CHEMISTRY OF SPACE

CHEMISTRY OF SPACE

Editor

Manas Korrapati

scitus academics

Chemistry of Space

Edited by **Manas Korrapati**

ISBN: 9781681178899

Library of Congress Control Number: 2017948917

SCITUS Academics LLC,

616, Corporate Way, Suite 2, 4766,

Valley Cottage, NY 10989

www.scitusacademics.com

Printed in United States of America.

Preface

Astronomy, space science and astrophysics allow us to see the universe and our place in it. Through studying these subjects, mankind has continually enlarged its horizons and explored the cosmos. The subjects continually evolve and change every year based on discoveries by researchers around the world. Space science is very often associated with physics. Physicists have been asking questions about the origins of the universe and using their specialized knowledge to make educated guesses about the past, the present and the future. The study of chemistry includes finding out about atoms, molecules and chemical reactions. Elements are made up of atoms of one kind. Atoms of elements can react to form molecules, sometimes of the same element (for example Cl_2) or sometimes with different elements forming compounds (for example HCl). Bonds form between the different atoms; these can be covalent or ionic, depending on the atoms involved. An element is made up of the same particles (sometimes the atoms are different isotopes) wherever it is found; iron (Fe) is iron whether you find it on Earth or in space, but conditions in space are different from those on Earth and different reactions may occur. Reactions of elements to form more complex compounds, such as amino acids, are important to the study of chemistry in space. Atoms and molecules are pretty much everywhere. From the iron-heavy center of Earth to the hydrogen-rich envelope of distant stars, nuclei and their attendant electrons are ubiquitous. It makes sense, then, that astronomers would want to sort out what atoms and molecules make up which astronomical objects and swim in the space between them. But they cannot just reach out, grab a hunk of star, and bring it back to the lab. They must do the analysis remotely and, as always in astronomy, using only light that telescopes passively receive.

'Chemistry of Space' discusses current research and advances in the field of space chemistry, including the origins of the universe, the chemical composition of planets and meteors, and stellar evolution. It looks at the way that the Earth's present atmosphere has evolved from possible earlier atmospheres. Some of the available evidence for different scenarios is presented and critically discussed.

Table of Contents

CHAPTER 1

Exoplanetary Atmospheres—Chemistry, Formation Conditions, and Habitability

Nikku Madhusudhan[1], Marcelino Agúndez[2], Julianne I. Moses[3], Yongyun Hu[4]

1.Institute of AstronomyUniversity of Cambridge Cambridge, UK

2.Instituto de Ciencia de Materiales de Madrid,CSICC antoblanco, Spain

3. Space Science Institute, Boulder, USA

4.Laboratory for Climate and Ocean-Atmosphere Sciences, Department of Atmospheric and Oceanic Sciences, School of PhysicsPeking University Beijing, China

ABSTRACT

Characterizing the atmospheres of extrasolar planets is the new frontier in exoplanetary science. The last two decades of exoplanet discoveries have revealed that exoplanets are very common and extremely diverse in their orbital and bulk properties. We now enter a new era as we begin to investigate the chemical diversity of exoplanets, their atmospheric and interior processes, and their formation conditions. Recent developments in the field have led to unprecedented advancements in our understanding of atmospheric chemistry of exoplanets and the implications for their formation conditions. We review these developments in the present work. We review in detail the theory of atmospheric chemistry in all classes of exoplanets discovered to date, from highly irradiated gas giants, ice giants, and super-Earths, to directly imaged giant planets at large orbital separations. We then review the observational detections of chemical species in exoplanetary atmospheres of these various types using different methods, including transit spectroscopy, Doppler spectroscopy, and direct imaging. In addition to chemical detections, we discuss the advances in determining chemical abundances in these atmospheres and how such abundances are being used to constrain exoplanetary formation conditions and migration mechanisms. Finally, we review recent theoretical work on the atmospheres of habitable exoplanets, followed by a discussion of future outlook of the field.

Keywords: Exoplanets, Exoplanetary, atmospheres, Atmospheric chemistry, Planet formation, Habitability

1. INTRODUCTION

The study of extrasolar planets (or 'exoplanets') is one of the most dynamic frontiers of modern astronomy. A few thousand exoplanets are now known. The numerous exoplanet discoveries via different methods have shown that exoplanetary systems are very common and are extremely diverse in their macroscopic properties. Recent statistics from observational surveys are suggesting that almost every star in the solar neighborhood hosts at least one planet around it, and that terrestrial-size exoplanets are amongst the most numerous (Howard et al. 2012; Fressin et al. 2013). The exoplanets detected to date span a diverse range in masses, radii, temperatures, and orbital parameters (orbital periods, separations, eccentricities, inclinations, etc.). The extreme diversity of these macroscopic properties of exoplanets has for the first time placed the solar system in cosmic context, thereby opening a plethora of new questions with far reaching implications.

We are now entering a new era in exoplanetary science. Going beyond planet detections, the field is now moving towards detailed characterization of exoplanetary atmospheres through spectroscopic observations. Encoded within a spectrum of an exoplanet is information about the chemical composition and manifold physical processes in its atmosphere. State-of-the-art observations are now beginning to provide both the high sensitivity and long spectral baseline required to place detailed constraints on the various physicochemical properties of exoplanetary atmospheres. Such observations in recent years are already providing good constraints on the chemical compositions of exoplanetary atmospheres. The most observed exoplanets to date are hot giant planets whose large scale-heights and high temperatures (~800–3000 K) make them particularly conducive to atmospheric observations. Several prominent molecules of carbon and oxygen and several atomic species (e.g. Na, K) are expected to be abundant and observable in these atmospheres, making giant exoplanets rich laboratories for understanding atmospheric chemistry. The atmospheric chemical compositions are in turn beginning to provide the first insights into the possible formation conditions and migration mechanisms of exoplanets.

Atmospheric observations have been reported for a variety of exoplanets detected via transits, direct imaging, as well as the radial velocity method. Spitzer observations have been obtained for about 20 transiting exoplanets in at least four photometric bands (3.6, 4.5, 5.8, 8 μm) and about 50 in two (3.6 and 4.5 μm), primarily for close-in hot Jupiters (Madhusudhan et al. 2014c). Photometric observations have also been obtained in the near infrared using large ground-based telescopes. Recently, high S/N near-infrared transit spectroscopy has become possible thanks to the HST Wide Field Camera 3 (WFC3) leading to the first high-confidence detections of H_2O in several transiting exoplanets (e.g. Deming et al. 2013). Multi-wavelength datasets are also providing the long spectral baseline and high precision required to derive joint constraints on the chemical compositions and temperature profiles of exoplanetary atmospheres. Over 400 HST orbits have been recently allocated for high S/N spectra of over a dozen transiting exoplanets using the HST instruments in the visible and near-infrared. High resolution spectroscopic observations have also been reported for several directly imaged planets in the near-infrared, and dedicated surveys are expected to pursue the same for dozens more. Finally, it has also now become possible to detect molecules in the atmospheres of close-in planets using very high resolution (R=105R=105) Doppler spectroscopy (Snellen et al. 2010). These observational advancements have led to the detection of chemical species in a sizable ensemble of planets and have motivated rapid progress in our theoretical understanding of atmospheric chemistry in exoplanets.

In the present work, we review the latest developments in our understanding of atmospheric chemistry of exoplanets. We begin with a brief introduction to observational methods for exoplanet detection and atmospheric characterization in Sect. 2. We then discuss the theoretical developments in our understanding of atmospheric chemistry in exoplanets (in Sect. 3), with particular emphasis on large gas giant planets (in Sect. 4) as well as ice giants and super-Earths (in Sect. 5). We then review, in Sect. 6, observational inferences of chemical species in exoplanetary atmospheres of various class using different observational methods. In Sect. 7, we review developments towards a theoretical framework for using atmospheric chemical abundances to constrain exoplanetary formation conditions and migration pathways. In Sect. 8, we discuss theoretical developments in our expectations for the habitability of terrestrial exoplanets. We conclude with a discussion of the future outlook for the field.

2. OBSERVATIONAL METHODS

The thousands of exoplanets known to date have been detected via a variety of observational methods (see Fischer et al. 2014, for a review). By far the majority of the planets have been detected using the transit method and the radial velocity method, followed by some detections using various other methods such as pulsar timing, gravitational microlensing, and direct imaging. On the other hand, observations of exoplanetary atmospheres present much more stringent requirements on the instrumental capabilities than planet detection. A combination of significantly higher sensitivity and higher spectral resolution are required for reliable atmospheric detections. Amongst the various planet finding methods, those amenable for atmospheric characterization are the transit method, direct imaging, and radial velocity method, with the transit method being the most successful to date. A detailed review of all these observational methods for atmospheric characterization can be found in Madhusudhan et al. (2014c). Here we briefly review the key attributes of each method.

One of the most successful methods for observing exoplanetary atmospheres has been the transit method. When the planet transits in front of the host star, part of the star light traverses through the day-night terminator region of the planetary atmosphere before reaching the observer. The resulting 'transmission spectrum', obtained by subtracting the in-transit spectrum from the out-of-transit spectrum, contains absorption features imprinted on the starlight by chemical species in the planetary atmosphere. Therefore, transmission spectra probe the chemical composition and temperature structure at the day-night terminator of the planet. On the other hand, when the planet is at full phase before being occulted by the star (i.e. 'secondary eclipse') the thermal emission and reflection spectrum from the planet is observed along with the stellar spectrum which can be subtracted out later; during secondary eclipse only the stellar spectrum is observed. Thus, secondary eclipse spectra probe composition and thermal structure of the dayside atmosphere of the planet. Spectra for exoplanets have been obtained in both transmission and at secondary eclipse, and in a wide range of wavelengths. While transmission spectra have been obtained from UV to mid-infrared wavelengths, thermal emission spectra have been obtained predominantly in the near to mid infrared. Besides the transit method, atmospheric observations have also been obtained from direct imaging of planets as well as by high-resolution Doppler spectroscopy, which are both discussed briefly in their respective sections below.

3. FACTORS INFLUENCING ATMOSPHERIC CHEMISTRY OF EXOPLANETS

The advent of observational techniques able to characterize atmospheres of extrasolar planets has been a major breakthrough in astronomy. The discovery of a great diversity of exoplanets, many of them with no analogue in the solar system, suggests that there is an exoplanet zoo out there with a continuum of possible sizes (from Jupiter-sized planets to sub-Earth bodies), atmospheric temperatures (from thousands of degrees Kelvin down to a few tens), and elemental compositions (from H/He-dominated atmospheres with a solar composition to atmospheres dominated by heavy molecules such as H_2O, CO_2, or N_2). This diversity poses a great challenge to theorists who seek an understanding of the variety of existing planetary climates.

A theoretical approach to the chemistry of exoplanet atmospheres must start by posing a question of the following type. How can we infer the chemical composition of a planetary atmosphere from a reduced set of parameters related to the planet? Of course it is first necessary to identify such parameters and to understand how each of them affect the atmospheric chemistry. Some of these parameters can be obtained in a straightforward way from observations, although there are some others for which we can just have guesses, albeit in some cases with formidable uncertainties. Let's have a look to each of these parameters.

Gravity—This parameter follows directly from the mass and radius of the planet, which in turn can be derived from radial velocity and transit techniques, respectively. Whether or not a planet can retain a substantial amount of atmosphere is largely related to its gravity and the X-ray and EUV flux of the host star (see e.g. chapter on protoatmospheres in Massol et al. 2016, this issue). Moreover, depending on the mass and radius of the planet, light elements such as hydrogen and helium may or not escape from the planet, with strong implications for the elemental composition of the atmosphere. The gravity is also important in that, together with the mean mass of particles and temperature, it sets the scale height of the atmosphere, i.e., how compact or extended the atmosphere is.

Elemental composition—The relative abundances of the different elements is one of the most important aspects that determine the atmospheric composition, although unfortunately these are rather difficult to infer from observations. Based on the study of solar system planets and on existing theories of planet formation (Lissauer 1993; Hubbard et al. 2002; Goldreich et al. 2004; Alibert et al. 2005; Papaloizou and Terquem 2006; Blum and Wurm 2008; Chiang and Youdin 2010; Morbidelli et al. 2012), it is expected that giant planets will from by core accretion, becoming massive enough at an early stage of planet formation to efficiently capture the nebular gas, and thus will retain a thick H/He-dominated atmosphere (see e.g. chapter on protoatmospheres in Massol et al. 2016, this issue). Terrestrial planets, on the other hand, will not become massive enough to efficiently accrete or retain H_2/He from the nebular gas, and their atmospheres, whether thick or thin, will depend on their mass and evolutionary history (e.g., orbital evolution, stellar/disk evolution, impact history, interior outgassing, atmospheric escape, climate evolution, surface-atmosphere interactions, magnetospheric interactions). Terrestrial-planet atmospheres are expected to be typically dominated by secondary products such as H_2O, CO_2, CO, N_2, Ne, Ar, Kr, SO_2, SiO_2 that were outgassed from the solid planetesimals that formed the planets or that were supplied by later impacts by solid bodies (see e.g. chapter on planet formation in Bai et al. 2016, this issue). In between giant and terrestrial planets there is a regime of super-Earths/sub-Neptunes whose elemental composition is largely unknown (e.g., Haghighipour 2013; Fortney et al. 2013; Lopez and Fortney 2014; Marcy et al. 2014; Lee et al. 2014; Rogers 2015). By comparing the outcomes of chemical models and observations, it is in principle possible to put constraints on the elemental composition of the planetary atmosphere (e.g., Madhusudhan et al. 2014c), which in turn can provide clues to the formation and evolution of the planet itself.

Insolation—The amount of energy received at the top of the atmosphere per unit time and unit area depends essentially on the luminosity of the host star and the orbital distance, two properties that can be relatively well constrained by observations. Of course, the way the stellar luminosity is divided across the electromagnetic spectrum, i.e., the spectral type of the star, has a great importance in affecting atmospheric characteristics. Incoming visible and infrared photons take care of the heating of the atmosphere, especially at pressures greater than ~1 microbar (e.g., Seager and Sasselov 1998; Sudarsky et al. 2003; Barman et al. 2005; Iro et al. 2005; Fortney et al. 2005; Burrows et al. 2006), while ultraviolet photons heat the high-altitude thermosphere and lead to ionization and dissociation of atmospheric constituents and induce photochemistry at a variety of altitudes (e.g., Liang et al. 2003; Yelle 2004; García Muñoz 2007; Zahnle et al. 2009a; Line et al. 2010; Moses et al. 2011; Venot et al. 2012; Agúndez et al. 2012; Hu et al. 2012; Miller-Ricci Kempton et al. 2012; Koskinen et al. 2013; Lavvas et al. 2014; Shaikhislamov et al. 2014; Rimmer and Helling 2015).

Internal heating—The interiors of planets can provide an important source of heat for the atmosphere, depending on the planet's age and tidal interactions with the host star or other planets in the system. The internal heating can be estimated from theoretical models of the evolution of the planet interior and its reaction to tidal forces (Guillot 2005; Baraffe et al. 2010; Ogilvie 2014). However, uncertainties in some key magnitudes, such as the dissipation properties of the bulk material of the planet interior and the age of the star (and thus of the planetary system), keep the internal heating poorly constrained in most cases.

A self-consistent model of a planetary atmosphere should in principle be able to describe the physical and chemical state of the atmosphere from the parameters described above. The situation resembles that of stellar atmospheres (e.g., Kurucz 1979; Hauschildt et al. 1999), where models need just three parameters (gravity, metallicity, and effective temperature)—external insolation being not important. Of course, in certain types of planets, parameters or processes different from those listed above, such as, e.g., exchange of matter with the surface in the case of terrestrial planets (e.g., Leconte et al. 2014) or grain formation and resulting opacity effects (e.g., Tsuji et al. 1996; Marley et al. 1996; Allard et al. 1997), may be also of great relevance in establishing the physical and chemical atmospheric properties.

In practice, models of planetary atmospheres lack a full self-consistency and tend to focus on some particular aspects, adopting approximations for some others. On the one hand, there are one-dimensional models that concentrate on solving the radiative transfer and thermal balance in the vertical direction. These models result in a pressure-temperature (PP-TT) profile for which the temperature gradient (dT/dPdT/dP) is subadiabatic in the radiative zone of the atmosphere and adiabatic in the deeper convective regions (e.g., McKay et al. 1989; Burrows et al. 1997, 2008; Seager et al. 2005). These radiative-convective equilibrium models usually make some assumptions about the chemical composition of the atmosphere, either adopting a particular one suitable for the modeled planet or assuming that chemical equilibrium holds in the atmosphere, and neglecting any fluid dynamics. For an assumed composition, in addition to the PP-TT profiles these non-gray numerical models also calculate the line-by-line thermal emission and reflectance spectra of the planetary atmospheres. Recent studies have also reported analytic and semi-analytic PP-TT profiles derived for irradiated planetary atmospheres in radiative equilibrium under the assumptions of gray/non-gray opacities (Hansen 2008; Guillot 2010; Robinson and Catling 2012; Heng et al. 2012; Parmentier and Guillot 2014; Parmentier et al. 2015).

Other type of models, the so-called general circulation models (GCMs), were originally developed to study the climate of the Earth (e.g., Phillips

1956; Mellor and Yamada 1982) and are now routinely used to study atmospheres of other solar system planets (e.g., Forget et al. 1999) as well as exoplanets such as hot Jupiters (e.g., Showman and Guillot 2002; Showman et al. 2009), hot Neptunes (e.g., Lewis et al. 2010), super-Earths (e.g., Castan and Menou 2011) and terrestrial planets (e.g., Showman et al. 2013). These are three-dimensional models which solve the Navier-Stokes equations or a reduced set of "primitive equations" (see e.g. Showman et al. 2010; Heng and Showman 2015) and end up with a three-dimensional view of the circulation and thermal structure of the atmosphere. Some of the limitations of GCMs have to do with the approximations used to deal with the radiative transfer (e.g., Showman et al. 2009) and the chemical processes (Cooper and Showman 2006). Another type of models are those focused on the atmospheric chemistry, in which we concentrate hereafter.

Chemical models of planetary atmospheres aim at describing how the atmospheric constituents are distributed, usually in the vertical direction. To build such a model it is necessary to have information on some basic parameters, among them three of the four aforementioned:

1) *gravity*,
2) *elemental composition*, and
3) *ultraviolet/X-ray irradiation*. It is also necessary to have information on a couple of additional parameters.
4) *Thermal structure*—The spatial distribution of the atmospheric temperature, at least in the vertical direction, can be retrieved from radiative-convective models and GCMs (see above). In the case of these latter models it is possible to get a three-dimensional map of the temperature. The atmospheric temperature has an enormous influence on the chemical composition because depending on whether the atmosphere is hot or cold the major constituents and the type of condensates formed (if any) are completely different.
5) *Strength of transport processes*—An atmosphere is in essence a fluid and therefore a variety of processes such as advection, diffusion, and turbulent motions can occur at different scales, having as consequence the transport and mixing of material between different regions. These processes are an important source of disequilibrium because they can transport molecules that were originally formed in a given location to other atmospheric regions. In particular, the vertical chemical structure can be strongly affected by transport processes, the strength of which is usually parameterized through an eddy diffusion coefficient (KzzKzz if referred to the vertical direction). It is, however, difficult to obtain realistic estimates of the coefficient KzzKzz. In the case of solar system planets such as Jupiter and Saturn, if observations are able to determine the abundance of certain species

at a given altitude, it is possible to put constraints on the eddy diffusion coefficient (e.g., Atreya 1986). In the case of extrasolar planets, information on the strength of mixing processes must be obtained from theoretical models such as GCMs (e.g., Parmentier et al. 2013).

If one wants to build a chemical model of a planetary atmosphere, these five ingredients (*gravity, elemental composition, ultraviolet/X-ray irradiation, thermal structure*, and *strength of transport processes*) should be in principle enough to provide a good description of the chemical structure of the atmosphere.

The characterization of exoplanet atmospheres through observations has been to date restricted to hot Jupiters, some hot Neptunes, and a few super-Earths, most of them characterized by transit techniques, and to a few young and self-luminous gas giant planets with an orbital distance sufficiently large to be accessible by direct imaging. It is therefore not surprising that theoretical studies of exoplanet atmospheres carried out to date have been to a large extent biased toward these types of extrasolar planets.

A general theoretical characterization of habitable-zone exoplanet atmospheres has also been a popular topic, due to the high intrinsic interest in the potential for life (and its detectability) outside of Earth.

4. THEORY OF ATMOSPHERIC CHEMISTRY OF GAS GIANT EXOPLANETS

Gas giant exoplanets are expected to have H/He-dominated atmospheres based on their formation mechanism and on the bulk densities inferred from observations (Lissauer and Stevenson 2007; Baraffe et al. 2010). Their chemical composition depends on various parameters, perhaps the most important being the temperature. A sketch of the structure of such an atmosphere is depicted in Fig. 1, where various important processes at work are indicated.

The chemical composition is controlled by chemical equilibrium in deep atmospheric layers, by transport-induced quenching in upper layers, and by photochemistry in still upper layers. Cloud formation becomes increasingly important as the atmosphere is cooler. Atmospheric escape can be an important issue for highly irradiated giant planets (e.g., Koskinen et al. 2014), while sequestration of refractory elements plays an important role in cool atmospheres (e.g., Lodders 2010).

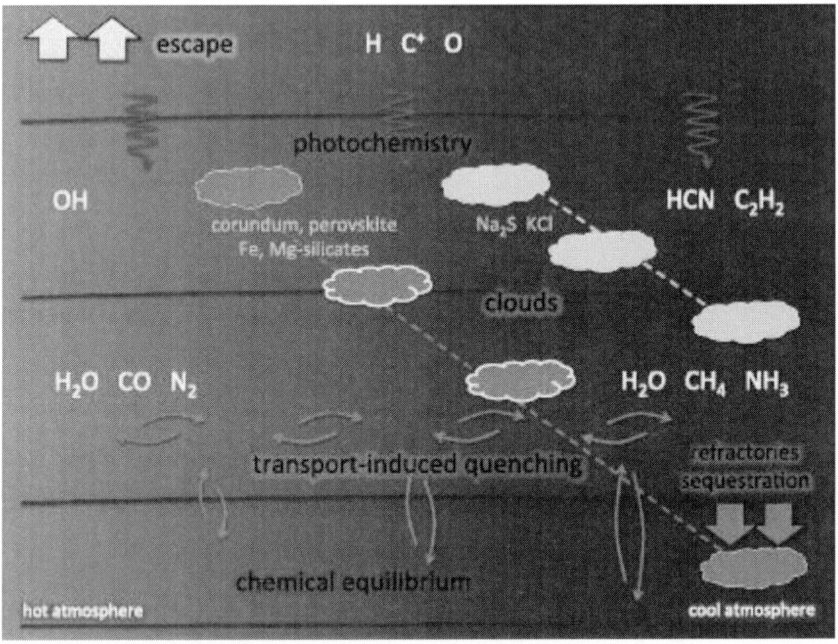

Figure 1. Sketch of the structure and processes at work in H/He-dominated atmospheres. The chemical composition in deep layers is controlled by chemical equilibrium, by transport-induced quenching in upper layers, and by photochemistry in still upper layers. Formation of clouds is an issue of increasing importance as the atmosphere becomes cooler. Escape can also be an important process in highly irradiated atmospheres (i.e., those with hot thermospheres), while sequestration of refractory elements becomes important in the cooler atmospheres

4.1 Chemical Equilibrium

Under the chemical equilibrium assumption, the chemical composition does only depend on temperature, pressure, and elemental abundances, and can be calculated in a relatively straightforward way by minimizing the Gibbs free energy of the system. Chemical equilibrium can accurately describe the composition of an atmosphere provided it is sufficiently hot and dense to ensure that chemical reactions occur faster than any other process at work (dynamics, interaction with energetic radiation, etc.). This can be the case for atmospheres of cool stars and brown dwarfs, where chemical equilibrium has been routinely applied (Tsuji 1973; Fegley and Lodders 1996; Allard et al. 1996; Burrows and Sharp 1999; Lodders and Fegley 2002), and of hot exoplanets with atmospheric temperatures in excess of ~2000 K (Moses et al. 2011; Venot et al. 2012). In the case of cooler planets, it can still be a good starting point to have a general idea of which could be the main atmospheric constituents.

The atmospheric temperatures of gas giant planets may span over a very broad range, from a few hundreds of kelvins up to some thousands (see Fig. 2). Gas giant planets are expected to have a nearly solar elemental composition, although some elements heavier than helium could be enriched or depleted by a factor of a few relative to the solar composition, as occurs in Jupiter (Mahaffy et al. 2000; Wong et al. 2004). In any case, gas giant atmospheres are expected to be dominated by hydrogen and helium, with heavier elements like oxygen, carbon, and nitrogen being present at a lower level. Figure 2 shows various curves, calculated under chemical equilibrium for a solar elemental composition, which delimitate different regions of interest from a chemical view point.

The red dashed curve indicates where CO and CH_4 have the same abundance. To the right of this curve (at high temperatures/low pressures) we have the region of stability of CO while to the left (at low temperatures/high pressures) methane is the main carbon reservoir. Similarly, in the case of nitrogen chemical equilibrium indicates that to the right of the green dashed curve N_2 is the main nitrogen reservoir while to the left it is ammonia that dominates.

Thus, a first lesson to be learnt from chemical equilibrium is that at high temperatures and/or low pressures CO and N_2 are the main reservoirs of carbon and nitrogen, respectively, while at low temperatures and/or high pressures the hydrides CH_4 and NH_3 dominate. In the case of oxygen, water vapor remains a major reservoir over most of the temperature regime of interest. This molecule locks either the excess of oxygen not locked into CO if the atmosphere resides in the stability region of CO (oxygen has a solar abundance about twice than carbon), or directly locks most of the available oxygen if the atmosphere resides in the stability region of CH_4.

Only at very low temperatures (see blue dot-dashed line) can water vapor be depleted from the gas phase because of condensation, something that can occur also for ammonia, albeit at even lower temperatures (see green dot-dashed line). That is, according to chemical equilibrium, depending on the thermal profile, the atmosphere of a giant planet may be dominated by (apart from H_2 and He) a mixture of $H_2O/CO/N_2$, $H_2O/CH_4/N_2$, $H_2O/CH_4/NH_3$, or a modification of the latter in which first H_2O, then NH_3, and finally $CH_4$are progressively removed from the gas phase as the atmosphere gets cooler (e.g., Burrows et al. 1997; Sudarsky et al. 2000; Lodders and Fegley 2002).

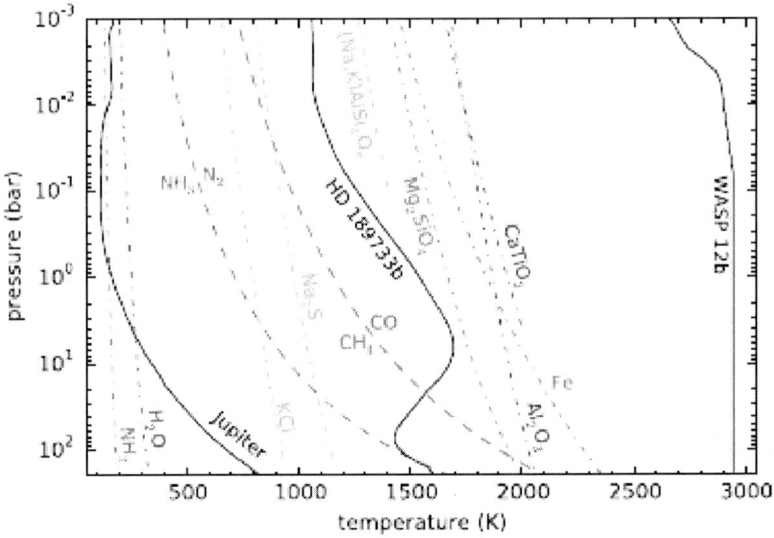

Figure 2. Pressure-temperature profiles for three gas giant planets covering a broad range of atmospheric temperatures: Jupiter (Lindal 1992), the widely studied hot Jupiter HD 189733b (Agúndez et al. 2014a), and the highly irradiated gas giant WASP-12b (Stevenson et al. 2014a). *Dashed lines* delimitate the regions where either CH_4 or CO (*red curve*) or NH_3 or N_2 (*green curve*) are the major carbon or nitrogen reservoirs, respectively. *Dot-dashed lines* indicate the condensation curves for water, ammonia, and various refractory species containing Ca, Ti, Al, Fe, Mg, Na, and K

4.2 Disequilibrium Chemistry Driven by Transport Processes

The atmospheric composition of gas giant exoplanets is governed by chemical equilibrium only in the hottest regions, usually in the deepest, densest regions (or the bulk of the atmosphere for very hot, strongly irradiated planets). In cooler upper regions, when chemical reactions become slower than dynamic processes (advection, diffusion, or turbulence), the bulk composition can be significantly driven out of equilibrium. The transport of material between different regions has as a consequence a redistribution (homogenization) of heat and chemical composition, with a net flow from hot to cooler regions.

The vertical distribution of the atmospheric constituents can be strongly affected by the so-called chemical quenching, in which as material moves upwards chemical reactions become slower, and at a certain level the abundances of the different species are progressively quenched. This phenomenon was first described by Prinn and Barshay (1977) to explain the presence of CO in the troposphere of Jupiter as a result of upward mixing from deeper and hotter levels, where CO is thermochemically stable. In gas giant exoplanets, chemical quenching regulates to a large extent the composition of the layers probed by observations, especially if their pressure-temperature profiles cross the boundaries of stability of the pairs CH_4/CO or NH_3/N_2 (see Fig. 2). A quantitative evaluation of chemical quenching (i.e., the quench level and resulting quenched abundance) requires a good knowledge of both the chemical kinetics of the interconversion scheme of interest and the vertical eddy mixing coefficient.

The main issue when dealing with quenching and chemical conversion schemes is the identification of the elementary reactions involved and the "limiting step", whose rate controls the kinetics of the overall conversion. The interconversion CO $\rightleftharpoons$ CH_4 is of great importance in cool to moderately warm gas giant planets because exoplanet temperature-pressure profiles often cross the stability boundary between CO and CH_4 (red dashed curve in Fig. 2). Understanding how the conversion between CO $\rightleftharpoons$ CH_4 proceeds kinetically and which reaction is the rate-limiting step is thus of primordial importance (see Visscher and Moses 2011 for more details). In the same vein, the interconversion $N_2 \rightleftharpoons NH_3$ is also of great relevance for the atmospheres of gas giant planets, N_2 and NH_3 being the main reservoirs of nitrogen at high and low temperatures (or low and high pressures), respectively. Although there are still important uncertainties in the reaction schemes and the rate-limiting step, the conversion $N_2 \rightarrow NH_3$ is likely intrinsically slower than the conversion CO $\rightarrow$ CH_4, resulting in a deeper quench level for the nitrogen species (see Moses 2014 for more details). The identification of interconversion reaction schemes is interesting in that it permits implementation of simple prescriptions to deal with chemical quenching in high-demanding computational models of atmospheres such as GCMs (Cooper and Showman 2006).

For 1D exoplanetary atmosphere models, the eddy diffusion coefficient must be quantified from 3D theoretical models like GCMs, or constrained from the observations themselves. Although mixing in planetary atmospheres can occur through large-scale advection, atmospheric waves, eddies of a variety of scales, and other transport processes that are not diffusive in a rigorous sense, vertical mixing can typically be well represented by a diffusion equation (e.g. Lindzen 1981; Strobel 1981; Brasseur et al. 1999). A common practice has been to estimate the vertical diffusion coefficient KzzKzz as the root mean square of the vertical velocity (as extracted from a GCM) times the vertical scale height (Line et al. 2010; Moses et al. 2011). A better approach when 3D circulation models are available, as outlined by Parmentier et al. (2013), is (1) to follow the behavior of passive tracers in a GCM simulation and fit the resulting planet-averaged vertical tracer profiles via a 1D diffusion equation with an effective diffusion coefficient and/or (2) to determine the diffusive flux and resulting KzzKzz profile that best matches the horizontally averaged vertical flux in the GCMs. These latter methods have been shown to result in KzzKzzvalues significantly lower than those obtained by representing KzzKzz as the root-mean-square vertical velocity times the scale height (Parmentier et al. 2013; Agúndez et al. 2014a). The exact value of the eddy diffusion coefficient, especially at pressures greater than ~0.1 bar, has a direct impact on the location of the quench level of each species (the higher the KzzKzz, the deeper the quench level) and the abundance at which each species gets quenched in the upper atmosphere. If the quench point falls within the radiative region of the atmosphere, KzzKzz is expected to vary with the inverse square root of atmospheric pressure near this quench point (Lindzen 1981; Parmentier et al. 2013). If the quench point falls within the deeper convective region of the atmosphere, mixing-length theory and free-convection theory can be used to estimate the magnitude of KzzKzz near the quench point (Visscher et al. 2010b), or expressions based on laboratory studies of turbulent rotating convection can be developed (Wang et al. 2015). However, the appropriate mixing length to use for the KzzKzz estimates is not straightforward (e.g., Smith 1998), and observations of quenched species may themselves provide our best means of estimating the magnitude of convective mixing at depth.

Chemical quenching in the vertical direction of hot Jupiter atmospheres has been studied through timescale arguments (Visscher et al. 2006; Line et al. 2010; Bilger et al. 2013) and using more robust chemical networks (in which endothermic reactions are included and reverse reaction rates are computed from thermochemical grounds) suitable to model high temperature environments (Zahnle et al. 2009a,b; Moses et al. 2011, 2013b; Venot et al. 2012; Kopparapu et al. 2012; Agúndez et al. 2014a; Miguel and Kaltenegger 2014; see also Fig. 3). Most of these studies focus on the atmospheres of the widely observed hot Jupiters HD 189733b and HD 209458b, where it is predicted that, assuming nearly solar elemental abundances, CO and N_2 are the major reservoirs of carbon and nitrogen, respectively, H_2O is very abundant (it takes up most of the excess of oxygen not locked into CO), and methane and ammonia are present at a lower level, with CO/CH_4 and N_2/NH_3 abundance ratios in the ranges 50–500 and 5–200, respectively, in HD 189733b, and in the ranges 1000–10000 and 100–1000, respectively, in the hotter HD 209458b (Moses et al. 2011; Venot et al. 2012; Agúndez et al. 2014a; Miguel and Kaltenegger 2014). The abundance of carbon dioxide is found to be somewhat low for atmospheres with an assumed solar elemental composition (see Fig. 3). It is interesting to note that when CO is the dominant carbon component, CO_2 is much less affected by chemical quenching than many other molecules because the various reactions involved in the interconversion between H_2O-CO-CO_2 are rapid enough (Moses et al. 2011; Venot et al. 2012). The chemistry of elements other than C, N, and O, although more difficult to constrain through observations and more uncertain due to a lack of robust reaction rate measurements under appropriate conditions, has been also studied theoretically. It has been pointed out, for example, that chemical quenching le to PH_3 becoming a major reservoir of phosphorus in hot Jupiter atmospheres (Visscher et al. 2006).

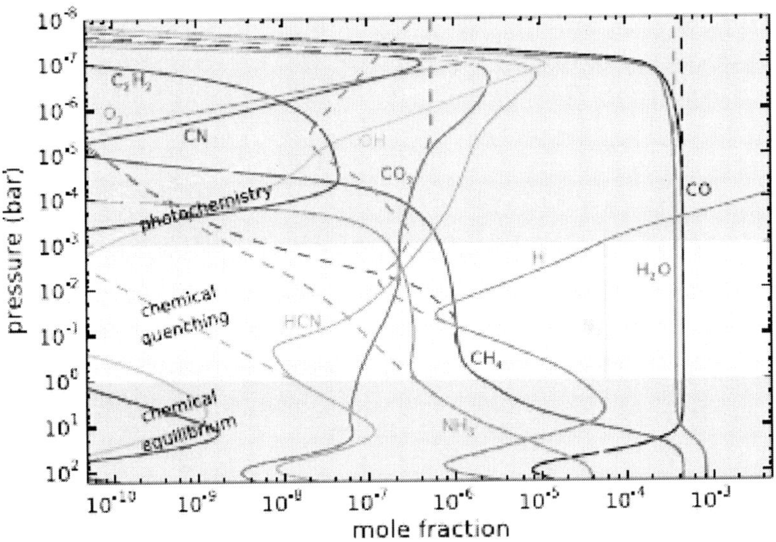

Figure 3. Vertical distribution of abundances at the substellar point of HD 189733b. *Dashed lines*correspond to chemical equilibrium and *solid lines* to a one-dimensional vertical model including thermochemical kinetics and photochemistry (Agúndez et al. 2014a). The atmosphere can be schematically divided into three regions where the composition is controlled by either chemical equilibrium, chemical quenching, or photochemistry

Chemical quenching can also occur horizontally driven by winds that move material from hot to cooler atmospheric regions. In hot Jupiters, which are usually tidally locked (with permanent day and night sides), the uneven heating of the planet may result in important temperature contrasts, and thus in possible variations in the composition, between the day and night sides. On the other hand, circulation dominated by a strong superrotating equatorial jet tends to homogenize both the temperature and chemical composition between the different planetary sides (Showman and Guillot 2002; Cooper and Showman 2006; Knutson et al. 2007). The main effect of horizontal chemical quenching is that molecular abundances are quenched horizontally to values typical of the hottest dayside regions, making the cooler nightside to be highly contaminated by the warmer dayside regions (Cooper and Showman 2006; Agúndez et al. 2012, 2014a). The effect is however more marked for molecules such as H_2O, CO, and N_2, which show rather uniform abundances, than for others such as CH_4, NH_3, CO_2, and HCN, which may still show important abundance gradients among the different planetary sides (see Fig. 4). The horizontal homogenization of abundances becomes more important as the atmosphere gets cooler. The distribution of the atmospheric constituents has implications for transit observations, which probe the terminator region, and for phase curves probing the different planetary sides.

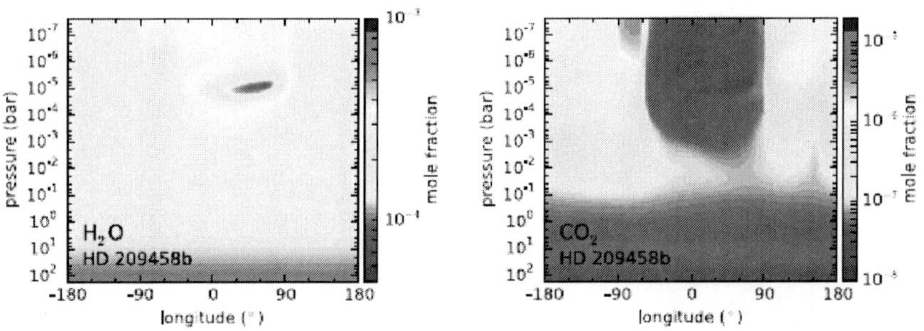

Figure 4. Calculated distributions of H_2O and CO_2 as a function of longitude and pressure in the atmosphere of HD 209458b, from Agúndez et al. (2014a). Note that H_2O maintains a rather homogeneous distribution while CO_2 experiences important abundance variations with longitude and height

4.3 Photochemistry

Hot Jupiters are strongly irradiated by their host stars and thus receive a high ultraviolet flux, which is absorbed in the upper atmosphere and induces a more or less rich photochemistry. That is, the parent molecules transported from deep atmospheric regions are photodissociated in upper layers producing radicals that react to form new species. The study of the photochemistry of exoplanet atmospheres has benefited from decades of study of solar system atmospheres such as those of Jupiter (Gltone et al.

1996), Saturn (Moses et al. 2000), and Titan (Yung et al. 1984; Lara et al. 1996). However, some of the chemical kinetics and photoabsorption cross section data still need to be adapted to the high temperatures of hot Jupiters (Venot et al. 2013). The type of molecules formed by photochemistry depends on the composition of the precursor material transported from deeper layers, which in turn depends on the elemental composition (see Sect. 4.4) and thermal atmospheric structure (see Sects. 4.1and 4.2). The cooler the planet the larger the extent of photochemistry because, on the one hand, high temperature chemistry counterbalances the action of photochemistry, and on the other, low temperatures favor the presence of CH_4 and NH_3, which are more active photochemical precursors than their respective high-temperature counterparts CO and N_2(Moses 2014).

There are various species produced by photochemistry in the atmospheres of gas giant planets (see Fig. 3). These atmospheres being hydrogen-rich, a major species produced by photochemistry is atomic hydrogen, which is formed by photolysis and thermal decomposition of H_2 as well as by catalytic photolysis of H_2O (Liang et al. 2003). The higher the irradiation of the planet the more abundant becomes H. Another important photochemical species is HCN, which is efficiently formed in the photochemical layer through schemes starting with the photodissociation of CH_4 and NH_3 (provided the atmosphere is not too hot to exclude these two hydrides as important constituents), and is able to diffuse to deeper regions (Zahnle et al. 2009a; Moses et al. 2011). The photodissociation of CH_4 also triggers the formation of larger hydrocarbons such as C_2H_2, which may then polymerize to form soots (Zahnle et al. 2009a; Bilger et al. 2013). Another important photochemical product is the radical OH, which results from the photodissociation of water and act as a key intermediate in the synthesis of other O-bearing molecules such as O_2 and NO (Moses et al. 2011). As concerns sulfur compounds, the major photochemical products of the dominant equilibrium sulfur species, H_2S, are expected to be S, HS, S_2, SO, and SO_2 (Zahnle et al. 2009b). Lavvas et al. (2014) show that for planets hot enough that refractory elements are not tied up in condensates, atomic neutrals and ions such as Mg, Mg^+, Fe, Fe^+, Ca, Ca^+, Na, Na^+, K, K^+, Si^+, Al^+ could be important photochemical and equilibrium species, although some dominant equilibrium molecular forms (e.g., SiO, SiS, AlOH, FeH) can also survive in the photosphere on hot planets (see also Visscher et al. 2010a), and molecular species like NaCl and KCl may form from both photochemistry and thermochemistry.

On highly irradiated hot Jupiters, most photochemical products are confined to relatively high altitudes (e.g., above ~0.1 mbar, see Fig. 3). Therefore, photochemical products are likely to have a limited impact on the infrared emission spectrum of the planet, whose photosphere is located at deeper layers, although they may leave their imprints on the transmission spectrum, which can probe higher atmospheric layers. The cooler the planet, however, the deeper the photochemical "layers" extend, and the higher the likelihood that the photochemical products affect the infrared emission spectra. On our own solar-system giant planets, for example, methane photochemical products such as C_2H_2 and C_2H_6 survive throughout the stratosphere, dominating the mid-IR spectra and enabling more effective cooling of the stratospheres of these planets. In the case of HCN, its extended vertical distribution and relatively high abundance in cool to moderately warm gas giant planets makes it a very likely candidate for detection (e.g., Moses et al. 2011).

One-dimensional models indicate that in still upper layers, molecules are completely photodestroyed and atoms become the major constituents. In the absence of a hydrodynamic escape flow, molecular diffusion at a high-altitude homopause layer (typically in the 10–610–6 to 10–910–9 bar range for hot Jupiters) will cause the atmospheric constituents to become layered according to their weight, following their own density scale heights. Because of both this molecular diffusion and the strong X-ray/ultraviolet irradiation from the host star (which strongly heats the high-altitude thermosphere), atomic hydrogen becomes the only remaining dominant neutral constituent at high altitudes on close-in giant planets (e.g., Moses et al. 2011). If the upper atmosphere were to remain cool, the molecular region could extend to very high altitudes (e.g., Agúndez et al. 2014a). However, thermospheric photochemical models that include X-ray and extreme ultraviolet photolysis and photodissociative ionization of H_2, ion chemistry, and realistic atmospheric escape processes (García Muñoz 2007; Koskinen et al. 2013) demonstrate that hydrodynamic winds fueled by the escaping gas can drag heavy atoms and ions to regions as high as the exosphere in strongly irradiated planets, but that molecules (including H_2) cannot survive the large thermospheric temperatures on such planets. In fact, various types of atoms, from the lightest hydrogen to heavy metals, have been detected in the vertically extended, escaping atmospheres of hot Jupiters (Vidal-Madjar et al. 2003, 2004, 2013; Lecavelier Des Etangs et al. 2010; Fossati et al. 2010; Linsky et al. 2010).

Ionization of atmospheric constituents by galactic cosmic rays provides an additional source of disequilibrium chemistry on exoplanets. Rimmer and Helling (2013) and Rimmer et al. (2014) have explored the effects of cosmic rays on the atmospheric compositions of extrasolar giants planets. Charging of mineral grains, and potential subsequent electrical discharges, are another possible source of disequilibrium chemistry on extrasolar giant planets (see Helling et al. 2013; Bailey et al. 2014; Stark et al. 2014).

4.4 Sensitivity to Bulk Elemental Composition

Gas giant planets are expected to have nearly solar elemental compositions. However some may show a different metallicity, i.e., the abundances of elements heavier than helium scaled up or down by a similar factor, and/or elemental abundance ratios different from those found in the Sun. These deviations from the solar elemental composition, whatever their origin (see Sect. 6), can have a significant impact on the atmospheric composition.

In general, an enhancement of metallicity in H/He-dominated atmospheres of gas giant planets favors an increase in the abundances of molecules that contain multiple heavy atoms, i.e., CO and N_2 are favored over CH_4 and NH_3. Molecules with more than two heavy atoms are even more favored, as occurs in the case of CO_2, whose abundance increases as the square of metallicity (Zahnle et al. 2009a,b). See more details on this subject in Sects. 5.1 and 5.2.

In warm H/He-dominated atmospheres, one of the most critical elemental ratios is C/O. The issue is well known in the study of evolved stars, some of which can bring out to the surface enough carbon to revert the C/O abundance ratio, which in the Sun is 0.55 (Asplund et al. 2009). The high stability of CO under these conditions makes it lock almost all the limiting reactant, either C or O, allowing for the reactant in excess to form O-bearing molecules such as H_2O when $C/O < 1$ and C-bearing molecules such as HCN and C_2H_2 when $C/O > 1$ (Tsuji 1973).

The observational claim of a carbon-rich atmosphere in the hot Jupiter WASP-12b (Madhusudhan et al. 2011b), although still subjected to debate (Cowan et al. 2012a; Crossfield et al. 2012; Madhusudhan 2012; Swain et al. 2013; Mandell et al. 2013; Stevenson et al. 2014a,b; Kreidberg et al. 2015), has opened a window on the possible existence of carbon-rich giant extrasolar planets. Whatever their origin (see Sect. 6), such planets would show a chemical composition dramatically different from the more traditional gas giants with a nearly solar C/O ratio (i.e., oxygen-rich). In moderately warm and hot carbon-rich atmospheres, water is no longer an abundant constituent, the C-bearing molecules C_2H_2, CH_4, and HCN become major constituents, and CO_2 vanishes to a negligible level (Madhusudhan 2012; Kopparapu et al. 2012; Moses et al. 2013a,b; Venot et al. 2015; Heng and Lyons 2016).

If the atmosphere is cool enough (<1000 K) then CH_4 and H_2O become the main reservoirs of carbon and oxygen, respectively, at the expense of other carbon- and oxygen-bearing molecules (Madhusudhan 2012; Moses et al. 2013a,b; Venot et al. 2015).

4.5 Clouds and Hazes

The terms *cloud* and *haze* are often used interchangeably, but here we follow the spirit of Marley et al. (2013) and use these terms to refer to two fundamentally different types of condensates that may appear in the atmospheres of gas giant planets. By *cloud* we refer to the typical "cooling clouds" (e.g., Rossow 1978) that form when a volume of gas is cooled dynamically or radiatively, such that the partial pressure of a constituent or constituents exceeds its saturation vapor pressure and condensation ensues. The typical equilibrium condensates expected along the cloud-condensation sequence in a hydrogen-dominated atmosphere (e.g., Lodders 2010; Marley et al. 2013; Morley et al. 2013) are thus considered clouds, regardless of their optical depth, horizontal extent, etc. We restrict the term "hazes" to refer to aerosols that form in situ by the action of photochemistry or any other disequilibrium chemical process (e.g., the so-called "soots" discussed by Zahnle et al. 2009a) or from photochemically produced gases flowing dynamically into cooler regions, where they can condense.

In the rainout scenario of cloud formation, a condensate forms and gravitationally settles in the atmospheric layer where the temperature equals its condensation temperature, and the atmosphere above the cloud becomes depleted in the elements that take part in that condensate (Lewis 1969; Fegley and Lodders 1994, 1996; Burrows and Sharp 1999; Lodders 1999; Lodders and Fegley 2002; Sudarsky et al. 2003). Thus, as one moves from the deep and hot atmosphere to upper and cooler regions, elements are progressively removed from the gas phase according to their refractory character. Chemical equilibrium calculations including gas and condensed species are very useful to identify the most plausible cloud-forming species (see Fig. 2). In the rainout approach, chemical equilibrium is solved in order of decreasing temperature and, when a given species is found to condense the elements which take part in that condensate are removed (in the corresponding stoichiometric proportions) before continuing to solve chemical equilibrium at lower temperatures. Thus, it is not only necessary to know the possible condensates but also the condensation sequence and the main condensates that deplete each element.

Unlike the case of Jupiter, whose atmosphere is so cold that water and ammonia condense to form tropospheric clouds, most extrasolar giant planets characterized to date are sufficiently hot to ensure the survival of water vapor and other volatiles in their atmospheres. In hot Jupiters only the most refractory species can condense to form clouds. Although the exact temperature-pressure profile of the planet in question controls which refractory species will condense first (i.e., at the deepest pressures;

see Lodders 2010), aluminum, titanium, and calcium are expected to be the first elements to be removed from the gas phase, condensing as corundum (Al_2O_3), Ca-aluminates such as hibonite ($CaAl_{12}O_{19}$), and Ca-titanates such as perovskite ($CaTiO_3$). Therefore, atmospheres hotter than about 2000 K can maintain titanium in the gas phase to form oxides such as TiO, while cooler atmospheres would deplete most of this element in the form of perovskite or other Ca-titanate (Burrows and Sharp 1999; Lodders 2002). In fact, it has been proposed that in the atmospheres of very hot giant exoplanets, the survival in the gas phase of species such as TiO and VO can provide a sufficiently high opacity at optical wavelengths as to induce a thermal inversion (Hubeny et al. 2003; Fortney et al. 2008a; Spiegel et al. 2009). The detections of TiO and VO in brown dwarf atmospheres at high temperatures (Kirkpatrick 2005) provide support to this proposition. However, TiO and VO have not yet been unambiguously identified in the atmospheres of strongly irradiated hot Jupiters (Désert et al. 2008; Huitson et al. 2013; Mancini et al. 2013; Swain et al. 2013; Sing et al. 2013; Gibson et al. 2013; Stevenson et al. 2014b; Schlawin et al. 2014; Hoeijmakers et al. 2015), though recently a potential detection of TiO has been reported for one of the most irradiated hot Jupiters WASP-33b (Haynes et al. 2015).

At temperatures below ~1800 K, iron is expected to condense homogeneously, and magnesium, together with a good fraction of silicon, are expected to deplete in the form of silicates such as Mg_2SiO_4 and $MgSiO_3$. The presence of such type of clouds has been inferred from transit observations of the hot Jupiter HD 189733b (Pont et al. 2008, 2013; Lecavelier Des Etangs et al. 2008a; Sing et al. 2009; Gibson et al. 2012). The alkali metals Na and K can be depleted in the form of the aluminosilicates $NaAlSi_3O_8$ and $KAlSi_3O_8$ at temperatures around 1500 K (Burrows and Sharp 1999), although in the rainout scenario it is likely that at these temperatures most of the aluminum is no longer available in the gas phase, in which case Na and K would condense at ~1000 K as Na_2S and KCl, respectively (Lodders 1999). In sufficiently hot atmospheres neutral Na and K atoms can survive in the gas phase and can be readily detected through the Na I doublet at 589.0 nm and 589.6 nm and the K I doublet at 766.5 nm and 769.9 nm (Charbonneau et al. 2002; Redfield et al. 2008; Wood et al. 2011; Sing et al. 2011a, 2012; Nikolov et al. 2014; Murgas et al. 2014).

The rainout scenario provides a useful methodology to predict the composition of the clouds that may be present in a given atmosphere and the base level at which each cloud forms. However, the computation of the size and concentration of particles above this level and the horizontal distribution of the clouds, i.e., how particles nucleate and grow and are affected by atmospheric dynamics, still remains a formidable challenge. Nevertheless, efforts to this end have been undertaken by various groups

(Ackerman and Marley 2001; Helling et al. 2008; Freytag et al. 2010; Parmentier et al. 2013; Morley et al. 2013; Lee et al. 2015).

The photochemical formation of organic hazes in the atmospheres of hot Jupiters has been addressed in a couple of theoretical studies (Liang et al. 2004; Zahnle et al. 2009a). The general view provided by these studies is that in hot atmospheres, most of the carbon remains locked into CO, while in cooler atmospheres the larger amount of carbon stored into CH_4 can be photochemically driven to larger hydrocarbons, eventually producing hydrocarbon aerosols or soots. Cooler planets would therefore be expected to have more prevalent photochemical hazes. This scenario agrees with the presence of hazes in the cold atmospheres of Jupiter, Saturn, and Titan (e.g., Tomasko et al. 2005; Zhang et al. 2013), although it is in contradiction with the inference of aerosols in the very hot atmosphere of WASP-12b (Sing et al. 2013), unless these aerosols result from the most refractory species along the cloud-condensation sequence (e.g., Al_2O_3) at the cooler limb of the planet (Sing et al. 2013), or unless the planet is indeed carbon-rich (and thus contains the more photochemically active molecules HCN and C_2H_2, e.g., Kopparapu et al. 2012 and Moses et al. 2013b). It is clear that more work is needed in both the theoretical and observational sides to better understand the formation and role of hazes in extrasolar giant planets.

4.6 Hot Jupiters Versus Directly Imaged Planets

One of the most recent and formidable successes achieved in the field of exoplanets has been the detection by direct imaging of young and self-luminous gas giant planets, opening the way to characterize their atmospheres by direct spectroscopy (Janson et al. 2010, 2013; Hinz et al. 2010; Bowler et al. 2010; Currie et al. 2011; Barman et al. 2011; Skemer et al. 2012; Oppenheimer et al. 2013; Konopacky et al. 2013; Chilcote et al. 2015). The few planets that have been characterized by this method are more massive than Jupiter and have effective temperatures in the range 600–1700 K. Thus, they have some similarities with brown dwarfs and free-floating planets (Zapatero Osorio et al. 2000; Burrows et al. 2001, 2003). Directly imaged planets share also some characteristics with hot Jupiters, as both are gas giants and hot. However, unlike hot Jupiters, directly imaged planets orbit far from their host star and are young, so that they are heated predominantly from the interior rather than irradiated by the star. This difference le to qualitative differences in the dynamical and thermal structure of the atmosphere. In hot Jupiters, the high irradiation causes the atmospheres to be radiative in almost the entire observable atmosphere, and the temperature profile is isothermal in the lower atmosphere, at pressures above $\sim$1 bar (see e.g. Burrows et al. 2008), before convection dominates in the deep atmosphere ($\gtrsim$100 bar). On the other hand, the atmospheres of self-luminous planets and brown dwarfs are largely driven by convection, leading to adiabatic temperature gradients even in the observable atmospheres.

The atmospheres of hot Jupiters and directly imaged planets have temperatures of the same order and are expected to have a nearly solar elemental composition, so that one would expect a similar atmospheric chemistry in both types of planets. There are, however, a couple of major differences. First, the atmospheres of directly imaged planets are likely less affected by photochemistry than in the case of hot Jupiters because of their much larger orbital distances; however, directly imaged planets tend to orbit young stars, and young stars tend to have high ultraviolet output, so photochemistry will not be negligible on these planets. And second, the source of heat being located in the interior of the planet rather than outside imprints differences in the atmospheric thermal profile. According to the recent theoretical study by Zahnle and Marley (2014), the higher temperatures in the deep atmosphere of self-luminous planets favor CO over CH_4 in deep layers, but also in the upper observable atmosphere as a consequence of upward mixing. Methane is therefore predicted to be a minor atmospheric constituent in most self-luminous planets characterized to date, except for a couple of planets with cool atmospheres whose near-infrared spectra show evidence for CH_4 absorption, such as GJ 504b (Janson et al. 2013), 51 Eri b (Macintosh et al. 2015), and HR 8799b (Barman et al. 2015). Years of study of brown dwarf atmospheres (see e.g., Marley and Robinson 2015), which share many similarities with the atmospheres of self-luminous planets, provide an invaluable basis to aid in the understanding of atmospheric chemistry of self-luminous planets, although further observations are still needed.

5. THEORY OF ATMOSPHERIC CHEMISTRY OF EXO-NEPTUNES AND SUPER-EARTHS

The same chemical processes discussed above for giant planets—thermochemical equilibrium, disequilibrium quenching due to transport, and disequilibrium photochemistry—also affect smaller planets, but the basic ingredients available to the atmospheres of smaller planets differ from those of gas giants. These differences emerge early on in the planet's evolution, with such factors as the dust/planetesimal surface density distribution within the protoplanetary disk (see chapters on protoplanetary disk evolution in Oberg et al. 2016, this issue; Fang et al. 2016, this issue), the planet's initial formation and feeding-zone location within the disk, and the planet's migration history being important parameters that control how massive the planet becomes, what atmospheric volatiles are collected, and how much atmospheric hydrogen and helium are retained (e.g., Pollack et al. 1996; Lissauer and Stevenson 2007; D'Angelo et al. 2010; Dodson-Robinson and Bodenheimer 2010; Marboeuf et al. 2014; Ali-Dib et al. 2014; Madhusudhan et al. 2014a; Helling et al. 2014). In the gravitational-instability theory for giant-planet formation (e.g., Boss 1997), giant planets are expected to end up with a metallicity similar to the host star. In the core-accretion theory of giant-planet formation (e.g. Pollack et al. 1996), protoplanetary cores form from the accretion of solid planetesimals, and gas accretion rates are initially slow. As the solid protoplanetary core grows, it can accrete more and more of the surrounding hydrogen-rich nebular gas (Lammer et al. 2014; Stökl et al. 2015, also see chapter by Massol et al. 2016, this issue). When the core becomes massive enough, with a critical mass of order ~10 Earth masses (but cf. Venturini et al. 2015), the protoplanet can experience a runaway gas-accretion phase in which the nebular gas is rapidly accreted onto the protoplanetary core. Planets that reach this runaway gas-accretion stage become hydrogen- and helium-rich gas-giant planets. Planets that don't reach this stage contain less H and He.

Whether a protoplanet can reach this runaway gas-accretion stage or not depends largely on the accretion rate of solids in comparison to the time scale for dissipation of gas from the disk. Just beyond condensation fronts in the disk, such as the water-ice "snow line", the surface density of solid material becomes large, leading to more rapid accretion of solids and a reduced time scale for formation of massive protoplanetary cores (e.g., Stevenson and Lunine 1988; Ciesla and Cuzzi 2006; Dodson-Robinson et al. 2009). The formation of giant planets is thus expected to be particularly efficient near the disk snow line. In lower-surface-density regions of the disk farther out from the snow line, the solid accretion rate is slower, and a protoplanet may never reach this runaway gas-accretion phase, leading to less accumulation of the nebular gas. In this classical picture, the formation of Uranus and Neptune occurred too slowly to allow these planets to fully reach the runaway gas-accretion phase, leading to "ice giant" planets whose total mass is dominated by heavier volatiles such H_2O rather than hydrogen and helium (e.g., Lissauer and Stevenson 2007; Fortney and Nettelmann 2010). The overall initial mass fraction of H/He in a planet's atmosphere is a sensitive function of the disk characteristics and lifetime, with small changes in disk properties or evolutionary history having a significant impact on the resulting mass and composition of planets that are intermediate in size between gas giants and solid terrestrial planets (e.g., Helled and Bodenheimer 2014; Lammer et al. 2014; Luger et al. 2015). While generally atmospheres of ice-giants are expected to be H_2-rich, recent theoretical studies have also suggested the possibility of He-dominated atmospheres (Hu et al. 2015b).

In this section, we discuss the atmospheric chemistry of these intermediate-sized planets, the so-called super-Earths, mini-Neptunes, and Neptune-sized planets, with radii spanning ~1–6 R⊕R⊕, that are observed to constitute a large percentage of the known planets in our galaxy (e.g., Borucki et al. 2011; Mayor et al. 2011; Howard et al. 2012; Batalha et al. 2013; Fressin et al. 2013; Burke et al. 2014; Rowe et al. 2015). These planets will typically contain less hydrogen and helium (and thus a higher atmospheric metallicity) than giant planets because of the less efficient accretion of nebular gas during their formation and evolution, as well as the higher likelihood of the escape of light gases over time. Outgassing from the planetary interior is expected to contribute additional volatiles—a component of the atmosphere that will become increasingly important for smaller planets—and other evolutionary processes such as atmospheric escape, impact delivery or erosion, atmosphere-surface exchange, weathering, and sequestration of volatiles into the interior can have and major influence on atmospheric composition and chemistry. The stochastic nature of the different possible evolutionary pathways is expected to lead to highly diverse atmospheric properties for intermediate and small planets (e.g. Pepin 2006; Dodson-Robinson and Bodenheimer 2010; Rogers et al. 2011; Lopez and Fortney 2013; Fortney et al. 2013; Haghighipour 2013; Leconte et al. 2014). Here, we consider the atmospheric chemistry of exo-Neptunes and super-Earths with widely diverse volatile contents, as well as the chemistry of outgassed atmospheres of smaller, hot, rocky planets. The atmospheric chemistry of terrestrial planets near the habitable zone is briefly discussed in Sect. 8.

5.1 Chemical Equilibrium in Exo-Neptune and Super-Earth Atmospheres

Thermochemical equilibrium can be maintained in high-temperature, high-pressure regions of exoplanet atmospheres (see Sect. 4.1), so equilibrium conditions are appropriate to consider to first order for any super-Earths or exo-Neptunes with thick, hot atmospheres. This statement is true whether the planet is a low-density fluid "ice giant", like Neptune itself, or has a thick atmosphere overlying a solid surface, like Venus. Even secondary atmospheres produced from the outgassing of interior volatiles during and after the accretion phase can lead to thick, high-pressure atmospheres that remain hot at depth due to accretional energy, radioactive decay in the interior, tidal heating, strong stellar irradiation, and/or a greenhouse effect.

Our own terrestrial planets have demonstrated the importance of secondary outgassing of interior volatiles in shaping the atmospheric properties of solid-surface planets (Pepin 2006). The theoretical equilibrium composition of super-Earth atmospheres dominated by such an outgassing source has been explored by several investigators. For example, Elkins-Tanton and Seager (2008) have examined the wide range of atmospheric masses and compositions that can result from the degassing of different meteoritic compositions during the planet's accretion phase. Schaefer and Fegley (2010) have performed similar more detailed calculations, albeit with a focus on the early Earth, to investigate the chemistry of potential steam atmospheres, such as those predicted to be associated with magma oceans (e.g., Abe and Matsui 1985) thought to form during the accretion of the Earth and other terrestrial (exo-)planets. Both Schaefer and Fegley (2010) and Elkins-Tanton and Seager (2008) find that the resulting atmospheric composition is a sensitive function of the assumed composition of the meteoritic material being accreted—water-dominated steam atmospheres occur only for CI and CM chondritic starting material. For other assumed meteoritic starting compositions, the atmospheres can be dominated by CO_2, N_2, H_2, CH_4, or CO, depending on the starting material composition and atmospheric temperatures (Schaefer and Fegley 2010). Schaefer and Fegley (2011) have also pursued the potential atmosphere-surface buffering of hot atmospheres in equilibrium with planetary surfaces.

The possible formation of exotic (by solar-system standards) silicate atmospheres on hot super-Earths, in which volatile elements such as H, C, N, S, and Cl have already escaped from the planet, is explored by Schaefer and Fegley (2009) and Miguel et al. (2011). In these calculations, the atmosphere is assumed to be in gas-melt equilibrium with a volatile-free magma ocean or partially molten lithosphere. Their results indicate that such atmospheres can be composed largely of atomic Na, O_2, O, with SiO, Fe, and/or Mg, depending on the planet's orbital distance and resulting atmospheric temperature. Ito et al. (2015) have performed a similar set of equilibrium calculations for super-Earth atmospheric compositions over a volatile-free magma ocean of various assumed compositions; they discuss the resulting thermal structure, spectroscopy, and detectability of such atmospheres. They find that SiO, in particular, affects the atmospheric opacity, causing thermal inversions and notable infrared emission signatures.

Schaefer et al. (2012) have extended their earlier 2010 investigation to consider a wider range of parameter space that might be relevant to outgassed atmospheres of hot terrestrial exoplanets and super-Earths, including considerations of variations in the relative abundances of elements such as H, C, and O. Again, these authors emphasize that the equilibrium atmospheric composition of such secondary outgassed atmospheres depends strongly on the source material and temperatures.

Outgassing from accreted material with the composition of the bulk silicate Earth or the terrestrial crust would result in a water-dominated atmosphere under conditions relative to GJ 1214b, with CO_2 as an important secondary component (see Fig. 5), whereas O_2 and Na could dominate in the hotter, lower-density atmosphere of CoRoT-7b (Schaefer et al. 2012; see also Léger et al. 2011). The chemistry of possible hot equilibrium atmospheres of Earth-like planets after giant impact events is explored by Lupu et al. (2014), who find that the main consituents are water and CO_2, along with smaller amounts of HCl, H_2, HF, CO, N_2, alkali halides, SO_2, and H_2S.

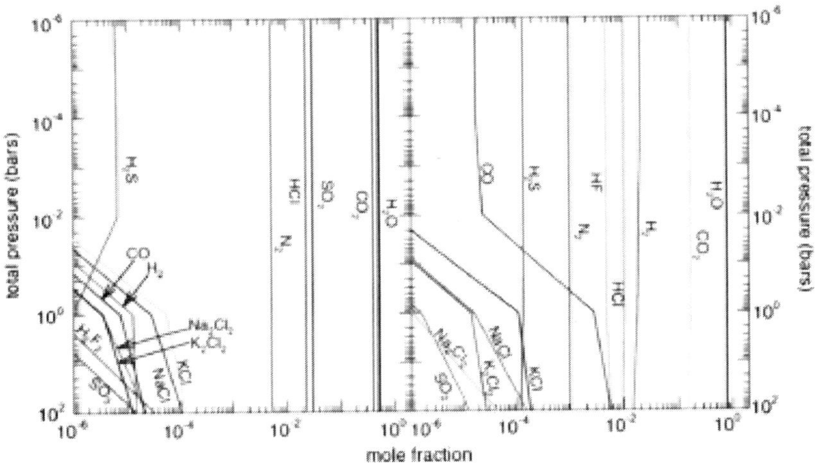

Figure 5. Equilibrium atmospheric composition for a GJ 1214b-like exoplanet, assuming the atmosphere results from outgassing from a high-temperature felsic silicate terrestrial crustal composition (*left*) or a more mafic bulk-Earth silicate composition (*right*). Figure from Schaefer et al. (2012)

The equilibrium atmospheric composition of warm super-Earths like GJ 581c is investigated by Miller-Ricci et al. (2009), for an assumed cometary-like complement of volatiles with variable hydrogen content. For hydrogen-rich atmospheres, the dominant constituents are H_2, H_2O, CH_4, and NH_3. For the hydrogen-poor situation, the resulting Venus-like atmosphere has dominant constituents CO_2 and N_2, with much less H_2O. For the "intermediate" hydrogen case, the atmosphere has an interesting mixture of dominant constituents like H_2O, H_2, CO_2, and CH_4. The overall hydrogen content therefore has a strong influence on the resulting composition.

Observations of exoplanets for which both mass and radius have been determined show a transition in planetary bulk densities near radii of ~1.5–2 R⊕R⊕; the smallest planets tend to be dense and presumably rocky, with at best a tiny mass fraction of light volatiles like hydrogen and helium, and larger planets tend to require increasingly larger mass fractions of H/He envelopes to explain the bulk density (cf. Weiss et al. 2013; Weiss and Marcy 2014; Marcy et al. 2014; Wu and Lithwick 2013; Hadden and Lithwick 2014; Rogers 2015). Population-synthesis models and other theoretical arguments suggest that the trend of decreasing hydrogen and helium content with decreasing planetary size is a natural consequence of planetary formation and evolution (Miller and Fortney 2011; Rogers et al. 2011; Lopez et al. 2012; Fortney et al. 2013; Owen and Wu 2013; Lopez and Fortney 2013, 2014; Benz et al. 2014; Wolfgang and Lopez 2015). Regardless of whether close-in exo-Neptunes and super-Earths with comparatively large H/He contents formed farther out in the disk and migrated/scattered inward (e.g., Ida and Lin 2005; Alibert et al. 2006; Terquem and Papaloizou 2007; Kennedy and Kenyon 2008; Mordasini et al. 2012; Inamdar and Schlichting 2015) or whether they formed in situ (Hansen and Murray 2012, 2013; Chiang and Laughlin 2013; Chatterjee and Tan 2014), the amount of hydrogen in the atmospheric envelope has a strong influence on the resulting atmospheric composition. The metallicity (or bulk mole fraction of H, in general) of the atmosphere is therefore an important parameter controlling the chemistry of low-density super-Earths and exo-Neptunes.

The effect of metallicity on the equilibrium composition, temperature structure, and/or spectra of exoplanet atmospheres has been studied by Lodders and Fegley (2002), Visscher et al. (2006, 2010a), and Fortney et al. (2008b) for metallicities up to 3–5× solar; Zahnle et al. (2009a,b), Spiegel et al. (2010), Fortney et al. (2010), Lewis et al. (2010), Line et al. (2011) for metallicities up to 30–50× solar; Agúndez et al. (2014b) and Venot et al. (2014) for metallicities up to 100× solar; and Moses et al. (2013a), Hu and Seager (2014), and Miguel et al. (2015) for very high metallicities (e.g., up to and beyond 1000× solar). These investigations reach concensus on several general trends. First, high-metallicity planets will have higher temperatures at lower pressures than otherwise similar low-metallicity planets, due to greater atmospheric opacity from the heavy (i.e., non-H_2) molecular constituents; that is, the photosphere will shift to higher altitudes. Second, the CO/CH_4 and N_2/NH_3 equal-abundance curves shift to lower temperatures with higher metallicities, leading to a higher likelihood that CO and N_2 will be important carbon and nitrogen components, respectively, of a high-metallicity planetary atmosphere. Third, molecules with multiple heavy elements, such as CO, CO_2, N_2, CS, S_2become favored at the expense of molecules that just contain one heavy element, like CH_4, NH_3, H_2S, as the metallicity increases—an effect that is particularly notable for molecules that contain more than two heavy elements, such as CO_2 and OCS (see Fig. 6). At sufficiently high metallicities, CO_2 will even replace H_2 as the dominant atmospheric constituent for an otherwise solar-composition atmosphere. Molecular hydrogen remains a major constituent of the atmosphere under the conditions studied in Fig. 6, but it ceases to dominate at the highest metallicities considered (e.g., several thousand times solar). Water increases roughly linearly with metallicity until metallicities of ~1000 times solar, at which point the lower H mole fraction begins to adversely impact all H-bearing molecules.

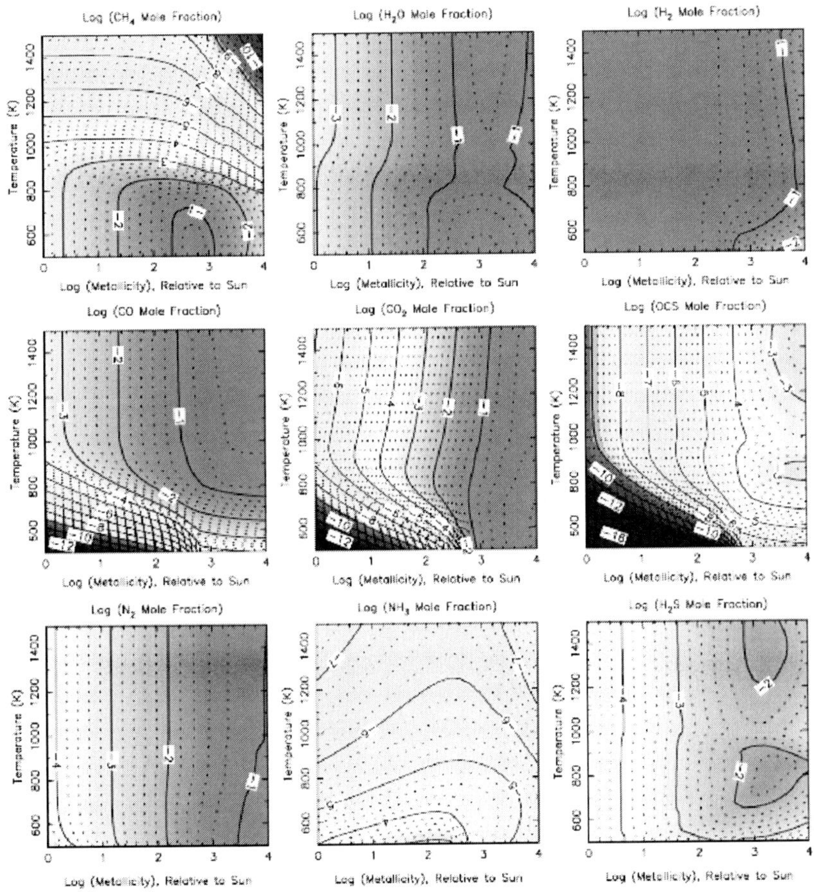

Figure 6. Equilibrium mixing ratios for various atmospheric constituents as a function of temperature and metallicity for an assumed typical photospheric pressure of 0.1 bar. Figure adapted from Moses et al. (2013a)

The expected depletion of hydrogen in super-Earths and exo-Neptunes due to inefficient accretion of the nebular gas and/or efficient escape of hydrogen at small orbital distances is therefore expected to lead to increased atmospheric mean molecular weights and a wide variety of interesting atmospheric compositions.

The sensitivity-to-metallicity calculations in Fig. 6 and in most of the investigations described in the previous paragraph have been performed assuming that the relative abundances of all elements other than H, He (and sometimes Ne) remain in solar proportion. Of course, that is not likely going to be true for exoplanetary atmospheres, given the different formation scenarios and evolutionary process at work. The sensitivity of the chemical equilibrium composition to changes in the C/O ratio at near-solar metallicities is discussed in Sect. 4.4. Moses et al. (2013a) examine the more general case of the sensitivity of the equilibrium atmospheric composition to both the C/O ratio and metallicity (i.e., to the relative abundances of H, C, and O) as a function of temperature, and Hu and Seager (2014) perform a similar general analysis considering disequilibrium processes like photochemistry (see also Sect. 5.2). Here is where the diversity of potential heavy-element-rich super-Earths and exo-Neptunes really stands out. The dominant equilibrium atmospheric constituent on intermediate-sized planets will typically be H_2 at low-enough metallicities (e.g., less than several hundred times solar) but can become H_2O at moderately high metallicities and subsolar C/O ratios, can become CO_2 at solar-like and subsolar C/O ratios and high metallicities, can potentially become CO at high metallicities and C/O ratios near unity (depending on graphite stability), and can even become O_2 at very low C/O ratios and high metallicities. At high C/O ratios and low metallicities, CH_4 is an important atmospheric component at low temperatures, while HCN and C_2H_2 become more important carbon phases at high temperatures. Hu and Seager (2014) suggest that hydrocarbons like C_2H_2 and C_2H_4 can even become dominant atmospheric constituents at bulk C/O ratios greater than ~2, but they have ignored graphite formation in their calculations. At high C/O ratios, the graphite stability field for equilibrium conditions is greatly expanded (see Moses et al. 2013a), and for a large range of temperature and metallicity conditions, graphite will condense out and sequester a significant fraction of the carbon when bulk C/O ratios are greater than ~1 (or even C/O <1 for high metallicities), leaving the remaining gas much less carbon-rich. Therefore, high-metallicity atmospheres can achieve an unusual state where the dominant gas is CO_2, even for bulk C/O ratios greatly exceeding unity (Moses et al. 2013a). It is also possible that atmospheric O may be lost preferentially by photodissociation of H_2O and subsequent hydrodynamic escape of O, thereby changing the C/O ratio (Chassefière 1996; Luger and Barnes 2015).

Between the potential for hot, silicate- and metal-rich outgassed atmospheres and the variety of volatile compositions available from inefficient accretion of nebular gas, the super-Earths and exo-Neptune population can be expected to have a rich diversity of atmospheric compositions from thermochemical equilibrium considerations, and disequilibrium processes (below) simply augment this possible diversity.

5.2 Disequilibrium Chemistry in Exo-Neptune and Super-Earth Atmospheres

Both photochemistry and transport-induced quenching can affect the atmospheric composition of intermediate-size planets, just as on giant planets (see Sects. 4.2 and 4.3). The first photochemical models specifically applied to intermediate-sized planets were those of Line et al. (2011) for the exo-Neptune GJ 436b and Miller-Ricci Kempton et al. (2012) for the super-Earth GJ 1214b, although the generic, higher-metallicity, hot-Jupiter models studied by Zahnle et al. (2009a,b) should also have relevance to some exoplanets in the intermediate-size range. Thermochemistry, photochemistry, and transport-induced quenching are considered in the Miller-Ricci Kempton et al. (2012) and Line et al. (2011) models, and the planets are assumed to have H_2-rich atmospheres with metallicities up to 30–50× solar. Given that both GJ 1214b and GJ 436b are expected to be relatively cool transiting planets, the results from both models are qualitatively similar. Methane is expected to be the dominant equilibrium carbon species for both planets up to 30–50× solar metallicity for the other conditions considered. At the quench point where interconversion between CH_4 and CO shuts down, methane is the main carbon component. However, transport-induced quenching causes CO to be more abundant than it otherwise would have been in equilibrium. This situation represents the opposite of the case for hotter planets, where quenching in the CO-dominated regime causes CO to be the major carbon component, with methane then being a less-abundant, but still important, quenched component (see Sect. 4.2). Photolysis of methane at high altitudes le to the production of C_2H_x hydrocarbons and, because of interactions with water photolysis products, the photochemical production of CO and CO_2. Coupled methane-ammonia photochemistry causes the production of HCN. However, methane is not removed from the photospheric region of either planet due to photochemistry (as was suggested as a possibility by Madhusudhan and Seager 2011), which is problematic, given that cloudless H_2-rich models with equilibrium methane abundances do not reproduce transit and eclipse observations of these planets (e.g., Stevenson et al. 2010; Désert et al. 2011; Bean et al. 2011; Berta et al. 2012; Kreidberg et al. 2014a).

This model-data mismatch led Moses et al. (2013a) to suggest that intermediate-sized planets could have much higher metallicities than previously considered, as long as their overall H content remains consistent with constraints supplied by the planet's bulk density. As discussed above, higher metallicities lead to hotter photospheres at lower pressures and shift the atmosphere toward the CO and CO_2 stability fields and away from the CH_4 stability field. Figure 7 shows how the atmospheric composition and photochemistry change for GJ 436b as the metallicity is assumed to increase from 1× solar to 10000× solar. The atmosphere transitions from being hydrogen-dominated, with abundant hydrogen-saturated components like H_2O, CH_4, and NH_3 and photochemically produced hydrocarbons and nitriles at low metallicities, to becoming CO_2-, CO-, H_2O-, and N_2-rich, with more oxidized photochemical products like O_2 and NO at high metallicities. Condensed graphite is a likely cloud component in the very-high-metallicity scenarios.

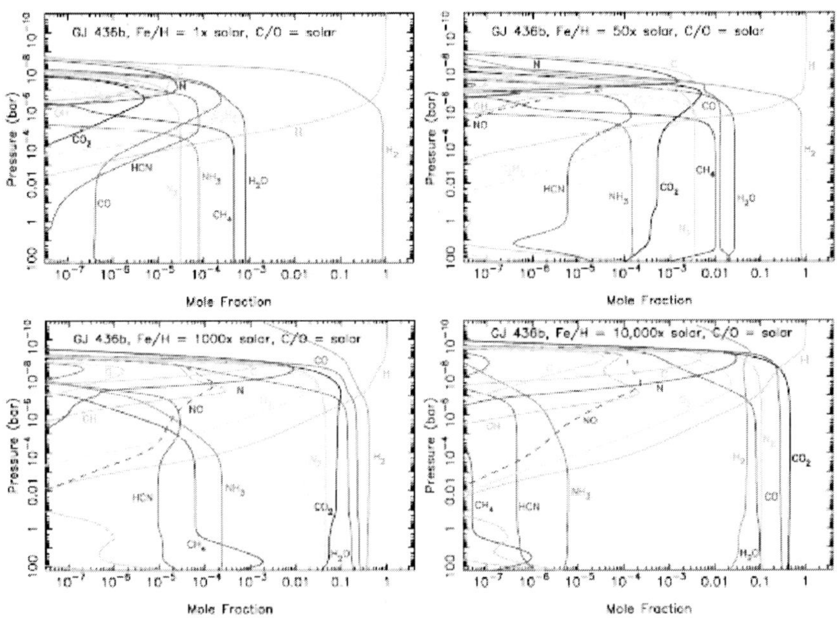

Figure 7. Vertical mixing-ratio profiles for several major atmospheric constituents in disequilibrium thermo/photochemical models of GJ 436b, for assumed metallicities of 1× solar (*top left*), 50× solar (*top right*), 1000× solar (*bottom left*), and 10000× solar (*bottom right*). Note the major change in the abundance of CO_2, CO, and CH_4 with the increase in metallicity (figure from Moses et al. 2013a)

Moses et al. (2013a) find that GJ 436b models with metallicities in the 230–2000× solar range provide the best overall consistency with both the planet's inferred interior structure (see Nettelmann et al. 2010, and references therein) and its apparent CO-rich, CH_4-poor dayside atmosphere and relatively flat transmission spectrum (e.g., Stevenson et al. 2010; Lanotte et al. 2014; Knutson et al. 2014a).

Venot et al. (2013) also consider the thermo/photochemistry of a presumed 100× solar metallicity atmosphere for GJ 436b, focusing in particular on how their new measurements of the CO_2 ultraviolet absorption cross sections at high temperatures (up to 800 K) affect the model abundances, and how stellar type affects the results. Venot et al. (2013) find that their results for the predicted abundances of NH_3, CO_2, and CO are especially affected by the temperature-sensitive CO_2 cross sections, with models that are irradiated by hotter stars (with their corresponding higher near-ultraviolet fluxes) exhibiting the biggest change in abundance due to use of high-temperature CO_2 cross sections.

Thermo/photochemical models for GJ 436b are also presented by Agúndez et al. (2014b) and Miguel et al. (2015). Agúndez et al. (2014b) explore the influence of tidal heating and metallicity on the thermal structure and resulting disequilibrium chemistry of planet, calculating the thermal structure self-consistently. Both greater tidal heating at depth and higher metallicities help shift the thermal structure into the CO dominated regime, again helping favor higher CO/CH_4 ratios and a greater CO_2 abundance. Miguel et al. (2015) examine the sensitivity of the atmospheric composition of GJ 436b to the flux in the stellar Lyman alpha line for both low-metallicity and high-metallicity models. They find that constituent abundances in the upper stratosphere at pressures less than ~0.1 mbar are affected by Lyman alpha, with stronger fluxes favoring the destruction of molecules and the resulting dominance of atomic species. Note, however, that when one considers the formation and presence of an extended hot thermosphere (García Muñoz 2007; Koskinen et al. 2013), which is the likely consequence of strong X-ray and EUV radiation received by close-in extrasolar planets—including GJ 436b (see Sanz-Forcada et al. 2011; Koskinen et al. 2014)—much or all of the Lyman alpha flux itself could be absorbed within the thermosphere (Lavvas et al. 2011), never reaching the stratosphere. On the other hand, other strong X-ray and EUV lines could have a similar effect to that described in Miguel et al. (2015), ultimately moving the base of the atomic-dominated hot thermosphere to deeper pressures for stronger EUV and X-ray fluxes.

Hu and Seager (2014) have examined the sensitivity of the disequilibrium atmospheric composition of several intermediate-sized exoplanets (e.g., GJ 1214b, HD 97658b, and 55 Cnc e) to the general variation of H, O, and C elemental abundances, using a thermo/photochemistry model that self-consistently calculates temperatures. Figure 8 shows their results for GJ 1214b. These results are similar to the equilibrium results shown in Fig. 6 in that the atmospheric composition can be highly diverse, depending on the relative abundances of the key volatile elements H, C, and O. At low metallicities, H_2 dominates, but GJ 1214b's atmosphere could also be dominated by H_2O, CO_2, CO, or O_2, depending on the overall H mole fraction and C/O ratio. Hu and Seager (2014) suggest that the atmosphere could be dominated by hydrocarbons such as C_2H_2 and C_2H_4 for high C/O ratios and moderately high metallicities; however, graphite condensation has been neglected in their calculations. Unless graphite formation is somehow kinetically inhibited (see Moses et al. 2013a), condensation of graphite will tie up much of the excess carbon, causing the remaining vapor to have lower C/O ratios, disfavoring hydrocarbons as the dominant constituents. Hu and Seager (2014) also show how the results for intermediate-sized planets can change with stellar flux (and thus temperature), and they suggest classification schemes for super-Earths with thick atmospheres based on the dominant constituents that appear for different H mole fractions and C/O ratios. In an earlier study, Hu et al. (2013) examine the fate of sulfur species from photochemistry in atmospheres of different dominant compositions relevant to super-Earths.

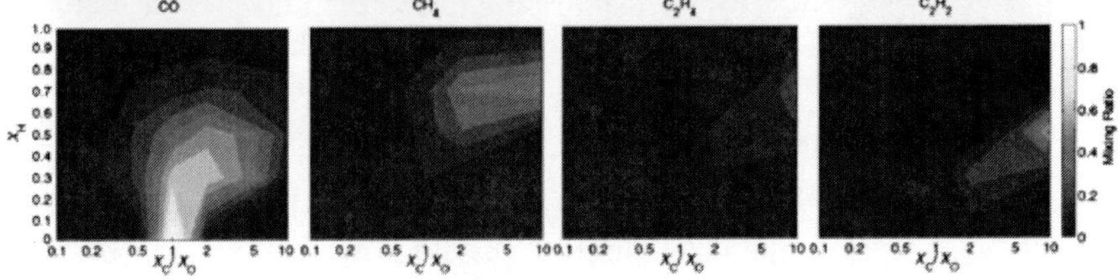

Figure 8. Mixing ratios for important atmospheric constituents vertically averaged over the pressure range 1–100 mbar from a disequilibrium thermo/photochemistry model of the super-Earth GJ 1214b, as a function of the bulk atmospheric H mole fraction (XHXH, where smaller values correspond to higher metallicities) and the C/O ratio (XC/XOXC/XO). Figure from Hu and Seager (2014)

Venot et al. (2014) have investigated the disequilibrium chemistry of the warm exo-Neptune GJ 3470b, using a grid of disequilibrium models with various metallicities, thermal profiles, incident ultraviolet fluxes, and eddy diffusion coefficients (specifying the strength of atmospheric mixing) to determine the possible sensitivity of the atmospheric composition to these parameters. They find that, like GJ 1214b and GJ 436b, methane is favored over CO as the dominant atmospheric constituent, except when both metallicities (up to 100× solar) and temperatures are high.

Conspicuously absent from the current literature is a study of how ion chemistry affects the gas-phase composition and possible formation of hazes on super-Earths and exo-Neptunes. Given that ion chemistry initiates the formation of high-molecular-weight organics in the N_2- and CH_4-rich upper atmosphere of Titan (Waite et al. 2007), it seems likely that ion chemistry would be interesting on intermediate-sized exoplanets (especially the cooler ones).

5.3 Clouds and Hazes in Exo-Neptune and Super-Earth Atmospheres

Basic concepts of cloud and haze formation in extrasolar-planet atmospheres are discussed in Sect. 4.5. The same physics and chemistry that was described for giant planets is relevant to intermediate-sized planets, but the different starting ingredients can lead to different possible aerosol compositions. Morley et al. (2013) provide a good discussion of the possible equilibrium condensates along the standard cloud condensation sequence for H_2-rich super-Earths and exo-Neptunes. In order from the hottest to the coldest dominant condensates, H_2-rich atmospheres could typically contain equilibrium clouds of Al-Ca-Ti oxides and silicates, Fe metal, Mg silicates, Cr metal, MnS, Na_2S, ZnS, KCl, $NH_4H_2PO_4$, H_2O, NH_4SH, NH_3, and CH_4. For a solar-metallicity atmosphere, the total available mass for some of these clouds (e.g., ZnS, KCl) is pretty sparse, suggesting that they would not be very optically thick in the vertical; however, Morley et al. (2013) demonstrate that this conclusion changes as the metallicity is increased. They find that ZnS and KCl clouds could obscure the transit spectra of a metal-rich GJ 1214b if the particles are lofted to sufficiently high altitudes and have sedimentation times that are sufficiently long (e.g., due to small particle sizes).

Other possibilities for aerosols on moderately H_2-rich planets include high-molecular-weight organics or "soots" that form through photochemical processes (e.g., Zahnle et al. 2009a; Moses et al. 2011; Miller-Ricci Kempton et al. 2012; Morley et al. 2013), or graphite that is stable in equilibrium for metal-rich atmospheres or ones with high C/O ratios (Moses et al. 2013a). Graphite can also be stable for a range of other compositions relevant to outgassed atmospheres of intermediate-sized planets (Schaefer and Fegley 2009, 2010), as can various other equilibrium condensates, such as alkali salts (Schaefer et al. 2012). Formation of sulfuric acid (H_2SO_4) clouds is likely through photochemical processes under a wide variety of conditions for high-metallicity or CO_2-rich atmospheres that have sufficient SO_2 (e.g., Schaefer and Fegley 2011; Schaefer et al. 2012; Hu et al. 2013), and elemental sulfur aerosols (e.g. S8) can be photochemically produced (e.g., Hu et al. 2013). However, the detailed formation pathways of photochemical hazes within the diverse atmospheres of intermediate-size planets has received relatively little attention and is still poorly understood.

6. OBSERVATIONAL INFERENCES OF CHEMICAL COMPOSITIONS

The last decade has witnessed substantial progress in observational inferences of chemical species in the atmospheres of giant exoplanets through a variety of methods. The planets for which such detections have been made are mostly hot giant planets, either in close-in orbits (i.e. 'hot Jupiters') or directly-imaged young objects at large orbital separations. The methods employed include differential photometry and spectroscopy of transiting hot Jupiters, high-resolution Doppler spectroscopy of transiting and non-transiting hot Jupiters, and high-resolution spectroscopy of directly-imaged planets. Whereas atomic species have been detected across the ultraviolet and visible, molecular species have been detected primarily in the near infrared. These detections were made thanks to pioneering observations using *Spitzer*, *HST*, and ground-based facilities. In what follows, we review the developments in each of these various areas.

6.1 Chemical Detections via Transit Spectroscopy

Atmospheric observations have been reported for over 50 transiting exoplanets to date, mostly as broadband/narrowband photometry or low resolution spectra. But, given the limited spectral resolution of the observations of most planets, molecular compositions have been inferred for only a few exoplanets to date. The inferred molecules typically include the most abundant and spectroscopically dominant molecules expected in hot atmospheres e.g. H_2O, CO, CH_4, and CO_2. On the other hand, inferences have also been made of several atomic species and the presence of clouds/hazes in some atmospheres.

6.1.1 Detections of Atomic Species

In high-temperature gas giant atmospheres (T≳1000 KT≳1000 K) in chemical equilibrium, alkali metals sodium (Na) and potassium (K) are expected to exist predominantly in atomic form thereby causing significant absorption in the visible via strong resonance lines at 589 nm and 770 nm, respectively (Seager and Sasselov 2000; Brown 2001). The strong line cores and wide pressure broadened wings are observable in optical transmission spectra of hot Jupiters. Thus the first Na detection was made using a visible band transmission spectrum of the hot Jupiter HD 209458b obtained using the HST STIS spectrograph (Charbonneau et al. 2002). However, while unambiguous, the spectrum revealed lower Na absorption than predicted by equilibrium models, indicating either lower atomic abundances than assumed or non-equilibrium processes such as rainout of condensed species and photoionization of Na/K (Fortney et al. 2003; Barman 2007).

The Na resonance doublet was also detected from ground in the optical transmission spectrum of the hot Jupiter HD 189733b (Redfield et al. 2008; Wyttenbach et al. 2015). Additionally, HST STIS observations of the transmission spectrum of HD 189733 also detected the Na absorption line core (Huitson et al. 2012). However, the ensemble of observations spanning UV to visible showed that the Na line is significantly weaker than that observed for HD 209458b, with a clear lack of broad line wings. Besides the weak Na I line, the spectrum was found to be largely featureless with a blue-ward slope which was consistent with the presence of strong scattering due to a thick haze of condensate grains (Vidal-Madjar et al. 2011; Pont et al. 2013). In recent years, Na I has been inferred in visible transmission spectra of a few other hot Jupiters, e.g. XO-2b (Sing et al. 2012), WASP-17b (Wood et al. 2011; Zhou and Bayliss 2012), HAT-P-1b (Nikolov et al. 2014), and WASP-12b (Burton et al. 2015).

Visible transmission spectra of hot Jupiters have also revealed other atomic species. Similar to Na, several studies have also detected the K resonance double at 770 nm using transmission spectra of hot Jupiters from both space and ground-based instruments, e.g., XO-2b (Sing et al. 2011a), and HD 80606 (Colón et al. 2012), HAT-P-1b (Wilson et al. 2015), WASP-31b (Sing et al. 2015). Other atomic species inferred in the visible include H αα at 656 nm (Jensen et al. 2012), Ca and Sc (Astudillo-Defru and Rojo 2013).

Several atomic species have also been detected in the exospheres of hot Jupiters using UV transmission spectroscopy. The first such observations in Ly αα revealed an extended envelope of escaping H in the hot Jupiter HD 209458b (Vidal-Madjar et al. 2003). Subsequent observations have revealed a rich population of atomic species in several hot Jupiter exospheres, e.g. H (Lecavelier Des Etangs et al. 2010; Bourrier et al. 2013), O and C (Vidal-Madjar et al. 2003), Si (Linsky et al. 2010; Schlawin et al. 2010), Mg (Fossati et al. 2010; Vidal-Madjar et al. 2013). Most recently, Ehrenreich et al. (2015) reported a detection of escaping atomic hydrogen in the exosphere of the hot Neptune GJ 436b.

6.1.2 Early Molecular Inferences

Early inferences of molecules in exoplanetary atmospheres were based on few channels of photometry or low-resolution spectra obtained using then available instruments on *Spitzer* and *HST*. For example, some early studies used 2–3 near-infrared photometric observations in transmission to infer the presence of H_2O at the day-night terminator regions of HD 209458b (Barman 2007) and HD 189733b Tinetti et al. (2007), but cf. Beaulieu et al. (2008) and Désert et al. (2009). Early attempts were also made to detect molecules in a handful of hot Jupiters using infrared spectroscopy with *HST* and *Spitzer*. For example, several studies used near-infrared spectra of hot Jupiters in transmission and/or emission obtained with the *HST*NICMOS instrument (1.8–2.3 μm) to report detections of H_2O, CH_4, CO, and/or CO_2 in the hot Jupiters HD 189733b (Swain et al. 2008, 2009b; Madhusudhan and Seager 2009), HD 209458b (Swain et al. 2009a), and XO-1b (Tinetti et al. 2010).

However, the uncertainties on the NICMOS observations have since been extensively debated in the literature leading to different molecular detection significances claimed by different teams, ranging from confident detections to no detections at the nominal 3-σσ significance (Gibson et al. 2011, 2012; Waldmann et al. 2013; Swain et al. 2014).

Similar molecular inferences were also made using observations over a longer spectral baseline using *Spitzer* photometry and spectroscopy. Grillmair et al. (2008) used *Spitzer IRS*spectroscopy to infer the presence of H_2O in the dayside atmosphere of HD 189733b using chemical equilibrium models (also see Todorov et al. 2014), whereas (Madhusudhan and Seager 2009) were able to place only an upper-limit on the possible abundances of H_2O using the same dataset. Using six-channel photometric and/or spectroscopic observations from *Spitzer* and/or *HST*, Madhusudhan and Seager (2009) also inferred the presence of H_2O, CH_4, CO, and/or CO_2 in the hot Jupiters HD 189733b and HD 209458b.

However, several of the *Spitzer* photometric observations have also since been revised drastically (Knutson et al. 2012; Diamond-Lowe et al. 2014), thanks to our improved understanding of the instrumental systematics and, hence, the observational uncertainties (but cf. Hansen et al. 2014). Therefore, early inferences based on such observations are not currently substantiated.

The above early inferences, were followed by more molecular inferences from multi-band photometry for smaller and/or more distant planets. Stevenson et al. (2010) and Madhusudhan and Seager (2011) used six-channel *Spitzer* photometry in thermal emission to infer the presence of CO and/or CO_2 and the absence of CH_4 in the dayside atmosphere of the hot Neptune GJ 436b. Given the relatively lower equilibrium temperature of the planet, the presence of CO and absence of CH_4 was suggested to be indicative of strong chemical disequilibrium and high metallicity in the atmosphere. While these data, and hence the molecular inferences, were originally contested by Beaulieu et al. (2011), recent comprehensive and independent analyses by Lanotte et al. (2014) support the original inferences of a methane poor and CO/CO_2 rich atmosphere in GJ 436b.

Another major observational advancement that followed was the possibility of detecting thermal emission from hot Jupiters in the near-infrared from ground (e.g. Sing and López-Morales 2009; Croll et al. 2010). The advantage of these measurements was that their spectral range, between ~0.9 μm and 2.3 μm, complemented that typically available from *Spitzer*photometry (3.6–8 μm) to provide a long spectral baseline to retrieve chemical species from the combined data. In the first such instance, observations of thermal emission in seven photometric bandpasses (between ~1.2–8 μm) were used to infer the presence of significant CO and CH_4 and lack of H_2O, and hence a C/O ratio ≥1, in the hot Jupiter WASP-12b (Madhusudhan et al. 2011a). Again, the *Spitzer* photometric observations were a subject of intense debate initially (Crossfield et al. 2012; Cowan et al. 2012a). However, subsequent multi-epoch *Spitzer* observations in the same bandpasses have reinstated the original data and the conclusions (Stevenson et al. 2014a) for the dayside atmosphere of WASP-12b.

6.1.3 Molecular Detections with HST WFC3

In recent years, near-infrared *HST* spectroscopy has led to detections of H_2O in the atmospheres of several transiting hot Jupiters. This has been made possible by the *HST* Wide Field Camera 3 (WFC3) spectrograph (McCullough and MacKenty 2012) which operates in the near-infrared range 1.1–1.8 μm, where water vapor has a strong absorption band, as shown in Figs. 9 and 10. Since H_2O is one of the most abundant molecules expected in giant exoplanetary atmospheres, the HST WFC3 instrument provides a unique opportunity to constrain H_2O abundances in such atmospheres. As shown in Fig. 9, Deming et al. (2013) reported the first WFC3 detection of H_2O in the transiting hot Jupiter HD 209458b using transmission spectroscopy, which probed the atmosphere at the day-night terminator region of the planet. The observations led to a clear 10-σσ detection of H_2O in a single transit thanks to the high brightness of the host star (V=7.6V=7.6).

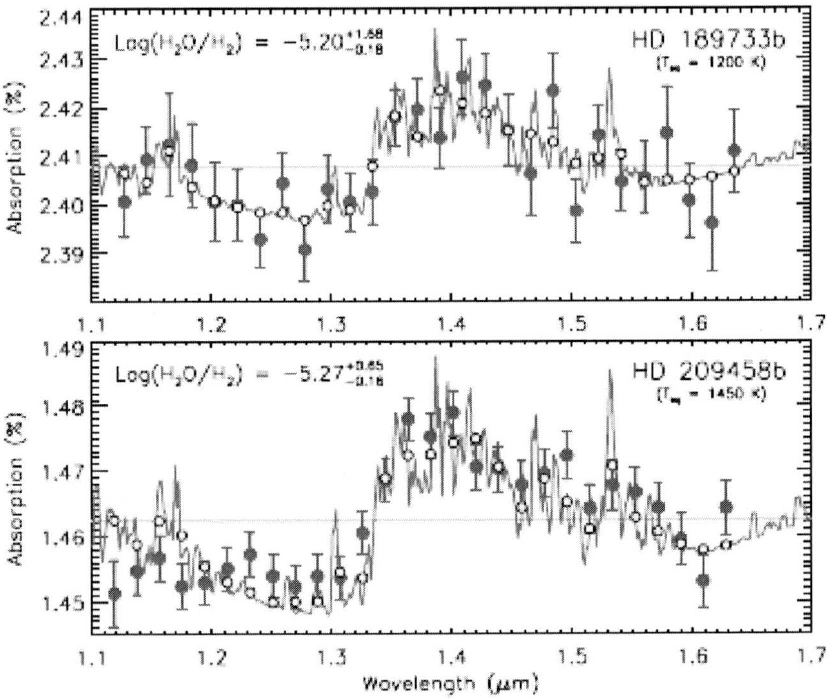

Figure 9. Transmission spectra of two hot Jupiters observed with HST WFC3 (adapted from Madhusudhan et al. 2014b). The *vertical axis* shows absorption (transit depth). The *blue circles* show the data: HD 209458b from Deming et al. (2013) and HD 189733b from McCullough et al. (2014). The *red curves* shows the best-fit model spectra, and the *cyan circles* show the models binned to the same resolution as the data. The *peaks* around 1.4 μm show H_2O absorption

Several subsequent studies reported single-event transmission spectra of transiting hot Jupiters orbiting less brighter stars with varied levels of success in H_2O detections. Some of the notable H_2O detections, albeit less significant than that for HD 209458b, were reported for hot Jupiters WASP-17b (Mandell et al. 2013), WASP-19b (Huitson et al. 2013), HAT-P-1b (Wakeford et al. 2013), WASP-12b (Stevenson et al. 2014a,b), and HD 189733b (McCullough et al. 2014), and a hot Neptune HAT-P-11b (Fraine et al. 2014), whereas similar observations of several other planets resulted in featureless spectra within the observed uncertainties (e.g. Swain et al. 2013; Mandell et al. 2013; Ranjan et al. 2014). These efforts demonstrate that single transits are inadequate to make high-confidence H_2O detections for most hot Jupiters currently known with the exception of those transiting the brightest stars such as HD 209458b. Consequently, more recent observations using multiple visits to reduce the observational uncertainties have led to good quality transmission spectra with clear H_2O detections for hot Jupiters orbiting even moderate brightness stars (Kreidberg et al. 2014a, 2015).

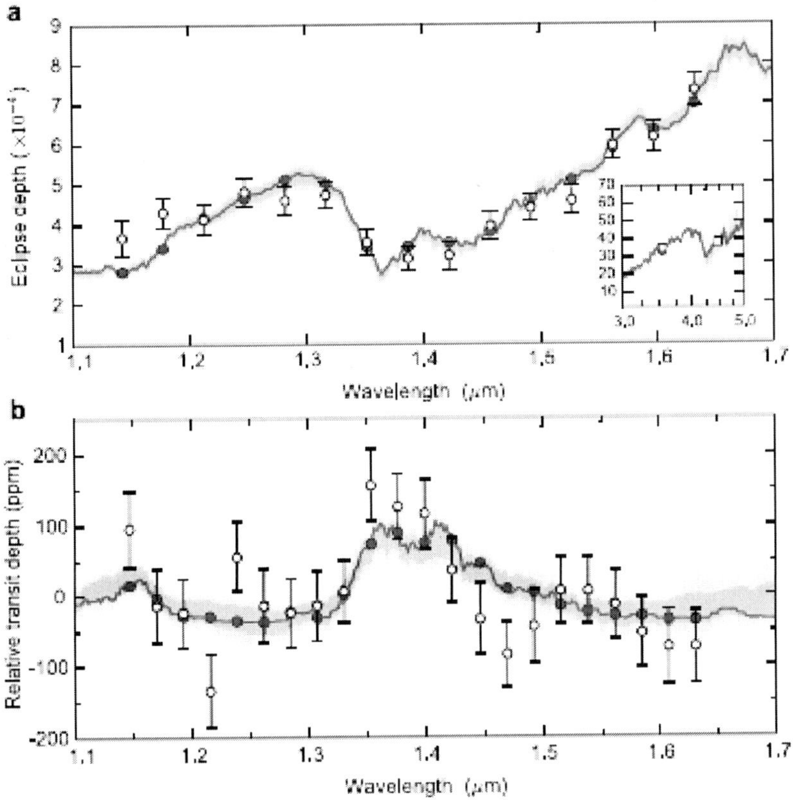

Figure 10. Spectra of the hot Jupiter WASP-43b observed with HST WFC3 (from Kreidberg et al. 2014b). *Top*: eclipse spectrum. *Bottom*: transmission spectrum. In both cases, the *black circles with error bars* show the data and *solid curves* show best-fit model spectra. The *filled dark blue circles* show the best-fit model binned to the same resolution as the data

Several studies have also used *HST* WFC3 to observe thermal emission spectra of hot Jupiters at secondary eclipse. In addition to the presence of molecular abundances, thermal emission spectra also provide constraints on the disk-averaged temperature profile of the dayside atmosphere.

However, with the exception of a few cases WFC3 observations to date have generally revealed thermal spectra with subdued molecular features, if any. The first such spectrum was observed for the very hot Jupiter WASP-12b (Teq~2500 KTeq~2500 K) which revealed a featureless thermal spectrum consistent with a blackbody spectrum (Swain et al. 2013; Madhusudhan 2012; Stevenson et al. 2014a). Considered on its own, the spectrum is consistent with an isothermal temperature structure in the atmosphere. On the other hand, considering together with existing photometric observations in the range 0.9–8 μm the observations require an atmosphere with a temperature profile decreasing outward and significantly depleted in H_2O as expected for a carbon-rich atmosphere, $C/O \geq 1$ (Madhusudhan 2012; Stevenson et al. 2014a). Subsequent studies have observed WFC3 thermal emission spectra for several other hot Jupiters (Wilkins et al. 2014; Ranjan et al. 2014; Stevenson et al. 2014a; Crouzet et al. 2014; Haynes et al. 2015).

Robust detections of molecular features in thermal emission spectra using HST WFC3 have been reported for the dayside atmospheres of only two hot Jupiters to date. Stevenson et al. (2014c) and Kreidberg et al. (2014b) reported unambiguous detection of H_2O in the dayside atmosphere of WASP-43b, as shown in Fig. 10. Most recently, Haynes et al. (2015) reported a detection of H_2O and the first detection of TiO in the dayside atmosphere of the extremely irradiated hot Jupiter WASP-33b.

6.1.4 Abundance Estimations Using Retrieval Methods

Going beyond detecting the presence of molecules, the exoplanetary spectra discussed above have also been used to derive the molecular abundances which in turn can be used to constrain atmospheric processes, elemental abundances, bulk compositions, and formation conditions. Molecular abundances are derived from exoplanetary spectra using detailed atmospheric retrieval methods which lead to joint statistical constraints on the chemical composition and temperature profile of an exoplanetary atmosphere given an observed spectrum; see e.g. Madhusudhan et al. (2014c) for a detailed review of atmospheric retrieval techniques for exoplanets. Atmospheric retrieval methods for exoplanets typically comprise of a 1-D atmospheric model coupled with an optimization algorithm to estimate the free parameters of the model given the data. The molecular abundances and the pressure-temperature (PP-TT) profile are free parameters in the model; typically there are over ten free parameters depending on the number of molecules included and the adopted parametrization of the temperature profile. For the optimization algorithm, a number of methods have been tried over time ranging from grid-search in the early days (Madhusudhan and Seager 2009) to Bayesian approaches such as the Markov Chain Monte Carlo (MCMC) method in subsequent years (Madhusudhan et al. 2011a; Line et al. 2012; Benneke and Seager 2013; Waldmann et al. 2015) and gradient descent methods (Lee et al. 2012).

Chemical abundances have been retrieved for several giant exoplanets to date. The majority of these planets are transiting hot Jupiters for which high-precision transmission and/or emission spectra have been obtained using the HST WFC3 spectrograph and, in some cases, Spitzer and/or ground-based photometry. Initial statistical constraints on atmospheric abundances (e.g. Madhusudhan and Seager 2009; Lee et al. 2012; Line et al. 2012) in the pre-WFC3 era were based on photometric data and/or low-resolution spectra, the data and uncertainties of which have since been revised as discussed above. On the other hand, while robust HST WFC3 spectra have been observed for over a dozen planets now, most of them have uncertainties large enough to be consistent with featureless spectra (Mandell et al. 2013; Ranjan et al. 2014). Therefore, in what follows we review only a selection of observed transiting exoplanets for which non-flat spectra have been observed at ≥ 3-$\sigma\sigma$ significance and for which abundance constraints have been reported. This includes five transiting hot Jupiters (HD 209458b, HD 189733b, WASP-12b, WASP-33b, WASP-43b), and a hot Neptune (HAT-P-11b).

Currently, the molecule with the best abundance estimates available is H_2O, thanks to high-precision spectra with HST WFC3. As discussed above, the HST WFC3 G141 grism with a spectral range of 1.1–1.8 μm covers a strong H_2O absorption band near 1.4 μm thereby allowing H_2O measurements from exoplanetary spectra. H_2O has been detected in all the six giant exoplanets listed above, albeit with different abundances. One important feature in almost all the transmission spectra with robust H_2O detections to date is that the amplitude of the H_2O absorption feature is typically smaller than model predictions for a clear (i.e. free of clouds/hazes) solar-abundance atmosphere. This was first noted for the hot Jupiter with the most-precise transmission spectrum, HD 209458b (Deming et al. 2013). For such spectra, assuming a clear atmosphere results in an H_2O abundance at the day-night terminator that is markedly sub-solar: 20–100× sub-solar for HD 209458b and 3–200× sub-solar for HD 189733b (Madhusudhan et al. 2014b). However, the derived low H_2O abundances are degenerate with the possibility of clouds or hazes at the terminator in these atmospheres (Sing et al. 2013; Benneke 2015), i.e. the same spectra can be explained with solar or super-solar abundances if high altitude clouds or hazes are allowed in the atmosphere. Transmission spectra of several other hot Jupiters were also found to be consistent with solar abundances when clouds/hazes are invoked or when the uncertainties are large enough to preclude high-precision abundance estimates (Huitson et al. 2013; Line et al. 2013; Barstow et al. 2014; Benneke 2015). Therefore, the possibility of clouds present a fundamental challenge in deriving chemical abundances from transmission spectra, which may be mitigated if optical transmission spectra are available (discussed in the following section) and/or where abundance estimates are available for the dayside atmospheres using thermal spectra.

Constraints on molecular abundances have also been reported in the dayside atmospheres of hot Jupiters using thermal emission spectra. An early example in this regard is the hot Jupiter WASP-12b, for which broadband photometric observations originally suggested a 3-σσ upper-limit on the H_2O abundance that was ~100× sub-solar, and implied a C/O ratio of ≥1 (Madhusudhan et al. 2011a). While the veracity of the Spitzer photometric observations of WASP-12b were a subject of substantial debate in the intervening years (Cowan et al. 2012a; Crossfield et al. 2012), latest multi-epoch Spitzer observations together with high-precision HST WFC3 spectra are consistent with the lack of significant H_2O in the dayside atmosphere of WASP-12b (Madhusudhan 2012; Stevenson et al. 2014b). Most recently, H_2O was detected at the day-night terminator region of the planetary atmosphere using a HST WFC3 transmission spectrum, however the abundance of H_2O detected was subject to model assumptions as discussed above (Kreidberg et al. 2015). Similar to the other datasets discussed above, a cloud-free atmosphere required a sub-solar H_2O abundance whereas the inclusion of clouds/hazes and/or the imposition of chemical equilibrium allowed the data to be consistent with solar abundance H_2O.

In another example, joint constraints on the H_2O abundance at both the terminator as well as the dayside atmosphere were obtained using transmission and emission spectra for the hot Jupiter WASP-43b resulting in an H_2O abundance of 0.3–3× solar, i.e. both sub-solar and marginally super-solar abundances are allowed by the data (Kreidberg et al. 2014b). Finally, abundance estimates for the dayside atmosphere of WASP-33b, the most irradiated hot Jupiter observed with HST WFC3 to date, also revealed a marginally sub-solar (~0.5×) H_2O abundance (Haynes et al. 2015). HST WFC3 thermal spectra for a few other hot Jupiters have also revealed relatively low-amplitude or non-existent H_2O features, hinting at the possibility of low H_2O abundances (e.g. Wilkins et al. 2014; Crouzet et al. 2014). While in principle clouds or hazes could also cause featureless (blackbody-like) spectra in thermal emission, the observed brightness temperatures of the spectra are quite high, indicative of the lower atmospheres, implying that they were unlikely to have been impeded by clouds/hazes in the atmosphere.

The observed molecular abundances can be used to constrain the O/H, C/H, and C/O ratio of the atmosphere (e.g. Madhusudhan et al. 2011a), which in turn are important to constrain exoplanetary formation conditions, as discussed in Sect. 6. However, estimating C/O ratios is non-trivial. Reliably estimating the C/O ratio for an exoplanetary atmosphere requires that the observations are able to constrain the molecular abundances of all the dominant O and C bearing molecules in a given atmosphere. As discussed in Sect. 3, for hot giant exoplanets, the dominant molecules are H_2O, CO, CH_4, and to a lesser extent CO_2, C_2H_2, and HCN, depending on the C/O ratio. Simultaneously constraining the abundances of all these molecules requires a long spectral baseline with observable bandpasses where these molecules have strong spectral features. This is, in principle, possible with the combination of HST WFC3 and the Spitzer IRAC Channels 3.6–8 μm (Madhusudhan 2012). Indeed, such constraints were placed for the few planets where such data are available in thermal emission allowing constraints on the C/O ratios of their dayside atmospheres. While a carbon-rich dayside atmosphere (C/O ≥ 1) has been reported for the hot Jupiter WASP-12b (Madhusudhan 2012; Stevenson et al. 2014b), oxygen-rich dayside atmospheres consistent with a solar C/O of 0.5 were reported for the hot Jupiters WASP-43b (Kreidberg et al. 2014b) and WASP-33b (Haynes et al. 2015).

On the other hand, determining the C/O ratio at the day-night terminator from transmission spectra is challenging and prone to misinterpretation. One of the main reason is that typically only WFC3 spectra are available in transmission, implying that stringent constraints are possible only on the H_2O abundance. In principle, the abundances of CH_4 and HCN may also be constrained in the WFC3 spectral range, however the most dominant O and C bearing molecule in hot Jupiters is CO, which is unconstrained by WFC3 observations. Consequently, with WFC3 data alone the C/H and hence the C/O ratio cannot be reliably constrained. For example, if a sub-solar H_2O abundance is observed in a hot Jupiter (assuming a cloud-free atmosphere), the low H_2O can be caused either by a low overall metallicity (i.e. low O/H) with a solar-like C/O ratio or by a high metallicity but a high C/O ratio (~1), depending on the temperature. Therefore, studies which attempt to derive C/O ratios from WFC3 transmission spectra alone do so by enforcing constraints of equilibrium chemistry or temperature profiles in radiative equilibrium or both (Benneke 2015; Kreidberg et al. 2015), neither of which is strictly justified at the day-night terminator regions of strongly irradiated atmospheres with strong atmospheric circulation. On the other hand, the presence of clouds only makes the inference more degenerate. Therefore, multi-wavelength observations besides WFC3 spectra are essential for reliable constraints on C/O ratios from transmission spectra.

6.1.5 Degeneracy with Clouds and Hazes

The presence of high-temperature condensates as hazes and clouds in hot exoplanetary atmospheres may contribute significantly to the optical depth in transmission spectra (Seager and Sasselov 2000; Fortney 2005). Therefore, determinations of chemical compositions from transmission spectra can be strongly degenerate with the presence of clouds or hazes in the atmospheres. In transmission spectra, the amplitude of a spectral feature is directly proportional to the height of the atmospheric annulus through which the star light traverses. The presence of an opaque cloud deck at a given height in the atmosphere means that only the layers above the cloud deck contribute to the transmission spectrum. Thus for a cloud deck located high enough in the atmosphere the amplitude of a spectral feature can be significantly diminished thereby confounding estimates of the corresponding molecular abundances.

High-precision observations of several transiting exoplanets have revealed such spectra with diminished features. At the extreme end are recent HST WFC3 observations of remarkably flat spectra, despite extremely high precisions, for super-Earths GJ 1214b (Kreidberg et al. 2014a) and HD 97658b (Knutson et al. 2014b), exo-Neptune GJ 436b (Knutson et al. 2014a), and exo-Uranus GJ 3470b (Ehrenreich et al. 2014). The featureless transmission spectra in these low-mass, low-temperature (TT below ~800 K), planets have been attributed to the presence of thick high-altitude clouds with cloud-top pressures below a mbar. On the other hand, even for hot Jupiters (T~1200--3000 KT~1200--3000 K) which show unambiguous H_2O features the amplitude of the spectral features are significantly diminished, corresponding to only ~1–2 scale heights (Deming et al. 2013; Kreidberg et al. 2015), compared to expectations of 5–8 scale heights (Madhusudhan and Redfield 2015). These diminished spectral features may be attributed to the presence of clouds/hazes in these hot Jupiters, but as discussed above the same features may also be attributed to significantly lower H_2O abundances than previously expected (Madhusudhan et al. 2014c; Benneke 2015).

Scattering due to haze particles have also been inferred from visible transmission spectra of hot Jupiters which show an inverse power-law dependence on wavelength with parameters different from that due to pure gaseous Rayleigh scattering. The archetype for such inferences is the hot Jupiter HD 189733b (Lecavelier Des Etangs et al. 2008b; Pont et al. 2008; Sing et al. 2011b) where observations with HST revealed a power law slope in the UV and optical indicating the presence of hazes, as shown in Fig. 11.

Similar inferences have been made for several other hot Jupiters in recent years (Sing et al. 2013, 2015; Nikolov et al. 2015). A potential degeneracy in such inferences arises from the fact that the presence of star spots may also cause similar optical spectra with inverse power law slopes (e.g. McCullough et al. 2014). However, in addition to the power-law slopes observations of partially muted alkali (Na and/or K) absorption lines in the optical have also indicated the presence of cloud decks in some atmospheres (Sing et al. 2013, 2015).

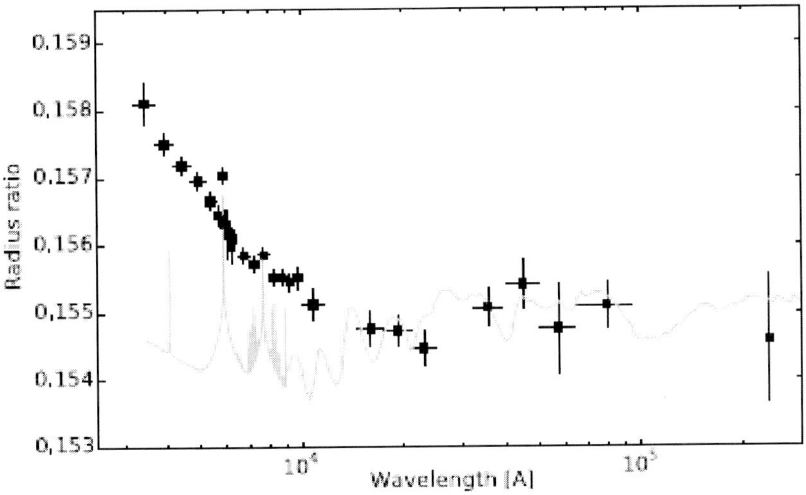

Figure 11. Transmission spectrum of the hot Jupiter HD 189733b from UV to near-infrared (from Pont et al. 2013). The *black circles with error bars* show the data available at that time, observed using multiple instruments on *HST* and *Spitzer*. The *grey curve* shows a model spectrum for a dust-free atmosphere. The *steep blue-ward rise* in the spectrum was indicative of strong scattering due to haze/dust in the atmosphere which was not consistent with the expected model spectrum for a haze-free atmosphere

6.1.6 Albedos and Optical Phase Curves

The presence of clouds/hazes have also been inferred using reflection spectra of hot Jupiters obtained at occultation. Early observations presented little evidence for clouds/hazes and suggested that hot Jupiters are dark, potentially due to strong absorption in the optical due to alkali metals Na and K. Such studies reported rather low geometric albedos for several hot Jupiters in the optical, e.g. 0.038±0.0450.038±0.045 for HD 209458b (Rowe et al. 2008) and 0.0136±0.00270.0136±0.0027 for TrES-2 (Kipping and Spiegel 2011). However, recent new observations are providing evidence for high geometric albedos and evidence for clouds/hazes in dayside spectra of hot Jupiters. Demory et al. (2011, 2013) reported a high visible geometric albedo of 0.35±0.020.35±0.02 for the hot Jupiter Kepler-7b observed in the Kepler photometric bandpass (~0.4–1.0 μm). Furthermore, based on the asymmetry in the visible phase curve they also suggested the presence of an inhomogeneous cloud cover in the atmosphere of Kepler-7b. In another study, Evans et al. (2013) reported a low-resolution albedo spectrum of the hot Jupiter HD 189733b in the 0.29–0.57 μm range using HST which nominally showed the albedo increasing blue-ward and inferred the presence of reflective clouds in the atmosphere; also see Barstow et al. (2014). Most recently, Martins et al. (2015) reported a high-resolution spectroscopic detection of reflected light for the non-transiting planet 51 Peg b which suggests a high geometric albedo (~0.5) of the planet. Several other recent studies have used photometric observations to understand the trends and/or report constraints on albedos and/or optical phase curves for a sizeable sample of exoplanets (e.g. Heng and Demory 2013; Esteves et al. 2015; Hu et al. 2015a; Schwartz and Cowan 2015).

While it is apparent from the above observations that clouds/hazes likely exist in at least a subset of irradiated atmospheres, the compositions of cloud forming species in currently unconstrained. However, as discussed in Sects. 3–4, theoretical studies over the past decade show that a wide range of high-temperature refractory condensates can form in such atmospheres in the 500–2000 K range. Such condensates range from NaCl, KCl, ZnS for TTbelow 1000 K to silicates, Fe, and Al_2O_3 at ~1000–2000 K temperatures (e.g. Sudarsky et al. 2003; Marley et al. 2013; Morley et al. 2013).

6.2 Molecular Detections via High-Resolution Doppler Spectroscopy

Robust detection of molecules in exoplanetary atmospheres have also been made using high-resolution Doppler spectroscopy in the near infrared. This technique involves the detection of molecular lines in the planetary spectrum that are shifted in wavelength due to the radial velocity of the planet (Brogi et al. 2012; Birkby et al. 2013). For close-in hot Jupiters, the orbital velocities are ~ kms−1~ kms−1 whereas the stellar orbital velocities are significantly lower (e.g. below ~100 ms−1~100 ms−1). Thus the spectrally shifted molecular lines of the planetary atmosphere are easily identifiable compared to those in the stellar spectrum as well as those in the telluric spectrum which is static. A template planetary spectrum including the sought after molecule is cross-correlated with the observed spectrum to detect the Doppler shift in the molecular lines with phase thereby revealing the presence of the molecule. Critical to this method, however, is the high resolution of the observed spectrum so that individual molecular lines can be resolved. Consequently, most of the success in this area has been achieved using the CRIRES instrument on the Very Large Telescope (VLT) with a spectral resolving power of 100000 in the near-infrared (Snellen et al. 2010). Moreover, as the planetary signal diminishes with the increased resolution the method has been successfully applied only to planets orbiting the brightest stars.

This method has been used to robustly detect CO and H_2O in several hot Jupiter atmospheres. The detections of these particular molecules are favored by the fact that they are expected to be the most dominant O and C bearing molecules in hot Jupiter atmospheres, especially for T>1300 KT>1300 K, and also have detectable absorption lines in the range ~2–3 μm where these atmospheres are most conducive to observe from ground. This wavelength range offers the optimal conditions because in the near-infrared the planet-star flux ratio increases with wavelength, however the background noise also increases with wavelength; therefore the K-band (around 2.1 μm) typically offers an optimal choice for ground based transit/eclipse spectroscopy. Snellen et al. (2010) reported the first detection of CO using this technique in the hot Jupiter HD 209458b in transit which in turn also led to a constraint on the day-night wind velocity in the planetary atmosphere. Subsequent observations have led to the detection of CO and H_2O in the dayside atmospheres of several transiting and non-transiting planets: ττBoo (Brogi et al. 2012; Rodler et al. 2012; Lockwood et al. 2014), 51 Peg b (Brogi et al. 2013), HD 189733b (de Kok et al. 2013; Birkby et al. 2013; Rodler et al. 2013), and HD 179949b (Brogi et al. 2014).

6.3 Directly Imaged Gas Giants

Direct imaging offers another avenue to characterize atmospheric compositions of exoplanets and works preferentially for a complementary region in planetary parameter space. Exoplanets detected to date via direct imaging are all young gas giants at large orbital separations. Their young ages (below ~100 Myr) implies high effective temperatures (~1000 K) and, hence, high planet-star flux contrasts in the near-infrared. On the other hand, their large orbital separations (beyond ~10 AU), which are set by the inner working angles of the instruments used, help in minimizing the stellar glare. Together these factors make young gas giants particularly favorable to near-infrared spectroscopy via direct imaging.

The advantage of direct imaging is that the detection of a planet simultaneously results in observation of its atmospheric thermal emission spectrum, i.e. atmospheric characterization can be pursued simultaneously with detection. This is in contrast to transiting exoplanets where the detections are generally made using photometric transit surveys and RV confirmations, while atmospheric spectra are obtained for optimal targets using follow-up observations with specialized instruments. On the other hand, the challenge with directly imaged planets is that generally only the atmospheric spectrum is available with little information about any other planet property; the mass, radius, and hence gravity, temperature, age, are all unknowns in modeling the planets. Therefore, typically, planetary evolution models are required along with atmospheric models to robustly constrain the atmospheric and physical parameters of the planets. Moreover, spectra are available at only one point of the orbital phase which means thermal phase curves are not observable precluding constraints on atmospheric properties with longitude. Nevertheless, the possibility of high-resolution absolute near-infrared spectroscopy for directly-imaged planets means that chemical signatures can be robustly detected in emission spectra using ground-based instruments.

Chemical detections have been reported for a few directly-imaged exoplanets in recent years. Given the expected temperature range of young giant planets, the dominant molecular species expected are H_2O, CO, and CH_4. All these molecules have been detected in one or more directly-imaged planets. Most of the detections have been reported for planets in the well characterized and nearby HR 8799 system. H_2O, CO, and CH_4 have been unambiguously detected in the planet HR 8799b using high-resolution spectroscopy in the H and K band (Barman et al. 2011, 2015), though the CH_4 absorption is weaker than expected from chemical equilibrium (Bowler et al. 2010; Barman et al. 2015). H_2O and CO have also been detected in high-resolution spectra of HR 8799c (Konopacky et al. 2013). The presence of CH_4 and other candidate molecules (e.g. NH_3, C_2H_2, CO_2, and/or HCN) have also been suggested using lower resolution spectra for the four planets in the HR 8799 system (Oppenheimer et al. 2013). Janson et al. (2013) reported an H-band detection of CH_4 in GJ 504b. More recently, CO was also detected unambiguously in ββ Pic b (Snellen et al. 2014) which also provided the first measurement of the spin period of an exoplanet.

In addition to molecular detections, spectra of directly imaged planets have also been use to place constraints on their elemental abundance ratios just as have been pursued for transiting hot Jupiters. However, determining elemental abundance ratios for directly imaged planets is challenging since most of the planetary properties (e.g. mass, radius, gravity, temperature) are not known a priori and hence need to be fit for, often requiring evolutionary models, in addition to the usual complexities introduced by the presence of clouds. Nevertheless, several studies have attempted to fit static models to spectra of directly-imaged planets to report nominal constraints on the elemental abundances. Using a model grid to fit spectroscopic observations of HR 8799 c, Konopacky et al. (2013) suggested a C/O ratio of 0.65+0.10−0.050.65−0.05+0.10 which is slightly higher than a solar value of 0.54. Lee et al. (2013) performed an atmospheric retrieval analysis for HR 8799b to suggest a high metallicity, with a mean molecular weight of 3.8 compared to 2.4 for solar abundance, and a C/O ratio ~1 in its atmosphere. More recently, Todorov et al. (2015) performed an atmospheric retrieval analysis to constrain the molecular abundances in the sub-stellar companion κκ and b and reported an H_2O abundance of ~10–4~10–4, nearly consistent with a solar abundance composition and those of several transiting hot Jupiters.

7. IMPLICATIONS FOR FORMATION CONDITIONS OF EXOPLANETS

Atmospheric elemental abundances of solar-system giant planets have led to important constraints on the origins of the solar system. In Jupiter's atmosphere, the abundances of C, N, S, Ar, Kr, and Xe, relative to H have been measured to be 2–3× solar values (Owen et al. 1999; Atreya and Wong 2005) which suggest substantial accretion of solids during its formation, and have been used as evidence for its metal-rich interior and its formation by core accretion. However, the abundance of O is not known for Jupiter. Given the low temperatures ($\lesssim$125 K) in Jupiter's observable troposphere, H_2O is condensed down to deeper layers of the atmosphere at pressures greater than ~10 bar. As the Galileo probe descended into Jupiter's atmosphere, its last measurement showed an H_2O abundance, and hence an O abundance, of 0.3× solar, which is 10× lower than the other elements (Atreya and Wong 2005). However, this measurement of Jupiter's H_2O abundance is considered to be a lower-limit since the region the probe entered is thought to be an anomalously dry spot in Jupiter's atmosphere. On the other hand, indirect constraints based on kinetics models explaining the observed CO abundance suggest an O abundance of 0.3–7.3× solar (Visscher et al. 2010b), which allows for a wide range of C/O ratios.

Taking the currently available lower-limit on O at its true abundance, implying more carbon than oxygen, would require unusual formation conditions in the early solar system. Lodders (2004) studied the hypothetical possibility and suggested that tarry planetesimals may have dominated Jupiter's accretion history, instead of planetesimals dominant in water-ice as expected from compositions of minor bodies in the solar system. Instead, using core accretion models Mousis et al. (2012) suggested that considering a solar-composition nebula predicts an O abundance of 3–7× solar. Therefore, a significantly lower observed O abundance would require substantial depletion of H_2O ice in Jupiter's formation region, unless the planetesimals are unusually carbon-rich. Thus, accurately measuring the O abundance is critical to constrain the formation conditions of Jupiter, and of the outer solar system in general. The H_2O abundance is similarly unknown for any other giant planet in the solar system. The upcoming Juno mission to Jupiter (Matousek 2007), therefore, aims to constrain its O abundance by measuring the H_2O abundance in Jupiter's atmosphere. In particular, the H_2O abundances play a central role in constraining exoplanetary formation conditions. Since O is cosmically the most abundant heavy element it is expected that H_2O is one of the most dominant volatile in interstellar and planet-forming environments (van Dishoeck et al. 2014). Thus, the H_2O abundances in giant planetary atmospheres are of fundamental importance to constrain the inventory of O that was available in planet's formation environments.

The O/H and C/O ratios are easier to measure for hot giant exoplanets than they are for solar-system giant planets (Madhusudhan 2012). The vast majority of extrasolar gas giants known have T~600--3000 KT~600--3000 K, thus hosting gaseous H_2O in their observable atmospheres accessible to spectroscopic observations. Other detectable gases include CH_4, CO, CO_2, and NH_3, depending on the temperature and incident irradiation. Measurements of such molecular abundances allow estimations of elemental abundances ratios involving H, C, O, and N. Such elemental abundances can in turn provide crucial clues regarding exoplanetary atmospheric processes, interior compositions, and formation mechanisms, just as pursued for solar system planets. This fortuitous opportunity makes hot giant exoplanets the perfect laboratories to investigate the origins of giant planets and the diversity of their atmospheres and interiors. Indeed, as discussed above, recent observations are already leading to unprecedented detections of chemical species in giant exoplanetary atmospheres. Such spectra have already led to clear H_2O detections in atmospheres of several transiting hot Jupiters, as discussed above (e.g. Deming et al. 2013; Kreidberg et al. 2014b; Madhusudhan et al. 2014b). H_2O, CO, and/or CH_4 have also been detected using high-resolution ground-based spectroscopy of hot Jupiters (e.g. Brogi et al. 2014), as well as directly imaged planets (e.g. Konopacky et al. 2013). Upcoming large facilities such as the *JWST* and *E-ELT* will further revolutionize the field.

New studies are beginning to investigate the influence of formation and migration histories of giant exoplanets on their observable chemical compositions. As discussed in previous sections, observations are suggesting the possibility of carbon-rich ($C/O \geq 1$) as well as oxygen-rich ($C/O < 1$) atmospheres in giant exoplanets (Madhusudhan et al. 2011a; Madhusudhan 2012) which are in turn motivating new ideas on the formation conditions of planetary systems. One of the key questions in this regard is about how C-rich gas giants can form around O-rich sun-like stars; the solar C/O is 0.5 (Asplund et al. 2009). As discussed above, an early investigation into this question was pursued in the context of Jupiter in the solar-system for which, as discussed above, only a lower limit on the O/H is known, which allows for a C/O > 1 in Jupiter. More recently, following the inference of C/O ≥ 1 in WASP-12b (Madhusudhan et al. 2011a), Öberg et al. (2011) suggested that C/O ratios in giant planetary envelopes depend on the formation location of the planets in the disk relative to the snow lines of major C and O bearing volatile species, such as H_2O, CO, and CO_2, since the C/O ratio of the gas approaches 1 outside the CO and CO_2 snow lines. By predominantly accreting such C-rich gas, more so than O-rich planetesimals, gas giants could, in principle, host C-rich atmospheres even when orbiting O-rich stars. On the other hand, it may also be possible that inherent inhomogeneities in the C/O ratios of the disk itself may contribute to higher C/O ratios of the planets relative to the host stars (Kuchner and Seager 2005; Madhusudhan et al. 2011b; Mousis et al. 2012; Moses et al. 2013b; Ali-Dib et al. 2014). The compositions of gas and solids accreted also depend on the physicochemical properties of the disk at the given location which are time-dependent (Helling et al. 2014; Marboeuf et al. 2014). Furthermore, some of the volatile elements can also be depleted due to dust formation and settling in the atmosphere (Moses et al. 2013b; Helling et al. 2014). These various scenarios predict different limits on the metallicities and C/O ratios of the giant planets, which high-precision observations of their atmospheres will be able to test in the near future.

Besides the various factors discussed above, the main formation mechanisms also contribute significantly to the planetary composition. Giant planets are thought to form via one of two primary mechanisms, core accretion (CI) versus gravitational instability (GI). In the CI model (Pollack et al. 1996), the planetary embryos start out as 10 Earth-mass cores in the protoplanetary disk that subsequently undergo runaway accretion of a large volume of gas and planetesimals to form a massive gaseous envelope. On the other hand, a GI in a young disk can cause rapid collapse of a large volume of ambient gas and solids to form a giant planet (Boss 2000). Both scenarios occur in planet-forming disks, but at different orbital separations. While CA is favored closer to the snow-line (within ~2–10 AU) because cores take too long to form at larger distances and only reach large masses after the disk has dispersed, GI is favored at larger distances ($\gtrsim$10 AU) where the disk can cool sufficiently on orbital timescales to fragment. In this regard, GI may be the favored mechanism for the formation of distant gas giants detected via direct imaging. However, because the different formation mechanisms are favorable at different orbital separations in the disk they may be expected to probe different chemical compositions in the disk. In principle, both CA and GI can cause significant metallicity enhancements or depletions depending on the specific accretion history (Helled and Bodenheimer 2010), but the relative elemental abundances (e.g. C, O, N, etc.) may be different depending on the formation location corresponding to the respective snow lines.

The existence of hot Jupiters in very close-in orbits (e.g. $\lesssim$0.1 AU) presents a key challenge to theories of giant planet formation. Neither GI nor CA discussed above is thought to operate in such a way that allows hot Jupiters to form in situ at their current locations close to the host stars. The disk cannot fragment at those distances (Gammie 2001; Rafikov 2005), and cores with sufficient mass to attract significant envelopes cannot form. Therefore, the existence of hot Jupiters requires some form of "migration" from their original formation locations to their present orbits. Two competing hypotheses suggest that the planets migrated either through interaction with the protoplanetary disk during their formation (Papaloizou et al. 2007), or by 'disk-free' mechanisms such as gravitational interactions with a third body (e.g. Rasio and Ford 1996; Fabrycky and Tremaine 2007). Measurements of spit-orbit misalignment (or stellar obliquity) for a large sample of hot Jupiters over the past decade (e.g. Triaud et al. 2010; Albrecht et al. 2012) have been advocated as possible metrics to distinguish between the two hypotheses for migration. In a simplistic view, migration through the disk was thought to likely align the orbital angular momentum vector of the planet with the stellar spin axis whereas disk-free migration via dynamical encounters could lead to spin-orbit misalignment. A significant number of large spin-orbit misalignments (or stellar obliquity) observed in hot Jupiter systems initially supported disk-free migration mechanisms (Winn et al. 2010). However, recent studies show that spin-orbit misalignments may also be caused by planet migration through disks that are themselves misaligned and due to star-disk interactions (Crida and Batygin 2014; Lai 2014). Consequently, observed dynamical properties of hot Jupiters have been unable to conclusively constrain their migration pathways.

Instead, recent studies suggest that chemical abundances of hot Jupiters could provide stronger constraints on their formation and migration pathways. As discussed above, the O/H, C/H, and C/O ratios can change substantially with their formation locations in the protoplanetary disk. Madhusudhan et al. (2014a) suggested that atmospheric metallicities of hot Jupiter atmospheres could potentially constrain their migration mechanisms as migration through the disk is more likely to cause metal enrichment due to planetesimal accretion compared to disk-free migration mechanisms. The results lead to three key predictions as evident from Fig. 12. Firstly, planets migrating through the disk always accrete solids efficiently enough to result in solar or super-solar abundances (top-right quadrant in Fig. 12). Secondly, the C and O abundances are enhanced or depleted together, i.e. no cases in the top-left or bottom-right quadrants. Thirdly, planets with sub-stellar O and C abundances could not result from disk-migration. These findings imply that elemental abundances of hot Jupiters could potentially provide important constraints not only on their local formation environments but also on the migration pathways.

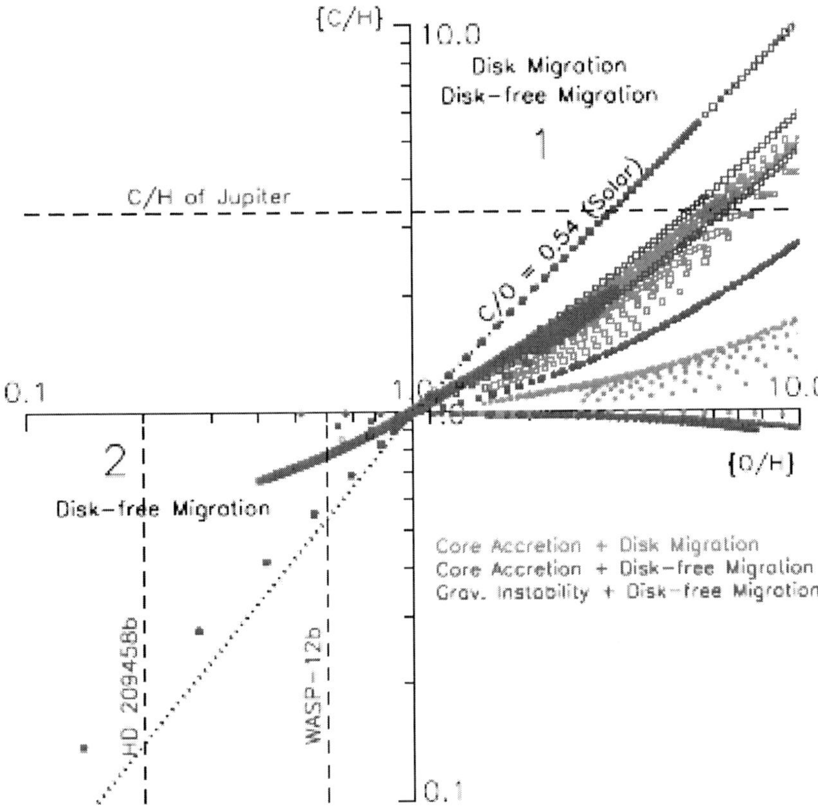

Figure 12. Predicted elemental O and C abundances in hot Jupiters from semi-analytic chemical models of hot Jupiter formation and migration (from Madhusudhan et al. 2014a). The *axes* show O/H and C/H abundances relative to solar values, and the *colored symbols*represent models with different formation-migration histories. *Each dot* represents a different model realization and shows the final O/H and C/H of the planet resulting from that model. The final elemental abundances in a planet depend critically on the amount of gas and solids accreted by the planet as well the region from which that material is accreted relative to the H_2O, CO_2, and CO *snow lines*, giving rise to the distinct loci of models in the O-C plane as shown. The results suggest that different migration mechanisms can cause different patterns of elemental enhancements in hot Jupiters

Current observations are already providing the first constraints on the O and C abundances which in turn are being used to constrain formation and migration mechanisms (Madhusudhan et al. 2014a). These efforts will be further bolstered in the future as the spectral range of JWST instruments will contain strong molecular features of several molecules which will allow precise abundance estimates for a wider range of elements (e.g. O, C, N, P, S, Si, etc.).

8. THEORY: CLIMATES OF HABITABLE EXOPLANETS

8.1 Habitability of Exoplanets

Existence of life requires many kinds of conditions. One of the critically necessary conditions for a planet to harbor life is permanent existence of liquid water on the planet's surface. To maintain liquid water, the planetary surface temperature has to be between the freezing point of water (0 °C) and the boiling point (100 °C), i.e., at 1 bar,

$$0\ ^\circ\mathrm{C} < T_s < 100\ ^\circ\mathrm{C}. \tag{1}$$

In reality, the upper limit of surface temperature has to be lower than the boiling point of water. Otherwise, the planet would fall into the runaway greenhouse state. We will return to this issue in the next section. To retain a fairly dense atmosphere and large amount of water against escape processes, the planetary mass has to be large enough so that the planet has sufficiently strong gravity. However, the planetary mass cannot be too large. Otherwise, it could have accreted a massive H_2-He envelope whose greenhouse effect along with the high pressure of the envelope would warm the surface and prevent water from being liquid, or it could be a Neptune-like ice-giant planet (Selsis et al. 2007; Venturini et al. 2015). Specifically, a habitable planet has to be a solid planet, and its mass should be in an approximate range given by:

$$0.5M_\oplus < M < 10M_\oplus, \tag{2}$$

where, M⊕M⊕ denotes Earth mass (Selsis et al. 2007).

Based on the above two conditions, an Earth Similarity Index (ESI) has been proposed (Schulze-Makuch et al. 2011) which attempts to quantify the similarity of a given property of an exoplanet to that of the Earth and ranges between 0.0 and 1.0. There are about 30 exoplanets that have ESI greater than 0.5. In other words, there are about 30 potentially habitable exoplanets among nearly 2000 discovered exoplanets. Gl 581c is the first exoplanet that had ever been thought to be habitable (Udry et al. 2007). However, radiative-convective model simulations indicated that Gl 581c is too hot to maintain liquid water (Selsis et al. 2007; Hu and Ding 2011). Especially, Hu and Ding (2011) showed that the climate of Gl 581c can readily fall into runaway greenhouse state even if the CO_2 level is extremely low and cloud albedo is extremely high. Gl 581d had also been considered a habitable super-Earth (Udry et al. 2007; Selsis et al. 2007; Von Bloh et al. 2007), and it is probably the first exoplanet whose habitability has drawn intensive studies. Wordsworth et al. (2010), Hu and Ding (2011), and Von Paris et al. (2011) all showed that at least 7 bars of CO_2 are required to maintain the surface temperature of Gl 581d above the freezing point of water. Kepler 186f, which has a size similar to Earth's, was also suggested to be a habitable exoplanet. However, calculations indicate that the surface temperature of Kepler 186f would be much colder than that of Gl 581d for same CO2 concentration because Kepler 186f receives less stellar radiation than Gl 581d does. So far, none of these potentially habitable exoplanets has been confirmed to be habitable.

8.2 Habitable Zones

Condition (1) defines a circumstellar zone in which a terrestrial planet can hold permanent liquid water on its surface. Such a zone is the so-called Habitable Zone (HZ). The HZ was first proposed by Shapley (1953) as the region around a star in which liquid water could exist on a planet's surface. The term 'habitable zone' was first introduced by Huang (1960), in the context of planetary habitability and extraterrestrial life. The HZ was more precisely defined in later works (Kasting et al. 1993; Kopparapu 2013), based on results from one-dimensional radiative-convective climate models. In these works, various factors that influence the width of the HZ, such as surface albedo, atmospheric compositions, and stellar radiation spectra, were considered. Especially, Kasting et al. (1993) extensively addressed how the inner and outer edges of HZ are constrained by the positive water-vapor feedback and by the saturation limit of the maximum CO_2 concentration, respectively. One might expect that the inner edge of HZ is at the distance where planetary surface temperature is equal the boiling point of water (100 °C). However, as shown by Ingersoll (1969), all liquid water on the planetary surface would completely evaporate into the atmosphere due to the positive water-vapor feedback once the surface temperature reaches about 70 °C, and the planet falls into the runaway greenhouse climate state. Then, water is lost throughout photolysis and hydrogen escape. The outer edge is constrained by the limit of CO_2 saturation, with which CO_2 condensation begins to take place. Thus, further increasing CO_2 does not enhance the greenhouse effect. Instead, latent heat release by CO_2 condensation even causes weakened greenhouse effect due to the decrease of lapse rate. Using the one-dimensional climate model, Kasting et al. (1993) found that conservative estimates of the inner and outer edges of HZ in our own Solar System is about 0.95 and 1.37 AU, respectively. Thus, the width of the solar HZ is about 0.42 AU. It is wider than the mean spacing between the four terrestrial planets in the Solar System, which is about 0.35 AU. It suggests that, statistically speaking, at least one of the terrestrial planets ought to be in the solar HZ. Kasting (2010) pointed out that the possibility that at least one rocky planet will be in the HZ should be high if such a statistical result can be applied to other exoplanetary systems. For darker stars like M dwarfs that have effective temperatures of about 3500 K, the HZ is much closer to the stars, at about 0.1 AU.

One uncertainty of the HZ width calculated from radiative-convective models is cloud radiative effects that are not included in these one-dimensional models. It was thought that the inner edge could be pushed closer to stars as the negative feedback of cloud reflection of stellar radiation is considered (Kasting et al. 1993; Selsis et al. 2007). Three-dimensional general circulation models (GCMs) include cloud radiative effects because GCMs have self-adjusted dynamical and physical processes of cloud formation. Several recent works have studied the HZ width with GCMs. However, Leconte et al. (2013) demonstrated that the inner edge of the solar HZ calculated from their GCM is nearly the same as that predicted by one-dimensional models. The reason why the inner edge does not move closer to Sun in their GCM is because of two important factors. One is that water-vapor absorption of solar radiation is higher in their model than previously assumed. The other one is that for high surface temperatures and water-rich atmosphere conditions high cirrus clouds increase in fraction faster than low stratus clouds. The former consists of ice particles and have greenhouse effects, and the later mainly reflects solar radiation and cools the surface. On the other hand, Yang et al. (2013) used a GCM to show that deep convective clouds develop around the substellar point for slow-rotating or tidally-locked exoplanets. These clouds largely reflect stellar radiation and cool the surface. As a result, the inner edge of the HZ is pushed much closer to the star, and the HZ width is nearly doubled.

It is worthwhile to point out that the cloud radiative effect on the inner edge of the HZ has large uncertainties. Parameterized clouds in GCMs are the primary source of uncertainty even in simulating modern climates where we have a wealth of direct observations. Thus, it is not clear to what extent cloud parameterizations in Earth GCMs can be applied to studying exoplanetary atmospheres. Using Earth GCMs to study the outer edge of the HZ also has great challenges. There is also lack of parameterizations of CO_2 clouds. Radiative effects of CO_2 clouds are not well understood (Forget and Pierrehumbert 1997).

8.3 Climates and Habitability of Tidal-Locking Exoplanets in the HZ of M Dwarfs

M dwarfs are the most common stars in the Universe (Rodono 1986). Thus, the probability of finding habitable exoplanets around M dwarfs is much higher than around other types of stars. Moreover, it is much easier to find habitable exoplanets closer to stars than that with greater distances with current observation techniques. Therefore, it is very likely to first discover habitable exoplanets around M dwarfs. Since M dwarfs have much weaker luminosities than Sun-like stars, the HZ around an M dwarf is typically about 0.1 AU, much closer than that for a solar-type star. However, the short distance of the HZ could also cause exoplanets in the HZ around M dwarfs to be uninhabitable, as discussed below.

First, exoplanets within such short distance to M dwarfs can be exposed to high levels of X-ray and extreme ultraviolet radiation (X-EUV) and strong particle fluxes from stellar winds or coronal mass ejections (Lammer et al. 2011). Such strong emissions are due to M dwarfs magnetic activity. It can generate significant atmospheric escapes to space and can even potentially cause atmosphere erosion or strip the whole atmosphere. For solar-type stars, such strong X-EUV emissions may be limited to the first few hundred millions years. However, the extreme irradiation could last several Gyr for M-type stars. Thus, exoplanets in the HZ of M-type stars may receive X-EUV fluxes that are 10–100 times higher than those in the HZ of solar-type stars of the same age (Selsis et al. 2007). X-EUV flux could cause a CO_2 dominant atmosphere with modest amount of free oxygen (O_2). Rapid escape of carbon from a CO_2dominant atmosphere could lead to the formation of O_2 even without the presence of biological O_2 production (Tian 2009). Photochemical models also suggest that O_3 and CH_4 can also be produced abiotically for different stellar characteristics and levels of volcanic outgassing (Domagal-Goldman et al. 2014). On the other hand, several other compounds, e.g. of sulfur, could act as potential biosignatures (Domagal-Goldman et al. 2011). At the current stage, more observational and quantitative modeling works are needed for understanding the effects of the extreme irradiation on atmospheres of exoplanets in the HZ around M dwarfs (see e.g. Segura et al. 2010).

Second, habitable exoplanets orbiting M-type stars are very likely to be tidally locked to their primaries due to strong forcing of gravitational gradients. Tidal-locking exoplanets receive very uneven stellar heating because their one side permanently faces their stars and the other side remains dark. While the dayside can be warm enough to sustain liquid water, the nightside could be so cold that any gases condense out there. If there exists atmosphere collapse, exoplanets in the HZ of M dwarfs would be uninhabitable. There were works that studied whether atmospheric heat transports could prevent atmosphere from collapse on the nightside using simplified or sophisticate models (Haberle et al. 1996; Joshi et al. 1997; Joshi 2003; Pierrehumbert 2011). They all showed that for a sufficiently dense atmosphere (surface air pressure greater than 0.1 bar) atmospheric heat transports would be able to warm the nightside and maintain nightside surface temperature above the condensation point of CO_2 (−78.5 °C at 1 bar of CO_2 partial pressure, and much lower at lower partial pressures). Therefore, atmosphere collapse can hardly happen as long as exoplanets are able to attract a fairly dense atmosphere envelope. However, more recent works suggest that the threshold of air pressure to prevent atmosphere collapse requires further investigation (Heng and Kopparla 2012; Wordsworth 2015).

The above studies have not taken ocean into account. If an exoplanet has an extensive ocean, its habitability also involves ocean heat transports and the positive sea-ice albedo feedback (Hu and Yang 2014). On the one hand, it is well known that ocean heat transports are equally important in Earth's climate (Peixoto and Oort 1992). In the presence of sea ice, ocean heat transports are likely to be especially important, since it is known from studies of the Snowball Earth phenomenon in Earth-like conditions that ocean heat transports are very effective in holding back the advance of the sea-ice margin (Pierrehumbert 2011; Yang et al. 2012a,b,c). On the other hand, in the presence of sea ice there exists the possibility that the exoplanet could be locked in a globally glaciated Snowball state due to the positive ice-albedo feedback. Note that the distribution of ice on tidally locked exoplanets is a particularly interesting issue only for M-type stars, since exoplanets orbiting hotter stars in orbits close enough to yield tidal-locking are likely to be too hot to permit ice, and may even be too hot to retain liquid water. Using a fully coupled atmospheric-oceanic GCM (AOGCM), Hu and Yang (2014) carried out simulations and demonstrated that ocean heat transports substantially extend the area of open water along the equator, and that the open-ocean area shows a lobster-like spatial pattern (Fig. 13), instead of an eyeball as shown in Pierrehumbert (2011). Figure 13 also shows that as CO_2 concentration increases, the dayside temperature does not increase very much, but the nightside temperature increases a lot, indicating that most heat is transported from the dayside to the nightside by ocean currents. They also showed that ocean heat transports can even lead to complete deglaciation of the nightside as greenhouse gas concentration is sufficiently high or stellar radiation is sufficiently strong. By contrast, the open-ocean area is not expanded very much by atmospheric heat transports alone for the same high level of CO_2. They also showed that it is more ready for aqua-planets to fall into runaway greenhouse and runaway freezing at the inner and outer edges, respectively, compared with simulation results without a dynamic ocean. It implies that the HZ would be narrower for aqua-planets or exoplanets with extensive oceans. These results indicate that ocean heat transports play critically important roles in determining the climate state and habitability of exoplanets.

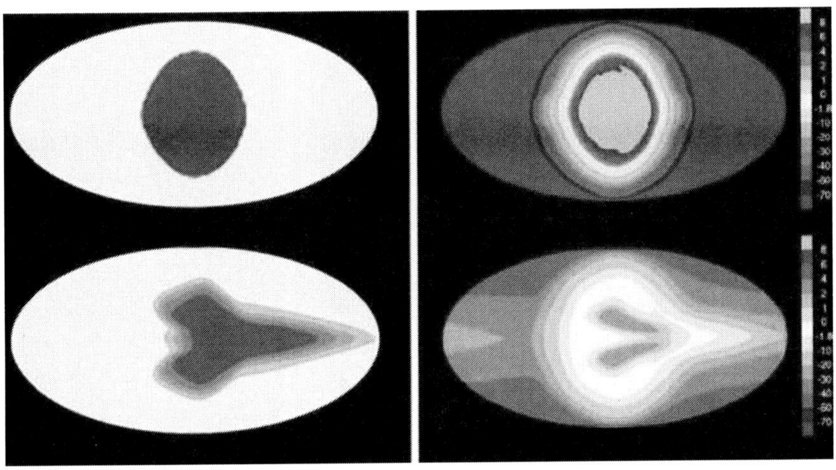

Figure 13. Spatial patterns of open-ocean areas and surface temperatures. *Left panels*: sea-ice fraction. *Blue colors* indicate open-ocean areas, and *light-gray* indicates ice coverage. *Right panels*: surface temperatures, *color bar* unit is °C. *Top panels*: with only atmospheric heat transport, and *bottom panels*: heat transports by both atmosphere and ocean. The substellar point is *at the center of each panel*. After Hu and Yang Hu and Yang (2014)

The third problem is whether water could be completely trapped on the nightside of tidal-locking exoplanets in the HZ of M dwarfs (Menou 2013). Although Fig. 13 shows that the lowest surface temperature on the nightside is about −40 °C, well above the condensation point of CO_2, it is still far below the freezing point of water. Therefore, there is the possibility of water trap on the nightside. Hu and Yang (2014) showed that as long as the ocean is not completely frozen, wind stresses transport sea ice toward the dayside and the ocean carries heat toward the nightside. As a result, sea-ice thickness on the nightside remains thin, less than 10 meters. Yang et al. (2014) studied that a case with a supercontinent on the nightside and an ocean on the dayside. In this case, ocean heat transports from the dayside to the nightside are ceased. It is found that ice sheets over the nightide supercontinent can grow 2000 m thick if geothermal heat flux is close to Earth's or smaller (Fig. 14). It suggests that if the dayside ocean is deeper than 2000 m, water will not be completely trapped on the nightside continent. Only exoplanets with a geothermal heat flux lower than Earths, and much of their surface covered by continents would be susceptible to complete water trapping.

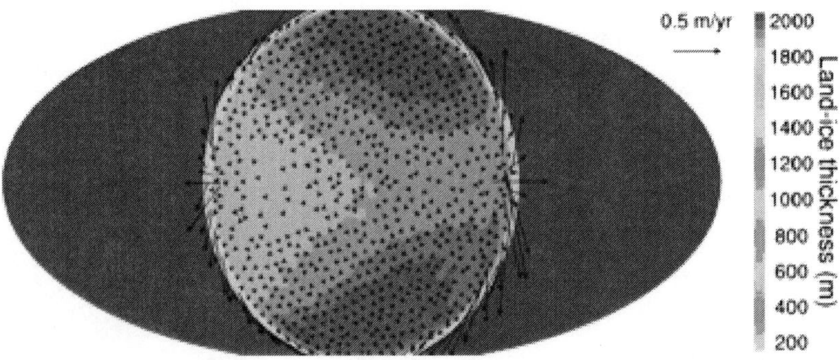

Figure 14. Ice sheets over a supercontinent spanning the entire nightside. *Color shading* indicates ice-sheet thickness, and *arrows* indicate ice-sheet flow velocity. Geothermal flux is $0.1\ Wm{-}20.1\ Wm{-}2$ in the simulation. Adapted from Yang et al. (2014)

8.4 Observable Signatures of Tidally-Locked Habitable Exoplanets

The above climate patterns have important implications for future observations. Showman et al. (2009), Showman and Polvani (2011), Heng and Showman (2015) showed that an equatorial superrotating jet stream develops in the atmosphere of tidal-locking exoplanets. The hot spot of tidal-locking exoplanets is advected to the downstream of the substellar point by the equatorial superrotating jet stream. Their simulation results are confirmed by observations for hot Jupiters (Knutson et al. 2009). It was shown that the hot spot is shifted 20°--$30^{\circ}20^{\circ}$--30° east of the substellar point of hot Jupiter HD 189733b. The equatorial superrotating flow is a common feature for tidal-locking exoplanets (Heng and Vogt 2011; Hu and Ding 2013). Thus, the hot-spot shift shall also be observed from the thermal phase curve of terrestrial exoplanets. In fact, the climate pattern in Fig. 13 also suggests eastward shift of the hot spot. Especially, as a dynamic ocean is considered, the equatorial ocean current, together with the equatorial atmospheric jet stream, can cause further downstream shift of the hot spot. Cowan et al. (2012b) studied how surface thermal inertia of Earth-like exoplanets le to distinguishable thermal phase curve for observations. They studied two types of climate states. One is the temperate climate which is just like the modern Earth with low- and middle-latitude oceans and polar ice-caps. The other one is the Snowball state. They found that the former has a relatively flat thermal phase curve because its relatively large surface thermal inertia tends to damp out the amplitude of the thermal phase curve. Wang et al. (2014) argued that terrestrial exoplanets around M dwarfs could have different climate states. They use an atmospheric GCM coupled with a slab ocean to show that exoplanets with nonzero eccentricities could have different spin-orbit resonance states different from the synchronous rotation state, and these exoplanets should have a striped-ball climate pattern, with a global belt of open ocean at low and middle latitudes and ice caps over both polar regions. This is in contrast to synchronous rotating habitable exoplanets around M dwarfs that have an eyeball climate pattern a limited region of open water on the dayside and ice on the rest of the planet (Pierrehumbert 2011; Hu and Yang 2014). They pointed out that the striped-ball climate state should be common for habitable exoplanets in large eccentric orbits around M dwarfs. They further suggested that these different climate patterns can be observed by future exoplanet detection missions. Overall, the features of climate patterns provide future observations of habitable exoplanets with additional constraints and information.

9. FUTURE PROSPECTS

Major observational advancements are happening in two key directions: (a) detections of exoplanets around bright stars, and (b) high precision observations of exoplanetary spectra. Firstly, bright exoplanet host stars are important to obtain precise measurements of exoplanetary masses, radii, and atmospheric spectra. Upcoming space missions, CHEOPS (Broeg et al. 2014) and TESS (Ricker et al. 2015), as well as various ground-based surveys (e.g. NGTS, SPECULOOS, MEARTH, etc.) are expected to find thousands of transiting exoplanets, including hundreds of super-Earths, orbiting nearby bright stars within the next five years (also see chapter on future landscape by Fridlund et al. 2016, this issue). During the same time, major ground-based direct imaging platforms (e.g. SPHERE, GPI, etc.) are also expected to find at least dozens of exoplanets at wide orbital separations orbiting nearby young stars. On a longer run, early 2020s, the recently selected ESA mission, PLATO (Rauer et al. 2014), will discover numerous transiting exoplanets in habitable zones of nearby stars. Secondly, major parallel efforts are also being pursued for spectroscopic observations of exoplanetary atmospheres. In this regard, enormous amount of time is currently being dedicated on the Hubble and Spitzer space telescopes, as well as major ground-based telescopes (e.g. VLT, Keck, Gemini, Magellan, CFHT, etc.). In addition, the James Webb Space Telescope (*JWST*) scheduled for launch in 2018 will revolutionize exoplanetary spectroscopy. By early 2020s, next generation ground-based telescopes, such as the European-Extremely Large Telescope (E-ELT), will further revolutionize the field. Future spectra with *JWST* would be of unprecedented precision and resolution which will enable us to derive precise chemical abundances for transiting exoplanets. In addition, JWST would also be able to observe atmospheres of much cooler low-mass planets that are possible to observe today.

ACKNOWLEDGEMENTS

This review was conceived during the 2014 ISSI workshop in Beijing on "The Disk in Relation to the Formation of Planets and their Protoatmospheres". N. Madhusudhan and M. Agúndez thank the International Space Science Institute (ISSI) and ISSI-Beijing for the hospitality during the workshop and the participants of the workshop for the lively discussions on exoplanetary atmospheres. M. Agúndez thanks K. Lodders for interesting discussions on atmospheric condensates. M. Agúndez acknowledges funding support from Spanish MINECO through grants CSD2009-00038, AYA2009-07304, and AYA2012-32032 and from the European Research Council (ERC Grant 610256: NANOCOSMOS). J. Moses thanks the NASA Exoplanet Research program NNX15AN82G for support. Y. Hu is supported by the National Natural Science Foundation of China 435 (NSFC) under grants 41375072 and 41530423.

REFERENCES

1. Y. Abe, T. Matsui, The formation of an impact-generated H_2O atmosphere and its implications for the early thermal history of the Earth. J. Geophys. Res. **90**, 545 (1985).
2. A.S. Ackerman, M.S. Marley, Precipitating condensation clouds in substellar atmospheres. Astrophys. J. **556**, 872–884 (2001).
3. M. Agúndez, O. Venot, N. Iro, F. Selsis, F. Hersant, E. Hébrard, M. Dobrijevic, The impact of atmospheric circulation on the chemistry of the hot Jupiter HD 209458b. Astron. Astrophys. **548**, 73 (2012).
4. M. Agúndez, V. Parmentier, O. Venot, F. Hersant, F. Selsis, Pseudo 2D chemical model of hot-Jupiter atmospheres: Application to HD 209458b and HD 189733b. Astron. Astrophys. **564**, 73 (2014a).

5. M. Agúndez, O. Venot, F. Selsis, N. Iro, The puzzling chemical composition of GJ 436b's atmosphere: Influence of tidal heating on the chemistry. Astrophys. J. **781**, 68 (2014b).
6. S. Albrecht, J.N. Winn, J.A. Johnson, A.W. Howard, G.W. Marcy, R.P. Butler, P. Arriagada, J.D. Crane, S.A. Shectman, I.B. Thompson, T. Hirano, G. Bakos, J.D. Hartman, Obliquities of hot Jupiter host stars: Evidence for tidal interactions and primordial misalignments. Astrophys. J. **757**, 18 (2012).
7. M. Ali-Dib, O. Mousis, J.-M. Petit, J.I. Lunine, Carbon-rich planet formation in a solar composition disk. Astrophys. J. **785**, 125 (2014).
8. Y. Alibert, C. Mordasini, W. Benz, C. Winisdoerffer, Models of giant planet formation with migration and disc evolution. Astron. Astrophys. **434**, 343–353 (2005).
9. Y. Alibert, I. Baraffe, W. Benz, G. Chabrier, C. Mordasini, C. Lovis, M. Mayor, F. Pepe, F. Bouchy, D. Queloz, S. Udry, Formation and structure of the three Neptune-mass planets system around HD 69830. Astron. Astrophys. **455**, 25–28 (2006).
10. F. Allard, P.H. Hauschildt, I. Baraffe, G. Chabrier, Synthetic spectra and mass determination of the brown dwarf Gl 229B. Astrophys. J. Lett. **465**, 123 (1996).
11. F. Allard, P.H. Hauschildt, D.R. Alexander, S. Starrfield, Model atmospheres of very low mass stars and brown dwarfs. Annu. Rev. Astron. Astrophys. **35**, 137–177 (1997).
12. M. Asplund, N. Grevesse, A.J. Sauval, P. Scott, The chemical composition of the Sun. Annu. Rev. Astron. Astrophys. **47**, 481–522 (2009).
13. N. Astudillo-Defru, P. Rojo, Ground-based detection of calcium and possibly scandium and hydrogen in the atmosphere of HD 209458b. Astron. Astrophys. **557**, 56 (2013).
14. S.K. Atreya, *Atmospheres and Ionospheres of the Outer Planets and Their Satellites* (1986)
15. S.K. Atreya, A.-S. Wong, Coupled clouds and chemistry of the giant planets—A case for multiprobes. Space Sci. Rev. **116**, 121–136 (2005).
16. Bai et al., Space Sci. Rev. (2016, this issue)
17. R.L. Bailey, C. Helling, G. Hodosán, C. Bilger, C.R. Stark, Ionization in atmospheres of brown dwarfs and extrasolar planets VI: Properties of large-scale discharge events. Astrophys. J. **784**, 43 (2014).
18. Baraffe, G. Chabrier, T. Barman, The physical properties of extra-solar planets. Rep. Prog. Phys. **73**(1), 016901 (2010).
19. T. Barman, Identification of absorption features in an extrasolar planet atmosphere. Astrophys. J. Lett. **661**, 191–194 (2007).
20. T.S. Barman, P.H. Hauschildt, F. Allard, Phase-dependent properties of extrasolar planet atmospheres. Astrophys. J. **632**, 1132–1139 (2005).
21. T.S. Barman, B. Macintosh, Q.M. Konopacky, C. Marois, Clouds and chemistry in the atmosphere of extrasolar planet HR8799b. Astrophys. J. **733**, 65 (2011).
22. T.S. Barman, Q.M. Konopacky, B. Macintosh, C. Marois, Simultaneous detection of water, methane, and carbon monoxide in the atmosphere of exoplanet HR8799b. Astrophys. J. **804**, 61 (2015).
23. J.K. Barstow, S. Aigrain, P.G.J. Irwin, T. Hackler, L.N. Fletcher, J.M. Lee, N.P. Gibson, Clouds on the hot Jupiter HD189733b: Constraints from the reflection spectrum. Astrophys. J. **786**, 154 (2014).
24. N.M. Batalha, J.F. Rowe, S.T. Bryson, T. Barclay, C.J. Burke, D.A. Caldwell, J.L. Christiansen, F. Mullally, S.E. Thompson, T.M. Brown, A.K. Dupree, D.C. Fabrycky, E.B. Ford, J.J. Fortney, R.L. Gilliland, H. Isaacson, D.W. Latham, G.W. Marcy, S.N. Quinn, D. Ragozzine, A. Shporer, W.J. Borucki, D.R. Ciardi, T.N. Gautier III, M.R. Haas, J.M. Jenkins, D.G. Koch, J.J. Lissauer, W. Rapin, G.S. Basri, A.P. Boss, L.A. Buchhave, J.A. Carter, D. Charbonneau, J. Christensen-Dalsgaard, B.D. Clarke, W.D. Cochran, B.-O. Demory, J.-M. Desert, E. Devore, L.R. Doyle, G.A. Esquerdo, M. Everett, F. Fressin, J.C. Geary, F.R. Girouard, A. Gould, J.R. Hall, M.J. Holman, A.W. Howard, S.B. Howell, K.A. Ibrahim, K. Kinemuchi, H. Kjeldsen, T.C. Klaus, J. Li, P.W. Lucas, S. Meibom, R.L. Morris, A. Prša, E. Quintana, D.T. Sanderfer, D. Sasselov, S.E. Seader, J.C. Smith, J.H. Steffen, M. Still, M.C. Stumpe, J.C. Tarter, P. Tenenbaum, G. Torres, J.D. Twicken, K. Uddin, J. Van Cleve, L. Walkowicz, W.F. Welsh, Planetary candidates observed by Kepler. III. Analysis of the first 16 months of data. Astrophys. J. Suppl. Ser. **204**, 24 (2013).

25. J.L. Bean, J.-M. Désert, P. Kabath, B. Stalder, S. Seager, E. Miller-Ricci Kempton, Z.K. Berta, D. Homeier, S. Walsh, A. Seifahrt, The optical and near-infrared transmission spectrum of the super-Earth GJ 1214b: Further evidence for a metal-rich atmosphere. Astrophys. J. **743**, 92 (2011).
26. J.P. Beaulieu, S. Carey, I. Ribas, G. Tinetti, Primary transit of the planet HD 189733b at 3.6 and 5.8 μm. Astrophys. J. **677**, 1343–1347 (2008).
27. J.-P. Beaulieu, G. Tinetti, D.M. Kipping, I. Ribas, R.J. Barber, J.Y.-K. Cho, I. Polichtchouk, J. Tennyson, S.N. Yurchenko, C.A. Griffith, V. Batista, I. Waldmann, S. Miller, S. Carey, O. Mousis, S.J. Fossey, A. Aylward, Methane in the atmosphere of the transiting hot Neptune GJ436B? Astrophys. J. **731**, 16 (2011).
28. B. Benneke, Strict upper limits on the carbon-to-oxygen ratios of eight hot Jupiters from self-consistent atmospheric retrieval. ArXiv e-prints (2015)
29. B. Benneke, S. Seager, How to distinguish between cloudy mini-Neptunes and water/volatile-dominated super-Earths. Astrophys. J. **778**, 153 (2013).
30. W. Benz, S. Ida, Y. Alibert, D. Lin, C. Mordasini, Planet population synthesis, in *Protostars and Planets VI*, ed. by H. Beuther, R.S. Klessen, C.P. Dullemond, T. Henning (2014), pp. 691–713.
31. Z.K. Berta, D. Charbonneau, J.-M. Désert, E. Miller-Ricci Kempton, P.R. McCullough, C.J. Burke, J.J. Fortney, J. Irwin, P. Nutzman, D. Homeier, The flat transmission spectrum of the super-Earth GJ1214b from wide field camera 3 on the Hubble Space Telescope. Astrophys. J. **747**, 35 (2012).
32. C. Bilger, P. Rimmer, C. Helling, Small hydrocarbon molecules in cloud-forming brown dwarf and giant gas planet atmospheres. Mon. Not. R. Astron. Soc. **435**, 1888–1903 (2013).
33. J.L. Birkby, R.J. de Kok, M. Brogi, E.J.W. de Mooij, H. Schwarz, S. Albrecht, I.A.G. Snellen, Detection of water absorption in the day side atmosphere of HD 189733 b using ground-based high-resolution spectroscopy at 3.2 μm. Mon. Not. R. Astron. Soc. **436**, 35–39 (2013).
34. J. Blum, G. Wurm, The growth mechanisms of macroscopic bodies in protoplanetary disks. Annu. Rev. Astron. Astrophys. **46**, 21–56 (2008).
35. W.J. Borucki, D.G. Koch, G. Basri, N. Batalha, T.M. Brown, S.T. Bryson, D. Caldwell, J. Christensen-Dalsgaard, W.D. Cochran, E. DeVore, E.W. Dunham, T.N. Gautier III, J.C. Geary, R. Gilliland, A. Gould, S.B. Howell, J.M. Jenkins, D.W. Latham, J.J. Lissauer, G.W. Marcy, J. Rowe, D. Sasselov, A. Boss, D. Charbonneau, D. Ciardi, L. Doyle, A.K. Dupree, E.B. Ford, J. Fortney, M.J. Holman, S. Seager, J.H. Steffen, J. Tarter, W.F. Welsh, C. Allen, L.A. Buchhave, J.L. Christiansen, B.D. Clarke, S. Das, J.-M. Désert, M. Endl, D. Fabrycky, F. Fressin, M. Haas, E. Horch, A. Howard, H. Isaacson, H. Kjeldsen, J. Kolodziejczak, C. Kulesa, J. Li, P.W. Lucas, P. Machalek, D. McCarthy, P. MacQueen, S. Meibom, T. Miquel, A. Prsa, S.N. Quinn, E.V. Quintana, D. Ragozzine, W. Sherry, A. Shporer, P. Tenenbaum, G. Torres, J.D. Twicken, J. Van Cleve, L. Walkowicz, F.C. Witteborn, M. Still, Characteristics of planetary candidates observed by Kepler. II. Analysis of the first four months of data. Astrophys. J. **736**, 19 (2011).
36. A.P. Boss, Giant planet formation by gravitational instability. Science **276**, 1836–1839 (1997).
37. A.P. Boss, Possible rapid gas giant planet formation in the solar nebula and other protoplanetary disks. Astrophys. J. Lett. **536**, 101–104 (2000).
38. V. Bourrier, A. Lecavelier des Etangs, H. Dupuy, D. Ehrenreich, A. Vidal-Madjar, G. Hébrard, G.E. Ballester, J.-M. Désert, R. Ferlet, D.K. Sing, P.J. Wheatley, Atmospheric escape from HD 189733b observed in H I Lyman-αα: Detailed analysis of HST/STIS September 2011 observations. Astron. Astrophys. **551**, 63 (2013).
39. B.P. Bowler, M.C. Liu, T.J. Dupuy, M.C. Cushing, Near-infrared spectroscopy of the extrasolar planet HR 8799 b. Astrophys. J. **723**, 850–868 (2010).
40. G.P. Brasseur, J.J. Orlando, G.S. Tyndall, *Atmospheric Chemistry and Global Change* (1999)
41. C. Broeg, W. Benz, N. Thomas (Cheops Team), The CHEOPS mission. Contrib. Astron. Obs. Skaln. Pleso **43**, 498 (2014)

42. M. Brogi, I.A.G. Snellen, R.J. de Kok, S. Albrecht, J. Birkby, E.J.W. de Mooij, The signature of orbital motion from the dayside of the planet ττ Boötis b. Nature **486**, 502–504 (2012).
43. M. Brogi, I.A.G. Snellen, R.J. de Kok, S. Albrecht, J.L. Birkby, E.J.W. de Mooij, Detection of molecular absorption in the dayside of exoplanet 51 Pegasi b? Astrophys. J. **767**, 27 (2013).
44. M. Brogi, R.J. de Kok, J.L. Birkby, H. Schwarz, I.A.G. Snellen, Carbon monoxide and water vapor in the atmosphere of the non-transiting exoplanet HD 179949 b. Astron. Astrophys. **565**, 124 (2014).
45. T.M. Brown, Transmission spectra as diagnostics of extrasolar giant planet atmospheres. Astrophys. J. **553**, 1006–1026 (2001).
46. C.J. Burke, S.T. Bryson, F. Mullally, J.F. Rowe, J.L. Christiansen, S.E. Thompson, J.L. Coughlin, M.R. Haas, N.M. Batalha, D.A. Caldwell, J.M. Jenkins, M. Still, T. Barclay, W.J. Borucki, W.J. Chaplin, D.R. Ciardi, B.D. Clarke, W.D. Cochran, B.-O. Demory, G.A. Esquerdo, T.N. Gautier III, R.L. Gilliland, F.R. Girouard, M. Havel, C.E. Henze, S.B. Howell, D. Huber, D.W. Latham, J. Li, R.C. Morehead, T.D. Morton, J. Pepper, E. Quintana, D. Ragozzine, S.E. Seader, Y. Shah, A. Shporer, P. Tenenbaum, J.D. Twicken, A. Wolfgang, Planetary candidates observed by Kepler IV: Planet sample from q1–q8 (22 months). Astrophys. J. Suppl. Ser. **210**, 19 (2014).

A. Burrows, C.M. Sharp, Chemical equilibrium abundances in brown dwarf and extrasolar giant planet atmospheres. Astrophys. J. **512**, 843–863 (1999).

47. Burrows, M. Marley, W.B. Hubbard, J.I. Lunine, T. Guillot, D. Saumon, R. Freedman, D. Sudarsky, C. Sharp, A nongray theory of extrasolar giant planets and brown dwarfs. Astrophys. J. **491**, 856–875 (1997)
48. Burrows, W.B. Hubbard, J.I. Lunine, J. Liebert, The theory of brown dwarfs and extrasolar giant planets. Rev. Mod. Phys. **73**, 719–765 (2001).
49. Burrows, D. Sudarsky, J.I. Lunine, Beyond the T dwarfs: Theoretical spectra, colors, and detectability of the coolest brown dwarfs. Astrophys. J. **596**, 587–596 (2003).
50. Burrows, D. Sudarsky, I. Hubeny, Theory for the secondary eclipse fluxes, spectra, atmospheres, and light curves of transiting extrasolar giant planets. Astrophys. J. **650**, 1140–1149 (2006).
51. Burrows, J. Budaj, I. Hubeny, Theoretical spectra and light curves of close-in extrasolar giant planets and comparison with data. Astrophys. J. **678**, 1436–1457 (2008).
52. J.R. Burton, C.A. Watson, P. Rodríguez-Gil, I. Skillen, S.P. Littlefair, S. Dhillon, D. Pollacco, Defocused transmission spectroscopy: A potential detection of sodium in the atmosphere of WASP-12b. Mon. Not. R. Astron. Soc. **446**, 1071–1082 (2015).
53. T. Castan, K. Menou, Atmospheres of hot super-Earths. Astrophys. J. Lett. **743**, 36 (2011).
54. D. Charbonneau, T.M. Brown, R.W. Noyes, R.L. Gilliland, Detection of an extrasolar planet atmosphere. Astrophys. J. **568**, 377–384 (2002).
55. E. Chassefière, Hydrodynamic escape of oxygen from primitive atmospheres: Applications to the cases of Venus and Mars. Icarus **124**, 537–552 (1996).
56. S. Chatterjee, J.C. Tan, Inside-out planet formation. Astrophys. J. **780**, 53 (2014).
57. E. Chiang, G. Laughlin, The minimum-mass extrasolar nebula: In situ formation of close-in super-Earths. Mon. Not. R. Astron. Soc. **431**, 3444–3455 (2013).
58. E. Chiang, A.N. Youdin, Forming planetesimals in solar and extrasolar nebulae. Annu. Rev. Earth Planet. Sci. **38**, 493–522 (2010).
59. J. Chilcote, T. Barman, M.P. Fitzgerald, J.R. Graham, J.E. Larkin, B. Macintosh, B. Bauman, A.S. Burrows, A. Cardwell, R.J. De Rosa, D. Dillon, R. Doyon, J. Dunn, D. Erikson, D. Gavel, S.J. Goodsell, M. Hartung, P. Hibon, P. Ingraham, P. Kalas, Q. Konopacky, J. Maire, F. Marchis, M.S. Marley, C. Marois, M. Millar-Blanchaer, K. Morzinski, A. Norton, R. Oppenheimer, D. Palmer, J. Patience, M. Perrin, L. Poyneer, L. Pueyo, F.T. Rantakyrö, N. Sadakuni, L. Saddlemyer, D. Savransky, A. Serio, A. Sivaramakrishnan, I. Song, R. Soummer, S. Thomas, J.K. Wallace, S. Wiktorowicz, S. Wolff, The first H-band spectrum of the giant planet ββPictoris b. Astrophys. J. Lett. **798**, 3 (2015).
60. F.J. Ciesla, J.N. Cuzzi, The evolution of the water distribution in a viscous protoplanetary disk. Icarus **181**, 178–204 (2006).

61. K.D. Colón, E.B. Ford, S. Redfield, J.J. Fortney, M. Shabram, H.J. Deeg, S. Mahadevan, Probing potassium in the atmosphere of HD 80606b with tunable filter transit spectrophotometry from the Gran Telescopio Canarias. Mon. Not. R. Astron. Soc. **419**, 2233–2250 (2012).
62. C.S. Cooper, A.P. Showman, Dynamics and disequilibrium carbon chemistry in hot Jupiter atmospheres, with application to HD 209458b. Astrophys. J. **649**, 1048–1063 (2006).
63. N.B. Cowan, P. Machalek, B. Croll, L.M. Shekhtman, A. Burrows, D. Deming, T. Greene, J.L. Hora, Thermal phase variations of WASP-12b: Defying predictions. Astrophys. J. **747**, 82 (2012a).
64. N.B. Cowan, A. Voigt, D.S. Abbot, Thermal phases of Earth-like planets: Estimating thermal inertia from eccentricity, obliquity, and diurnal forcing. Astrophys. J. **757**(1), 80 (2012b)

A. Crida, K. Batygin, Spin-orbit angle distribution and the origin of (mis)aligned hot Jupiters. Astron. Astrophys. **567**, 42 (2014).

B. Croll, L. Albert, D. Lafreniere, R. Jayawardhana, J.J. Fortney, Near-infrared thermal emission from the hot Jupiter TrES-2b: Ground-based detection of the secondary eclipse. Astrophys. J. **717**, 1084–1091 (2010).

65. I.J.M. Crossfield, T. Barman, B.M.S. Hansen, I. Tanaka, T. Kodama, Re-evaluating WASP-12b: Strong emission at 2.315 μm, deeper occultations, and an isothermal atmosphere. Astrophys. J. **760**, 140 (2012).
66. N. Crouzet, P.R. McCullough, D. Deming, N. Madhusudhan, Water vapor in the spectrum of the extrasolar planet HD 189733b. II. The eclipse. Astrophys. J. **795**, 166 (2014).
67. T. Currie, A. Burrows, Y. Itoh, S. Matsumura, M. Fukagawa, D. Apai, N. Madhusudhan, P.M. Hinz, T.J. Rodigas, M. Kasper, T.-S. Pyo, S. Ogino, A combined Subaru/VLT/MMT 1-5 μm study of planets orbiting HR 8799: Implications for atmospheric properties, masses, and formation. Astrophys. J. **729**, 128 (2011).
68. G. D'Angelo, R.H. Durisen, J.J. Lissauer, Giant planet formation, in *Exoplanets*, ed. by S. Seager (2010), pp. 319–346
69. R.J. de Kok, M. Brogi, I.A.G. Snellen, J. Birkby, S. Albrecht, E.J.W. de Mooij, Detection of carbon monoxide in the high-resolution day-side spectrum of the exoplanet HD 189733b. Astron. Astrophys. **554**, 82 (2013).
70. D. Deming, A. Wilkins, P. McCullough, A. Burrows, J.J. Fortney, E. Agol, I. Dobbs-Dixon, N. Madhusudhan, N. Crouzet, J.-M. Desert, R.L. Gilliland, K. Haynes, H.A. Knutson, M. Line, Z. Magic, A.M. Mandell, S. Ranjan, D. Charbonneau, M. Clampin, S. Seager, A.P. Showman, Infrared transmission spectroscopy of the exoplanets HD 209458b and XO-1b using the wide field camera-3 on the Hubble Space Telescope. Astrophys. J. **774**, 95 (2013).
71. B.-O. Demory, S. Seager, N. Madhusudhan, H. Kjeldsen, J. Christensen-Dalsgaard, M. Gillon, J.F. Rowe, W.F. Welsh, E.R. Adams, A. Dupree, D. McCarthy, C. Kulesa, W.J. Borucki, D.G. Koch, The high albedo of the hot Jupiter Kepler-7 b. Astrophys. J. Lett. **735**, 12 (2011).
72. B.-O. Demory, J. de Wit, N. Lewis, J. Fortney, A. Zsom, S. Seager, H. Knutson, K. Heng, N. Madhusudhan, M. Gillon, T. Barclay, J.-M. Desert, V. Parmentier, N.B. Cowan, Inference of inhomogeneous clouds in an exoplanet atmosphere. Astrophys. J. Lett. **776**, 25 (2013).
73. J.-M. Désert, A. Vidal-Madjar, A. Lecavelier Des Etangs, D. Sing, D. Ehrenreich, G. Hébrard, R. Ferlet, TiO and VO broad band absorption features in the optical spectrum of the atmosphere of the hot-Jupiter <ASTROBJ>HD 209458b</ASTROBJ>. Astron. Astrophys. **492**, 585–592 (2008).
74. J.-M. Désert, A. Lecavelier des Etangs, G. Hébrard, D.K. Sing, D. Ehrenreich, R. Ferlet, A. Vidal-Madjar, Search for carbon monoxide in the atmosphere of the transiting exoplanet HD 189733b. Astrophys. J. **699**, 478–485 (2009).
75. J.-M. Désert, J. Bean, E. Miller-Ricci Kempton, Z.K. Berta, D. Charbonneau, J. Irwin, J. Fortney, C.J. Burke, P. Nutzman, Observational evidence for a metal-rich atmosphere on the super-Earth GJ1214b. Astrophys. J. Lett. **731**, 40 (2011).
76. H. Diamond-Lowe, K.B. Stevenson, J.L. Bean, M.R. Line, J.J. Fortney, New analysis indicates no thermal inversion in the atmosphere of HD 209458b. Astrophys. J. **796**, 66 (2014).
77. S.E. Dodson-Robinson, P. Bodenheimer, The formation of Uranus and Neptune in solid-rich feeding zones: Connecting chemistry and dynamics. Icarus **207**, 491–498 (2010).

78. S.E. Dodson-Robinson, K. Willacy, P. Bodenheimer, N.J. Turner, C.A. Beichman, Ice lines, planetesimal composition and solid surface density in the solar nebula. Icarus **200**, 672–693 (2009).
79. S.D. Domagal-Goldman, V.S. Meadows, M.W. Claire, J.F. Kasting, Using biogenic sulfur gases as remotely detectable biosignatures on anoxic planets. Astrobiology **11**, 419–441 (2011).
80. S.D. Domagal-Goldman, A. Segura, M.W. Claire, T.D. Robinson, V.S. Meadows, Abiotic ozone and oxygen in atmospheres similar to prebiotic Earth. Astrophys. J. **792**, 90 (2014).
81. D. Ehrenreich, X. Bonfils, C. Lovis, X. Delfosse, T. Forveille, M. Mayor, V. Neves, N.C. Santos, S. Udry, D. Ségransan, Near-infrared transmission spectrum of the warm-Uranus GJ 3470b with the wide field camera-3 on the Hubble Space Telescope. Astron. Astrophys. **570**, 89 (2014).
82. D. Ehrenreich, V. Bourrier, P.J. Wheatley, A. Lecavelier des Etangs, G. Hébrard, S. Udry, X. Bonfils, X. Delfosse, J.-M. Désert, D.K. Sing, A. Vidal-Madjar, A giant comet-like cloud of hydrogen escaping the warm Neptune-mass exoplanet GJ 436b. Nature **522**, 459–461 (2015).
83. L.T. Elkins-Tanton, S. Seager, Ranges of atmospheric mass and composition of super-Earth exoplanets. Astrophys. J. **685**, 1237–1246 (2008).
84. L.J. Esteves, E.J.W. De Mooij, R. Jayawardhana, Changing phases of alien worlds: Probing atmospheres of Kepler planets with high-precision photometry. Astrophys. J. **804**, 150 (2015).
85. T.M. Evans, F. Pont, D.K. Sing, S. Aigrain, J.K. Barstow, J.-M. Désert, N. Gibson, K. Heng, H.A. Knutson, A. Lecavelier des Etangs, The deep blue color of HD 189733b: Albedo measurements with Hubble Space Telescope/space telescope imaging spectrograph at visible wavelengths. Astrophys. J. Lett. **772**, 16 (2013).
86. D. Fabrycky, S. Tremaine, Shrinking binary and planetary orbits by Kozai cycles with tidal friction. Astrophys. J. **669**, 1298–1315 (2007).
87. Fang et al., Space Sci. Rev. (2016, this issue)
88. Fegley Jr., K. Lodders, Chemical models of the deep atmospheres of Jupiter and Saturn. Icarus **110**, 117–154 (1994).
89. Fegley Jr., K. Lodders, Atmospheric chemistry of the brown dwarf Gliese 229B: Thermochemical equilibrium predictions. Astrophys. J. Lett. **472**, 37 (1996).
90. D.A. Fischer, A.W. Howard, G.P. Laughlin, B. Macintosh, S. Mahadevan, J. Sahlmann, J.C. Yee, Exoplanet detection techniques, in *Protostars and Planets VI* (2014), pp. 715–737.
91. F. Forget, R.T. Pierrehumbert, Warming early Mars with carbon dioxide clouds that scatter infrared radiation. Science **278**(5341), 1273–1276 (1997)
92. F. Forget, F. Hourdin, R. Fournier, C. Hourdin, O. Talagrand, M. Collins, S.R. Lewis, P.L. Read, J.-P. Huot, Improved general circulation models of the Martian atmosphere from the surface to above 80 km. J. Geophys. Res. **104**, 24155–24176 (1999).
93. J.J. Fortney, The effect of condensates on the characterization of transiting planet atmospheres with transmission spectroscopy. Mon. Not. R. Astron. Soc. **364**, 649–653 (2005).
94. J.J. Fortney, N. Nettelmann, The interior structure, composition, and evolution of giant planets. Space Sci. Rev. **152**, 423–447 (2010).
95. J.J. Fortney, D. Sudarsky, I. Hubeny, C.S. Cooper, W.B. Hubbard, A. Burrows, J.I. Lunine, On the indirect detection of sodium in the atmosphere of the planetary companion to HD 209458. Astrophys. J. **589**, 615–622 (2003).
96. J.J. Fortney, M.S. Marley, K. Lodders, D. Saumon, R. Freedman, Comparative planetary atmospheres: Models of TrES-1 and HD 209458b. Astrophys. J. Lett. **627**, 69–72 (2005).
97. J.J. Fortney, K. Lodders, M.S. Marley, R.S. Freedman, A unified theory for the atmospheres of the hot and very hot Jupiters: Two classes of irradiated atmospheres. Astrophys. J. **678**, 1419–1435 (2008a).
98. J.J. Fortney, M.S. Marley, D. Saumon, K. Lodders, Synthetic spectra and colors of young giant planet atmospheres: Effects of initial conditions and atmospheric metallicity. Astrophys. J. **683**, 1104–1116 (2008b).

99. J.J. Fortney, M. Shabram, A.P. Showman, Y. Lian, R.S. Freedman, M.S. Marley, N.K. Lewis, Transmission spectra of three-dimensional hot Jupiter model atmospheres. Astrophys. J. **709**, 1396–1406 (2010).
100. J.J. Fortney, C. Mordasini, N. Nettelmann, E.M.-R. Kempton, T.P. Greene, K. Zahnle, A framework for characterizing the atmospheres of low-mass low-density transiting planets. Astrophys. J. **775**, 80 (2013).
101. L. Fossati, C.A. Haswell, C.S. Froning, L. Hebb, S. Holmes, U. Kolb, C. Helling, A. Carter, P. Wheatley, A. Collier Cameron, B. Loeillet, D. Pollacco, R. Street, H.C. Stempels, E. Simpson, S. Udry, Y.C. Joshi, R.G. West, I. Skillen, D. Wilson, Metals in the exosphere of the highly irradiated planet WASP-12b. Astrophys. J. Lett. **714**, 222–227 (2010).
102. J. Fraine, D. Deming, B. Benneke, H. Knutson, A. Jordán, N. Espinoza, N. Madhusudhan, A. Wilkins, K. Todorov, Water vapour absorption in the clear atmosphere of a Neptune-sized exoplanet. Nature **513**, 526–529 (2014).
103. F. Fressin, G. Torres, D. Charbonneau, S.T. Bryson, J. Christiansen, C.D. Dressing, J.M. Jenkins, L.M. Walkowicz, N.M. Batalha, The false positive rate of Kepler and the occurrence of planets. Astrophys. J. **766**, 81 (2013).
104. Freytag, F. Allard, H.-G. Ludwig, D. Homeier, M. Steffen, The role of convection, overshoot, and gravity waves for the transport of dust in M dwarf and brown dwarf atmospheres. Astron. Astrophys. **513**, 19 (2010).
105. Fridlund et al., Space Sci. Rev. (2016, this issue)
106. C.F. Gammie, Nonlinear outcome of gravitational instability in cooling, gaseous disks. Astrophys. J. **553**, 174–183 (2001).
A. García Muñoz, Physical and chemical aeronomy of HD 209458b. Planet. Space Sci. **55**, 1426–1455 (2007).
107. N.P. Gibson, F. Pont, S. Aigrain, A new look at NICMOS transmission spectroscopy of HD 189733, GJ-436 and XO-1: No conclusive evidence for molecular features. Mon. Not. R. Astron. Soc. **411**, 2199–2213 (2011).
108. N.P. Gibson, S. Aigrain, F. Pont, D.K. Sing, J.-M. Désert, T.M. Evans, G. Henry, N. Husnoo, H. Knutson, Probing the haze in the atmosphere of HD 189733b with Hubble Space Telescope/WFC3 transmission spectroscopy. Mon. Not. R. Astron. Soc. **422**, 753–760 (2012).
109. N.P. Gibson, S. Aigrain, J.K. Barstow, T.M. Evans, L.N. Fletcher, P.G.J. Irwin, The optical transmission spectrum of the hot Jupiter HAT-P-32b: Clouds explain the absence of broad spectral features? Mon. Not. R. Astron. Soc. **436**, 2974–2988 (2013).
110. G.R. Gltone, M. Allen, Y.L. Yung, Hydrocarbon photochemistry in the upper atmosphere of Jupiter. Icarus **119**, 1–52 (1996).
111. P. Goldreich, Y. Lithwick, R. Sari, Planet formation by coagulation: A focus on Uranus and Neptune. Annu. Rev. Astron. Astrophys. **42**, 549–601 (2004).
112. C.J. Grillmair, A. Burrows, D. Charbonneau, L. Armus, J. Stauffer, V. Meadows, J. van Cleve, K. von Braun, D. Levine, Strong water absorption in the dayside emission spectrum of the planet HD189733b. Nature **456**, 767–769 (2008).
113. T. Guillot, The interiors of giant planets: Models and outstanding questions. Annu. Rev. Earth Planet. Sci. **33**, 493–530 (2005).
114. T. Guillot, On the radiative equilibrium of irradiated planetary atmospheres. Astron. Astrophys. **520**, 27 (2010).
115. R.M. Haberle, C.P. McKay, D. Tyler, R.T. Reynolds, Can synchronously rotating planets support an atmosphere?, in *Circumstellar Habitable Zones*, vol. 1, 1996, p. 29
116. S. Hadden, Y. Lithwick, Densities and eccentricities of 139 Kepler planets from transit time variations. Astrophys. J. **787**, 80 (2014).
117. N. Haghighipour, The formation and dynamics of super-Earth planets. Annu. Rev. Earth Planet. Sci. **41**, 469–495 (2013).
118. B.M.S. Hansen, On the absorption and redistribution of energy in irradiated planets. Astrophys. J. Suppl. Ser. **179**, 484–508 (2008).
119. B.M.S. Hansen, N. Murray, Migration then assembly: Formation of Neptune-mass planets inside 1 AU. Astrophys. J. **751**, 158 (2012).

120. B.M.S. Hansen, N. Murray, Testing in situ assembly with the Kepler planet candidate sample. Astrophys. J. **775**, 53 (2013).
121. C.J. Hansen, J.C. Schwartz, N.B. Cowan, Features in the broad-band eclipse spectra of exoplanets: Signal or noise? Mon. Not. R. Astron. Soc. **444**, 3632–3640 (2014).
122. P.H. Hauschildt, F. Allard, E. Baron, The NextGen model atmosphere grid for 3000≤Teff≤10000 K3000≤Teff≤10000 K. Astrophys. J. **512**, 377–385 (1999).
123. K. Haynes, A.M. Mandell, N. Madhusudhan, D. Deming, H. Knutson, Spectroscopic evidence for a temperature inversion in the dayside atmosphere of hot Jupiter WASP-33b. Astrophys. J. **806**, 146 (2015).
124. R. Helled, P. Bodenheimer, Metallicity of the massive protoplanets around HR 8799 if formed by gravitational instability. Icarus **207**, 503–508 (2010).
125. R. Helled, P. Bodenheimer, The formation of Uranus and Neptune: Challenges and implications for intermediate-mass exoplanets. Astrophys. J. **789**, 69 (2014).
126. Helling, A. Ackerman, F. Allard, M. Dehn, P. Hauschildt, D. Homeier, K. Lodders, M. Marley, F. Rietmeijer, T. Tsuji, P. Woitke, A comparison of chemistry and dust cloud formation in ultracool dwarf model atmospheres. Mon. Not. R. Astron. Soc. **391**, 1854–1873 (2008).
127. Helling, M. Jardine, C. Stark, D. Diver, Ionization in atmospheres of brown dwarfs and extrasolar planets. III. Breakdown conditions for mineral clouds. Astrophys. J. **767**, 136 (2013).
128. Helling, P. Woitke, P.B. Rimmer, I. Kamp, W.-F. Thi, R. Meijerink, Disk evolution, element abundances and cloud properties of young gas giant planets. Life **4**, 142–173 (2014).
129. K. Heng, B.-O. Demory, Understanding trends associated with clouds in irradiated exoplanets. Astrophys. J. **777**, 100 (2013).
130. K. Heng, P. Kopparla, On the stability of super-Earth atmospheres. Astrophys. J. **754**(1), 60 (2012).
131. K. Heng, J.R. Lyons, Carbon dioxide in exoplanetary atmospheres: Rarely dominant compared to carbon monoxide and water in hot, hydrogen-dominated atmospheres. Astrophys. J. **817**, 149 (2016)
132. K. Heng, A.P. Showman, Atmospheric dynamics of hot exoplanets. Annu. Rev. Earth Planet. Sci. **43**, 509–540 (2015).
133. K. Heng, S.S. Vogt, Gliese 581g as a scaled-up version of Earth: Atmospheric circulation simulations. Mon. Not. R. Astron. Soc. **415**(3), 2145–2157 (2011)
134. K. Heng, W. Hayek, F. Pont, D.K. Sing, On the effects of clouds and hazes in the atmospheres of hot Jupiters: Semi-analytical temperature-pressure profiles. Mon. Not. R. Astron. Soc. **420**, 20–36 (2012).
135. P.M. Hinz, T.J. Rodigas, M.A. Kenworthy, S. Sivanandam, A.N. Heinze, E.E. Mamajek, M.R. Meyer, Thermal infrared MMTAO observations of the HR 8799 planetary system. Astrophys. J. **716**, 417–426 (2010).
136. H.J. Hoeijmakers, R.J. de Kok, I.A.G. Snellen, M. Brogi, J.L. Birkby, H. Schwarz, A search for TiO in the optical high-resolution transmission spectrum of HD 209458b: Hindrance due to inaccuracies in the line database. Astron. Astrophys. **575**, 20 (2015).
137. A.W. Howard, G.W. Marcy, S.T. Bryson, J.M. Jenkins, J.F. Rowe, N.M. Batalha, W.J. Borucki, D.G. Koch, E.W. Dunham, T.N. Gautier III, J. Van Cleve, W.D. Cochran, D.W. Latham, J.J. Lissauer, G. Torres, T.M. Brown, R.L. Gilliland, L.A. Buchhave, D.A. Caldwell, J. Christensen-Dalsgaard, D. Ciardi, F. Fressin, M.R. Haas, S.B. Howell, H. Kjeldsen, S. Seager, L. Rogers, D.D. Sasselov, J.H. Steffen, G.S. Basri, D. Charbonneau, J. Christiansen, B. Clarke, A. Dupree, D.C. Fabrycky, D.A. Fischer, E.B. Ford, J.J. Fortney, J. Tarter, F.R. Girouard, M.J. Holman, J.A. Johnson, T.C. Klaus, P. Machalek, A.V. Moorhead, R.C. Morehead, D. Ragozzine, P. Tenenbaum, J.D. Twicken, S.N. Quinn, H. Isaacson, A. Shporer, P.W. Lucas, L.M. Walkowicz, W.F. Welsh, A. Boss, E. Devore, A. Gould, J.C. Smith, R.L. Morris, A. Prsa, T.D. Morton, M. Still, S.E. Thompson, F. Mullally, M. Endl, P.J. MacQueen, Planet occurrence within 0.25 AU of solar-type stars from Kepler. Astrophys. J. Suppl. Ser. **201**, 15 (2012).
138. Y. Hu, F. Ding, Radiative constraints on the habitability of exoplanets Gliese 581c and Gliese 581d. Astron. Astrophys. **526**, 135 (2011)
139. Y. Hu, F. Ding, Atmospheric circulations and climate of tidal-locking exoplanets. Sci. Sinica Phys. Mech. Astron. **43**(10), 1356 (2013). R. Hu, S. Seager, Photochemistry in terrestrial

exoplanet atmospheres. III. Photochemistry and thermochemistry in thick atmospheres on super Earths and mini Neptunes. Astrophys. J. **784**, 63 (2014).

140. Y. Hu, J. Yang, Role of ocean heat transport in climates of tidally locked exoplanets around m dwarf stars. Proc. Natl. Acad. Sci. **111**(2), 629–634 (2014)
141. R. Hu, S. Seager, W. Bains, Photochemistry in terrestrial exoplanet atmospheres. I. Photochemistry model and benchmark cases. Astrophys. J. **761**, 166 (2012).
142. R. Hu, S. Seager, W. Bains, Photochemistry in terrestrial exoplanet atmospheres. II. H_2S and SO_2 photochemistry in anoxic atmospheres. Astrophys. J. **769**, 6 (2013).
143. R. Hu, B.-O. Demory, S. Seager, N. Lewis, A.P. Showman, A semi-analytical model of visible-wavelength phase curves of exoplanets and applications to Kepler- 7 b and Kepler- 10 b. Astrophys. J. **802**, 51 (2015a).
144. R. Hu, S. Seager, Y.L. Yung, Helium atmospheres on warm Neptune- and sub-Neptune-sized exoplanets and applications to GJ 436b. Astrophys. J. **807**, 8 (2015b).
145. S.-S. Huang, Life-supporting regions in the vicinity of binary systems, in *Publications of the Astronomical Society of the Pacific* (1960), pp. 106–114
146. W.B. Hubbard, A. Burrows, J.I. Lunine, Theory of giant planets. Annu. Rev. Astron. Astrophys. **40**, 103–136 (2002).

I. Hubeny, A. Burrows, D. Sudarsky, A possible bifurcation in atmospheres of strongly irradiated stars and planets. Astrophys. J. **594**, 1011–1018 (2003).

147. C.M. Huitson, D.K. Sing, A. Vidal-Madjar, G.E. Ballester, A. Lecavelier des Etangs, J.-M. Désert, F. Pont, Temperature-pressure profile of the hot Jupiter HD 189733b from HST sodium observations: Detection of upper atmospheric heating. Mon. Not. R. Astron. Soc. **422**, 2477–2488 (2012).
148. C.M. Huitson, D.K. Sing, F. Pont, J.J. Fortney, A.S. Burrows, P.A. Wilson, G.E. Ballester, N. Nikolov, N.P. Gibson, D. Deming, S. Aigrain, T.M. Evans, G.W. Henry, A. Lecavelier des Etangs, A.P. Showman, A. Vidal-Madjar, K. Zahnle, An HST optical-to-near-IR transmission spectrum of the hot Jupiter WASP-19b: Detection of atmospheric water and likely absence of TiO. Mon. Not. R. Astron. Soc. **434**, 3252–3274 (2013).
149. S. Ida, D.N.C. Lin, Toward a deterministic model of planetary formation. III. Mass distribution of short-period planets around stars of various masses. Astrophys. J. **626**, 1045–1060 (2005).
150. N.K. Inamdar, H.E. Schlichting, The formation of super-Earths and mini-Neptunes with giant impacts. Mon. Not. R. Astron. Soc. **448**, 1751–1760 (2015).
151. A.P. Ingersoll, The runaway greenhouse: A history of water on Venus. J. Atmos. Sci. **26**(6), 1191–1198 (1969)
152. N. Iro, B. Bézard, T. Guillot, A time-dependent radiative model of HD 209458b. Astron. Astrophys. **436**, 719–727 (2005).
153. Y. Ito, M. Ikoma, H. Kawahara, H. Nagahara, Y. Kawashima, T. Nakamoto, Theoretical emission spectra of atmospheres of hot rocky super-Earths. Astrophys. J. **801**, 144 (2015).
154. M. Janson, C. Bergfors, M. Goto, W. Brandner, D. Lafrenière, Spatially resolved spectroscopy of the exoplanet HR 8799 c. Astrophys. J. Lett. **710**, 35–38 (2010).
155. M. Janson, T.D. Brandt, M. Kuzuhara, D.S. Spiegel, C. Thalmann, T. Currie, M. Bonnefoy, N. Zimmerman, S. Sorahana, T. Kotani, J. Schlieder, J. Hashimoto, T. Kudo, N. Kusakabe, L. Abe, W. Brandner, J.C. Carson, S. Egner, M. Feldt, M. Goto, C.A. Grady, O. Guyon, Y. Hayano, M. Hayashi, S. Hayashi, T. Henning, K.W. Hodapp, M. Ishii, M. Iye, R. Kandori, G.R. Knapp, J. Kwon, T. Matsuo, M.W. McElwain, K. Mede, S. Miyama, J.-I. Morino, A. Moro-Martín, T. Nakagawa, T. Nishimura, T.-S. Pyo, E. Serabyn, T. Suenaga, H. Suto, R. Suzuki, Y. Takahashi, M. Takami, N. Takato, H. Terada, D. Tomono, E.L. Turner, M. Watanabe, J. Wisniewski, T. Yamada, H. Takami, T. Usuda, M. Tamura, Direct imaging detection of methane in the atmosphere of GJ 504 b. Astrophys. J. Lett. **778**, 4 (2013).
156. A.G. Jensen, S. Redfield, M. Endl, W.D. Cochran, L. Koesterke, T. Barman, A detection of H$\alpha\alpha$ in an exoplanetary exosphere. Astrophys. J. **751**, 86 (2012).
157. M. Joshi, Climate model studies of synchronously rotating planets. Astrobiology **3**(2), 415–427 (2003)

158. M. Joshi, R. Haberle, R. Reynolds, Simulations of the atmospheres of synchronously rotating terrestrial planets orbiting m dwarfs: Conditions for atmospheric collapse and the implications for habitability. Icarus **129**(2), 450–465 (1997)
159. J. Kasting, *How to Find a Habitable Planet*. Science Essentials (Princeton University Press, Princeton, 2010). 9780691138053
160. J.F. Kasting, D.P. Whitmire, R.T. Reynolds, Habitable zones around main sequence stars. Icarus **101**(1), 108–128 (1993)
161. G.M. Kennedy, S.J. Kenyon, Planet formation around stars of various masses: Hot super-Earths. Astrophys. J. **682**, 1264–1276 (2008).
162. D.M. Kipping, D.S. Spiegel, Detection of visible light from the darkest world. Mon. Not. R. Astron. Soc. **417**, 88–92 (2011).
163. J.D. Kirkpatrick, New spectral types L and T. Annu. Rev. Astron. Astrophys. **43**, 195–245 (2005).
164. H.A. Knutson, D. Charbonneau, L.E. Allen, J.J. Fortney, E. Agol, N.B. Cowan, A.P. Showman, C.S. Cooper, S.T. Megeath, A map of the day-night contrast of the extrasolar planet HD 189733b. Nature **447**, 183–186 (2007).
165. H.A. Knutson, D. Charbonneau, N.B. Cowan, J.J. Fortney, A.P. Showman, E. Agol, G.W. Henry, The 8 μm phase variation of the hot Saturn hd 149026b. Astrophys. J. **703**(1), 769 (2009)
166. H.A. Knutson, N. Lewis, J.J. Fortney, A. Burrows, A.P. Showman, N.B. Cowan, E. Agol, S. Aigrain, D. Charbonneau, D. Deming, J.-M. Désert, G.W. Henry, J. Langton, G. Laughlin, 3.6 and 4.5 μm phase curves and evidence for non-equilibrium chemistry in the atmosphere of extrasolar planet HD 189733b. Astrophys. J. **754**, 22 (2012).
167. H.A. Knutson, B. Benneke, D. Deming, D. Homeier, A featureless transmission spectrum for the Neptune-mass exoplanet GJ436b. Nature **505**, 66–68 (2014a).
168. H.A. Knutson, D. Dragomir, L. Kreidberg, E.M.-R. Kempton, P.R. McCullough, J.J. Fortney, J.L. Bean, M. Gillon, D. Homeier, A.W. Howard, Hubble Space Telescope near-IR transmission spectroscopy of the super-Earth HD 97658b. Astrophys. J. **794**, 155 (2014b).
169. Q.M. Konopacky, T.S. Barman, B.A. Macintosh, C. Marois, Detection of carbon monoxide and water absorption lines in an exoplanet atmosphere. Science **339**, 1398–1401 (2013).
170. R.K. Kopparapu, A revised estimate of the occurrence rate of terrestrial planets in the habitable zones around Kepler m-dwarfs. Astrophys. J. Lett. **767**(1), 8 (2013)
171. R.k. Kopparapu, J.F. Kasting, K.J. Zahnle, A photochemical model for the carbon-rich planet WASP-12b. Astrophys. J. **745**, 77 (2012).
172. T.T. Koskinen, M.J. Harris, R.V. Yelle, P. Lavvas, The escape of heavy atoms from the ionosphere of HD209458b. I. A photochemical-dynamical model of the thermosphere. Icarus **226**, 1678–1694 (2013).
173. T.T. Koskinen, P. Lavvas, M.J. Harris, R.V. Yelle, Thermal escape from extrasolar giant planets. Philos. Trans. R. Soc. A, Math. Phys. Eng. Sci. **372**, 30089 (2014).
174. L. Kreidberg, J.L. Bean, J.-M. Désert, B. Benneke, D. Deming, K.B. Stevenson, S. Seager, Z. Berta-Thompson, A. Seifahrt, D. Homeier, Clouds in the atmosphere of the super-Earth exoplanet GJ1214b. Nature **505**, 69–72 (2014a).
175. L. Kreidberg, J.L. Bean, J.-M. Désert, M.R. Line, J.J. Fortney, N. Madhusudhan, K.B. Stevenson, A.P. Showman, D. Charbonneau, P.R. McCullough, S. Seager, A. Burrows, G.W. Henry, M. Williamson, T. Kataria, D. Homeier, A precise water abundance measurement for the hot Jupiter WASP-43b. Astrophys. J. Lett. **793**, 27 (2014b).
176. L. Kreidberg, M.R. Line, J.L. Bean, K.B. Stevenson, J.-M. Désert, N. Madhusudhan, J.J. Fortney, J.K. Barstow, G.W. Henry, M.H. Williamson, A.P. Showman, A detection of water in the transmission spectrum of the hot Jupiter WASP-12b and implications for its atmospheric composition. Astrophys. J. **814**, 66 (2015).
177. M.J. Kuchner, S. Seager, Extrasolar carbon planets. ArXiv Astrophysics e-prints (2005)
178. R.L. Kurucz, Model atmospheres for G, F, A, B, and O stars. Astrophys. J. Suppl. Ser. **40**, 1–340 (1979).
179. Lai, Star-disc-binary interactions in protoplanetary disc systems and primordial spin-orbit misalignments. Mon. Not. R. Astron. Soc. **440**, 3532–3544 (2014).

180. H. Lammer, V. Eybl, K. Kislyakova, J. Weingrill, M. Holmström, M. Khodachenko, Y.N. Kulikov, A. Reiners, M. Leitzinger, P. Odert, et al., Uv transit observations of euv-heated expanded thermospheres of Earth-like exoplanets around m-stars: Testing atmosphere evolution scenarios. Astrophys. Space Sci. **335**(1), 39–50 (2011)
181. H. Lammer, A. Stökl, N.V. Erkaev, E.A. Dorfi, P. Odert, M. Güdel, Y.N. Kulikov, K.G. Kislyakova, M. Leitzinger, Origin and loss of nebula-captured hydrogen envelopes from 'sub'- to 'super-Earths' in the habitable zone of Sun-like stars. Mon. Not. R. Astron. Soc. **439**, 3225–3238 (2014).
182. A.A. Lanotte, M. Gillon, B.-O. Demory, J.J. Fortney, N. Astudillo, X. Bonfils, P. Magain, X. Delfosse, T. Forveille, C. Lovis, M. Mayor, V. Neves, F. Pepe, D. Queloz, N. Santos, S. Udry, A global analysis of Spitzer and new HARPS data confirms the loneliness and metal-richness of GJ 436 b. Astron. Astrophys. **572**, 73 (2014).
183. L.M. Lara, E. Lellouch, J.J. López-Moreno, R. Rodrigo, Vertical distribution of Titan's atmospheric neutral constituents. J. Geophys. Res. **101**, 23261–23283 (1996).
184. P. Lavvas, T. Koskinen, R.V. Yelle, Chemical composition of extrasolar giant planets, in *EPSC-DPS Joint Meeting*, 2011, 2011, p. 645
185. P. Lavvas, T. Koskinen, R.V. Yelle, Electron densities and alkali atoms in exoplanet atmospheres. Astrophys. J. **796**, 15 (2014).
A. Lecavelier Des Etangs, F. Pont, A. Vidal-Madjar, D. Sing, Rayleigh scattering in the transit spectrum of HD 189733b. Astron. Astrophys. **481**, 83–86 (2008a).
186. Lecavelier Des Etangs, A. Vidal-Madjar, J.-M. Désert, D. Sing, Rayleigh scattering by H_2 in the extrasolar planet HD 209458b. Astron. Astrophys. **485**, 865–869 (2008b).
187. Lecavelier Des Etangs, D. Ehrenreich, A. Vidal-Madjar, G.E. Ballester, J.-M. Désert, R. Ferlet, G. Hébrard, D.K. Sing, K.-O. Tchakoumegni, S. Udry, Evaporation of the planet HD 189733b observed in H I Lyman-αα. Astron. Astrophys. **514**, 72 (2010).
188. J. Leconte, F. Forget, B. Charnay, R. Wordsworth, A. Pottier, Increased insolation threshold for runaway greenhouse processes on Earth-like planets. Nature **504**(7479), 268–271 (2013)
189. J. Leconte, F. Forget, H. Lammer, On the (anticipated) diversity of terrestrial planet atmospheres. Exp. Astron. (2014).
190. J.-M. Lee, L.N. Fletcher, P.G.J. Irwin, Optimal estimation retrievals of the atmospheric structure and composition of HD 189733b from secondary eclipse spectroscopy. Mon. Not. R. Astron. Soc. **420**, 170–182 (2012).
191. J.-M. Lee, K. Heng, P.G.J. Irwin, Atmospheric retrieval analysis of the directly imaged exoplanet HR 8799b. Astrophys. J. **778**, 97 (2013).
192. E.J. Lee, E. Chiang, C.W. Ormel, Make super-Earths, not Jupiters: Accreting nebular gas onto solid cores at 0.1 AU and beyond. Astrophys. J. **797**, 95 (2014).
193. Lee, C. Helling, I. Dobbs-Dixon, D. Juncher, Modelling the local and global cloud formation on HD 189733b. Astron. Astrophys. **580**, 12 (2015).
194. Léger, O. Grasset, B. Fegley, F. Codron, A.F. Albarede, P. Barge, R. Barnes, P. Cance, S. Carpy, F. Catalano, C. Cavarroc, O. Demangeon, S. Ferraz-Mello, P. Gabor, J.-M. Grießmeier, J. Leibacher, G. Libourel, A.-S. Maurin, S.N. Raymond, D. Rouan, B. Samuel, L. Schaefer, J. Schneider, P.A. Schuller, F. Selsis, C. Sotin, The extreme physical properties of the CoRoT-7b super-Earth. Icarus **213**, 1–11 (2011).
195. J.S. Lewis, Observability of spectroscopically active compounds in the atmosphere of Jupiter. Icarus **10**, 393–409 (1969).
196. N.K. Lewis, A.P. Showman, J.J. Fortney, M.S. Marley, R.S. Freedman, K. Lodders, Atmospheric circulation of eccentric hot Neptune GJ436b. Astrophys. J. **720**, 344–356 (2010).
197. M.-C. Liang, C.D. Parkinson, A.Y.-T. Lee, Y.L. Yung, S. Seager, Source of atomic hydrogen in the atmosphere of HD 209458b. Astrophys. J. Lett. **596**, 247–250 (2003).
198. M.-C. Liang, S. Seager, C.D. Parkinson, A.Y.-T. Lee, Y.L. Yung, On the insignificance of photochemical hydrocarbon aerosols in the atmospheres of close-in extrasolar giant planets. Astrophys. J. Lett. **605**, 61–64 (2004).
199. G.F. Lindal, The atmosphere of Neptune—An analysis of radio occultation data acquired with Voyager 2. Astron. J. **103**, 967–982 (1992).

200. R.S. Lindzen, Turbulence and stress owing to gravity wave and tidal breakdown. J. Geophys. Res. **86**, 9707–9714 (1981).
201. M.R. Line, M.C. Liang, Y.L. Yung, High-temperature photochemistry in the atmosphere of HD 189733b. Astrophys. J. **717**, 496–502 (2010).
202. M.R. Line, G. Vasisht, P. Chen, D. Angerhausen, Y.L. Yung, Thermochemical and photochemical kinetics in cooler hydrogen-dominated extrasolar planets: A methane-poor GJ436b? Astrophys. J. **738**, 32 (2011).
203. M.R. Line, X. Zhang, G. Vasisht, V. Natraj, P. Chen, Y.L. Yung, Information content of exoplanetary transit spectra: An initial look. Astrophys. J. **749**, 93 (2012).
204. M.R. Line, A.S. Wolf, X. Zhang, H. Knutson, J.A. Kammer, E. Ellison, P. Deroo, D. Crisp, Y.L. Yung, A systematic retrieval analysis of secondary eclipse spectra. I. A comparison of atmospheric retrieval techniques. Astrophys. J. **775**, 137 (2013).
205. J.L. Linsky, H. Yang, K. France, C.S. Froning, J.C. Green, J.T. Stocke, S.N. Osterman, Observations of mass loss from the transiting exoplanet HD 209458b. Astrophys. J. **717**, 1291–1299 (2010).
206. J.J. Lissauer, Planet formation. Annu. Rev. Astron. Astrophys. **31**, 129–174 (1993).
207. J.J. Lissauer, D.J. Stevenson, Formation of giant planets, in *Protostars and Planets V* (2007), pp. 591–606
208. A.C. Lockwood, M.E. Brown, J. Stansberry, The size and shape of the oblong dwarf planet Haumea. Earth Moon Planets **111**, 127–137 (2014).
209. K. Lodders, Alkali element chemistry in cool dwarf atmospheres. Astrophys. J. **519**, 793–801 (1999).
210. K. Lodders, Titanium and vanadium chemistry in low-mass dwarf stars. Astrophys. J. **577**, 974–985 (2002).
211. K. Lodders, Jupiter formed with more tar than ice. Astrophys. J. **611**, 587–597 (2004).
212. K. Lodders, in *Exoplanet Chemistry*, ed. by R. Barnes (2010), p. 157.
213. K. Lodders, B. Fegley, Atmospheric chemistry in giant planets, brown dwarfs, and low-mass dwarf stars. I. Carbon, nitrogen, and oxygen. Icarus **155**, 393–424 (2002).
214. E.D. Lopez, J.J. Fortney, The role of core mass in controlling evaporation: The Kepler radius distribution and the Kepler-36 density dichotomy. Astrophys. J. **776**, 2 (2013).
215. E.D. Lopez, J.J. Fortney, Understanding the mass-radius relation for sub-Neptunes: Radius as a proxy for composition. Astrophys. J. **792**, 1 (2014).
216. E.D. Lopez, J.J. Fortney, N. Miller, How thermal evolution and mass-loss sculpt populations of super-Earths and sub-Neptunes: Application to the Kepler-11 system and beyond. Astrophys. J. **761**, 59 (2012).
217. R. Luger, R. Barnes, Extreme water loss and abiotic O2Buildup on planets throughout the habitable zones of M dwarfs. Astrobiology **15**, 119–143 (2015).
218. R. Luger, R. Barnes, E. Lopez, J. Fortney, B. Jackson, V. Meadows, Habitable evaporated cores: Transforming mini-Neptunes into super-Earths in the habitable zones of M dwarfs. Astrobiology **15**, 57–88 (2015).
219. R.E. Lupu, K. Zahnle, M.S. Marley, L. Schaefer, B. Fegley, C. Morley, K. Cahoy, R. Freedman, J.J. Fortney, The atmospheres of earthlike planets after giant impact events. Astrophys. J. **784**, 27 (2014).
220. Macintosh, J.R. Graham, T. Barman, R.J. De Rosa, Q. Konopacky, M.S. Marley, C. Marois, E.L. Nielsen, L. Pueyo, A. Rajan, J. Rameau, D. Saumon, J.J. Wang, J. Patience, M. Ammons, P. Arriaga, E. Artigau, S. Beckwith, J. Brewster, S. Bruzzone, J. Bulger, B. Burningham, A.S. Burrows, C. Chen, E. Chiang, J.K. Chilcote, R.I. Dawson, R. Dong, R. Doyon, Z.H. Draper, G. Duchêne, T.M. Esposito, D. Fabrycky, M.P. Fitzgerald, K.B. Follette, J.J. Fortney, B. Gerard, S. Goodsell, A.Z. Greenbaum, P. Hibon, S. Hinkley, T.H. Cotten, L.-W. Hung, P. Ingraham, M. Johnson-Groh, P. Kalas, D. Lafreniere, J.E. Larkin, J. Lee, M. Line, D. Long, J. Maire, F. Marchis, B.C. Matthews, C.E. Max, S. Metchev, M.A. Millar-Blanchaer, T. Mittal, C.V. Morley, K.M. Morzinski, R. Murray-Clay, R. Oppenheimer, D.W. Palmer, R. Patel, M.D. Perrin, L.A. Poyneer, R.R. Rafikov, F.T. Rantakyrö, E.L. Rice, P. Rojo, A.R. Rudy, J.-B. Ruffio, M.T. Ruiz, N. Sadakuni, L. Saddlemyer, M. Salama, D. Savransky, A.C. Schneider, A. Sivaramakrishnan, I. Song, R. Soummer, S. Thomas, G. Vasisht, J.K. Wallace, K. Ward-Duong, S.J. Wiktorowicz,

S.G. Wolff, B. Zuckerman, Discovery and spectroscopy of the young Jovian planet 51 Eri b with the Gemini Planet Imager. Science **350**, 64–67 (2015).

221. N. Madhusudhan, C/O ratio as a dimension for characterizing exoplanetary atmospheres. Astrophys. J. **758**, 36 (2012).
222. N. Madhusudhan, S. Redfield, Optimal measures for characterizing water-rich super-Earths. Int. J. Astrobiol. **14**, 177–189 (2015).
223. N. Madhusudhan, S. Seager, A temperature and abundance retrieval method for exoplanet atmospheres. Astrophys. J. **707**, 24–39 (2009).
224. N. Madhusudhan, S. Seager, High metallicity and non-equilibrium chemistry in the dayside atmosphere of hot-Neptune GJ 436b. Astrophys. J. **729**, 41 (2011).
225. N. Madhusudhan, A. Burrows, T. Currie, Model atmospheres for massive gas giants with thick clouds: Application to the HR 8799 planets and predictions for future detections. Astrophys. J. **737**, 34 (2011a).
226. N. Madhusudhan, J. Harrington, K.B. Stevenson, S. Nymeyer, C.J. Campo, P.J. Wheatley, D. Deming, J. Blecic, R.A. Hardy, N.B. Lust, D.R. Anderson, A. Collier-Cameron, C.B.T. Britt, W.C. Bowman, L. Hebb, C. Hellier, P.F.L. Maxted, D. Pollacco, R.G. West, A high C/O ratio and weak thermal inversion in the atmosphere of exoplanet WASP-12b. Nature **469**, 64–67 (2011b).
227. N. Madhusudhan, M.A. Amin, G.M. Kennedy, Toward chemical constraints on hot Jupiter migration. Astrophys. J. Lett. **794**, 12 (2014a).
228. N. Madhusudhan, N. Crouzet, P.R. McCullough, D. Deming, C. Hedges, H_2O abundances in the atmospheres of three hot Jupiters. Astrophys. J. Lett. **791**, 9 (2014b).
229. N. Madhusudhan, H. Knutson, J.J. Fortney, T. Barman, Exoplanetary atmospheres, in *Protostars and Planets VI* (2014c), pp. 739–762.
230. P.R. Mahaffy, H.B. Niemann, A. Alpert, S.K. Atreya, J. Demick, T.M. Donahue, D.N. Harpold, T.C. Owen, Noble gas abundance and isotope ratios in the atmosphere of Jupiter from the Galileo probe mass spectrometer. J. Geophys. Res. **105**, 15061–15072 (2000).
231. L. Mancini, S. Ciceri, G. Chen, J. Tregloan-Reed, J.J. Fortney, J. Southworth, T.G. Tan, M. Burgdorf, S. Calchi Novati, M. Dominik, X.-S. Fang, F. Finet, T. Gerner, S. Hardis, T.C. Hinse, U.G. Jørgensen, C. Liebig, N. Nikolov, D. Ricci, S. Schäfer, F. Schönebeck, J. Skottfelt, O. Wertz, K.A. Alsubai, V. Bozza, P. Browne, P. Dodds, S.-H. Gu, K. Harpsøe, T. Henning, M. Hundertmark, J. Jessen-Hansen, N. Kains, E. Kerins, H. Kjeldsen, M.N. Lund, M. Lundkvist, N. Madhusudhan, M. Mathiasen, M.T. Penny, S. Prof, S. Rahvar, K. Sahu, G. Scarpetta, C. Snodgrass, J. Surdej, Physical properties, transmission and emission spectra of the WASP-19 planetary system from multi-colour photometry. Mon. Not. R. Astron. Soc. **436**, 2–18 (2013).
232. A.M. Mandell, K. Haynes, E. Sinukoff, N. Madhusudhan, A. Burrows, D. Deming, Exoplanet transit spectroscopy using WFC3: WASP-12 b, WASP-17 b, and WASP-19 b. Astrophys. J. **779**, 128 (2013).
233. U. Marboeuf, A. Thiabaud, Y. Alibert, N. Cabral, W. Benz, From stellar nebula to planetesimals. Astron. Astrophys. **570**, 35 (2014).
234. G.W. Marcy, H. Isaacson, A.W. Howard, J.F. Rowe, J.M. Jenkins, S.T. Bryson, D.W. Latham, S.B. Howell, T.N. Gautier III, N.M. Batalha, L. Rogers, D. Ciardi, D.A. Fischer, R.L. Gilliland, H. Kjeldsen, J. Christensen-Dalsgaard, D. Huber, W.J. Chaplin, S. Basu, L.A. Buchhave, S.N. Quinn, W.J. Borucki, D.G. Koch, R. Hunter, D.A. Caldwell, J. Van Cleve, R. Kolbl, L.M. Weiss, E. Petigura, S. Seager, T. Morton, J.A. Johnson, S. Ballard, C. Burke, W.D. Cochran, M. Endl, P. MacQueen, M.E. Everett, J.J. Lissauer, E.B. Ford, G. Torres, F. Fressin, T.M. Brown, J.H. Steffen, D. Charbonneau, G.S. Basri, D.D. Sasselov, J. Winn, R. Sanchis-Ojeda, J. Christiansen, E. Adams, C. Henze, A. Dupree, D.C. Fabrycky, J.J. Fortney, J. Tarter, M.J. Holman, P. Tenenbaum, A. Shporer, P.W. Lucas, W.F. Welsh, J.A. Orosz, T.R. Bedding, T.L. Campante, G.R. Davies, Y. Elsworth, R. Handberg, S. Hekker, C. Karoff, S.D. Kawaler, M.N. Lund, M. Lundkvist, T.S. Metcalfe, A. Miglio, V. Silva Aguirre, D. Stello, T.R. White, A. Boss, E. Devore, A. Gould, A. Prsa, E. Agol, T. Barclay, J. Coughlin, E. Brugamyer, F. Mullally, E.V. Quintana, M. Still, S.E. Thompson, D. Morrison, J.D. Twicken, J.-M. Désert, J. Carter, J.R. Crepp, G. Hébrard, A. Santerne, C. Moutou, C. Sobeck, D. Hudgins, M.R. Haas, P. Robertson,

J. Lillo-Box, D. Barrado, Masses, radii, and orbits of small Kepler planets: The transition from gaseous to rocky planets. Astrophys. J. Suppl. Ser. **210**, 20 (2014).
235. M.S. Marley, T.D. Robinson, On the cool side: Modeling the atmospheres of brown dwarfs and giant planets. Annu. Rev. Astron. Astrophys. **53**, 279–323 (2015).
236. M.S. Marley, D. Saumon, T. Guillot, R.S. Freedman, W.B. Hubbard, A. Burrows, J.I. Lunine, Atmospheric, evolutionary, and spectral models of the brown dwarf Gliese 229 B. Science **272**, 1919–1921 (1996).
237. M.S. Marley, A.S. Ackerman, J.N. Cuzzi, D. Kitzmann, Clouds and hazes in exoplanet atmospheres, in *Comparative Climatology of Terrestrial Planets*, ed. by S.J. Mackwell, A.A. Simon-Miller, J.W. Harder, M.A. Bullock (2013), pp. 367–391.
238. J.H.C. Martins, N.C. Santos, P. Figueira, J.P. Faria, M. Montalto, I. Boisse, D. Ehrenreich, C. Lovis, M. Mayor, C. Melo, F. Pepe, S.G. Sousa, S. Udry, D. Cunha, Evidence for a spectroscopic direct detection of reflected light from <ASTROBJ>51 Pegasi b</ASTROBJ>. Astron. Astrophys. **576**, 134 (2015).
239. Massol et al., Space Sci. Rev. (2016, this issue)
240. S. Matousek, The Juno new frontiers mission. Acta Astronaut. **61**, 932–939 (2007).
241. M. Mayor, M. Marmier, C. Lovis, S. Udry, D. Ségransan, F. Pepe, W. Benz, J.-. Bertaux, F. Bouchy, X. Dumusque, G. Lo Curto, C. Mordasini, D. Queloz, N.C. Santos, The HARPS search for southern extra-solar planets XXXIV. Occurrence, mass distribution and orbital properties of super-Earths and Neptune-mass planets. ArXiv e-prints (2011)
242. P. McCullough, J. MacKenty, Considerations for using Spatial Scans with WFC3. Technical report, 2012
243. P.R. McCullough, N. Crouzet, D. Deming, N. Madhusudhan, Water vapor in the spectrum of the extrasolar planet HD 189733b. I. The transit. Astrophys. J. **791**, 55 (2014).
244. C.P. McKay, J.B. Pollack, R. Courtin, The thermal structure of Titan's atmosphere. Icarus **80**, 23–53 (1989).
245. G.L. Mellor, T. Yamada, Development of a turbulence closure model for geophysical fluid problems. Rev. Geophys. Space Phys. **20**, 851–875 (1982).
246. K. Menou, Water-trapped worlds. Astrophys. J. **774**(1), 51 (2013)
247. Y. Miguel, L. Kaltenegger, Exploring atmospheres of hot mini-Neptunes and extrasolar giant planets orbiting different stars with application to HD 97658b, WASP-12b, CoRoT-2b, XO-1b, and HD 189733b. Astrophys. J. **780**, 166 (2014).
248. Y. Miguel, L. Kaltenegger, B. Fegley, L. Schaefer, Compositions of hot super-Earth atmospheres: Exploring Kepler candidates. Astrophys. J. Lett. **742**, 19 (2011).
249. Y. Miguel, L. Kaltenegger, J.L. Linsky, S. Rugheimer, The effect of Lyman αα radiation on mini-Neptune atmospheres around M stars: Application to GJ 436b. Mon. Not. R. Astron. Soc. **446**, 345–353 (2015).
250. N. Miller, J.J. Fortney, The heavy-element masses of extrasolar giant planets, revealed. Astrophys. J. Lett. **736**, 29 (2011).
251. Miller-Ricci Kempton, K. Zahnle, J.J. Fortney, The atmospheric chemistry of GJ 1214b: Photochemistry and clouds. Astrophys. J. **745**, 3 (2012).
252. Miller-Ricci, S. Seager, D. Sasselov, The atmospheric signatures of super-Earths: How to distinguish between hydrogen-rich and hydrogen-poor atmospheres. Astrophys. J. **690**, 1056–1067 (2009).
253. Morbidelli, J.I. Lunine, D.P. O'Brien, S.N. Raymond, K.J. Walsh, Building terrestrial planets. Annu. Rev. Earth Planet. Sci. **40**, 251–275 (2012).
254. Mordasini, Y. Alibert, C. Georgy, K.-M. Dittkrist, H. Klahr, T. Henning, Characterization of exoplanets from their formation. II. The planetary mass-radius relationship. Astron. Astrophys. **547**, 112 (2012).
255. C.V. Morley, J.J. Fortney, E.M.-R. Kempton, M.S. Marley, C. Visscher, K. Zahnle, Quantitatively assessing the role of clouds in the transmission spectrum of GJ 1214b. Astrophys. J. **775**, 33 (2013).
256. J.I. Moses, Chemical kinetics on extrasolar planets. Philos. Trans. R. Soc. A, Math. Phys. Eng. Sci. **372**, 30073 (2014).

257. J.I. Moses, B. Bézard, E. Lellouch, G.R. Gltone, H. Feuchtgruber, M. Allen, Photochemistry of Saturn's atmosphere. I. Hydrocarbon chemistry and comparisons with ISO observations. Icarus **143**, 244–298 (2000).
258. J.I. Moses, C. Visscher, J.J. Fortney, A.P. Showman, N.K. Lewis, C.A. Griffith, S.J. Klippenstein, M. Shabram, A.J. Friedson, M.S. Marley, R.S. Freedman, Disequilibrium carbon, oxygen, and nitrogen chemistry in the atmospheres of HD 189733b and HD 209458b. Astrophys. J. **737**, 15 (2011).
259. J.I. Moses, M.R. Line, C. Visscher, M.R. Richardson, N. Nettelmann, J.J. Fortney, T.S. Barman, K.B. Stevenson, N. Madhusudhan, Compositional diversity in the atmospheres of hot Neptunes, with application to GJ 436b. Astrophys. J. **777**, 34 (2013a).
260. J.I. Moses, N. Madhusudhan, C. Visscher, R.S. Freedman, Chemical consequences of the C/O ratio on hot Jupiters: Examples from WASP-12b, CoRoT-2b, XO-1b, and HD 189733b. Astrophys. J. **763**, 25 (2013b).
261. O. Mousis, J.I. Lunine, N. Madhusudhan, T.V. Johnson, Nebular water depletion as the cause of Jupiter's low oxygen abundance. Astrophys. J. Lett. **751**, 7 (2012).
262. Murgas, E. Pallé, M.R. Zapatero Osorio, L. Nortmann, S. Hoyer, A. Cabrera-Lavers, The GTC exoplanet transit spectroscopy survey. I. OSIRIS transmission spectroscopy of the short period planet WASP-43b. Astron. Astrophys. **563**, 41 (2014).
263. N. Nettelmann, U. Kramm, R. Redmer, R. Neuhäuser, Interior structure models of GJ 436b. Astron. Astrophys. **523**, 26 (2010).
264. N. Nikolov, D.K. Sing, F. Pont, A.S. Burrows, J.J. Fortney, G.E. Ballester, T.M. Evans, C.M. Huitson, H.R. Wakeford, P.A. Wilson, S. Aigrain, D. Deming, N.P. Gibson, G.W. Henry, H. Knutson, A. Lecavelier des Etangs, A.P. Showman, A. Vidal-Madjar, K. Zahnle, Hubble Space Telescope hot Jupiter transmission spectral survey: A detection of Na and strong optical absorption in HAT-P-1b. Mon. Not. R. Astron. Soc. **437**, 46–66 (2014).
265. N. Nikolov, D.K. Sing, A.S. Burrows, J.J. Fortney, G.W. Henry, F. Pont, G.E. Ballester, S. Aigrain, P.A. Wilson, C.M. Huitson, N.P. Gibson, J.-M. Désert, A.L.d. Etangs, A.P. Showman, A. Vidal-Madjar, H.R. Wakeford, K. Zahnle, HST hot-Jupiter transmission spectral survey: Haze in the atmosphere of WASP-6b. Mon. Not. R. Astron. Soc. **447**, 463–478 (2015).
266. K.I. Öberg, R. Murray-Clay, E.A. Bergin, The effects of snowlines on C/O in planetary atmospheres. Astrophys. J. Lett. **743**, 16 (2011).
267. Oberg et al., Space Sci. Rev. (2016, this issue)
268. G.I. Ogilvie, Tidal dissipation in stars and giant planets. Annu. Rev. Astron. Astrophys. **52**, 171–210 (2014).
269. B.R. Oppenheimer, C. Baranec, C. Beichman, D. Brenner, R. Burruss, E. Cady, J.R. Crepp, R. Dekany, R. Fergus, D. Hale, L. Hillenbrand, S. Hinkley, D.W. Hogg, D. King, E.R. Ligon, T. Lockhart, R. Nilsson, I.R. Parry, L. Pueyo, E. Rice, J.E. Roberts, L.C. Roberts Jr., M. Shao, A. Sivaramakrishnan, R. Soummer, T. Truong, G. Vasisht, A. Veicht, F. Vescelus, J.K. Wallace, C. Zhai, N. Zimmerman, Reconnaissance of the HR 8799 exosolar system. I. Near-infrared spectroscopy. Astrophys. J. **768**, 24 (2013).
270. J.E. Owen, Y. Wu, Kepler planets: A tale of evaporation. Astrophys. J. **775**, 105 (2013).
271. T. Owen, P. Mahaffy, H.B. Niemann, S. Atreya, T. Donahue, A. Bar-Nun, I. de Pater, A low-temperature origin for the planetesimals that formed Jupiter. Nature **402**, 269–270 (1999).
272. J.C.B. Papaloizou, C. Terquem, Planet formation and migration. Rep. Prog. Phys. **69**, 119–180 (2006).
273. J.C.B. Papaloizou, R.P. Nelson, W. Kley, F.S. Masset, P. Artymowicz, Disk-planet interactions during planet formation, in *Protostars and Planets V* (2007), pp. 655–668
274. V. Parmentier, T. Guillot, A non-grey analytical model for irradiated atmospheres. I. Derivation. Astron. Astrophys. **562**, 133 (2014).
275. V. Parmentier, A.P. Showman, Y. Lian, 3D mixing in hot Jupiters atmospheres. I. Application to the day/night cold trap in HD 209458b. Astron. Astrophys. **558**, 91 (2013).
276. V. Parmentier, T. Guillot, J.J. Fortney, M.S. Marley, A non-grey analytical model for irradiated atmospheres. II. Analytical vs. numerical solutions. Astron. Astrophys. **574**, 35 (2015).
277. J. Peixoto, A.H. Oort, *Physics of Climate* (Am. Inst. of Phys., New York, 1992). 520 pp.

278. R.O. Pepin, Atmospheres on the terrestrial planets: Clues to origin and evolution. Earth Planet. Sci. Lett. **252**, 1–14 (2006).
279. N.A. Phillips, The general circulation of the atmosphere: A numerical experiment. Q. J. R. Meteorol. Soc. **82**, 123–164 (1956).
280. R.T. Pierrehumbert, A palette of climates for Gliese 581g. Astrophys. J. Lett. **726**(1), 8 (2011)
281. J.B. Pollack, O. Hubickyj, P. Bodenheimer, J.J. Lissauer, M. Podolak, Y. Greenzweig, Formation of the giant planets by concurrent accretion of solids and gas. Icarus **124**, 62–85 (1996).
282. Pont, H. Knutson, R.L. Gilliland, C. Moutou, D. Charbonneau, Detection of atmospheric haze on an extrasolar planet: The 0.55-1.05 μm transmission spectrum of HD 189733b with the Hubble Space Telescope. Mon. Not. R. Astron. Soc. **385**, 109–118 (2008).
283. F. Pont, D.K. Sing, N.P. Gibson, S. Aigrain, G. Henry, N. Husnoo, The prevalence of dust on the exoplanet HD 189733b from Hubble and Spitzer observations. Mon. Not. R. Astron. Soc. **432**, 2917–2944 (2013).
284. R.G. Prinn, S.S. Barshay, Carbon monoxide on Jupiter and implications for atmospheric convection. Science **198**, 1031–1034 (1977).
285. R.R. Rafikov, Can giant planets form by direct gravitational instability? Astrophys. J. Lett. **621**, 69–72 (2005).
286. S. Ranjan, D. Charbonneau, J.-M. Désert, N. Madhusudhan, D. Deming, A. Wilkins, A.M. Mandell, Atmospheric characterization of five hot Jupiters with the wide field camera 3 on the Hubble Space Telescope. Astrophys. J. **785**, 148 (2014).
287. F.A. Rasio, E.B. Ford, Dynamical instabilities and the formation of extrasolar planetary systems. Science **274**, 954–956 (1996).
288. Rauer, C. Catala, C. Aerts, T. Appourchaux, W. Benz, A. Brandeker, J. Christensen-Dalsgaard, M. Deleuil, L. Gizon, M.-J. Goupil, M. Güdel, E. Janot-Pacheco, M. Mas-Hesse, I. Pagano, G. Piotto, D. Pollacco, Ċ. Santos, A. Smith, J.-C. Suárez, R. Szabó, S. Udry, V. Adibekyan, Y. Alibert, J.-M. Almenara, P. Amaro-Seoane, M.A.-v. Eiff, M. Asplund, E. Antonello, S. Barnes, F. Baudin, K. Belkacem, M. Bergemann, G. Bihain, A.C. Birch, X. Bonfils, I. Boisse, A.S. Bonomo, F. Borsa, I.M. Brandão, E. Brocato, S. Brun, M. Burleigh, R. Burston, J. Cabrera, S. Cassisi, W. Chaplin, S. Charpinet, C. Chiappini, R.P. Church, S. Csizmadia, M. Cunha, M. Damasso, M.B. Davies, H.J. Deeg, R.F. Díaz, S. Dreizler, C. Dreyer, P. Eggenberger, D. Ehrenreich, P. Eigmüller, A. Erikson, R. Farmer, S. Feltzing, F. de Oliveira Fialho, P. Figueira, T. Forveille, M. Fridlund, R.A. García, P. Giommi, G. Giuffrida, M. Godolt, J. Gomes da Silva, T. Granzer, J.L. Grenfell, A. Grotsch-Noels, E. Günther, C.A. Haswell, A.P. Hatzes, G. Hébrard, S. Hekker, R. Helled, K. Heng, J.M. Jenkins, A. Johansen, M.L. Khodachenko, K.G. Kislyakova, W. Kley, U. Kolb, N. Krivova, F. Kupka, H. Lammer, A.F. Lanza, Y. Lebreton, D. Magrin, P. Marcos-Arenal, P.M. Marrese, J.P. Marques, J. Martins, S. Mathis, S. Mathur, S. Messina, A. Miglio, J. Montalban, M. Montalto, M.J.P.F.G. Monteiro, H. Moradi, E. Moravveji, C. Mordasini, T. Morel, A. Mortier, V. Nascimbeni, R.P. Nelson, M.B. Nielsen, L. Noack, A.J. Norton, A. Ofir, M. Oshagh, R.-M. Ouazzani, P. Pápics, V.C. Parro, P. Petit, B. Plez, E. Poretti, A. Quirrenbach, R. Ragazzoni, G. Raimondo, M. Rainer, D.R. Reese, R. Redmer, S. Reffert, B. Rojas-Ayala, I.W. Roxburgh, S. Salmon, A. Santerne, J. Schneider, J. Schou, S. Schuh, H. Schunker, A. Silva-Valio, R. Silvotti, I. Skillen, I. Snellen, F. Sohl, S.G. Sousa, A. Sozzetti, D. Stello, K.G. Strassmeier, M. Švanda, G.M. Szabó, A. Tkachenko, D. Valencia, V. Van Grootel, S.D. Vauclair, P. Ventura, F.W. Wagner, N.A. Walton, J. Weingrill, S.C. Werner, P.J. Wheatley, K. Zwintz, The PLATO 2.0 mission. Exp. Astron. **38**, 249–330 (2014).
289. S. Redfield, M. Endl, W.D. Cochran, L. Koesterke, Sodium absorption from the exoplanetary atmosphere of HD 189733b detected in the optical transmission spectrum. Astrophys. J. Lett. **673**, 87–90 (2008).
290. G.R. Ricker, J.N. Winn, R. Vanderspek, D.W. Latham, G.Á. Bakos, J.L. Bean, Z.K. Berta-Thompson, T.M. Brown, L. Buchhave, N.R. Butler, R.P. Butler, W.J. Chaplin, D. Charbonneau, J. Christensen-Dalsgaard, M. Clampin, D. Deming, J. Doty, N. De Lee, C. Dressing, E.W. Dunham, M. Endl, F. Fressin, J. Ge, T. Henning, M.J. Holman, A.W. Howard, S. Ida, J. Jenkins, G. Jernigan, J.A. Johnson, L. Kaltenegger, N. Kawai, H. Kjeldsen, G. Laughlin,

A.M. Levine, D. Lin, J.J. Lissauer, P. MacQueen, G. Marcy, P.R. McCullough, T.D. Morton, N. Narita, M. Paegert, E. Palle, F. Pepe, J. Pepper, A. Quirrenbach, S.A. Rinehart, D. Sasselov, B. Sato, S. Seager, A. Sozzetti, K.G. Stassun, P. Sullivan, A. Szentgyorgyi, G. Torres, S. Udry, J. Villasenor, Transiting Exoplanet Survey Satellite (TESS). J. Astron. Telesc. Instrum. Syst. **1**(1), 014003 (2015).

291. P.B. Rimmer, C. Helling, Ionization in atmospheres of brown dwarfs and extrasolar planets. IV. The effect of cosmic rays. Astrophys. J. **774**, 108 (2013).
292. P.B. Rimmer, C. Helling, A chemical kinetics network for lightning and life in planetary atmospheres. ArXiv e-prints (2015)
293. P.B. Rimmer, C. Helling, C. Bilger, The influence of galactic cosmic rays on ion-neutral hydrocarbon chemistry in the upper atmospheres of free-floating exoplanets. Int. J. Astrobiol. **13**, 173–181 (2014).
294. T.D. Robinson, D.C. Catling, An analytic radiative-convective model for planetary atmospheres. Astrophys. J. **757**, 104 (2012).
295. F. Rodler, M. Lopez-Morales, I. Ribas, Weighing the non-transiting hot Jupiter ττ Boo b. Astrophys. J. Lett. **753**, 25 (2012).
296. F. Rodler, M. Kürster, J.R. Barnes, Detection of CO absorption in the atmosphere of the hot Jupiter HD 189733b. Mon. Not. R. Astron. Soc. **432**, 1980–1988 (2013).
297. M. Rodono, The atmospheres of m dwarfs: Observations. NASA Spec. Publ. **492**, 409–453 (1986)
298. L.A. Rogers, Most 1.6 Earth-radius planets are not rocky. Astrophys. J. **801**, 41 (2015).
299. L.A. Rogers, P. Bodenheimer, J.J. Lissauer, S. Seager, Formation and structure of low-density exo-Neptunes. Astrophys. J. **738**, 59 (2011).
300. W.B. Rossow, Cloud microphysics—Analysis of the clouds of Earth, Venus, Mars, and Jupiter. Icarus **36**, 1–50 (1978).
301. J.F. Rowe, J.M. Matthews, S. Seager, E. Miller-Ricci, D. Sasselov, R. Kuschnig, D.B. Guenther, A.F.J. Moffat, S.M. Rucinski, G.A.H. Walker, W.W. Weiss, The very low albedo of an extrasolar planet: MOST space-based photometry of HD 209458. Astrophys. J. **689**, 1345–1353 (2008).
302. J.F. Rowe, J.L. Coughlin, V. Antoci, T. Barclay, N.M. Batalha, W.J. Borucki, C.J. Burke, S.T. Bryson, D.A. Caldwell, J.R. Campbell, J.H. Catanzarite, J.L. Christiansen, W. Cochran, R.L. Gilliland, F.R. Girouard, M.R. Haas, K.G. Hełminiak, C.E. Henze, K.L. Hoffman, S.B. Howell, D. Huber, R.C. Hunter, H. Jang-Condell, J.M. Jenkins, T.C. Klaus, D.W. Latham, J. Li, J.J. Lissauer, S.D. McCauliff, R.L. Morris, F. Mullally, A. Ofir, B. Quarles, E. Quintana, A. Sabale, S. Seader, A. Shporer, J.C. Smith, J.H. Steffen, M. Still, P. Tenenbaum, S.E. Thompson, J.D. Twicken, C. Van Laerhoven, A. Wolfgang, K.A. Zamudio, Planetary candidates observed by Kepler. V. Planet sample from Q1-Q12 (36 months). Astrophys. J. Suppl. Ser. **217**, 16 (2015).
303. Sanz-Forcada, G. Micela, I. Ribas, A.M.T. Pollock, C. Eiroa, A. Velasco, E. Solano, D. García-Álvarez, Estimation of the XUV radiation onto close planets and their evaporation. Astron. Astrophys. **532**, 6 (2011).
304. L. Schaefer, B. Fegley Jr., Atmospheric chemistry of Venus-like exoplanets. Astrophys. J. **729**, 6 (2011).
305. L. Schaefer, B. Fegley, Chemistry of silicate atmospheres of evaporating super-Earths. Astrophys. J. Lett. **703**, 113–117 (2009).
306. L. Schaefer, B. Fegley, Chemistry of atmospheres formed during accretion of the Earth and other terrestrial planets. Icarus **208**, 438–448 (2010).
307. L. Schaefer, K. Lodders, B. Fegley, Vaporization of the Earth: Application to exoplanet atmospheres. Astrophys. J. **755**, 41 (2012).
308. E. Schlawin, E. Agol, L.M. Walkowicz, K. Covey, J.P. Lloyd, Exoplanetary transits of limb-brightened lines: Tentative Si IV absorption by HD 209458b. Astrophys. J. Lett. **722**, 75–79 (2010).
309. E. Schlawin, M. Zhao, J.K. Teske, T. Herter, A 0.8-2.4 μm transmission spectrum of the hot Jupiter CoRoT-1b. Astrophys. J. **783**, 5 (2014).
310. Schulze-Makuch, A. Méndez, A.G. Fairén, P. Von Paris, C. Turse, G. Boyer, A.F. Davila, M.R.d.S. António, D. Catling, L.N. Irwin, A two-tiered approach to assessing the habitability of exoplanets. Astrobiology **11**(10), 1041–1052 (2011)

311. J.C. Schwartz, N.B. Cowan, Balancing the energy budget of short-period giant planets: Evidence for reflective clouds and optical absorbers. Mon. Not. R. Astron. Soc. **449**, 4192–4203 (2015).
312. S. Seager, D.D. Sasselov, Extrasolar giant planets under strong stellar irradiation. Astrophys. J. Lett. **502**, 157–161 (1998).
313. S. Seager, D.D. Sasselov, Theoretical transmission spectra during extrasolar giant planet transits. Astrophys. J. **537**, 916–921 (2000).
314. S. Seager, L.J. Richardson, B.M.S. Hansen, K. Menou, J.Y.-K. Cho, D. Deming, On the dayside thermal emission of hot Jupiters. Astrophys. J. **632**, 1122–1131 (2005).
315. Segura, L.M. Walkowicz, V. Meadows, J. Kasting, S. Hawley, The effect of a strong stellar flare on the atmospheric chemistry of an Earth-like planet orbiting an M dwarf. Astrobiology **10**, 751–771 (2010).
316. F. Selsis, J. Kasting, B. Levrard, J. Paillet, I. Ribas, X. Delfosse, Habitable planets around the star Gliese 581? Astron. Astrophys. **476**(3), 1373–1387 (2007)
317. I.F. Shaikhislamov, M.L. Khodachenko, Y.L. Sasunov, H. Lammer, K.G. Kislyakova, N.V. Erkaev, Atmosphere expansion and mass loss of close-orbit giant exoplanets heated by stellar XUV. I. Modeling of hydrodynamic escape of upper atmospheric material. Astrophys. J. **795**, 132 (2014).
318. Shapley, *Climatic Change: Evidence, Causes, and Effects* (1953)
319. A.P. Showman, T. Guillot, Atmospheric circulation and tides of "51 Pegasus b-like" planets. Astron. Astrophys. **385**, 166–180 (2002).
320. A.P. Showman, L.M. Polvani, Equatorial superrotation on tidally locked exoplanets. Astrophys. J. **738**(1), 71 (2011)
321. A.P. Showman, J.J. Fortney, Y. Lian, M.S. Marley, R.S. Freedman, H.A. Knutson, D. Charbonneau, Atmospheric circulation of hot Jupiters: Coupled radiative-dynamical general circulation model simulations of HD 189733b and HD 209458b. Astrophys. J. **699**, 564–584 (2009).
322. A.P. Showman, J.Y.-K. Cho, K. Menou, Atmospheric circulation of exoplanets, in *Exoplanets*, ed. by S. Seager (2010), pp. 471–516
323. A.P. Showman, R.D. Wordsworth, T.M. Merlis, Y. Kaspi, Atmospheric circulation of terrestrial exoplanets, in *Comparative Climatology of Terrestrial Planets*, ed. by S.J. Mackwell, A.A. Simon-Miller, J.W. Harder, M.A. Bullock (2013), pp. 277–326.
324. D.K. Sing, M. López-Morales, Ground-based secondary eclipse detection of the very-hot Jupiter OGLE-TR-56b. Astron. Astrophys. **493**, 31–34 (2009).
325. D.K. Sing, J.-M. Désert, A. Lecavelier Des Etangs, G.E. Ballester, A. Vidal-Madjar, V. Parmentier, G. Hebrard, G.W. Henry, Transit spectrophotometry of the exoplanet HD 189733b. I. Searching for water but finding haze with HST NICMOS. Astron. Astrophys. **505**, 891–899 (2009).
326. D.K. Sing, J.-M. Désert, J.J. Fortney, A. Lecavelier Des Etangs, G.E. Ballester, J. Cepa, D. Ehrenreich, M. López-Morales, F. Pont, M. Shabram, A. Vidal-Madjar, Gran Telescopio Canarias OSIRIS transiting exoplanet atmospheric survey: Detection of potassium in XO-2b from narrowband spectrophotometry. Astron. Astrophys. **527**, 73 (2011a).
327. D.K. Sing, F. Pont, S. Aigrain, D. Charbonneau, J.-M. Désert, N. Gibson, R. Gilliland, W. Hayek, G. Henry, H. Knutson, A. Lecavelier Des Etangs, T. Mazeh, A. Shporer, Hubble Space Telescope transmission spectroscopy of the exoplanet HD 189733b: High-altitude atmospheric haze in the optical and near-ultraviolet with STIS. Mon. Not. R. Astron. Soc. **416**, 1443–1455 (2011b).
328. D.K. Sing, C.M. Huitson, M. Lopez-Morales, F. Pont, J.-M. Désert, D. Ehrenreich, P.A. Wilson, G.E. Ballester, J.J. Fortney, A. Lecavelier des Etangs, A. Vidal-Madjar, GTC OSIRIS transiting exoplanet atmospheric survey: Detection of sodium in XO-2b from differential long-slit spectroscopy. Mon. Not. R. Astron. Soc. **426**, 1663–1670 (2012).
329. D.K. Sing, A. Lecavelier des Etangs, J.J. Fortney, A.S. Burrows, F. Pont, H.R. Wakeford, G.E. Ballester, N. Nikolov, G.W. Henry, S. Aigrain, D. Deming, T.M. Evans, N.P. Gibson, C.M. Huitson, H. Knutson, A.P. Showman, A. Vidal-Madjar, P.A. Wilson, M.H. Williamson, K. Zahnle, HST hot-Jupiter transmission spectral survey: Evidence for aerosols and lack of TiO in the atmosphere of WASP-12b. Mon. Not. R. Astron. Soc. **436**, 2956–2973 (2013).

330. D.K. Sing, H.R. Wakeford, A.P. Showman, N. Nikolov, J.J. Fortney, A.S. Burrows, G.E. Ballester, D. Deming, S. Aigrain, J.-M. Désert, N.P. Gibson, G.W. Henry, H. Knutson, A. Lecavelier des Etangs, F. Pont, A. Vidal-Madjar, M.W. Williamson, P.A. Wilson, HST hot-Jupiter transmission spectral survey: Detection of potassium in WASP-31b along with a cloud deck and Rayleigh scattering. Mon. Not. R. Astron. Soc. **446**, 2428–2443 (2015).
331. A.J. Skemer, P.M. Hinz, S. Esposito, A. Burrows, J. Leisenring, M. Skrutskie, S. Desidera, D. Mesa, C. Arcidiacono, F. Mannucci, T.J. Rodigas, L. Close, D. McCarthy, C. Kulesa, G. Agapito, D. Apai, J. Argomedo, V. Bailey, K. Boutsia, R. Briguglio, G. Brusa, L. Busoni, R. Claudi, J. Eisner, L. Fini, K.B. Follette, P. Garnavich, R. Gratton, J.C. Guerra, J.M. Hill, W.F. Hoffmann, T. Jones, M. Krejny, J. Males, E. Masciadri, M.R. Meyer, D.L. Miller, K. Morzinski, M. Nelson, E. Pinna, A. Puglisi, S.P. Quanz, F. Quiros-Pacheco, A. Riccardi, P. Stefanini, V. Vaitheeswaran, J.C. Wilson, M. Xompero, First light LBT AO images of HR 8799 bcde at 1.6 and 3.3 μm: New discrepancies between young planets and old brown dwarfs. Astrophys. J. **753**, 14 (2012).
332. M.D. Smith, Estimation of a length scale to use with the quench level approximation for obtaining chemical abundances. Icarus **132**, 176–184 (1998).
333. I.A.G. Snellen, R.J. de Kok, E.J.W. de Mooij, S. Albrecht, The orbital motion, absolute mass and high-altitude winds of exoplanet HD209458b. Nature **465**, 1049–1051 (2010).
334. I.A.G. Snellen, B.R. Brandl, R.J. de Kok, M. Brogi, J. Birkby, H. Schwarz, Fast spin of the young extrasolar planet ββ Pictoris b. Nature **509**, 63–65 (2014).
335. D.S. Spiegel, K. Silverio, A. Burrows, Can TiO explain thermal inversions in the upper atmospheres of irradiated giant planets? Astrophys. J. **699**, 1487–1500 (2009).
336. D.S. Spiegel, A. Burrows, L. Ibgui, I. Hubeny, J.A. Milsom, Models of Neptune-mass exoplanets: Emergent fluxes and albedos. Astrophys. J. **709**, 149–158 (2010).
337. C.R. Stark, C. Helling, D.A. Diver, P.B. Rimmer, Electrostatic activation of prebiotic chemistry in substellar atmospheres. Int. J. Astrobiol. **13**, 165–172 (2014).
338. D.J. Stevenson, J.I. Lunine, Rapid formation of Jupiter by diffuse redistribution of water vapor in the solar nebula. Icarus **75**, 146–155 (1988).
339. K.B. Stevenson, J. Harrington, S. Nymeyer, N. Madhusudhan, S. Seager, W.C. Bowman, R.A. Hardy, D. Deming, E. Rauscher, N.B. Lust, Possible thermochemical disequilibrium in the atmosphere of the exoplanet GJ 436b. Nature **464**, 1161–1164 (2010).
340. K.B. Stevenson, J.L. Bean, N. Madhusudhan, J. Harrington, Deciphering the atmospheric composition of WASP-12b: A comprehensive analysis of its dayside emission. Astrophys. J. **791**, 36 (2014a).
341. K.B. Stevenson, J.L. Bean, A. Seifahrt, J.-M. Désert, N. Madhusudhan, M. Bergmann, L. Kreidberg, D. Homeier, Transmission spectroscopy of the hot Jupiter WASP-12b from 0.7 to 5 μm. Astron. J. **147**, 161 (2014b).
342. K.B. Stevenson, J.-M. Désert, M.R. Line, J.L. Bean, J.J. Fortney, A.P. Showman, T. Kataria, L. Kreidberg, P.R. McCullough, G.W. Henry, D. Charbonneau, A. Burrows, S. Seager, N. Madhusudhan, M.H. Williamson, D. Homeier, Thermal structure of an exoplanet atmosphere from phase-resolved emission spectroscopy. Science **346**, 838–841 (2014c).
343. Stökl, E. Dorfi, H. Lammer, Hydrodynamic simulations of captured protoatmospheres around Earth-like planets. Astron. Astrophys. **576**, 87 (2015).
344. D.F. Strobel, Parameterization of linear wave chemical transport in planetary atmospheres by eddy diffusion. J. Geophys. Res. **86**, 9806–9810 (1981).
345. Sudarsky, A. Burrows, P. Pinto, Albedo and reflection spectra of extrasolar giant planets. Astrophys. J. **538**, 885–903 (2000).
346. Sudarsky, A. Burrows, I. Hubeny, Theoretical spectra and atmospheres of extrasolar giant planets. Astrophys. J. **588**, 1121–1148 (2003).
347. M.R. Swain, G. Vasisht, G. Tinetti, The presence of methane in the atmosphere of an extrasolar planet. Nature **452**, 329–331 (2008).
348. M.R. Swain, G. Tinetti, G. Vasisht, P. Deroo, C. Griffith, J. Bouwman, P. Chen, Y. Yung, A. Burrows, L.R. Brown, J. Matthews, J.F. Rowe, R. Kuschnig, D. Angerhausen, Water, methane, and carbon dioxide present in the dayside spectrum of the exoplanet HD 209458b. Astrophys. J. **704**, 1616–1621 (2009a).

349. M.R. Swain, G. Vasisht, G. Tinetti, J. Bouwman, P. Chen, Y. Yung, D. Deming, P. Deroo, Molecular signatures in the near-infrared dayside spectrum of HD 189733b. Astrophys. J. Lett. **690**, 114–117 (2009b).
350. M. Swain, P. Deroo, G. Tinetti, M. Hollis, M. Tessenyi, M. Line, H. Kawahara, Y. Fujii, A.P. Showman, S.N. Yurchenko, Probing the extreme planetary atmosphere of WASP-12b. Icarus **225**, 432–445 (2013).
351. M.R. Swain, M.R. Line, P. Deroo, On the detection of molecules in the atmosphere of HD 189733b using HST NICMOS transmission spectroscopy. Astrophys. J. **784**, 133 (2014).
352. Terquem, J.C.B. Papaloizou, Migration and the formation of systems of hot super-Earths and Neptunes. Astrophys. J. **654**, 1110–1120 (2007).
353. Tian, Thermal escape from super Earth atmospheres in the habitable zones of m stars. Astrophys. J. **703**(1), 905 (2009)
354. Tinetti, A. Vidal-Madjar, M.-C. Liang, J.-P. Beaulieu, Y. Yung, S. Carey, R.J. Barber, J. Tennyson, I. Ribas, N. Allard, G.E. Ballester, D.K. Sing, F. Selsis, Water vapour in the atmosphere of a transiting extrasolar planet. Nature **448**, 169–171 (2007).
355. Tinetti, P. Deroo, M.R. Swain, C.A. Griffith, G. Vasisht, L.R. Brown, C. Burke, P. McCullough, Probing the terminator region atmosphere of the hot-Jupiter XO-1b with transmission spectroscopy. Astrophys. J. Lett. **712**, 139–142 (2010).
356. K.O. Todorov, D. Deming, A. Burrows, C.J. Grillmair, Updated Spitzer emission spectroscopy of bright transiting hot Jupiter HD 189733b. Astrophys. J. **796**, 100 (2014).
357. K.O. Todorov, M.R. Line, J.E. Pineda, M.R. Meyer, S.P. Quanz, S. Hinkley, J.J. Fortney, The water abundance of the directly imaged substellar companion {\kappa}{\kappa} and b retrieved from a near infrared spectrum. ArXiv e-prints (2015)
358. M.G. Tomasko, B. Archinal, T. Becker, B. Bézard, M. Bushroe, M. Combes, D. Cook, A. Coustenis, C. de Bergh, L.E. Dafoe, L. Doose, S. Douté, A. Eibl, S. Engel, F. Gliem, B. Grieger, K. Holso, E. Howington-Kraus, E. Karkoschka, H.U. Keller, R. Kirk, R. Kramm, M. Küppers, P. Lanagan, E. Lellouch, M. Lemmon, J. Lunine, E. McFarlane, J. Moores, G.M. Prout, B. Rizk, M. Rosiek, P. Rueffer, S.E. Schröder, B. Schmitt, C. See, P. Smith, L. Soderblom, N. Thomas, R. West, Rain, winds and haze during the Huygens probe's descent to Titan's surface. Nature **438**, 765–778 (2005).
359. A.H.M.J. Triaud, A. Collier Cameron, D. Queloz, D.R. Anderson, M. Gillon, L. Hebb, C. Hellier, B. Loeillet, P.F.L. Maxted, M. Mayor, F. Pepe, D. Pollacco, D. Ségransan, B. Smalley, S. Udry, R.G. West, P.J. Wheatley, Spin-orbit angle measurements for six southern transiting planets. New insights into the dynamical origins of hot Jupiters. Astron. Astrophys. **524**, 25 (2010).
360. T. Tsuji, Molecular abundances in stellar atmospheres. II. Astron. Astrophys. **23**, 411–431 (1973)
361. T. Tsuji, K. Ohnaka, W. Aoki, T. Nakajima, Evolution of dusty photospheres through red to brown dwarfs: How dust forms in very low mass objects. Astron. Astrophys. **308**, 29–32 (1996)
362. S. Udry, X. Bonfils, X. Delfosse, T. Forveille, M. Mayor, C. Perrier, F. Bouchy, C. Lovis, F. Pepe, D. Queloz, et al., The harps search for southern extra-solar planets ⋆,⋆⋆. Astron. Astrophys. **469**, 43–47 (2007)
363. E.F. van Dishoeck, E.A. Bergin, D.C. Lis, J.I. Lunine, Water: From clouds to planets, in *Protostars and Planets VI* (2014), pp. 835–858.
364. O. Venot, E. Hébrard, M. Agúndez, M. Dobrijevic, F. Selsis, F. Hersant, N. Iro, R. Bounaceur, A chemical model for the atmosphere of hot Jupiters. Astron. Astrophys. **546**, 43 (2012).
365. O. Venot, N. Fray, Y. Bénilan, M.-C. Gazeau, E. Hébrard, G. Larcher, M. Schwell, M. Dobrijevic, F. Selsis, High-temperature measurements of VUV-absorption cross sections of CO_2 and their application to exoplanets. Astron. Astrophys. **551**, 131 (2013).
366. O. Venot, M. Agúndez, F. Selsis, M. Tessenyi, N. Iro, The atmospheric chemistry of the warm Neptune GJ 3470b: Influence of metallicity and temperature on the CH_4/CO ratio. Astron. Astrophys. **562**, 51 (2014).
367. O. Venot, E. Hébrard, M. Agúndez, L. Decin, R. Bounaceur, New chemical scheme for studying carbon-rich exoplanet atmospheres. Astron. Astrophys. **577**, A33 (2015)

368. Venturini, Y. Alibert, W. Benz, M. Ikoma, Critical core mass for enriched envelopes: The role of H_2O condensation. Astron. Astrophys. **576**, 114 (2015).
369. Vidal-Madjar, A. Lecavelier des Etangs, J.-M. Désert, G.E. Ballester, R. Ferlet, G. Hébrard, M. Mayor, An extended upper atmosphere around the extrasolar planet HD209458b. Nature **422**, 143–146 (2003).
370. Vidal-Madjar, J.-M. Désert, A. Lecavelier des Etangs, G. Hébrard, G.E. Ballester, D. Ehrenreich, R. Ferlet, J.C. McConnell, M. Mayor, C.D. Parkinson, Detection of oxygen and carbon in the hydrodynamically escaping atmosphere of the extrasolar planet HD 209458b. Astrophys. J. Lett. **604**, 69–72 (2004).
371. Vidal-Madjar, D.K. Sing, A. Lecavelier Des Etangs, R. Ferlet, J.-M. Désert, G. Hébrard, I. Boisse, D. Ehrenreich, C. Moutou, The upper atmosphere of the exoplanet HD 209458 b revealed by the sodium D lines. Temperature-pressure profile, ionization layer, and thermosphere. Astron. Astrophys. **527**, 110 (2011).
372. Vidal-Madjar, C.M. Huitson, V. Bourrier, J.-M. Désert, G. Ballester, A. Lecavelier des Etangs, D.K. Sing, D. Ehrenreich, R. Ferlet, G. Hébrard, J.C. McConnell, Magnesium in the atmosphere of the planet HD 209458 b: Observations of the thermosphere-exosphere transition region. Astron. Astrophys. **560**, 54 (2013).
373. Visscher, J.I. Moses, Quenching of carbon monoxide and methane in the atmospheres of cool brown dwarfs and hot Jupiters. Astrophys. J. **738**, 72 (2011).
374. Visscher, K. Lodders, B. Fegley Jr., Atmospheric chemistry in giant planets, brown dwarfs, and low-mass dwarf stars. II. Sulfur and phosphorus. Astrophys. J. **648**, 1181–1195 (2006).
375. Visscher, K. Lodders, B. Fegley Jr., Atmospheric chemistry in giant planets, brown dwarfs, and low-mass dwarf stars. III. Iron, magnesium, and silicon. Astrophys. J. **716**, 1060–1075 (2010a).
376. Visscher, J.I. Moses, S.A. Saslow, The deep water abundance on Jupiter: New constraints from thermochemical kinetics and diffusion modeling. Icarus **209**, 602–615 (2010b).
377. W. Von Bloh, C. Bounama, M. Cuntz, S. Franck, The habitability of super-earths in Gliese 581. Astron. Astrophys. **476**(3), 1365–1371 (2007)
378. P. Von Paris, S. Gebauer, M. Godolt, H. Rauer, B. Stracke, Atmospheric studies of habitability in the Gliese 581 system. Astron. Astrophys. **532**, 58 (2011)
379. J.H. Waite, D.T. Young, T.E. Cravens, A.J. Coates, F.J. Crary, B. Magee, J. Westlake, The process of tholin formation in Titan's upper atmosphere. Science **316**, 870–875 (2007).
380. H.R. Wakeford, D.K. Sing, D. Deming, N.P. Gibson, J.J. Fortney, A.S. Burrows, G. Ballester, N. Nikolov, S. Aigrain, G. Henry, H. Knutson, A. Lecavelier des Etangs, F. Pont, A.P. Showman, A. Vidal-Madjar, K. Zahnle, HST hot Jupiter transmission spectral survey: Detection of water in HAT-P-1b from WFC3 near-IR spatial scan observations. Mon. Not. R. Astron. Soc. **435**, 3481–3493 (2013).
381. I.P. Waldmann, G. Tinetti, P. Deroo, M.D.J. Hollis, S.N. Yurchenko, J. Tennyson, Blind extraction of an exoplanetary spectrum through independent component analysis. Astrophys. J. **766**, 7 (2013).
382. I.P. Waldmann, G. Tinetti, M. Rocchetto, E.J. Barton, S.N. Yurchenko, J. Tennyson, Tau-REx I: A next generation retrieval code for exoplanetary atmospheres. Astrophys. J. **802**, 107 (2015).
383. Y. Wang, F. Tian, Y. Hu, Climate patterns of habitable exoplanets in eccentric orbits around m dwarfs. Astrophys. J. Lett. **791**(1), 12 (2014)
384. Wang, P.J. Gierasch, J.I. Lunine, O. Mousis, New insights on Jupiter's deep water abundance from disequilibrium species. Icarus **250**, 154–164 (2015).
385. L.M. Weiss, G.W. Marcy, The mass-radius relation for 65 exoplanets smaller than 4 Earth radii. Astrophys. J. Lett. **783**, 6 (2014).
386. L.M. Weiss, G.W. Marcy, J.F. Rowe, A.W. Howard, H. Isaacson, J.J. Fortney, N. Miller, B.-O. Demory, D.A. Fischer, E.R. Adams, A.K. Dupree, S.B. Howell, R. Kolbl, J.A. Johnson, E.P. Horch, M.E. Everett, D.C. Fabrycky, S. Seager, The mass of KOI-94d and a relation for planet radius, mass, and incident flux. Astrophys. J. **768**, 14 (2013).

387. A.N. Wilkins, D. Deming, N. Madhusudhan, A. Burrows, H. Knutson, P. McCullough, S. Ranjan, The emergent 1.1-1.7 μm spectrum of the exoplanet CoRoT-2b as measured using the Hubble Space Telescope. Astrophys. J. **783**, 113 (2014).
388. P.A. Wilson, D.K. Sing, N. Nikolov, A. Lecavelier des Etangs, F. Pont, J.J. Fortney, G.E. Ballester, M. López-Morales, J.-M. Désert, A. Vidal-Madjar, GTC OSIRIS transiting exoplanet atmospheric survey: Detection of potassium in HAT-P-1b from narrow-band spectrophotometry. Mon. Not. R. Astron. Soc. **450**, 192–200 (2015).
389. J.N. Winn, D. Fabrycky, S. Albrecht, J.A. Johnson, Hot stars with hot Jupiters have high obliquities. Astrophys. J. Lett. **718**, 145–149 (2010).
390. Wolfgang, E. Lopez, How rocky are they? The composition distribution of Kepler's sub-Neptune planet candidates within 0.15 AU. Astrophys. J. **806**, 183 (2015)
391. M.H. Wong, P.R. Mahaffy, S.K. Atreya, H.B. Niemann, T.C. Owen, Updated Galileo probe mass spectrometer measurements of carbon, oxygen, nitrogen, and sulfur on Jupiter. Icarus **171**, 153–170 (2004).
392. P.L. Wood, P.F.L. Maxted, B. Smalley, N. Iro, Transmission spectroscopy of the sodium 'D' doublet in WASP-17b with the VLT. Mon. Not. R. Astron. Soc. **412**, 2376–2382 (2011).
393. R. Wordsworth, Atmospheric heat redistribution and collapse on tidally locked rocky planets. Astrophys. J. **806** (2015).
394. R. Wordsworth, F. Forget, F. Selsis, J.-B. Madeleine, E. Millour, V. Eymet, Is Gliese 581d habitable? Some constraints from radiative-convective climate modeling. Astron. Astrophys. **522**, 22 (2010)
395. Y. Wu, Y. Lithwick, Density and eccentricity of Kepler planets. Astrophys. J. **772**, 74 (2013).
396. Wyttenbach, D. Ehrenreich, C. Lovis, S. Udry, F. Pepe, Spectrally resolved detection of sodium in the atmosphere of HD 189733b with the HARPS spectrograph. Astron. Astrophys. **577**, 62 (2015).
397. Yang, W.R. Peltier, Y. Hu, The initiation of modern soft and hard snowball Earth climates in ccsm4. Clim. Past **8**(3), 907–918 (2012a).
398. Yang, W.R. Peltier, Y. Hu, The initiation of modern soft snowball and hard snowball climates in ccsm3. part i: The influences of solar luminosity, co2 concentration, and the sea ice/snow albedo parameterization. J. Climate **25**(8), 2711–2736 (2012b)
399. Yang, W.R. Peltier, Y. Hu, The initiation of modern soft snowball and hard snowball climates in ccsm3. part ii: Climate dynamic feedbacks. J. Climate **25**(8), 2737–2754 (2012c)
400. Yang, N.B. Cowan, D.S. Abbot, Stabilizing cloud feedback dramatically expands the habitable zone of tidally locked planets. Astrophys. J. Lett. **771**(2), 45 (2013)
401. Yang, Y. Liu, Y. Hu, D.S. Abbot, Water trapping on tidally locked terrestrial planets requires special conditions. Astrophys. J. Lett. **796**(2), 22 (2014)
402. R.V. Yelle, Aeronomy of extra-solar giant planets at small orbital distances. Icarus **170**, 167–179 (2004).
403. Y.L. Yung, M. Allen, J.P. Pinto, Photochemistry of the atmosphere of Titan—Comparison between model and observations. Astrophys. J. Suppl. Ser. **55**, 465–506 (1984).
404. Zahnle, M.S. Marley, Methane, carbon monoxide, and ammonia in brown dwarfs and self-luminous giant planets. Astrophys. J. **797**, 41 (2014).
405. Zahnle, M.S. Marley, J.J. Fortney, Thermometric soots on warm Jupiters? ArXiv e-prints (2009a)
406. Zahnle, M.S. Marley, R.S. Freedman, K. Lodders, J.J. Fortney, Atmospheric sulfur photochemistry on hot Jupiters. Astrophys. J. Lett. **701**, 20–24 (2009b).
407. M.R. Zapatero Osorio, V.J.S. Béjar, E.L. Martín, R. Rebolo, D. Barrado y Navascués, C.A.L. Bailer-Jones, R. Mundt, Discovery of young, isolated planetary mass objects in the σσ orionis star cluster. Science **290**, 103–107 (2000).
408. X. Zhang, R.A. West, D. Banfield, Y.L. Yung, Stratospheric aerosols on Jupiter from Cassini observations. Icarus **226**, 159–171 (2013).
409. G. Zhou, D.D.R. Bayliss, Detection of sodium absorption in WASP-17b with Magellan. Mon. Not. R. Astron. Soc. **426**, 2483–2488 (2012).

CHAPTER 2

Evidence for methane in Martian meteorites

Nigel J.F. Blamey1,2, John Parnell2, Sean McMahon2, Darren F. Mark3, Tim Tomkinson3,4, Martin Lee4, Jared Shivak5, Matthew R.M. Izawa5, Neil R. Banerjee5 & Roberta L. Flemming5

1Department of Earth Sciences, Brock University, 500 Glenridge Avenue, St Catharines, Ontario, Canada L2S 3A1.

2 School of Geosciences, University of Aberdeen, Aberdeen AB24 3UE, UK.

3 Scottish Universities Environmental Research Centre, Glasgow G75 0QF, UK.

4 School of Geographical and Earth Sciences, University of Glasgow, Glasgow G12 8QQ, UK.

5Department of Earth Sciences, University of Western Ontario, Ontario, Canada

ABSTRACT

The putative occurrence of methane in the Martian atmosphere has had a major influence on the exploration of Mars, especially by the implication of active biology. The occurrence has not been borne out by measurements of atmosphere by the MSL rover Curiosity but, as on Earth, methane on Mars is most likely in the subsurface of the crust. Serpentinization of olivine-bearing rocks, to yield hydrogen that may further react with carbon-bearing species, has been widely invoked as a source of methane on Mars, but this possibility has not hitherto been tested. Here we show that some Martian meteorites, representing basic igneous rocks, liberate a methane-rich volatile component on crushing. The occurrence of methane in Martian rock samples adds strong weight to models whereby any life on Mars is/was likely to be resident in a subsurface habitat, where methane could be a source of energy and carbon for microbial activity.

1. INTRODUCTION

On Earth, methane (CH_4) can have a microbial origin (methanogenesis) and/or is a source of carbon and metabolic energy for microbes (methanotrophs); hence, the putative occurrence of methane in the Martian atmosphere[1,2] attracted much attention for its possible biological significance[3,4,5,6]. However, alternative treatments of the data have raised uncertainty about the occurrence of methane[7,8]. Nevertheless, analysis of the local Martian atmosphere by the Mars Science Laboratory (MSL) rover Curiosity has detected transient methane anomalies[9]. Any life on Mars is likely to be in the subsurface[10,11,12], and the potential remains for a subsurface habitat based on methane derived inorganically from low-temperature (<100 °C) reactions of water or carbon-bearing fluids with basalt and other rocks[13]. The serpentinization of olivine and hydration of pyroxene in basalt and other rocks in the presence of water yields hydrogen, which in turn may interact with carbon-bearing species such as carbon monoxide (CO), carbon dioxide (CO_2) and formic acid (HCOOH) to yield methane[3,14,15,16].

Methane in submarine terrestrial basalts can support a microbial population[17]. There are also large volumes of basalt and other basic igneous rocks on Mars[18]. Olivine-rich volcanic rocks have been identified in each of the Noachian, Hesperian and Amazonian successions, and olivine is a significant component of Martian sediment, unlike on Earth[19]. Pyroxene is volumetrically greater than olivine within the Martian crust, and is potentially important as, in addition to a source of hydrogen[20], on Earth altered pyroxenes harbour life[21]. Widespread hydrated silicates on Mars imply extensive water–rock interaction[22], and the opportunities for gas-generating alteration in the Martian crust should, therefore, be widespread. The olivine in Martian meteorites, including all of the nakhlites and many of the shergottites, has experienced aqueous alteration including serpentinization[23]. Serpentinization of olivine at some stage in the past is also recorded on Mars through orbital measurements, using MRO-CRISM[24,25] and could be extensive[10]. Martian meteorites contain magmatic carbon, carbonate carbon from interaction with Martian atmospheric CO_2 and organic carbon from meteoritic infall[26]. We anticipate that all of these carbon-bearing components could become entrained in hydration reactions, which alter olivine and pyroxene, and contribute to methane generation.

There is an opportunity to assess whether Martian rocks could be releasing methane by the analysis of Martian meteorites. The volatile components of the meteorites must be derived from a mixture of sources, including Martian atmosphere, Martian magmatism, Martian crustal processes and terrestrial contamination27. All meteorites from Mars have a bulk basic or ultrabasic igneous composition18; hence, it is possible that some of them have generated methane upon aqueous alteration before ejection from Mars. A range of other fluid-related signatures obtained from Martian meteorites (for example, refs 28, 29, 30, 31) implies that methane generated from alteration on Mars could survive the journey to Earth and be amenable to extraction.

The analysis of entrapped gas has been undertaken on six Martian meteorites (Supplementary Table 1, Supplementary Note 1). These were nakhlites Nakhla, Y 000749, NWA 5790 and MIL 03346, and the shergottites Zagami and LA002. The nakhlites are known to have experienced extensive alteration in a highly oxidizing environment on Mars, and many shergottites show evidence of some interaction with oxidizing fluids in the subsurface28. Nakhla and the shergottites do not exhibit significant terrestrial weathering28,32; however, MIL 03346, NWA 5790 and Y 000749 show evidence of significant aqueous alteration on both Mars and Earth31,33,34,35. With the exception of NWA 5790, the samples were from meteorite interiors, to mitigate against terrestrial contamination. All samples excluded any fusion crust, so they should not have been thermally altered during atmospheric entry (Supplementary Note 2), and fluid inclusions should remain intact36.

The meteorites were analysed using the crush-fast scan technique, in which incremental crushing at room temperature liberates gases trapped in fluid inclusions and crystal boundaries, into a quadrupole mass spectrometer37. Each crush yields a burst of gas that is individually analysed. Gas compositions are calibrated against standard gas mixtures and fluid inclusion standards of known composition.

This technique has the advantage that it does not involve the liberation of gas through heating, in which carbon-bearing components could be converted into a different form. Thus, carbonates, which would yield a strongly CO_2-rich signature due to thermal breakdown upon heating, can yield a CH_4-rich signature if that is the predominant entrapped gas. Analyses of gases in meteorites following the heating of samples provide valuable information, including the isotopic composition of the carbon reservoir38, and liberation of entrained hydrocarbons28, but preclude detailed determination of the indigenous carbon-bearing gas species.

2. RESULTS

2.1 Gases in Martian meteorites

The gases released were dominated by CH_4, CO_2, H_2, N_2, with traces of O_2 and Ar. The highest yields were from Nakhla and LA 002 and lowest from NWA 5790 and MIL 03346 (Fig. 1; Supplementary Table 2). The CH_4/CO_2 ratio varies from about 3 (LA002, NWA 5790, MIL 03346) to 23 (Zagami). The H_2/CO_2 ratio varies from 0.5 to 7. These values are all in the same range as those for terrestrial basalts (Fig. 2). The lower CH_4/CO_2 ratio for Martian samples compared with terrestrial serpentinites39 may reflect an abundance of CO_2 in the Martian crust, evidenced by carbonation observed in meteorites40 and from orbit41. Variations in H_2/CH_4 ratio imply varying degrees of interaction between H_2 and available carbon, but would also reflect differential retention of H_2 and CH_4 in the rock. The artificial generation of CH_4during application of the steel crushing device was monitored using procedural blanks of heat-degassed basalt, as previous studies showed that crushing creates traces of CH_4 (refs 42, 43, 44). The contributions of the blanks to the methane found in the meteorite samples (Fig. 1) were 0.03 to 0.1% (mean 0.07%) for the mean blank, and 0.7 to 2.3% for the maximum blank of mean value plus s.d. The contribution of the blank to the methane measured in the Nakhla sample is 0.05%. Normalization of gases to CO_2 and comparison with the composition of the terrestrial and Martian atmosphere27 allows the plotting of data relative to mixing lines between the two atmospheres (Fig. 3). A plot of argon against oxygen shows that the meteorite data plots much closer to the Martian atmosphere composition with Ar/O_2 ratio ~13. Contamination by the terrestrial atmosphere, which has Ar/O_2 ratio ~0.05, would shift the data along the mixing line away from the Martian end member. The proximity of the meteorite data to the Martian end member indicates that terrestrial contamination is limited, although the composition of terrestrial soil gas lies closer to the meteorite data than the composition of the terrestrial atmosphere. However, a plot of methane against oxygen shows meteorite data well removed from the atmospheric mixing line (Fig. 3). There is substantially more methane than in the Martian atmosphere, strongly suggesting a mechanism for methane generation in the Martian crust. An alternative explanation could be that the methane composition of the atmosphere was much higher when the nakhlites were being altered about 600 Myr ago, but the similar enrichment in terrestrial basalts suggests that a crustal origin is more likely. The scatter in the data reflects variations in mineralogy including proportions of alteration products, and gas retention, and is comparable with the scatter found in terrestrial basalts45, as shown in Fig. 2.

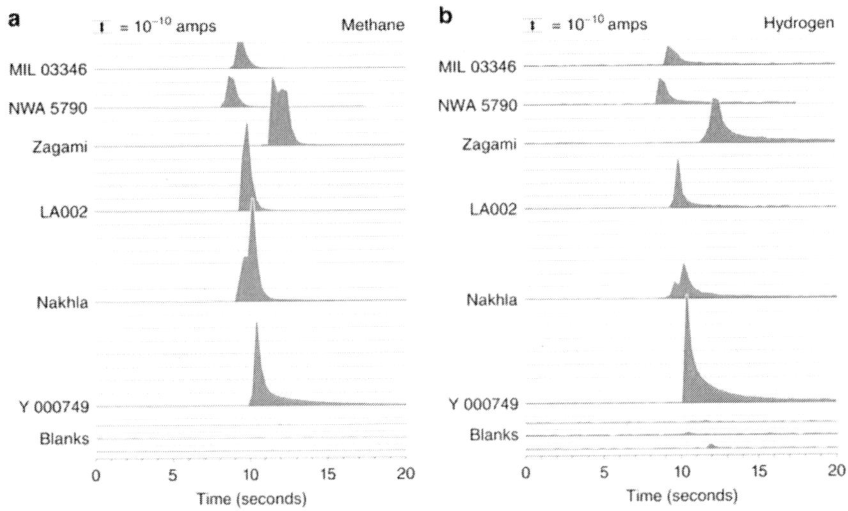

Figure 1. Output from crush fast scan-mass spectrometry analyses of Martian meteorites.

Output shown from during CH_4 (**a**) and H_2 (**b**) release. Output is over 80 cycles (~20–25 s) at *m/z*=15 and *m/z*=2, representing ionized CH_3+ and H_2+ during sample crushing. The ionized volatiles produce an instantaneous current in the detector proportional to their amount. The signals from Martian meteorites are shown above quartz blanks, in units of 10^{-10} amps above background.

Hydrogen generates more current than an equal volume of methane because of a difference in sensitivity factor. Doublet peaks may represent gas release from successively crushed inclusions. Peaks are displaced laterally for clarity only. The meteorite methane peak signals are 2 to 3 orders of magnitude higher than the highest signal from the blank.

The terrestrial samples are from basalts and terrestrial fluids. Each point corresponds to one burst of volatiles produced by an increment of crushing. Gases of interest are normalized to CO_2. All Martian meteorites plot within the data field for terrestrial basalts; both are more methane-rich than terrestrial fluids and carbonaceous chondrites.

Errors are obtained from molar volumes of individual species±one weighted s.d. (see Methods). One Tagish Lake analysis has large error bar, but in most cases error bars are smaller than data symbols. Data sources: terrestrial serpentinites39, and terrestrial basalts45, both crushed in same manner as the meteorites.

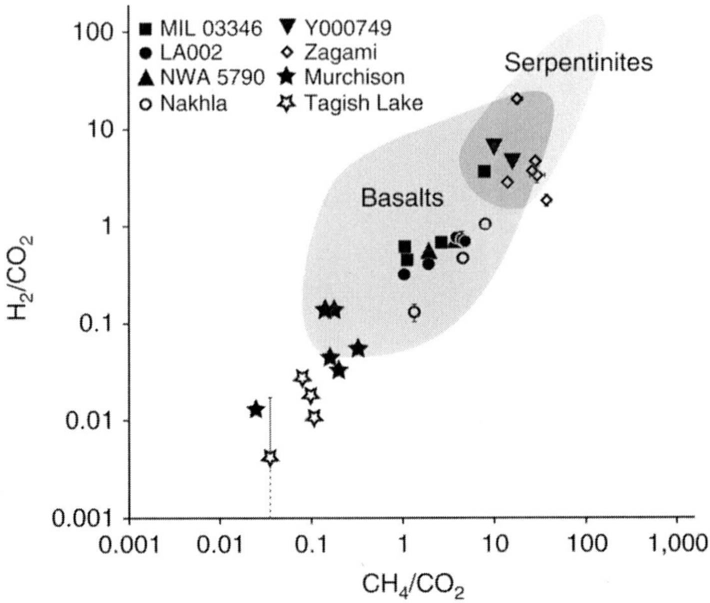

Figure 2. Comparison of gas compositions in Martian meteorites with those in terrestrial samples and carbonaceous chondrites.

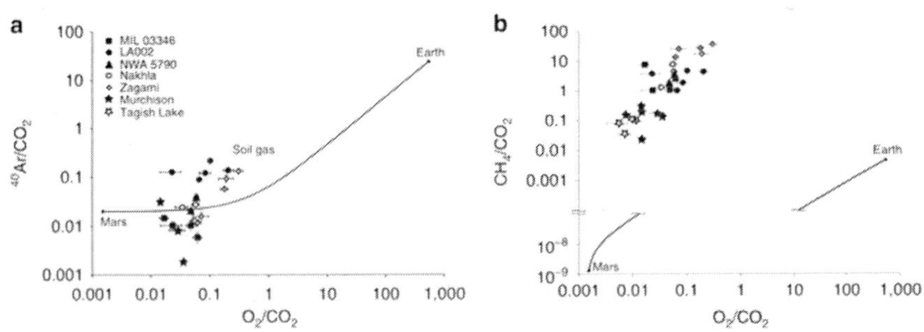

Figure 3. Comparison of Martian meteorite data with mixing lines between Martian and terrestrial atmospheres.

Data are from ref. 27. Each point corresponds to one burst of volatiles produced by an increment of crushing. Gases of interest are normalized to CO_2. Ar/O_2 plot (**a**) implies a primary Martian signature with limited contamination from the terrestrial atmosphere. CH_4/O_2 plot (**b**) implies substantial enrichment in CH_4 by a non-atmospheric source, most likely in the Martian crust. The Ar/O_2 meteorite data lie close to the composition of terrestrial soil gas, but the CH_4 enrichment excludes soil gas as a major contaminant. Together the plots show that contamination from terrestrial atmosphere was minimal, and that the gas signature is crustal rather than atmospheric. Errors are obtained from molar volumes of individual species±one weighted s.d. (see Methods).

3. DISCUSSION

The gases released by crushing are likely to be resident within fluid inclusions, which have been observed in nakhlite meteorites29, and crystal boundaries. The relatively high hydrogen and methane contents are consistent with the aqueous alteration of olivine and pyroxene in the meteorites. Nakhlites in particular are known to have experienced significant aqueous alteration on Mars, including the formation of carbonates, sulfates and hydrous silicates23. Olivines found within the nakhlites and other Martian meteorites are iron-rich compared with typical terrestrial olivines, and are therefore particularly susceptible to serpentinization reactions3,14. Indeed, traces of serpentine have been detected in the Yamato 000593 meteorite23. Therefore, hydrogen and methane should have been generated from these rocks on Mars, just as they are in comparable rocks on Earth. There is a strong implication that the gases measured in this study are predominantly Martian in origin.

Several other lines of evidence suggest that the gases had a Martian origin.

First, the samples that had experienced most alteration, both on Mars and Earth, have the lowest gas yields. This suggests that alteration had facilitated gas release, either during the alteration process or during subsequent history. This does not preclude the orption of gas after arrival on Earth, but the much greater gas yields from the less altered meteorites, including the shergottites, implies that their gas signatures are largely Martian in origin.

Second, the gas assemblages in the meteorites compare closely with those from terrestrial basalts, that is, the meteorites contain those gases expected from basaltic rocks (Fig. 2), suggesting a similar genesis by alteration of the basalt mineralogy. Measurements on terrestrial basalts and serpentinites show that they consistently yield gas with high CH_4/CO_2 ratios, commonly >1 (refs 39, 45). Many other data sets measured by the same technique yield very different assemblages37. These data sets underline the CH_4-rich signature ($CH_4/CO_2>5$) for the meteorite basalts. If the gas assemblage was modified on Mars, for example by uptake of atmospheric CO_2 in secondary carbonates, or during the shock of meteorite ejection26, the CH_4/CO_2 ratio would have been lowered from the original composition, further emphasizing the significance of the CH_4 as a primary signature of basalt alteration.

Third, samples of a group of meteorites with a very different composition, carbonaceous chondrites (Murchison, Tagish Lake), yield a distinct gas assemblage, characterized by 1 to 2 orders of magnitude lower CH_4/CO_2 and H_2/CO_2 (Fig. 2; data from this study, in Supplementary Table 2). Both groups of meteorite have experienced low-temperature alteration of silicates, probably at low water–rock ratios[46]. The carbonaceous chondrites differ in containing organic carbon at the percentage level, orders of magnitude greater than in the Martian meteorites. This suggests that the gas assemblages are a product of the meteorite mineralogy, rather than a characteristic of the terrestrial environment in which the meteorites arrived.

There is evidence for the orption of terrestrial air[47] and for a terrestrial component of carbon[26,28] in Martian meteorites. The incorporation of air in surface environments on Earth can result in measurable levels of oxygen in the gases liberated on crushing, including in samples of oxidized basalt[45]. Traces of oxygen occur in the meteorites, most abundantly in those meteorites known to have experienced terrestrial weathering, MIL 03346 and NWA 5790. This observation indicates that some interaction with terrestrial air may have occurred, but terrestrial air or soil gas would not contribute the methane-rich gas assemblage. It might be argued that some interaction with water could have occurred since arrival on Earth. However, in the gas-rich Nakhla and Zagami samples, it is unrealistic to envisage this gas generation on the timescales since they fell to Earth (1911, 1962). Traces of oxygen also occur in the Martian atmosphere, but the meteorites do not exhibit the high Ar content in the Martian atmosphere. We can estimate the possible degree of contamination, based on the oxygen contents of the meteorite gases. One meteorite (Y000749) has no oxygen, and effectively represents a Martian basalt end member, while the terrestrial atmosphere has ~21%. The percentage oxygen in the other meteorites (ranging up to 0.72%) would be the equivalent of up to 3.4% contamination by the terrestrial atmosphere.

The availability of methane and hydrogen is critical to the potential of the Martian crust as a habitat for microbial life. The hostile Martian surface is probably less habitable than the subsurface, and several scenarios have been proposed for deep Martian life[10,11,12]. The gases would be most concentrated in subsurface environments such as fracture systems and basalt lava vesicles, where they could support a deep biosphere, as envisaged in basalts and other crystalline basement rocks on Earth[14,48]. Methane supports the deep biosphere on Earth, including in basalt where, critically for a Martian analogue, the methane is used by anaerobic microbes and may be partially coupled to the microbial cycling of sulfur[17].

The evidence presented here indicates that a methane-bearing subsurface habitat is similarly available on Mars. Whether or not the habitat has been occupied remains to be determined.

4. METHODS

4.1 Sample processing

Extraction and analysis of volatiles was performed by the crush-fast scan (CFS) method[37,49], which offers very low detection limits[50]. Samples the size of a match head (that is, milligram to gram scale) were first dried below 100 °C. Samples were then loaded into a crushing chamber then held under vacuum (~10^{-8} Torr) for 72 h. The samples were crushed in swift increments, with each crush producing a burst of mixed volatiles which is individually analysed. A typical sample size of about 250 mg (one or two 3-mm grains) released 4–10 bursts of volatiles (up to ~2×10^{-11} mol) into the vacuum chamber, which remained there for 8–10 analyser scans (~2 s) before removal by the vacuum pump. This method does not require a carrier gas and volatiles are not separated from each other but released simultaneously into the chamber. Each individual crush was analysed for H_2, He, CH_4, N_2, O_2, Ar and CO_2, using a Pfeiffer Prisma 'residual-gas' quadrupole mass spectrometer operating in fast-scan, peak-hopping mode at room temperature. In multiple ion detection mode, both isotopic peaks and fragmentation peaks are observed; the major peaks for individual gases are generally selected but some exceptions occur. Methane is detected at mass 15 (that is, the CH_3+ fragment) to avoid interference by O- fragments from water (the methane peak at mass 15 is about 90% as intense as the peak at mass 16). CH_3+ fragments generated by other organic compounds are typically much less abundant and can be corrected for. It is likely that the meteorites host multiple fluid inclusions and, as the sample is incrementally crushed, inclusions can rupture several milliseconds apart to produce a doublet in some mass scans, as seen in Fig. 1. Ratios of gases are reported, as the crushing does not liberate the total amount of entrapped gas. The instrument was calibrated using Scott Gas Mini-mix commercial standard gas mixtures, synthetic inclusions filled with gas mixtures and three in-house fluid inclusion gas standards[49]. The Mini-mix gas mixtures have a possible 2% error in the value of the secondary gas mixed with hydrogen. Data are reported on a water-free basis. Instrumental blanks are also analysed routinely. The amount of each species was calculated by matrix methods[51,52] to provide a quantitative analysis, which is corrected for the instrumental background. Crushing does not liberate all the entrapped gas from samples, so data are generated as molar percentages rather than moles. To circumvent the dependence of individual percentages on irrelevant species, we report methane abundances as the ratio (mol% CH_4)/(mol% CO_2), hydrogen abundances as (mol% H_2)/(mol% CO_2) and oxygen abundances as (mol% O_2)/(mol% CO_2) (hereafter CH_4/CO_2, $H_2/CO_2$and O_2/CO_2 respectively). Analyses were only treated as valid if they yielded an order of magnitude more total gas than (that is, at least 20 ×) the blanks. The mass spectrometer has a range of linearity up to 10e−6 amps. The bursts from the meteorites are too low to reach this part of the current range where non-linearity begins, so linearity is assured. Before analysis, the crushing area and the bellows of the crusher were cleaned using potassium hydroxide. The potassium hydroxide was then removed by swabbing with deionized water, followed by air drying. The apparatus is also routinely cleaned with isopropanol. Thereafter, the crushing chamber is baked at about 150–200 °Celsius for 72 h before loading and analysing the samples at room temperature the next day. The crushing area is isolated from the main chamber so that the main chamber can be baked out every evening. This way the samples are not heated overnight and lose gas, but they are exposed to high vacuum to remove adhered gases. Meteorite samples were transferred directly onto hard metal disks for crushing to avoid handling contamination.

4.2 Calibration using blank analysis

A standard NB84, consisting of calcite containing a 25.0% carbon dioxide-water mix, was analysed 52 times. The weighted mean content of carbon dioxide measured was 25.18 mol%, within <1% of the standard composition.

This standard is used to calibrate water-gas ratios, but is not essential to measurement of the relative proportions of gases reported in this study. Three blank/reference sample measurements are considered in the measurement of gases.

First, a blank is measured from moving parts (bellows in crushing mechanism), without crushing a sample. Second, a procedural blank is measured from crushing a Cenozoic olivine basalt (Beinn Totaig Group, Loch Beag, Island of Skye, Scotland; ref. 53), after heating for 4 h at 750 °C in a vacuum, to cause complete degassing. Third, a blank is measured from crushing synthetic quartz free of visible inclusions (so none larger than 1 μm, if any are present).

The compositions of the quartz blank and basalt blank (mean and standard deviation) are given in Table 1.

The blank from the bellows is negligible compared with the blanks from crushing and is so small that we cannot quantify it. The meteorite yields are much greater than the basalt and quartz blanks, so we have confidence that the yields are predominantly gas from the meteorites rather than an artefact of crushing. The mean basalt blank gives yields 0.03 to 0.10% (mean 0.07%) of the meteorite yields of methane after blank correction (Table 2). The maximum basalt blank (uppermost value of the standard deviation) gives yields 0.68 to 2.30% of the meteorite yields.

The procedural blank was deducted from measurements, excepting for carbon dioxide that was measured from the quartz blank, as the heating-degassing process generates new carbon dioxide.

Table 1. Composition of quartz and basalt blanks.

Gas	Qtz blank mean (mols)	Qtz blank s.d. (mols)	Basalt blank mean (mols)	Basalt blank s.d. (mols)
H_2	5.1E − 15	1.31E − 14	3.01E − 15	4.09E − 15
He	Below detection	0	2.12E − 16	2.50E − 16
CH_4	4.14E − 14	2.08E − 14	6.62E − 16	1.42E − 15
N_2	2.84E − 15	3.2E − 14	1.14E − 12	6.81E − 13
O_2	1.1E − 15	4.7E − 16	0	0
Ar	1E − 16	1.6E − 16	3.01E − 16	5.37E − 16
CO_2	4.5E − 15	2.83E − 15	5.50E − 11	4.27E − 11

Table 2. Blank methane value as percentages of meteorite methane values.

Sample	Blank % of meteorite	Blank + s.d. % of meteorite
NWA5790	0.08	1.76
Zagami	0.04	0.90
LA002	0.10	2.30
Nakhla	0.05	1.04
Y000749	0.03	0.68
MIL03346	0.10	2.26

The blanks quantify the maximum possible contribution of trace gas produced by the crushing process, which we attribute to the metal crushing discs. The blank current is about double that of the background current. The s.d. of the mass spectrometer's signal to noise ratio at the mass 15 background is 1.196e−13.

4.3 Detection limits

The detection limit measurements are as described in ref. 50. The formula is the same as used by ref. 54 and slightly modified by later workers55,56. The s.d. of an individual determination of the background of a sample is given by the formula:

$$\sigma_{\text{individual}} = \sqrt{\left(\sum (X - X_i)^2/(n_b - 1)\right)} \qquad (1)$$

where X is the mean, and X_i is one of the values contributing to the mean.

Previous workers50,54 recognized for a MS transient sampling signal that the measurement error included both elements of the background (n_b is number of cycles obtained from the background) and the analyte (n_a is number of cycles used for the measurement of the analyte signal) where the error associated with the net (analyte minus background) signal (σ_{Rnet}) equals the 1σ from the background times the square root of $1/n_b$plus $1/n_a$ as shown in the formula below. For σ_{Rnet}, the s.d. of the mean net signal (gross—background), both n_b and n_a are 10 thus improving the σ_{Rnet} by 0.45 times the $\sigma_{individual}$.

$$\sigma_{\text{Rnet}} = \sigma_{\text{individual}} \sqrt{\left(n_b^{-1} + n_a^{-1}\right)} \qquad (2)$$

The term σ_{Rnet} is the s.d. of the mean net signal and while it is common practice by geochemists to make 2σ as the confidence limits or approximately a 95% confidence limit, we adopt a 99% or $\sim 3\sigma$ value is used to confirm that the gas burst contains more than zero concentration. As the above s.d. is in terms of signal count rates or other convenient units, it is more useful to convert this value to concentration units by dividing by the sensitivity, S (signal per unit concentration):

$$LOD = 3\sigma_{\text{Rnet}}/S \qquad (3)$$

The sensitivity factors, relative to N_2=1.0, determined at the time of analysis were:

$$CO_2 = 0.544;\ CH_4 = 0.862$$

For the mass spectrometer system in its current configuration the 2σdetection limit ranges from 0.8 to 1.2 femtomol (10^{-15} mol), depending on noise within the mass spectrometer.

4.4 Introduction to error calculations

Total errors are calculated from six components: blank compositions; uncertainty in gas composition; analytical errors in measuring current; interferences in measuring gas species; error in linear range; and background instrument noise. There is an additional, seventh, component, which only applies to the summary data table (Supplementary Table 2), where mols are reported, based on the conversion of the mass spectrometer current into mols.

4.5 Blank compositions

The s.d. (1 sigma) of blank compositions, as determined above, expressed as current (amps), are:

$$\begin{array}{rl} CO_2 & 2.98e-12 \text{ (based on 25 measurements)} \\ N_2 & 3.37e-11 \text{ (based on 25 measurements)} \\ CH_4 & 1.50e-12 \text{ (based on 15 measurements)} \end{array}$$

4.6 Uncertainty in gas composition

During calibration, Scott Gas standard gas mixtures are used for the gases H_2, He, CH_4 and CO_2, for which the manufacturer specifies a 2% uncertainty. The gases N_2, O_2 and Ar are sourced from the atmosphere for which no uncertainty is assigned. Our measurements of the composition of CH_4, CO_2 and N_2 have a s.d. of 0.061, 0.062 and 0.043% respectively, determined over 400 acquisitions.

4.7 Analytical errors

To quantify the error associated with the incremental crush method, we used atmosphere sealed in seven gas capillary tubes and analysed the N_2/Ar ratios.

The analytical mean ratio was 83.4 and s.d. 0.4; atmosphere is 83.6. The error therefore includes estimation of the area under the peak curve as well as signal noise within the mass spectrometer. Note that, although some analysts using LA-ICP-MS apply a Gaussian formula57, gas signals released from fluid inclusions are not Gaussian in shape (also recognized using LA-ICP-MS55,58,59) and our estimation of the area under the curve bears a close resemblance to that of ref. 55. Our 1σ precision is 0.5%.

4.8 Interferences in measuring gas species

Three interferences are potentially relevant. First, analysts may consider analysing for CH_4 at *m/e*=16 (which is the major peak); however, this mass corresponds to a minor interference by a fragmented water ion. We used *m/e*=15 instead of 16 for methane as the *m/e*=15 peak (CH_3+) is 90% of the *m/e*=16 peak and avoids interference corrections from water. The trade-off of 10% increased error at *m/e*=15 versus interference from water makes the *m/e*=15 much more attractive for methane analysis. Second, a very minor interference of 0.032% is detected at the *m/e*=32 (principle mass for oxygen) and is attributable to CO_2. No adjustment is made for this very low value. The principle interference that occurs is that of CO_2 fragmentation to produce CO+ and which corresponds to the N_2 *m/e*=28 channel. Of the total CO_2signal, 8.2% is due to CO+ at *m/e*=28. However, typically the CO_2 is half that of N_2 for the meteorite samples and therefore the contribution of CO+ to the *m/e*=28 signal is only about 4%. The CO_2 content at *m/e*=44 is used to remove 4% contribution to the N_2 value. The relative error added to N_2 includes a square function and therefore CO_2 only contributes about 0.16% to the relative N_2 error.

4.9 Error in linear range

The error across the linear range of the mass spectrometer is estimated from the s.d. for capillary tube measurements of the N_2/Ar ratio. This covers the noise and the area under the peak curve. The measurements indicate a maximum error of 1%.

4.10 Background instrument noise

The background instrument noise is measured when no sample is being analysed. The signal is measured with the sample exposed to the main chamber and represents all residual gases that comprise the background with the turbo pump operating at maximum speed of 90,000 r.p.m. The mass spectrometer is instructed to begin recording and cycle through the selected masses relating to the gases being analysed. Approximately 30 s later the incremental crush occurs. The background is measured for 10 cycles stopping at 1 s before the gas burst. Both the background signal mean and s.d. are recorded.

4.11 Conversion of current to moles

Conversion from mass spectrometer current (amps) to gas concentration (mols) is achieved by analysing the contents of a Millipore one microliter volume capillary tube containing atmosphere at standard temperature and pressure, sealed using a high-vacuum epoxy glue. The tube is crushed in the same manner as any other sample. The number of moles of nitrogen (obtained using PV=nRT) is equated to the measured current. The equivalence determined at the time of measurement of the meteorites was 1,054 amps per mol, calculated by conventional matrix methods[51,52]. The uncertainty in measurement is no more than 5%, based on the s.d. of measurements of the ratio of current to moles calculated from the area under the peak curve. In addition, the manufacturer reports a maximum 1% uncertainty in the volume of the capillary tube. Adding these two errors give 5.04% (square root of added squares).

4.12 Propagation of errors

The uncertainty matrix includes multiple components as described above. There will be different contributions of uncertainties for each gas. The formula to incorporate the additive uncertainties is:

$$u_{Add} = \sqrt{u_a^2 + u_b^2 + u_c^2 + \cdots} \tag{4}$$

The components used in the additive error calculation are given in Table 3.

Table 3. Components used in additive error calculation.

	CO_2	N_2	CH_4
Blank (current (amps), 1 sigma s.d.)	2.98e − 12	3.37e − 11	1.50e − 12
Uncertainty in gas composition from standard	2%	0	2%
Interference	0	4% of CO_2	0
Analytical error	0.5%	0.5%	0.5%
Error in linear range	1.0%	1.0%	1.0%
Background instrument noise (1-sigma)	Specific to analysis	Specific to analysis	Specific to analysis
Conversion current to mols	5.04%	5.04%	5.04%

An example of the errors is given in Table 4. This is for a crush of the Zagami meteorite and is the smallest burst of gas measured in the programme.

Table 4. Errors calculated for smallest crush of the Zagami meteorite.

Source of error/ uncertainty	CO_2	N_2	CH_4
Background instrument noise (1-sigma)	5.20e − 13	1.47e − 12	1.05e − 13
Blank (1-sigma)	2.98e − 12	3.37e − 11	1.50e − 12
Uncertainty in gas composition from standard	4.73e − 13	0	1.19e − 11
Interference	0	2.92e − 13	0
Analytical error (0.5%)	1.18e − 13	4.75e − 13	2.98e − 12
Error in linear range (1%)	2.36e − 13	9.50e − 13	5.96e − 12
Additive error before converting to mols	3.07e − 12	3.37e − 11	1.37e − 11
Conversion current to mols for this crush	3.13e − 15 mols	3.24e − 14 mols	3.14e − 14 mols

ACKNOWLEDGEMENTS

Meteorite samples were kindly provided by the University of New Mexico (Zagami), Caroline Smith, Natural History Museum, London (Nakhla), Robert Verish, Meteorite Recovery Laboratory, California (LA 002), Darryl Pitt, Macovich Collection, New York NWA 5790), Hideyasu Kojima, National Institute of Polar Research, Japan (Yamoto 000749) and the Meteorite Working Group, NASA Johnson Space Center (Miller 03346). D.M. is supported by the STFC (ST/H002472/1 & ST/K000918/1). S.McM. was supported by a STFC Aurora studentship.

REFERENCES

1. Formisano, V., Atreya, S., Encrenaz, T., Ignatiev, N. & Giuranna, M. Detection of methane in the atmosphere of Mars. Science 306, 1758–1761 (2004).
2. Mumma, M. J. et al. Strong release of methane on Mars in northern summer 2003. Science 323, 1041–1045 (2009).
3. Schulte, M., Blake, D., Hoehler, T. & McCollum, T. Serpentinization and its implications for life on the early Earth and Mars. Astrobiology 6, 364–376 (2006).
4. Atreya, S. K., Mahaffy, P. R. & Wong, A. S. Methane and related trace species on Mars: Origin, loss, implications for life, and habitability. Planet. Space Sci. 55, 358–369 (2007).

5. Blank, J. G. et al. An alkaline spring system within the Del Puerto Ophiolite (California, USA): A Mars analog site. Planet. Space Sci. 57, 533–540 (2009).
6. ten Kate, I. L. Organics on Mars? Astrobiology 10, 589–603 (2010).
7. Lefevre, F. & Forget, F. Observed variations of methane on Mars unexplained by known atmospheric chemistry and physics. Nature 460, 720–723 (2009).
8. Zahnle, K., Freedman, R. S. & Catling, D. C. Is there methane on Mars? Icarus 212, 493–503 (2011).
9. Webster, C. R. et al. Mars methane detection and variability at Gale crater. Science 347, 415–417 (2015).
10. Fisk, M. R. & Giovannoni, S. J. Sources of nutrients and energy for a deep biosphere on Mars. J. Geophys. Res. 104, E11805–E11815 (1999).
11. Weiss, B. P., Yung, Y. L. & Nealson, K. H. Atmospheric energy for subsurface life on Mars? Proc. Natl Acad. Sci. USA 97, 1395–1399 (2000).
12. Michalski, J. R. et al. Groundwater activity on Mars and implications for a deep biosphere. Nat. Geosci. 6, 133–138 (2013).
13. Etiope, G., Ehlmann, B. L. & Schoell, M. Low temperature production and exhalation of methane from serpentinized rocks on Earth: a potential analog for methane production on Mars. Icarus 224, 276–285 (2013).
14. Sleep, N. H., Meibom, A., Fridriksson, T., Coleman, R. G. & Bird, D. K. H2-rich fluids from serpentinization: geochemical and biotic implications. Proc. Natl Acad. Sci. USA 101, 12818–12823 (2004).
15. Chassefière, E. & Leblanc, F. Methane release and the carbon cycle on Mars. Planet. Space Sci. 59, 207–217 (2011).
16. McCollom, T. M., Sherwood Lollar, B., Lacrampe-Couloume, G. & Seewald, J. S. The influence of carbon source on abiotic organic synthesis and carbon isotope fractionation under hydrothermal conditions. Geochim. Cosmochim. Acta 74, 2717–2740 (2010).
17. Lever, M. A. et al. Evidence for microbial carbon and sulfur cycling in deeply buried ridge flank basalt. Science 339, 1305–1308 (2013).
18. McSween, H. Y. Jr., Taylor, G. J. & Wyatt, M. B. Elemental composition of the martian crust. Science 324, 736–739 (2009).
19. Hamilton, V. E. & Christensen, P. R. Evidence for extensive, olivine-rich bedrock on Mars. Geology 33, 433–436 (2005).
20. Mayhew, L. E., Ellison, E. T., McCollom, T. M., Trainor, T. P. & Templeton, A. S. Hydrogen generation from low-temperature water-rock reactions. Nat. Geosci. 6, 478–484 (2013).
21. Ménez, B., Pasini, V. & Brunelli, D. Life in the hydrated suboceanic mantle. Nat. Geosci. 5, 133–137 (2012).
22. Mustard, J. F. et al. Hydrated silicate minerals on Mars observed by the Mars Reconnaissance Orbiter CRISM instrument. Nature 454, 305–309 (2008).
23. Noguchi, T. et al. Laihunite and jarosite in the Yamato 00 nakhlites: alteration products on Mars? J. Geophys. Res. 114, E10004 (2009).
24. Ehlmann, B. L., Mustard, J. F. & Murchie, S. L. Geologic setting of serpentine deposits on Mars. Geophys. Res. Lett. 37, L06201 (2010).
25. Wray, J. J. & Ehlmann, B. L. Geology of possible Martian methane source regions. Planet Space Sci. 59, 196–202 (2011).
26. Grady, M. M., Verchovsky, A. B. & Wright, I. P. Magmatic carbon in Martian meteorites: attempts to constrain the carbon cycle on Mars. Int. J. Astrobiol. 3, 117–124 (2004).
27. Mahaffy, P. R. et al. Abundance and isotopic composition of gases in the Martian atmosphere from the Curiosity rover. Science 341, 263–266 (2013).
28. Gooding, J. L., Aggrey, K. E. & Muenow, D. W. Volatile compounds in shergottite and nakhlite meteorites. Meteoritics 25, 281–289 (1990).
29. Bridges, J. C., Smith, M. P. & Grady, M. M. Progressive evaporation and relict fluid inclusions in the nakhlites. Lunar Planet. Sci. Conf. XXXI, 1590 (2000).
30. Mathew, K. J. & Marti, K. Martian atmospheric and interior volatiles in the meteorite Nakhla. Earth Planet. Sci. Lett. 199, 7–20 (2002).

31. Cartwright, J. A., Gilmour, J. D. & Burgess, R. Martian fluid and Martian weathering signatures identified in Nakhla, NWA 998 and MIL 03346 by halogen and noble gas analysis. Geochim. Cosmochim. Acta 105, 255–293 (2013).
32. Jambon, A. et al. The basaltic shergottite Northwest Africa 856: Petrology and chemistry. Meteor. Planet. Sci. 37, 1147–1164 (2002).
33. Shih, C.-Y., Nyquist, L. E. & Jambon, A. Sm-Nd isotopic studies of two nakhlites, NWA 5790 and Nakhla. Lunar Planet. Sci. Conf. XLI, 1367 (2010).
34. Hallis, L. J. & Taylor, G. J. Comparisons of the four Miller Range nakhlites, MIL 03346, 090030, 090032 and 090136: Textural and compositional observations of primary and secondary mineral assemblages. Meteor. Planet. Sci. 46, 1787–1803 (2011).
35. Tomkinson, T., Lee, M. R., Mark, D. F., Dobson, K. J. & Franchi, I. A. The Northwest Africa (NWA) 5790 meteorite: a mesostasis-rich nakhlite with little or no Martian aqueous alteration. Meteor. Planet. Sci. 50, 287–304 (2015).
36. Parnell, J., Mark, D. & Brandstatter, F. Response of sandstone to atmospheric heating during the STONE-5 experiment: Implications for the palaeofluid record in meteorites. Icarus 197, 282–290 (2008).
37. Blamey, N. J. F. Composition and evolution of crustal fluids interpreted using quantitative fluid inclusion gas analysis. J. Geochem. Explor. 116-117, 17–27 (2012).
38. Wright, I. P., Grady, M. M. & Pillinger, C. T. Chassigny and the nakhlites: Carbon-bearing components and their relationship to Martian environmental conditions. Geochim. Cosmochim. Acta 56, 817–826 (1992).
39. Parnell, J., Boyce, A. J. & Blamey, N. J. F. Follow the methane: the search for a deep biosphere, and the case for sampling serpentinites, on Mars. Int. J. Astrobiol. 9, 193–200 (2010).
40. Tomkinson, T., Lee, M. R., Mark, D. F. & Smith, C. L. Sequestration of Martian CO2 by mineral carbonation. Nat. Commun. 4, 2662 (2013).
41. Michalski, J. R. & Niles, P. B. Deep crustal carbonate rocks exposed by meteor impact on Mars. Nat. Geosci. 3, 751–755 (2010).
42. Andrawes, F. F. & Gibson, E. K. Release and analysis of gases from geological samples. Am. Mineral 64, 453–463 (1979).
43. Stauffer, B., Fischer, G., Neftel, A. & Oeschger, H. Increase of atmospheric methane recorded in Antarctic ice core. Science 229, 1386–1388 (1985).
44. Higaki, S., Oya, Y. & Makide, Y. Emission of methane from stainless steel investigated by using tritium as a radioactive tracer. Chem. Lett. 35, 292–293 (2006).
45. McMahon, S., Parnell, J. & Blamey, N. J. F. Sampling methane in basalt on Earth and Mars. Int. J. Astrobiol. 12, 113–122 (2013).
46. Zolensky, M. E., Krot, A. N. & Benedix, G. Record of low-temperature alteration in asteroids. Revs. Mineral 68, 429–462 (2008).
47. Busemann, H. & Eugster, O. The trapped heavy noble gases in recently found Martian meteorites. Lunar Planet. Sci. Conf. XXXIII, 1823 (2002).
48. Schrenk, M. O., Brazelton, W. J. & Lang, S. Q. Serpentinization, carbon, and deep life. Revs. Mineral 75, 575–606 (2013).
49. Norman, D. I. & Blamey, N. J. F. Quantitative analysis of fluid inclusion volatiles by a two quadrupole mass spectrometer system. ECROFI XVI, 341–344 (2001).
50. Blamey, N. J. F., Parnell, J. & Longerich, H. P. Understanding detection limits in fluid inclusion analysis using an incremental crush fast scan method for planetary science. Lunar Planet. Sci. Conf. XLIII, 1035 (2012).
51. Isenhour, T. L. & Jurs, P. C. Introduction to computer programming for chemists 325Allyn & Baco (1972).
52. Norman, D. I. & Sawkins, F. J. Analysis of volatiles in fluid inclusions by mass spectrometry. Chem. Geol. 61, 1–10 (1987).
53. Thompson, R. N., Gibson, I. L., Marriner, G. F., Mattey, D. P. & Morrison, M. A. Trace-element evidence of multistage mantle fusion and polybaric fractional crystallization in the Palaeocene lavas of Skye, NW Scotland. J. Petrol. 21, 265–293 (1980).

54. Longerich, H. P., Jackson, S. E. & Günther, D. Laser ablation inductively coupled plasma mass spectrometric transient signal data acquisition and analyte concentration calculation. J. Anal. Atom. Spectrom. 11, 899–904 (1996).
55. Tanner, M. & Günther, D. In torch laser ablation sampling for inductively coupled plasma mass spectrometry. J . Anal. Atom. Spectrom. 20, 987–989 (2005).
56. Borovinskaya, O., Gschwind, S., Hattendorf, B., Tanner, M. & Günther, D. Simultaneous mass quantification of nanoparticles of different composition in a mixture by microdroplet generator-ICPTOFMS. Anal. Chem. 86, 8142–8148 (2014).
57. Gschwind, S., Hattendorf, B., Frick, D. A. & Günther, D. Mass quantification of nanoparticles by single droplet calibration using inductively coupled plasma mass spectrometry. Anal. Chem. 85, 5875–5883 (2013).
58. Cottle, J. M., Horstwood, M. S. A. & Parrish, R. R. A new approach to single shot laser ablation analysis and its application to in situ Pb/U geochronology. J. Anal. Atom. Spectrom. 24, 1355–1363 (2009).
59. Pettke, T. et al. Recent developments in element concentration and isotope ratio analysis of individual fluid inclusions by laser ablation single and multiple collector ICP-MS. Ore Geol. Revs. 44, 10–38 (2012).

CHAPTER 3

The Gas Disk: Evolution and Chemistry

Christian Rab1, Carla Baldovin-Saavedra1, Odysseas Dionatos1, Eduard Vorobyov1,2, Manuel Güdel1

1 Department of Astrophysics, University of Vienna, Türkenschanzstr. 17, 1180 Vienna, Austria

2 Research Institute of Physics, Southern Federal University, Stachki 194, Rostov-on-Don, 344090, Russia

ABSTRACT

Protoplanetary disks are the birthplaces of planetary systems. The evolution of the star-disk system and the disk chemical composition determines the initial conditions for planet formation. Therefore a comprehensive understanding of the main physical and chemical processes in disks is crucial for our understanding of planet formation. We give an overview of the early evolution of disks, discuss the importance of the stellar high-energy radiation for disk evolution and describe the general thermal and chemical structure of disks. Finally we provide an overview of observational tracers of the gas component and disk winds.

Keywords: Stars: pre-main sequence, Stars: formation, Protoplanetary disks, Accretion, accretion disks, Planet-disk, interactions, ISM: jets and outflow Astrochemistry

1 INTRODUCTION

Disks around young stellar objects are the birthplaces of planetary systems. A comprehensive knowledge of their evolution, structure and chemical composition is therefore crucial for our understanding of planet formation.

Low mass stars like our Sun are formed with a disk component. Observations show that the fraction of disk-bearing stars in clusters with an age of $\approx$1 Myr$\approx$1 Myr is usually $\gtrsim$80 %·However, the disk fraction drops quite rapidly with cluster age. In clusters with ages of 2--3 Myr2--3 Myr still %$\approx$50 %of the stars show disks, whereas in older ($\approx$10 Myr$\approx$10 Myr) clusters the fraction of disk-bearing stars is lower than 5 % (e.g. Hernández et al. 2008; Fedele et al. 2010). This tells us that disks have a typical lifetime of a few million years. The lifetime of disks defines the timescale on which formation of massive (gaseous) planets can occur.

Disks evolve on timescales of a few million years. The evolutionary stages are determined by the balance of accretion onto the disk and different disk dispersal/removal mechanisms (see Williams and Cieza 2011 for a review). There is observational evidence that disks form already within 104--105 yr after the gravitational collapse of the parental cloud-core (e.g. Murillo et al. 2013) at the phase where the star is still deeply embedded (Class-0 protostellar phase; see review of Li et al. 2014). After the formation disks are growing due to the continuous inflow of mass from their surrounding envelopes during the Class-0 and Class-I (the star becomes visible) protostellar phases. Once this reservoir is drained and the star reaches its Class-II phase (T Tauri star), disk erosion processes due to accretion onto the star, disk photoevaporation, jet acceleration, and planet formation lead to the disappearance of dust and gas disks within typically 2–3 Myr (although in some rare cases disks are still present at ages of ~10 Myr). Observations of disks with large inner holes/gaps, the so called transitional disks, indicate that disk dispersal works from inside out. From the fraction of observed transition disks a timescale of yr$\gtrsim$105 yr is derived for this period (see Espaillat et al. 2014 for a review). The final result of the gas disk dispersal are debris disks. These disks are mainly made up of asteroid/comet/planet like bodies and dust but contain no or only very little amount of gas. This evolutionary phase can last for Gyr. However, it is still unclear if every gaseous disks ends up as a debris disk (see Matthews et al. 2014 for a review).

During these evolutionary stages also the dust component of the disk evolves. Recent discoveries like the horseshoe-shaped structures of larger dust particles in transition disks (e.g. van der Marel et al. 2013; Pérez et al. 2014a) or the spectacular dark rings in the thermal dust emission of

HL Tau (ALMA Partnership et al. 2015; Pinte et al. 2016) and TW Hya (Andrews et al. 2016) clearly show this. The dust component also influences the gas disk. Dust is a strong opacity source in the optical and ultraviolet and therefore absorbs most of the stellar radiation. This significantly affects the disk thermal structure and chemistry (see Sects. 5 and 6). However, our focus here is on the gas disk and we refer the reader to Chap. 3 of this book and to the review by Testi et al. (2014) for more details on dust evolution.

We now briefly summarize the essential steps of disk evolution from formation to dispersal due to various erosion processes. Three major processes: accretion, winds/outflows/jets and planet formation contribute to the slow removal of disk material, determining the typical lifetime of disks.

The first and best studied process is the *continuous accretion* of gas and dust through the disk onto the star. There are several possible mechanisms which can drive matter flows in the disk, of which the most prominent ones are: the magneto-rotational instability (MRI), magnetically launched jet winds, magneto-centrifugally driven winds, gravitational instability and hydrodynamical instabilities (see Turner et al. 2014 for a review). The first mechanism has been the most promising one to explain increased disk viscosity required for the observed accretion rate. However, MRI requires a disk ionization fraction of ≈10−12≈10−12 which is probably not reached everywhere in the disk (e.g. Cleeves et al. 2015). The second mechanism le to fast jets and associated molecular outflows, but here the acceleration mechanism of disk material along magnetic fields is poorly understood, and it is also still a matter of debate whether the launching mechanism of molecular outflows is directly associated with jet acceleration (see review of Frank et al. 2014). The other processes (turbulence, winds and instabilities) are sensitive to heating and cooling processes in the disk. Therefore high quality observational data in combination with detailed thermo-chemical models are needed to derive quantities such as the ionization fraction and the gas temperature (see Sects. 4, 5 and 6). In summary, the efficiency of all above processes remains unclear. We also note that disk accretion might affect the chemical composition due to additional heating and transport of material from colder to warmer regions of the disk (see Sect. 6), and jet formation may lead to additional irradiation of disks from above (see Sect. 3.5). Generally, self-consistent thermo-chemical and (magneto-)hydrodynamic simulations including feedback are needed for a full understanding of these processes.

Especially in their earlier phases, most disks probably undergo more *violent accretion phases.*The so-called FU Orionis objects show strong luminosity outbursts (increase in luminosity by factors of 10 to 100s).

These outbursts typically last for decades or centuries. During this period the accretion rate rapidly increases from $\approx 10^{-7}$ to $\approx 10^{-4}$ $M_{\odot}$ yr^{-1} (see Audard et al. 2014 for a review). The cause of this phenomenon is still unclear. Proposed explanations are disk instabilities (thermal, magneto-rotational and/or gravitational instabilities) and the perturbation of the disk by an external body. Another possibility is disk fragmentation and the infall of the formed fragments onto the star. These outbursts also have an impact on the chemical evolution of the disk (see Sects. 2 and 6). Although rare, repeated FU Ori-type accretion events may be crucial in star and disk evolution as they may bring a significant amount of matter onto the star, and may at the same time dominate disk mass-loss due to accretion on long timescales.

The second major mass-loss processes are via the formation of *disk winds*, either magnetically driven or irradiation driven. We distinguish between fast, probably magnetically driven jets with (at least in the Class-I phase) associated molecular outflows, and radiation-driven slow photoevaporative winds (see Sect. 8, Chap. 6 of this book and the review of Alexander et al. 2014). In the T Tauri phase, mass-loss rates in outflows/jets and photoevaporative winds amount to no more than about 10 % of the stellar accretion rate (Hartigan et al. 1995; Cabrit 2002; Frank et al. 2014). However, in late evolutionary phases the photoevaporation rate may locally (at distances of several au from the star) exceed the disk accretion rate, which le to the rapid formation of a gap and the subsequent erosion of the entire disk on time scales of only a few hundred kyr or ≈10 %≈10 % of the disk lifetime. This rapid-erosion model is compatible with the relatively rare occurrence of transitional disks with such gaps and eventually inner holes, compared to classical, optically thick disks. An example of gas photoevaporation are winds driven by X-ray or extreme-ultraviolet irradiation of the disk (see Sects. 5 and 8), leading to strong surface heating and ionization and consequently escape of ions. Photoevaporation might have an impact on the final appearance of a planetary systems. In fact the models of Ercolano and Rosotti (2015) suggest that the distribution of the semi-major axes of exoplanet orbits is sensitive to photo-evaporation as it affects planet-migration.

The third process eroding a disk is *planet formation* itself. The formation of gas giants requires an intact gas reservoir still to be present, which suggests that giants form to their full mass during the disk phase. The combined mass of the giant planets in the solar system amounts to about 0.13 % of the solar mass, which provides a strict lower limit to the disk mass loss in the T Tauri phase of our Sun through planetary accretion. Giant planets may open gaps and form spiral-arm patterns, leading to a complex exchange of angular momentum between planet and disk, leading to planet migration. Meanwhile, solids also accumulate and grow up to Mars-sized planetary embryos, adding to the disk mass loss (for more details see the reviews by Baruteau et al. 2014 and Raymond et al. 2014). One spectacular observed example for this process might be the LkCa 15 transitional disks. Recent adaptive optics observations by Sallum et al. (2015) reveal up to three accreting giant proto-planets in the large inner cavity (rcav≈50 aurcav≈50 au, Andrews et al. 2011) of this disk. At the same time the outer disk of this system still remains intact and shows similar observational signatures as classical disks (Öberg et al. 2011; Punzi et al. 2015).

As mentioned above, the relative importance of the disk erosion processes are unknown although all of them are observed, and all of them are expected to remove disk material in the range of maybe one tenth and about one percent of the solar mass during a typical disk lifetime (accretion and ejection rates of order $10^{-8}\ M_\odot\ yr^{-1}$ for a duration of $\sim 10^6$ yr).

Another important constraint for disk evolution and planet formation is the actual mass of the disk. Disk masses derived from millimeter continuum observations are in the range of $M_{disk} \approx 0.1\ M_\odot$ to $\lesssim 10^{-3}\ M_\odot$ (Andrews and Williams 2005). However, disk masses derived from sub-millimeter fluxes can be underestimated by a factor of 2, and in some cases up to a factor of 10 (Dunham et al. 2014).

Williams and Best (2014) proposed a method for a more direct measurement of the disk gas mass using CO isotopologue line ratios. Their results indicate rather low disk gas masses in the range of 10^{-2} to $10^{-3}\ M_\odot$ and consequently, lower gas to dust mass ratios of the order of 10, in contrast to the canonical value of 100. However, direct disk gas mass measurements are difficult and are affected by complex chemical processes, which are not fully considered in Williams and Best (2014) (this is also true for CO, see Sect. 6 for more details). In contrast to the results of Williams and Best (2014), Bergin et al. (2013) reported a gas mass of $M_{disk} \gtrsim 5 \times 10^{-2}\ M_\odot$ for the disk around TW Hya, using the HD molecule as the mass tracer. This result is especially interesting as TW Hya is supposed to be a rather old disk (3--10 Myr3--10 Myr, Bergin et al. 2013). The derived disk mass would still allow to form a planetary system similar to our solar system.

To improve our knowledge of the gas disk structure and evolution, further observations are required. However, deriving disk properties directly from observations is not straightforward. Not all regions of the disk (e.g. the midplane) are directly observable and the chemical processing of disk material has to be considered for the interpretation of observations (see Sect. 6). However, the first results of the Atacama Large Millimeter Array (ALMA) have already significantly improved our knowledge of the gaseous disk (see Sect. 7), and most certainly will provide us with more interesting and surprising results in the near future.

From the above introduction of disk evolution and important physical processes it becomes clear that a review like this cannot cover this topic exhaustively. We rather focus on several important aspects and refer the interested reader to dedicated reviews for topics we do not cover in detail.

In Sect. 2 we start with a discussion about the early evolution of gravitationally unstable disks, their potential to form gas giants and their role for the episodic accretion scenario. Section 3 is about the radiative environment of the disk with a focus on stellar high-energy processes as they are in particular interesting for disk-ionization, heating and disk dispersal. In Sect. 4 we discuss in more detail disk ionization and the magneto-rotational instability as a potential driver of accretion. In Sect. 5 we briefly discuss the thermal structure of the disk, heating/cooling processes and disk photo-evaporation as a consequence of gas heating by high-energy radiation. The general chemical structure of classical T Tauri and Herbig Ae/Be disks is described in Sect. 6. In this Section we also discuss in more detail X-ray chemistry, ice-lines and present a simple model for the chemical evolution during episodic accretion events. In Sect. 7we give an overview of gas disk observations and in Sect. 8 we discuss observational constraints of photoevaporative winds. We conclude this chapter with a summary (Sect. 9).

2 THE EARLY EVOLUTION OF CIRCUMSTELLAR DISKS

Stars form from the gravitational collapse of dense molecular clouds, a larger fraction of which passes through a circumstellar disk owing to conservation of the net angular momentum of the collapsing cloud. Observational evidence and numerical simulations suggest that circumstellar disks can form as early as a few thousand years after the formation of the protostar (Vorobyov 2011b; Tobin et al. 2013) and sometimes even earlier than the protostar itself (Machida et al. 2010). In this early evolutionary phase, most of the mass reservoir is residing in the collapsing cloud, which lands onto the disk outer regions and drives the disk to the boundary of gravitational instability. Theoretical and numerical studies indicate that circumstellar disks can be prone to gravitational fragmentation in the outer regions if they form from cores of sufficient initial mass and angular momentum (Kratter et al. 2008; Vorobyov 2010; Stamatellos et al. 2011; Zhu et al. 2012). The infall of material from the collapsing core in the early stages of star formation triggers fragmentation in disks that otherwise might have been stable against fragmentation. The location of the fragmentation boundary (usually, at several tens of au from the central star) depends somewhat on the disk physics and is currently a subject of ongoing debate (Rafikov 2005; Boley 2009; Meru and Bate 2012).

In the subsequent protostellar phase of disk evolution, lasting until the parental cloud is accreted onto the star plus disk system or dissipated via star formation feedback, the complex interplay between mass infall onto the disk and disk transport processes, such as gravitational and viscous torques, le to the formation of gravitationally unstable protostellar disks with radial and azimuthal profiles of density and temperature that are very different from those of the Minimum Mass Solar Nebula (Vorobyov 2011b; Bitsch et al. 2015). Non-axisymmetric structures that emerge in the disk, such as spiral arms, clumps, and vortices, may serve as likely spots for dust accumulation and planetesimal formation, facilitating the formation of solid protoplanetary cores (Rice et al. 2004; Nayakshin 2011; Regály et al. 2012; Gibbons et al. 2015). Gravitational instability and fragmentation in the early evolution of protostellar disks brings about a wealth of phenomena, the most important of which are briefly reviewed below.

2.1 The Burst Mode of Protostellar Accretion

Numerical studies show that fragments forming in the outer regions of protostellar disks can be quickly driven into the inner regions and onto the star thanks to exchange of angular momentum with spiral arms and other fragments. This fast inward migration le to luminosity outbursts as fragments fall onto the star and release their gravitational energy (Vorobyov and Basu 2006, 2010b; Machida et al. 2011). The magnitude, duration and frequency of the luminosity outbursts is similar to those typically observed in FU-Ori-type eruptions (Vorobyov and Basu 2015). Figure 1 illustrates the burst phenomenon showing the gas surface density distribution in a gravitationally unstable disk.[1] In particular, each row of images presents a time sequence capturing individual luminosity bursts caused by fragments falling onto the star. The left, middle and right panels correspond to the time instances immediately preceding the burst, during the burst, and soon after the burst, respectively. The corresponding times are identified in each panel. The inserts in the left column zoom onto the fragments that are about to fall onto the star and the inserts in the right column present the time evolution of the total stellar luminosity during the shown time sequence. Strong luminosity bursts ranging from several tens to several hundreds of the solar luminosity are evident. Luminosity bursts in the model of disk gravitational instability and fragmentation are a robust phenomenon estimated to take place in at least 40 % of protostellar disks (Vorobyov and Basu 2015).

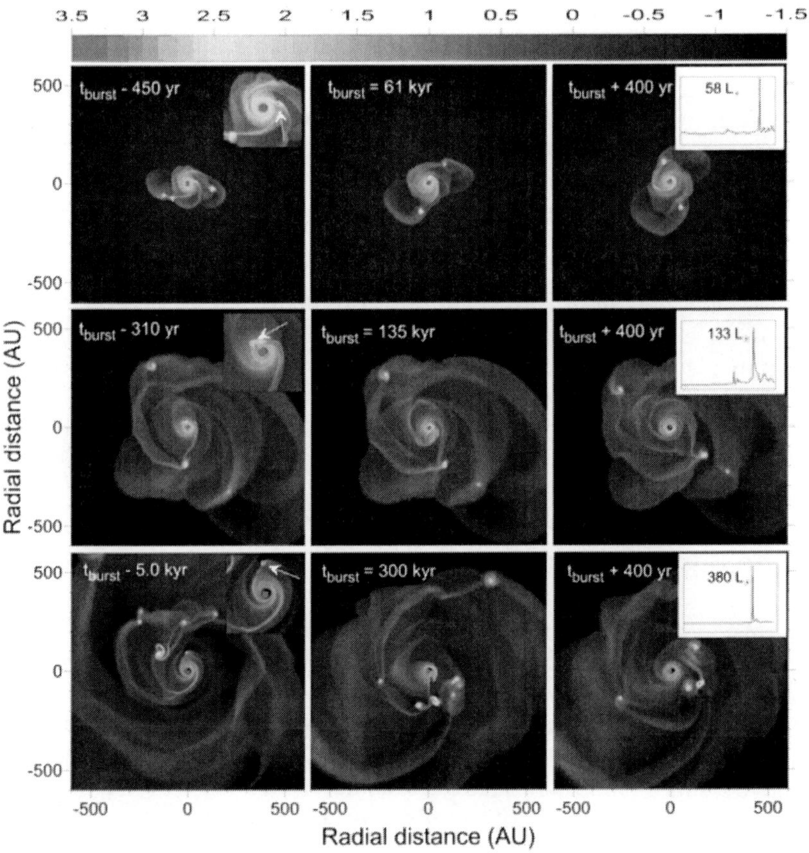

Figure 1. Sequence of disk images capturing the inward migration of a clump onto the central star. The *scale bar* shows the gas surface density in gcm−2gcm−2. See the text for more detail

2.2 Formation of Massive Gas Giants and Brown Dwarfs

Another interesting phenomenon associated with disk gravitational fragmentation is the possibility for the formation of massive gas giant planet and brown dwarf companions to solar-type stars (Boss 1997, 2011; Stamatellos and Whitworth 2009; Boley et al. 2010; Nayakshin 2010;

Kratter and Murray-Clay 2011; Rogers and Wley 2012). Numerical simulations capturing the long-term evolution of circumstellar disks (Vorobyov and Basu 2010a; Vorobyov 2013) show that under favorable conditions, i.e. in sufficiently massive disks that can experience fragmentation not only in the embedded but also T Tauri stages of star formation, some of the fragments may escape fast inward migration and mature into planetary or sub-stellar objects on wide orbits, similar to such systems as Fomalhaut or CT Chameleontis.

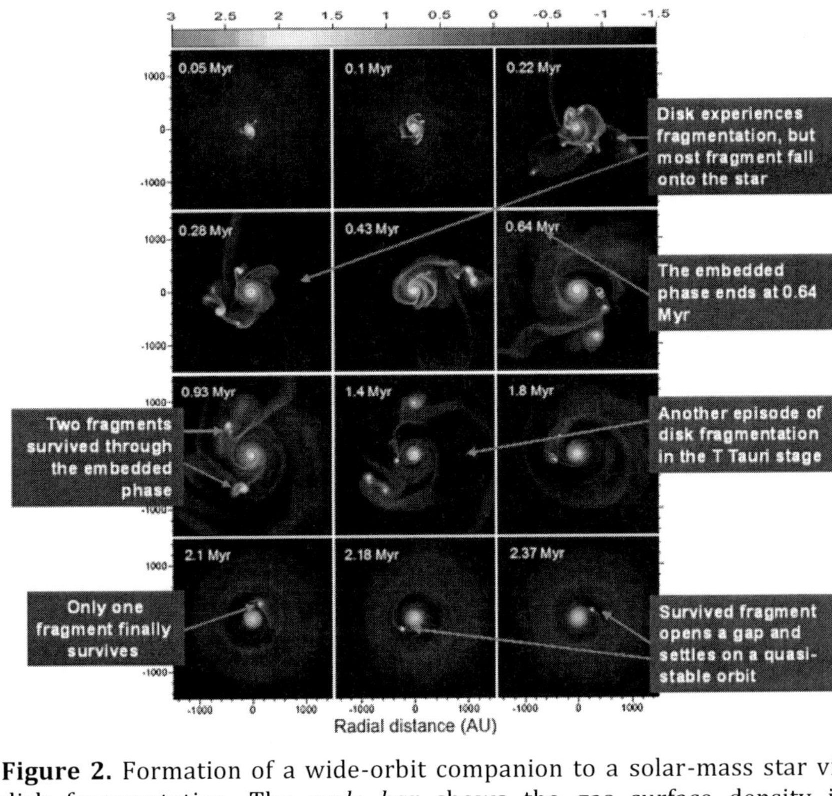

Figure 2. Formation of a wide-orbit companion to a solar-mass star via disk fragmentation. The *scale bar* shows the gas surface density in gcm−2gcm−2. See text for more detail

Figure 2 illustrates this process showing the evolution of a gravitationally unstable disk leading to fragmentation and ultimate survival of one of the fragments. The time elapsed since the formation of the central protostar is indicated in each panel. During the early phase $t \leq 0.65$ Myr$t \leq 0.65$ Myr, when the forming star and disk are both embedded in the parental cloud, the disk experiences vigorous fragmentation, but most fragments are torqued onto the star producing strong luminosity bursts. After the end of the embedded phase, only two fragments are left in the disk. But a new (and last) episode of disk fragmentation takes place around t=1.0 Myrt=1.0 Myr leading to the formation of two more fragments. However, only one fragment survives through the subsequent evolution and finally settles onto a wide-separation (~300 au~300 au) orbit around the host star. The mass of the surviving fragment is 11 MJup11 MJup. In most cases, however, companions in the brown dwarf mass regime are formed. The described process operates in massive disks, but has a rather stochastic nature and does not always lead to the formation of a stable companion, even for otherwise similar initial conditions.

2.3 Other Implications of Disk Fragmentation

a) *Ejection of fragments from the disk via many-body interaction.* Close encounters between fragments in a protostellar disk can lead to the ejection of the least massive one into the intracluster medium. The ejected fragments, upon cooling and contraction, can form freely floating brown dwarfs and very-low-mass stars (Basu and Vorobyov 2012; Vorobyov 2016). This phenomenon is distinct from previously suggested scenarios of brown dwarf formation via disk fragmentation, wherein finished point-sized objects are ejected (Bate 2009; Stamatellos and Whitworth 2009), and is more consistent with observations of isolated proto-brown dwarf clumps (André et al. 2012; Palau et al. 2014).

b) *Dust processing in the depths of massive fragments and formation of solid cores.* Theoretical considerations and numerical hydrodynamics simulations of fragmenting disks (Boley 2009; Vorobyov 2011a; Nayakshin et al. 2011) demonstrate that gas temperatures in the depths of massive fragments may exceed 800 K, thus initiating thermal annealing of amorphous silicates. If the tidal destruction timescale of these fragments is shorter than both the migration and dust sedimentation timescales, then the processed dust can be released at distances up to several hundred au, thus explaining the recent detection of crystalline silicates in the solar system comets. The fragments may also form solid cores in their interior before they are destroyed on their approach to the central star, providing a formation mechanism of protoplanetary cores that is alternative to the core accretion mechanism (Nayakshin 2010).

c) *Solution of the luminosity problem.* In the standard model of spherical core collapse (Shu 1977), the mass accretion rate is proportional to the cube of the sound speed. For typical gas temperatures in cloud cores on the order of 10–20 K, the corresponding accretion rate varies in the $(1\text{–}4)\times10^{-6}\ M_\odot$ range. A significant shortcoming of the standard model is the classic "luminosity problem", whereby accretion at the above rate produces accretion luminosities factors of 10–100 higher than typically observed for embedded protostars (Kenyon et al. 1990; Dunham et al. 2010). Accretion rates in gravitationally unstable disks can provide a solution to this problem. In the recent study, Dunham and Vorobyov (2012) calculated the radiative transfer of the collapsing cores throughout the full duration of the collapse using as inputs the disk, envelope, and stellar properties, as well as stellar mass accretion rates, predicted by the hydrodynamical simulations. They demonstrated that numerical hydrodynamics models that predict gradually declining accretion rates with episodic bursts caused by fragments migrating onto the star can account for the luminosity problem and explain a wide spread in luminosities of embedded sources.

d) *Chemical tracers of recent luminosity bursts*. As simple semi-analytic models show, luminosity outbursts similar in magnitude to those of FU-Orionis-type eruptions can raise the gas and dust temperature and drive significant chemical changes in the surrounding envelope and disk (Lee 2007; Visser and Bergin 2012; Kim et al. 2012; Visser et al. 2015). Numerical hydrodynamics simulations of disk-fragmentation-driven luminosity outbursts (Vorobyov et al. 2013) demonstrate that the CO ice can sublimate into the gas-phase in the inner envelope (a few ×103 au×103 au) during the luminosity burst, remaining in the gas phase long after the burst has subsided. For the typical conditions in the inner envelope, the e-folding time for CO freeze-out onto dust grains is about 2500 yr. The fact that the CO freeze-out time can be much longer than the burst duration (100–200 yr) opens up a possibility for the observational detection of recent luminosity outbursts (see also Sect. 6).

3 THE RADIATIVE ENVIRONMENT OF CIRCUMSTELLAR DISKS

Radiation from the central star is a crucial driver of disk physics, in particular heating processes, ionization processes, and chemical reactions. The dust component in a disk plays a special role because it adds significant opacity to radiation. The balance between heating mechanisms (e.g., due to chemical reactions, photoionization processes) and cooling processes (line radiation from the gas and continuum radiation from the dust) determines the vertical temperature profile in the upper layers of the disk. Close to the midplane, gas and dust temperatures are equal due to coupling interactions between the two components ("accommodation", Chiang and Goldreich 1997).

Both the chemical composition of the gas and the size distribution of dust grains are influencing the heating-cooling balance, so that complex radiation thermo-chemical simulations are required to model temperature profiles. Generally, the disk midplane is relatively cool. However, high-energy radiation le to excessive ionization and heating of the disk surface layers in which gas and dust thermally decouple due to the low densities. These mechanisms are at the origin of gas-disk erosion and chemical processing of the gas components. We therefore concentrate on high-energy radiation and its effects on the gas disk in the following, while we refer the reader to the chapter on the dust disk for mechanisms involving the dust component (Chap. 3).

Circumstellar disks are immersed in various types of high-energy radiation and particle fluxes. Among them figure most prominently extreme-ultraviolet and X-ray radiation from hot stellar coronae; even some collimated jets have revealed prominent X-ray emission (Güdel et al. 2008); stellar flares are expected to add hard X-ray and gamma radiation, apart from an enhanced flux of energetic particles ("stellar cosmic rays"). High-energy radiation and particles act as ionization agents of disk gas and consequently heat disk gas to several thousand K, thereby driving chemical reactions.

Ionization and heating also act toward disk dispersal: the former by possibly inducing the magneto-rotational instability (MRI) through coupling of ionized disk gas to magnetic fields, and the latter by photoevaporating gas from the uppermost layers of the disk. The disk lifetime may thus well be controlled by high-energy processes taking place on or near the central star.

In the subsequent sections, we first briefly review the relevant radiation sources in star-disk systems, and then discuss ionization and heating and their role in gas disk structure and evolution.

3.1 Stellar Coronal X-Rays

X-ray and extreme-ultraviolet radiation from T Tauri stars (TTS) originates predominantly from a magnetized corona, as in the case of the Sun. The X-ray radiation level is much higher than the Sun's, saturating at a maximum level of about 10−310−3 times the stellar bolometric luminosity (Telleschi et al. 2007), i.e., at levels of 1029--1031 ergs−11029--1031 ergs−1. Similar characteristics apply to Class I protostars. In the Orion region, the X-ray luminosities increase from protostars to TTS by about an order of magnitude, although the situation is unknown below 1–2 keV (Prisinzano et al. 2008).

3.2 Stellar X-Ray Flares

X-ray flares are also common in TTS, producing plasma temperatures of up to ≈108 K≈108 K (Stelzer et al. 2007; Imanishi et al. 2001). Like in main-sequence stars, the flare peak temperature, TpTp, correlates with the peak emission measure, EMpEMp (or, by implication, the peak X-ray luminosity), roughly as (Güdel 2004)

$$EM_p \propto T_p^{4.30\pm0.35}. \quad (1)$$

3.3 X-Ray Fluorescence

Photoionization of cool disk gas by X-rays above the Fe K edge at 7.11 keV produces a prominent line feature at 6.4 keV due to fluorescence. This feature has been detected in spectra of several classical TTS, in particular after intense X-ray flaring (Tsujimoto et al. 2005). But in at least one protostellar (Class I) case, strong fluorescence appears to be quasi-steady rather than related to strong X-ray flares (Favata et al. 2005); perhaps, the 6.4 keV line is excited by non-thermal electron impact in relatively dense, accreting magnetic loops.

3.4 Accretion-Induced X-Rays

Accretion streams falling from the circumstellar disk toward the central star reach a maximum velocity corresponding to the free-fall velocity,

$$v_{ff} = \left(\frac{2GM_*}{R}\right)^{1/2} \approx 620\left(\frac{M}{M_\odot}\right)^{1/2}\left(\frac{R}{R_\odot}\right)^{-1/2} \text{km s}^{-1}. \quad (2)$$

This velocity is an upper limit as the material starts only at the inner border of the disk; a more realistic terminal speed is vm≈0.8vffvm≈0.8vff (Calvet and Gullbring 1998). On the stellar surface, a shock develops with a temperature of

$$T_s = \frac{3}{16k} m_p \mu v_m^2 \approx 3.5 \times 10^6 \frac{M}{M_\odot}\left(\frac{R}{R_\odot}\right)^{-1} \text{K} \quad (3)$$

(mpmp is the proton mass and μμ is the mean molecular weight, i.e., μ≈0.62μ≈0.62 for ionized gas). For typical T Tauri stars, M=(0.1--1) $M = (0.1–1)\ M_\odot$, $R = (0.5–2)\ R_\odot$, and $M/R \approx (0.1–1)\ M_\odot/R_\odot$ and therefore

$T_s \approx (0.4\text{–}4) \times 10^6$ K.K. The ensuing X-rays may be absorbed in the shock gas itself, contributing to its heating, but suggestive evidence for accretion-related X-ray production is available in the very soft X-ray range (e.g., based on flux excess in the O VII and Ne IX line triplets or indications of very high densities; Güdel and Telleschi 2007, Kastner et al. 2002).

3.5 Jets and Herbig–Haro Objects

Terminal shocks between jets and the interstellar medium ("Herbig–Haro (HH) objects"), are obvious candidates for X-ray production. Faint X-ray sources at the shock fronts of HH objects have indeed been found (e.g., Pravdo et al. 2001). However, these sources are very distant and probably of little relevance for the disk. On the other hand, in a few cases X-ray sources have been detected in the launching jet regions no more than a few tens of au from the star; they are likely to form in shocks internal to the jet, and possibly also involve magnetic fields (e.g., Favata et al. 2002, Bally et al. 2003, Güdel et al. 2008).

The faint low-resolution spectra are soft and indicate temperatures of a few million K, compatible with shock velocities of a few 100 kms−1100 kms−1. These sources may be important as their X-ray luminosities compete with the stellar X-ray luminosity, but they are located vertically above the disk, providing their X-ray radiation relatively unobstructed access to the disk surfaces (in contrast to the grazing-incidence X-rays from the star), where they can drive heating, ionization and therefore photoevaporative flows or chemical reactions.

3.6 Stellar High-Energy Particles

Strong flares on the central star produce high fluxes of accelerated electrons and protons ("stellar cosmic rays") at planetary distances. From solar analogy and flare statistics, Feigelson et al. (2002) estimated a proton flux for the "T Tauri Sun" about 105105 times higher than at present (i.e., 107 protons cm−2s−1107 protons cm−2s−1 at 1 au).

This flux is probably present almost continuously given the high flare rate. Protostellar jets may also be excellent accelerators of high-energy particles (Padovani et al. 2015).

There is indeed some indirect evidence for high-energy particles ejected by young stars (Ceccarelli et al. 2014).

4 DISK IONIZATION AND THE MAGNETO-ROTATIONAL INSTABILITY

4.1 The Ionization Structure of the Gas Disk

The stellar X-ray spectrum is continuously being absorbed along the path from the star to the disk surface by the upper, tenuous layers of the disk gas and perhaps a disk wind. The dependence of the photon flux spectrum on photon energy εε and position (rr = radius of the disk segment) can be written as, for a geometrically thin, flat disk (Glassgold et al. 1997)

$$F_0(\varepsilon, r) = \underbrace{\frac{R}{r}\frac{1}{2}\frac{1}{4\pi r^2}\frac{L_X}{kT}}_{f(r)}\underbrace{\frac{1}{\varepsilon}e^{-\varepsilon/kT}}_{g(\varepsilon)}, \qquad \varepsilon > \varepsilon_0 \qquad (4)$$

where the X-ray source is located at the center of the disk but vertically displaced by R≪rR≪r; R/rR/r thus is the glancing angle at which the X-rays strike the disk; the 1/ε1/ε factor transforms energy flux to photon flux; the 1/(kT)1/(kT) factor normalizes the total luminosity to LXLX, and ε0ε0 is introduced to simulate the low-energy cutoff due to absorption along the way from the star to the disk. When arriving at the disk surface, this flux is continuously attenuated according to the photoionization cross section. Essentially all of the initial X-ray energy εε is used up to produce ionization, expending VΔε≈37 eV per ionization pair. Glassgold et al. (1997) and Igea and Glassgold (1999) performed numerical calculations of disk ionization for realistic disk profiles and shapes, also considering Compton scattering of energetic X-rays inside the disk. They essentially find that (see Fig. 3)

- the ionization rate by the softest X-rays at the disk surface at optical depth ≪1≪1 is approximately constant, owing to the spectral cutoff ε0ε0;
- the ionization rate rapidly decreases toward larger column depths where harder photons of the spectrum ionize predominantly heavier elements;
- harder spectra penetrate deeper into the disk and therefore ionize deeper layers;
- for a typical stellar coronal temperature of kT=1 keVkT=1 keV, the ionization degree corresponding to the adopted cosmic-ray ionization

degree of 2×10−17 s−12×10−17 s−1 is found at depths of $N_H \approx 10^{23-25}$ cm^{-2} at disk radii of ≈10–1 au.

- At column densities greater than $N_H \approx 1.5 \times 10^{24}$ cm^{-2}, the ionizing X-rays have energies >10 keV>10 keV where the Compton cross section becomes larger than the absorption cross section, and the ionization rate begins to be altered by Compton scattering.

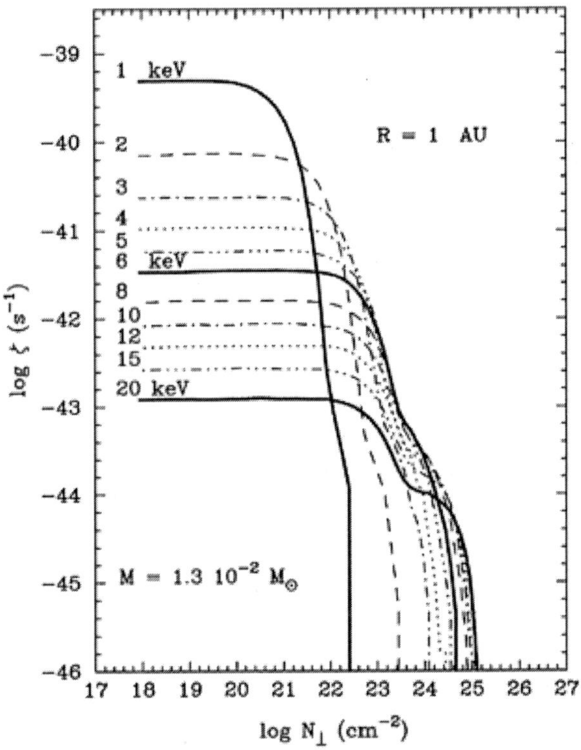

Figure 3. Ionization rates for monochromatic X-rays as a function of column density vertically into a disk for a gas with standard interstellar composition, at a radius of 1 au. The source X-ray luminosity has been normalized to 1 ergs−11 ergs−1. The shoulder at deep levels are due to Compton scattering (from Igea and Glassgold 1999, © AAS. Reproduced with permission)

Calculations of the resulting ionization fraction n(e)n(e) by Glassgold et al. (1997) and Igea and Glassgold (1999) (accounting for sedimentation of condensible heavy elements) show that the electron density n(e)n(e) decreases rapidly toward the disk midplane; for a given vertical column density N⊥N⊥, n(e)n(e) also decreases with distance rr; together, the disk ionization structure defines a wedge-shaped region around the inner disk mid-plane with its apex at the mid-plane at some outer radius, within which the ionization degree is very low ("dead zone"; Fig. 4).

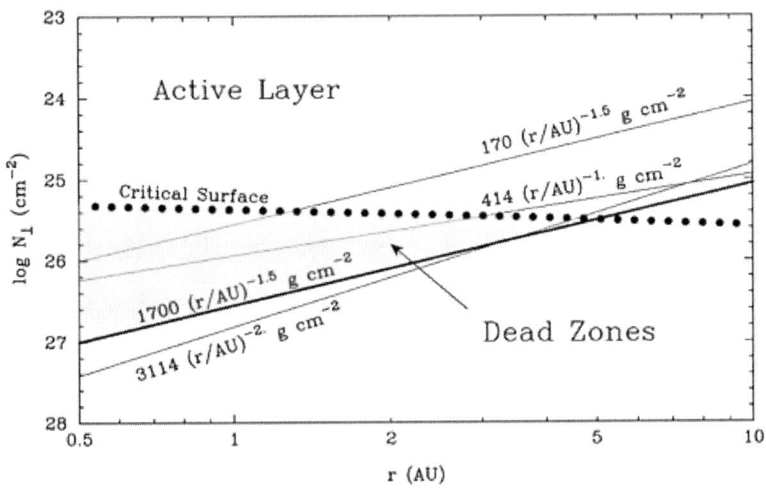

Figure 4. Electron fraction in a disk due to X-ray ionization, shown in the radius—N⊥N⊥ (*vertical column density*) plane. "Critical surface" marks the height below which the ionization fraction is too low to induce the magneto-rotational instability. The *solid lines* give disk mid-planes from different models (from Igea and Glassgold 1999; © AAS. Reproduced with permission)

4.2 The Magneto-Rotational Instability

Ionization of disk gas couples the disk gas to magnetic fields; this configuration in a differentially rotating disk is subject to the magneto-rotational instability (MRI; Balbus and Hawley 1991). The instability requires a minimum ionization degree; wherever this criterion is fulfilled, the disk is "active", other regions are called "dead zones" (Gammie 1996; see Fig. 4). MRI is held responsible for disk accretion. The criterion is given by the magnetic Reynolds number, ReMReM, expressing the relative importance between advection and magnetic diffusion,

$$Re_{\mathrm{M}} = \frac{Lv}{\eta} \approx 7.4 \times 10^{13} x_e \alpha^{1/2} \left(\frac{R}{1\ \mathrm{au}}\right)^{3/2} \left(\frac{T}{500\ \mathrm{K}}\right) \left(\frac{M}{\mathrm{M}_{\odot}}\right)^{-1/2} \quad (5)$$

where LL is the characteristic length of variations in BB, ηη is the magnetic diffusivity, vv is the plasma flow velocity, and the last numerical expression is due to Gammie (1996), with xexe being the electron fraction and αα the Shakura–Sunyaev viscosity parameter (α1/2α1/2 is of order unity). For MRI to be efficient, the critical ReMReM is about 100, and this corresponds to an ionization fraction of ≈10−12≈10−12 (see discussion in Fromang et al. 2002).

A significant part of the disk surface will couple sufficiently to magnetic fields to drive the instability, while the central layer around the midplane of the inner (few au) disk is insufficiently ionized and is therefore a "dead zone" (Fig. 4; Glassgold et al. 1997 and Igea and Glassgold 1999). However, considering also traces of metallic atoms, Fromang et al. (2002) find for a standard molecular αα disk and ReM,crit=100ReM,crit=100 a dead zone extending from a fraction of an au out to 10–100 au. On the other hand, a small admixture of heavy-metal atoms is rapidly ionized by charge exchange with the molecular ions, while atomic ions recombine only slowly by radiation. For standard disks, an atomic abundance of heavy metals of as little as 10−6--10−710−6--10−7of the cosmic value may make the entire disk active, without any dead zone. Newer calculations including complex chemical networks, however, show the presence of a dead zone even if metals are present (Walsh et al. 2012; Cleeves et al. 2013a).

Cosmic rays are also efficient in ionizing disk material. Their importance for driving MRI was recognized in particular by Gammie (1996) who developed a model of layered accretion in T Tauri disks. The crucial question is that of the ionization rate. In the interstellar space, $\zeta_{\rm CR} \approx 10^{-22}\text{–}10^{-20}\ {\rm s}^{-1}$. (Spitzer and Tomasko 1968). The propagation of cosmic rays onto the disk surface is, however, a very significant problem. In the present solar system, low-energy cosmic rays are suppressed because of diffusion, convection and adiabatic cooling in the solar wind. The low-energy cosmic-ray spectrum is therefore highly uncertain for energies ≲109 eV≲109 eV due to modulation of the interstellar CR spectrum by the magnetized solar wind (filling the "heliosphere"). Various extrapolations down to lower energies provide largely diverging spectra (see summary by Cleeves et al. 2013a). The interstellar ionization rate is therefore unlikely to be appropriate for a protoplanetary disk environment, especially because T Tauri stars are, like the Sun, producing magnetic fields and are very likely to produce ionized winds. Cleeves et al. (2013a) estimate an attenuation of the cosmic ray flux at, say, 1 au from the star by ≈3 orders of magnitude compared to the Sun and at least 6 orders of magnitude compared to interstellar conditions, resulting in a CR ionization rate of about $\zeta_{\rm CR} \approx 10^{-22}\text{–}10^{-20}\ {\rm s}^{-1}$. This may result in much larger "dead zones" in protoplanetary disks, possibly reaching out to distances of 25 au.

5 HEATING OF DISK GAS AND PHOTOEVAPORATION

5.1 Thermal Structure of a Gas Disk

As laid out in Sect. 3, high-energy radiation strongly modifies the disk surface layers due to their strong heating rates and the decoupling of gas and dust in these low-density regions. In lower regions, dust and chemical processes are pivotal for the temperature profile, but in the upper disk atmospheres most of the heating is due to irradiation by stellar (or jet) ultraviolet, extreme-ultraviolet and X-ray photons.

In Fig. 5 we show the main heating/cooling processes for an exemplary model of a disk around a T Tauri star. The main stellar and disk properties of the model are: stellar mass M∗=0.7 M⊙M∗=0.7 M⊙, photospheric luminosity L∗=1 L⊙, FUV luminosity LFUV=2.3×1031 ergs−1, X-ray luminosity LX=1030 ergs−1 and the disk mass is Mdisk=0.03 M⊙ (for more details on the model see Sect. 6 and Table 1).

In the upper layers of the disk the gas heating is dominated by radiation (e.g. X-ray Coulomb heating, C photo-ionisation due to FUV radiation). In the deeper layers the gas temperature is mainly controlled by the dust temperature due to collisions (heating/cooling by thermal accommodation on grains, Burke and Hollenbach 1983). The main cooling process in disks is line cooling. In the upper layers the cooling is dominated by emission of ionized or neutral atomic species like O or C+C+.

In the transition region where TgasTgas drops rapidly, at $N_\perp \approx (1\text{–}2) \times 10^{21}\ \mathrm{cm}^{-2}$ (see Fig. 6), molecular line cooling is important. Prominent examples are CO and in deeper layers H_2O. CO is formed where H still abounds in its atomic form, and accelerates cooling through CO ro-vibrational and CO rotational emission, generating a strong temperature gradient; corresponding emission lines provide important observational diagnostics of disk heating and cooling.

We want to emphasize that the results shown in Fig. 5 are model dependent. For example the model of Glassgold et al. (2004) includes a viscous heating process which can be dominant in the transition region from the hot upper layers to the cool midplane (see Fig. 6).

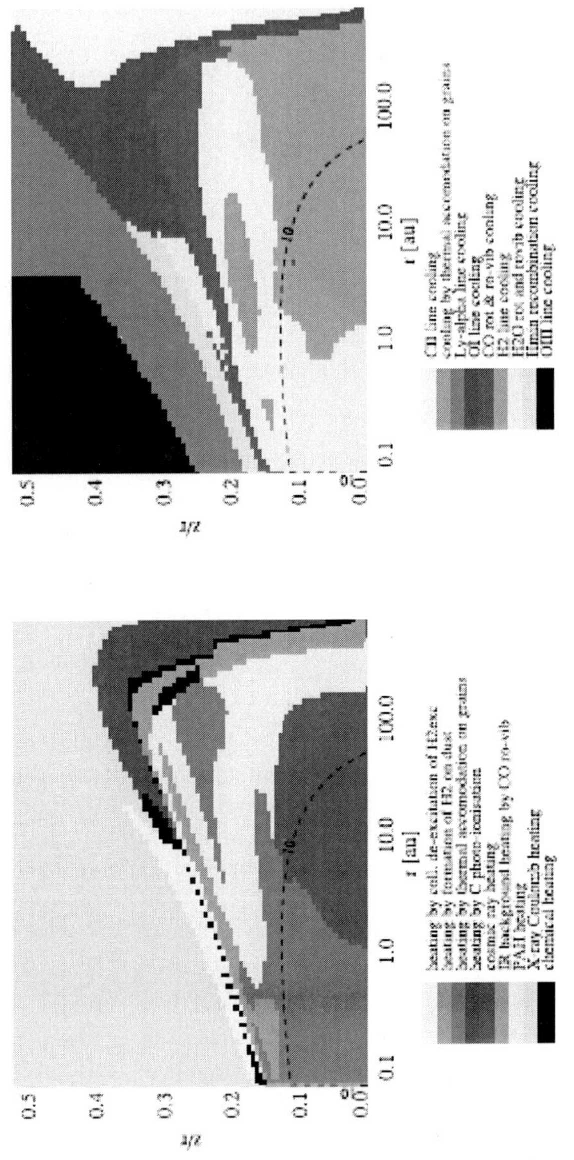

Figure 5. Dominant heating (*left panel*) and cooling (*right panel*) processes for an exemplary ProDiMo model (Protoplanetary Disk Model, see Sect. 6 for details) of a disk around a T Tauri star. The *plots* show the dominant heating/cooling process at every point in the disk (identified by the different colors and the legend). The height of the disk is scaled by the radius (z/rz/r) to show also the details in the inner region. The *dashed line* indicates a visual extinction of 10

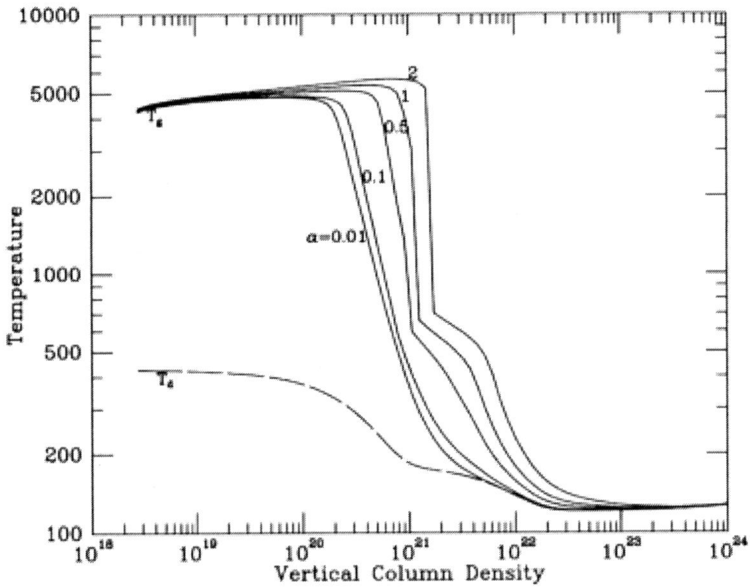

Figure 6. Gas and dust temperature profiles as a function of vertical column density into a disk, at 1 au from a central T Tauri star; apart from X-ray heating, viscous heating is present, characterized by the various αα values (α=0α=0 for no viscous heating; from Glassgold et al. 2004; © AAS. Reproduced with permission)

Table 1. Main stellar and disk properties for the T Tauri and Herbig Ae/Be (where different) ProDiMo disk models

	T Tauri	Herbig Ae/Be
Stellar mass	0.7 $M_\odot$	2.1 $M_\odot$
Stellar effective temperature	4000 K	6800 K
Stellar luminosity	1.0 $L_\odot$	32.0 $L_\odot$
FUV luminosity	2.3×10^{31} erg s^{-1}	5.2×10^{33} erg s^{-1}
X-ray luminosity	10^{30} erg s^{-1}	10^{29} erg s^{-1}
X-ray emission temperature	2×10^7 K	
Disk gas mass	0.03 $M_\odot$	
Gas/dust mass ratio	100	
Inner disk radius	0.07 au	0.4 au
Characteristic radius	100 au	
Column density power index	1.0	
Scale height at $r = 100$ au	10 au	
Flaring power index	1.15	

This process is not included in the model used for Fig. 5. However, all thermo-chemical models show the decoupling of the gas and dust temperature and the resulting high gas temperatures in the upper layers of the disk (e.g. Kamp and van Zadelhoff 2001; Gorti and Hollenbach 2004; Nomura and Millar 2005; Woitke et al. 2009). These results are qualitatively confirmed by Herschel observations (e.g. Bruderer et al. 2012; Tilling et al. 2012).

With the high spatial resolution of ALMA it now becomes possible to measure detailed radial temperature gradients in the disk. Schwarz et al. (2016) uses ALMA observations of several CO transitions (including isotopologues) to derive the radial temperature profile of the TW Hya disk. However, these observations mainly trace the warm molecular layer ($\approx$30--50 K$\approx$30--50 K see Sect. 6.1) and not necessarily the midplane of the disk. These kind of observations are required to improve and calibrate thermo-chemical disk models. However, it remains difficult to directly measure disk gas temperatures in particular in the deeper layers of the disk, and derived gas thermal structures of disks remain model dependent.

5.2 X-Ray Heating

X-ray irradiation le to substantial heating of the upper layers of protoplanetary disks. The heating energy per X-ray ionization event is about $\Delta\epsilon h\approx30$ eV$\Delta\epsilon h\approx30$ eV (Glassgold et al. 2004). Realistic gas disk calculations irradiated by stellar X-rays show strong heating of the upper, atomic layers of the disk to several thousand K at 1 au (Fig. 6). This surface heating le to a strong temperature inversion along the vertical direction through the disk, with a steep temperature decline at about $N\perp\approx1021$ cm$-2N\perp\approx1021$ cm-2. Very close to the star, much hotter, fully ionized “disk coronae” at $T\approx106$ K$T\approx106$ K are formed (Ercolano et al. 2008). X-rays are very important heating agents out to $\sim$35--40 au$\sim$35--40 au but disk ionization by stellar X-rays can be seen out to 190 au in such models.

5.3 Disk Photoevaporation

The heating of disk surface layers may produce temperatures at which individual particles escape from the stellar gravitational potential. This occurs once the mean particle energy, $\sim kT\sim kT$, exceeds the gravitational binding energy of the particles with mass mm, $GM*m/RGM*m/R$. For a given temperature, this process of *photoevaporation* becomes effective beyond the *gravitational radius*, RgRg, where the above criterion is fulfilled, i.e.,

$$R_g = \frac{GM_*}{c_s^2} \tag{6}$$

where cs=(kT/m)1/2cs=(kT/m)1/2 is the sound speed of the ionized gas.

Statistically, protoplanetary dust disks observed in the infrared disperse with a time scale of ≈3 Myr≈3 Myr (Mamajek 2009). However, the actual removal of an optically thick disk occurs rapidly at the end of this process, on a time scale of perhaps 0.5 Myr.

This "two-time scale problem" was addressed by the "UV switch" model by Clarke et al. (2001), assuming that at some point in time, the disk accretion rate falls below the wind mass loss rate induced by photoevaporation at a certain radius of the disk. Subsequently, the resupply of the inner disk is suppressed and a gap is forming at this distance. The inner disk now disperses on its own, short viscous time scale.

The EUV photoevaporation model of disks was first and extensively developed by Hollenbach et al. (1994) based on the diffuse recombination radiation field that ionizes the disk gas at large radii. The mass-loss rate depends on the flux of EUV photons, ΦΦ, and the mass of the central star, M∗M∗, as

$$\dot{M}_w \approx 4.4 \times 10^{-10}\left(\frac{\Phi}{10^{41}\ \mathrm{s}^{-1}}\right)^{1/2}\left(\frac{M_*}{1\mathrm{M}_\odot}\right)^{1/2} \mathrm{M}_\odot\ \mathrm{yr}^{-1}. \tag{7}$$

If the escaping gas is flowing at sound speed (cs≈10 kms−1cs≈10 kms−1), then the total mass loss rate at radius RR is

$$\dot{\Sigma}_w(R) = 2\mu m_{\mathrm{H}} n_0(R) c_s(R) \tag{8}$$

μμ being the mean molecular weight of the gas, ≈2.3≈2.3 for a molecular gas, mHmH the mass of a hydrogen atom, the factor of 2 accounting for the two sides of the disk, and n0n0 being the number density of the disk gas.

Once the inner disk has drained, there is a significant contribution of *direct* radiation supporting the evaporation process in particular at the inner edge of the disk so that the outer disk also rapidly disperses. Alexander et al. (2006a) find a typical escape speed of 0.4cs0.4cs and a mass-loss formula

$$\dot{M}_w(< R_{out}) \approx 2.2 \times 10^{-9} \left(\frac{\Phi}{10^{41}\ s^{-1}} \right)^{1/2} \left(\frac{R_{in}}{3\ au} \right)^{1/2} \left(\frac{H/R}{0.05} \right)^{-1/2} \times \left[1 - \left(\frac{R_{in}}{R_{out}} \right)^{0.42} \right] M_{\odot}\ yr^{-1} \quad (9)$$

where RinRin and RoutRout are the inner and outer disk radii, respectively, and H/RH/R is the ratio (assumed to be constant) between disk scale height HH and radius RR. The effect of direct radiation thus enhances the mass loss rate by an order of magnitude above the estimate in Eq. (7). Numerical simulations confirm this behavior and suggest that at the end of the disk lifetime (several Myr), the disk rapidly disperses within ≳105 yr≳105 yr (Alexander et al. 2006b).

EUV (Ly continuum) radiation is problematic because it may not reach the disk at all if disk winds absorb them early on. Also, Pascucci et al. (2014) claim that the EUV contribution is, based on observational evidence, too small to be effective for photoevaporation. Alternatively, soft X-rays may be relevant. Models based on Monte-Carlo methods for radiation show that X-rays may indeed dominate the photoevaporation process, with wind mass-loss rates of order 10−8 M⊙yr−110−8 M⊙yr−1 (Ercolano et al. 2008). These studies were qualitatively and quantitatively supported by full hydrodynamic simulations by Owen et al. (2010) and Owen et al. (2011), emphasizing that the soft X-ray part of a stellar spectrum is the dominant driving mechanism for photoevaporation (rather than EUV alone). They find the mass-loss rate to scale linearly with LXLX. The important advantage of X-rays here is their much larger penetration depths than for EUV, resulting in a much larger disk area subject to photoevaporation (out to ~40 au~40 au) compared to the EUV case (for which the inner cutoff at Rg≈5 auRg≈5 au for 104 K104 K gas limits the evaporation efficiency additionally). We will discuss observational evidence for photoevaporation of disks in Sect. 8.

6 CHEMICAL PROCESSING OF DISKS

Protoplanetary disks are considered to be the birthplaces of planets. It is likely that chemical processes occurring in those disks have an impact on the bodies formed in them. Molecular and atomic species are useful observational tracers of important physical properties of disks like mass, temperature, disk structure and turbulence. These are important quantities for planet formation theories. For the interpretation of such observations a detailed understanding of chemical processes is required.

This is a challenging task as at least from a chemical point of view, disks are very complex structures. In gas disks temperatures can vary from ≈10≈10 to several 1000 K and densities from ≈104--1016 cm−3≈104--1016 cm−3. Also external influences, in particular stellar radiation (e.g. UV, X-rays, see Sect. 3) but also cosmic rays, alter the chemical composition of the disk. Furthermore, the dust component plays an important role (see Chap. 3 on the dust disk). Dust is an important opacity source shielding the deep regions of the disk from stellar radiation. In addition, dust allows for freeze-out of molecules (ice formation) in cold regions and the dust surface can act as a chemical catalyst for the formation of molecules (surface chemistry). Last but not least also dynamical processes like viscous evolution, turbulence, planet-disk interaction (e.g. via spiral arms) and the evolution of the star likely alter the chemical composition of the disk. Figure 7gives an overview of the physical and chemical structure of a typical protoplanetary disk.

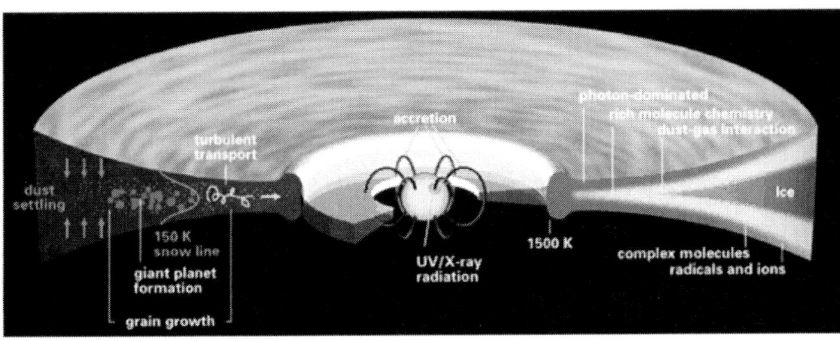

Figure 7. Sketch of the physical and chemical structure of a protoplanetary disk around a sunlike star (age ≈1--5 Myr≈1--5 Myr). Adapted with permission from Henning and Semenov (2013), ©(2013) American Chemical Society

In terms of chemical disk modeling the inclusion of all relevant processes remains challenging. Most chemical disk models commonly assume a fixed (steady state) two dimensional disk structure and fixed stellar properties. So-called *radiation thermo-chemical disk models* also consistently solve for the gas temperature (heating/cooling processes) and often include detailed dust radiative transfer (e.g. Aikawa et al. 2002; Gorti and Hollenbach 2008; Woitke et al. 2009; Walsh et al. 2010; Bruderer et al. 2012; Du and Bergin 2014).

Also the impact of turbulent mixing (e.g., Semenov et al. 2006; Furuya and Aikawa 2014) or dust evolution (e.g. Vasyunin et al. 2011; Akimkin et al.

2013) can be studied with such models. An extensive and complete review of this kind of models can be found in Henning and Semenov (2013) and Dutrey et al. (2014). Recently, chemical modeling has also been included in hydrodynamic simulations of disks.

Examples are: Ilee et al. (2011) and Evans et al. (2015) who study the chemistry in the context of gravitationally unstable disks (e.g., in spiral arms) or Vorobyov et al. (2013) who study the impact of episodic accretion on the CO gas phase transition (see Sect. 2.3).

Such models are computationally expensive and usually make simplifying assumptions about the radiative transfer and/or the chemistry. Nevertheless, such approaches are very powerful and give a first insight on the importance of dynamical processes on the chemical evolution of disks.

Here we do not discuss chemical models in detail but rather focus on the general chemical structure of the disk. For more complete reviews on disk chemical models and processes see Bergin et al. (2007), Henning and Semenov (2013), Dutrey et al. (2014). To qualitatively illustrate certain aspects discussed in this section, we use the results of two generic disk models, one T Tauri and one Herbig Ae/Be disk (see Table 1 for details). These two models should in particular illustrate the impact of the stellar properties on the chemical structure of the disk.

Therefore we use for both models the same parameters for the disk (e.g. disk mass) with the exception of the inner disk radius. The inner disk radius is given by the dust condensation radius, and is therefore larger for the Herbig Ae/Be disk due to the stronger stellar radiation field.

We apply the radiation thermo-chemical disk code ProDiMo (Protoplanetary Disk Model, Woitke et al. 2009, Kamp et al. 2010, Thi et al. 2011, Aresu et al. 2011, Meijerink et al. 2012, Woitke et al. 2016), to calculate the thermal and chemical structure of the disk. We use time-dependent chemistry and evolve the chemistry up to 2 million years (chemical age of the disk). For the chemical reaction network we use the KIDA 2014 (Kinetic Database for Astrochemistry, Wakelam et al. 2015) database supplemented by additional chemical reactions included in ProDiMo.

For the orption/desorption energies of the chemical species we use the data provided by the UMIST 2012 release (McElroy et al. 2013). More details on this particular model can be found in Woitke et al. (2016).

We use the T Tauri and Herbig Ae/Be ProDiMo disk models to discuss the general chemical structure of disks (Sect. 6.1, Figs. 8 and 9) and the importance of ice lines for planet formation (Sect. 6.3, Fig. 10).

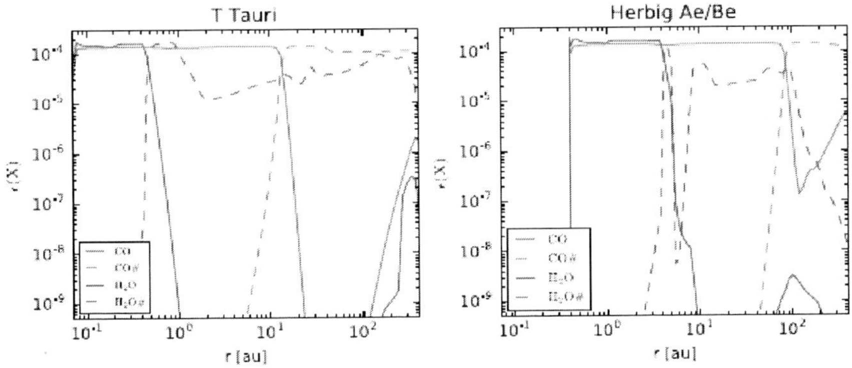

Figure 8. Abundance of CO and water (relative to the total hydrogen number density) in gas and solid phase (as ice on the dust grains) in the midplane (z=0 auz=0 au) of the disk for the two generic *ProDiMo* disk models (*left panel*: T Tauri, *right panel*: Herbig Ae/Be, see also Table 1). The *solid lines* show the gas phase, the *dashed lines* the ice-phase abundances. The ice-line for water is at ≈0.5(5) au≈0.5(5) au, the one for CO at ≈15(80) au≈15(80) au for this particular T Tauri (Herbig Ae/Be) disk model. We note that the inner radius of the Herbig Ae/Be disk is at 0.4 au. For easier comparison we use the same scales for both plots

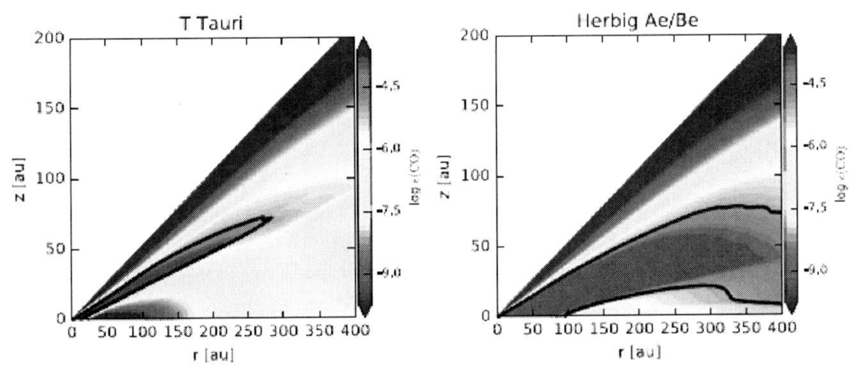

Figure 9. CO abundance ϵ(CO)ϵ(CO) (relative to the total hydrogen number density) for the two generic *ProDiMo* disk models (*left panel:* T Tauri, *right panel:* Herbig Ae/Be, see also Table 1). The black contour roughly encircles the warm molecular layer (ϵ(CO)=10−5ϵ(CO)=10−5). Above this layer CO is photo-dissociated, below this layer CO freezes-out on dust grains. In the outer region of the midplane (r≳200 aur≳200 au) CO is also released back into the gas phase due to photo-desorption. The CO layer in the Herbig Ae/Be disk is thicker and more extended compared to the T Tauri disk, as the disk is warmer due to the stronger stellar radiation field

6.1 Chemical Structure

The following description of the chemical structure is strictly speaking only valid for T Tauri disks. However, the main chemical processes described are also valid for Herbig Ae/Be disks. We discuss the main differences between T Tauri and Herbig Ae/Be disks at the end of this section.

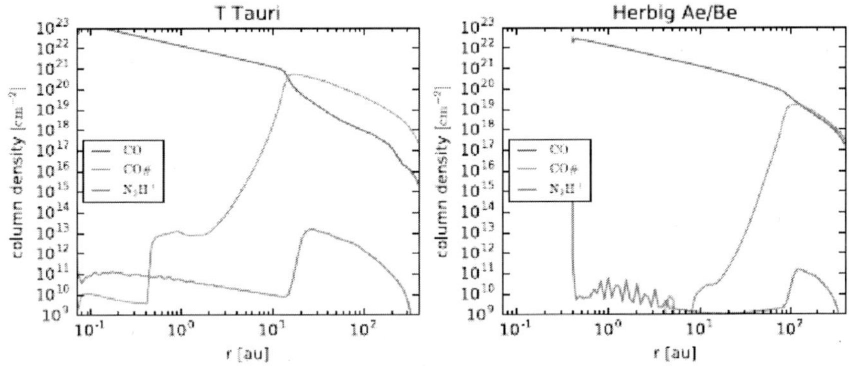

Figure 10. Vertical column densities for CO, CO ice (CO#) and N2H+N2H+ as a function of radius for the two generic *ProDiMo* disk models (*left panel:* T Tauri, *right panel:* Herbig Ae/Be, see also Table 1). The jump in the N2H+N2H+ column density at the CO ice lines (at ≈15 and 80 au≈15 and 80 au) is visible in both models but less pronounced in the Herbig Ae/Be disk

For the discussion of the chemical structure of the disk we divide the disk in different radial zones and vertical layers. From a chemical point of view the midplane of the disk contains three distinct zones:

- *inner zone:* extends up to a distance of a few au from the star; temperatures T>100 KT>100 K; mostly shielded from radiation due to high optical depths;
- *middle zone:* extends from a few au up to $100\ \mathrm{au};\ 100 \gtrsim T \gtrsim 20\ \mathrm{K}$:; most species are orbed onto dust grains (e.g. water ice); shielded from radiation;
- *outer zone:*$T \lesssim 20$ K; most species freeze-out on dust grains; only partly shielded from radiation.

As an example we show in Fig. 8 the abundance of CO and water (for the gas and solid phase) in the midplane as a function of radius. The different locations of the ice-lines due to the different orption energies for CO and water are clearly visible. Also the re-increase of the gas phase abundance in the outer disk, in particular for CO, is apparent.

In vertical direction it is common to define three distinct layers:

- *midplane*: see above; ice dominated chemistry (dust–gas interaction, surface chemistry);
- *rich molecular layer*: warm enough (K20$\lesssim$T$\lesssim$100 K) so that most molecules are in gas phase; partly shielded from radiation; rich ion-neutral chemistry;
- *photon-dominated layer*: photochemistry (ionization, photo-dissociation); most species are in atomic form and ionized.

These three layers are also indicated in Fig. 7. In Fig. 9 we show, as an example, the two dimensional distribution of the CO molecule. The vertically layered molecular structure is clearly visible and is similar for other molecules.

Due to the strongly varying physical conditions the chemistry can be very different in the zones described above. For a proper treatment of the chemistry in the *inner zone* chemical reactions with high activation barriers can also become important (e.g. Harada et al. 2010), different from the usually colder interstellar conditions. Therefore, it can be expected that the resulting chemical abundances are also different. Observations presented in Pontoppidan and Blevins (2014) show clear indications for this scenario. This work also shows that thermo-chemical models still have problems to explain the observed abundances. One reason might be that radial migration processes (e.g. large dust particles migrate inwards) are neglected. Migrating dust particles are probably coated in water ice. This ice is sublimates in the warmer conditions of the inner zone, enhancing the oxygen abundance which can subsequently also influence the chemistry. However, the thermo-chemical model of Walsh et al. (2014), for example, takes accretion flows into account, showing that also the abundance of complex molecules (e.g. methanol) can be strongly enhanced owing to the transport of ice-coated dust particles into the inner zone.

In the radiation shielded and much cooler *middle zone* of the midplane the important chemical drivers are cosmic rays and/or radionuclide ionization (and possibly also high-energy X-rays). The ionization rates of these processes are quite low, in the range of ≈10−20--10−17 s−1≈10−20--10−17 s−1 (e.g., Cleeves et al. 2013a,b) compared to the upper layers of the disk. Further photochemistry induced by the secondary UV field from cosmic rays (e.g. Gredel et al. 1989) becomes important in this region. A detailed study by Chaparro Molano and Kamp (2012a,b) has shown that for disks proper dust and gas opacities have to be taken into account for the cosmic-ray induced photo processes. Their derived chemical abundances are compatible with measurements of the chemical composition of solar-system comets. The gas phase abundances of chemical species probably also influences the composition of gas giants that might be formed in this zone. One interesting quantity is the resulting C/O gas phase abundance ratio which can be significantly altered, and therefore directly influences the chemical composition of giant planet atmospheres (e.g Helling et al. 2014).

Different from the *middle zone*, in the cold *outer zone* non-thermal desorption mechanisms (e.g. photo-desorption) may become important. Recent observations of DCO^+ in IM Lup suggest that non-thermal desorption (most likely photo-desorption due to stellar UV radiation) is efficient in the outer disk (Öberg et al. 2015; see also Sect. 7). In addition to the stellar UV radiation also the interstellar background field might play an important role for the DCO^+abundance (Teague et al. 2015). Surface chemistry processes produce complex molecules on the dust surface (e.g. hydrogenation of CO to form H_2CO, Qi et al. 2013), and these molecules may subsequently be photo-desorbed. One prominent example for this suggested formation process is methanol (Walsh et al. 2010; Dutrey et al. 2014). Gas phase methanol was recently detected in the disk of TW Hya (Walsh et al. 2016). The observationally derived average abundances of methanol are up to two orders or magnitude lower compared to model predictions. This observations probably indicate a surface chemistry dominated formation pathway of methanol, but existing models seem to overestimate the efficiency of methanol formation and/or the efficiency of various desorption processes (Walsh et al. 2016).

Vertical turbulent mixing can transport ice coated dust particles to higher layers where photo-desorption is more efficient, and as a consequence gas phase abundances of complex molecules formed on dust grains are enhanced (Semenov and Wiebe 2011). On the other hand, vertical mixing can also decrease the abundance of complex organic molecules. Enhanced hydrogenation of radicals due to transport of atomic hydrogen to deeper layers can significantly suppress complex molecule formation (Furuya and Aikawa 2014).

In the warm molecular layer above the midplane one finds a very rich chemistry driven by higher temperatures, UV and X-ray radiation (Aikawa et al. 2002; Semenov et al. 2004; Henning and Semenov 2013). In our generic T Tauri disk model the layer is clearly visible in CO (Fig. 9). However, the location and the thickness of this layer probably vary from disk to disk and also depend on the dust properties. Vasyunin et al. (2011) have shown that in chemical models including large dust grains, this layer becomes thicker (e.g. extends deeper into the disk) compared to models considering ISM like dust properties (e.g. only small grains). Similar results are found by Akimkin et al. (2013) who used a self-consistent chemical model including dust evolution. They also find that on average, column densities of neutral molecules (e.g. CO, H_2O) are enhanced in models considering dust evolution.

Another important chemical process in this layer is CO isotopologue chemistry (e.g. Visser et al. 2009), in particular isotopologue selective photo-dissociation. This chemical process may explain the anomalous ^{17}O and ^{18}O ratios measured in meteorites (Visser et al. 2009). CO isotope-selective photo-dissociation must also be considered if CO isotopologue line ratios are used to estimate disk gas masses. Thermo-chemical disk models of Miotello et al. (2014) show that using constant (throughout the disk) CO isotopologue ratios can lead to underestimating disk masses by up to an order of magnitude. Also, the possible depletion of gas phase CO in disks as suggested by Favre et al. (2013) and Bruderer et al. (2012) has to be taken into account. Although this depletion may be explained by isotope-selective photo-dissociation (in particular $C^{18}O$), chemical processes occurring closer to the midplane, like dissociation of CO by He+He+(Aikawa et al. 1998; Bergin et al. 2014; Furuya and Aikawa 2014) and/or surface chemistry (Reboussin et al. 2015) must also be considered if CO and its isotopologues are used as disk mass tracer.

Above the warm molecular layer direct ionization processes of atomic species and dissociation of molecules due to X-ray and UV photons are the main chemistry drivers. This layer shows typical properties of photo-dissociation regions (e.g. Hollenbach and Tielens 1999). A proper treatment of chemical processes in this region is important to determine the gas temperature, which is relevant for e.g. photo-evaporation in disks (see Sect. 5.3 and 8). Also X-rays play a major role in this layer ionizing noble gases like Ne and Ar and heating the gas to temperatures up to 10000 K10000 K (see Sect. 5).

6.1.1 T Tauri Versus Herbig Ae/Be Disk Chemistry

A disk around a Herbig Ae/Be disk is on average warmer simply due to its stronger radiation field in the UV and optical wavelength regime. This has an impact on both the radial and vertical chemical structure of the disk. The locations of midplane ice lines are on average at larger radii in a Herbig Ae/Be disk than in a T Tauri disk (see Fig. 8). But also the vertical structure is affected. Compared to the T Tauri disk the warm molecular layer starts at slightly lower height (due to enhanced photo-dissociation) but extends much deeper towards the midplane of the disk. The warm molecular layer is mainly thicker due to the higher temperatures which prevents freeze-out of molecules (see Fig. 9).

Also molecular observations indicate that T Tauri disks have a larger cold chemistry reservoir compared to Herbig Ae/Be disks. The main indication here is the lower detection rates of cold chemistry tracers like H_2CO (Öberg et al. 2011; Dutrey et al. 2014). Also the measured CO snowline location in HD 163296 of ≈90 au≈90 au (Qi et al. 2015) is a strong indication for warmer Herbig Ae/Be disks. As ice lines might play an important role for dust evolution (see Sect. 6.3), these chemical differences are most relevant for the study of early planet formation.

6.2 X-Ray Driven Disk Ion Chemistry

Given the importance of X-ray disk ionization, we briefly summarize chemical reactions directly induced by X-rays, in particular those illustrating some principal destruction routes of molecular ions. Among the principal ion destruction paths are the following, leading to "chemical heating" (Glassgold et al. 2012):

$$H_2^+ + e \rightarrow H + H$$
$$H_2^+ + H_2 \rightarrow H_3^+ + H$$
$$H_3^+ + e \rightarrow H_2 + H$$
$$H_3^+ + e \rightarrow H + H + H$$
$$H_3^+ + CO \rightarrow HCO^+ + H_2$$
$$H_3^+ + O \rightarrow OH^+ + H_2$$
$$H_3^+ + H_2O \rightarrow H_3O^+ + H$$
$$HCO^+ + e \rightarrow CO + H$$
$$OH^+ + H_2 \rightarrow H_3O^+$$
$$H_3O^+ + e \rightarrow H_2O + H$$
$$H_3O^+ + e \rightarrow OH + H_2$$
$$H_3O^+ + e \rightarrow OH + 2H$$
$$H^+ + H_2O \rightarrow H_2O^+ + H$$
$$H_2O^+ + H_2 \rightarrow H_3O^+ + H$$
$$He^+ + CO \rightarrow C^+ + O + He$$
$$C^+ + H_2O \rightarrow HCO^+ + H$$

Complex thermo-chemical models of irradiated disks show that the abundances of neutral molecules such as CO, OH, and H_2O are not overly sensitive to X-ray irradiation (a change by less than an order of magnitude over the range of many orders of magnitude in LXLX). In contrast, the X-ray sensitive ionization fraction in the upper layers of the disk strongly influences ion chemistry. The following reactions are important (Aresu et al. 2011; MM indicates the total mass of the species in the disk):

$H_2 + e \rightarrow H_2^+$	(secondary electrons from direct X-ray ionization)
$H_2^+ + H_2 \rightarrow H_3^+ + H$	
$H_3^+ + O \rightarrow OH^+ + H_2$	$M(OH)$ grows approximately linearly with L_X.
$H_3^+ + OH \rightarrow H_2O^+ + H_2$	$M(H_2O^+)$ grows somewhat less than OH^+.
$H_3^+ + H_2O \rightarrow H_3O^+ + H$	$M(H_3O^+)$ grows by 10–100× for 1000× L_X.

Also N+N+ reacts very sensitively to the presence of X-rays and its total mass grows about linearly with LXLX. These reactions also spread in disk area, being pushed to larger distances as LXLXgrows. A special case of interest is the fine-structure emission of [O I]. The models clearly predict a linear growth of its line flux with the sum of LX+LUVLX+LUV. Whichever of the two luminosities dominates determines the correlation. Because of the usually appreciable UV field around T Tauri stars, an X-ray dependence comes into play only above LX≈1030 ergs−1LX≈1030 ergs−1(Aresu et al. 2014). Observations support the flux levels approximately although the correlation is not evident, possibly due to some influence of other parameters such as grain size distribution, surface density distribution, etc. in real objects.

6.3 Ice Lines

Ice lines (or condensation fronts) probably play a significant role in planet formation processes (e.g. through dust growth) but are also relevant for the final composition of bodies formed in the disk.

Recently Qi et al. (2013) reported the imaging of the CO snowline in the TW Hya disk via observations of the N2H+N2H+ ion. N2H+N2H+ is efficiently destroyed by proton transfer to CO

$$N_2H^+ + CO \rightarrow HCO^+ + N_2. \qquad (10)$$

Therefore N2H+N2H+ can only be abundant in regions with low CO gas phase abundances. The resulting ring like structure visible in images of the N2H+N2H+ line emissions is also, at least qualitatively, reproduced by thermo-chemical disk models (e.g. Cleeves et al. 2015; Aikawa et al. 2015). The situation is illustrated in Fig. 10 for our generic T Tauri disk model. However, Aikawa et al. (2015) discuss other relevant chemical processes influencing the resulting N2H+N2H+distribution. One example is the so called sink effect, where CO is converted to less volatile species in regions with temperatures exceeding the sublimation temperature of CO (e.g. >20 K>20 K). Therefore, the peak of the N2H+N2H+ abundance may be slightly below the CO sublimation radius.

There is a similar mechanism involving water, where HCO^+ is destroyed by water vapor. This mechanism was proposed to explain the ring-like appearance of the H13CO+H13CO+ emission in young protostars experiencing a recent luminosity outburst (Jørgensen et al. 2013). Such outbursts significantly enhance the gas phase water abundance, leading to subsequent destruction of HCO+HCO+. This chemical process is also relevant for disks; however, more detailed studies are required to investigate its efficiency under disk conditions. Furthermore, direct imaging of the water ice line via HCO+HCO+ would require a much higher spatial resolution, as the water ice line is located much closer to the star (typically around 1 au1 au, see also Fig. 8).

The recent work of Zhang et al. (2015) suggests that the locations of the dark rings visible in ALMA continuum observations of HL Tau (ALMA Partnership et al. 2015) coincide with the expected ice line locations of several molecules. In particular, the locations of the H_2O, $NH_3$and CO ice lines agree well with the three most prominent dips in the radial surface brightness distribution. The same authors argue that around the location of the ice lines, enhanced pebble-growth (particles larger than a few cm) quickly reduces the number of mm-sized particles, which would explain the lack of emission inside the rings.

6.4 FU Orions Like Outburst Scenario

As an example for the impact of stellar/disk evolution on the chemical structure we discuss the impact of luminosity outbursts (see Sect. 2.1). During these outbursts the total luminosity of the system can increase by more than a factor of 100 compared to quiescent phases. The most prominent example for such outbursts is probably FU Orionis (e.g., Zhu et al. 2007). As already mentioned, Vorobyov et al. (2013) studied the impact of such outbursts on the gas phase transition of CO in a complex hydrodynamic model with a simplified chemistry (only thermal orption/desorption processes for CO and CO_2 were considered).

We present here a simple disk+envelope model using the radiation thermo-chemical disk code ProDiMo. The modeling approach is similar to Visser et al. (2015), who studied the impact of luminosity outbursts on the chemical structure of envelopes around young protostars. Different from the model of Visser et al. (2015), our model considers an outburst scenario at a later evolutionary stage when a prominent Keplerian disk has already formed (e.g. similar to FU Orionis). In addition to the disk component, a remnant envelope structure is included (see Table 2). For the outburst we assume an instantaneous increase of the stellar luminosity by a factor of 100 lasting for 100 years.

After the outburst CO freezes out again. In the inner disk this happens quite quickly, whereas in the outer disk ($r\approx100$ au$r\approx100$ au) the freeze-out timescales are longer (≈1000 yr≈1000 yr) due to lower densities. In the envelope structure the gas phase CO abundance is still enhanced after 24 000 yr after the outburst. Approximately 100 000 yr after the outburst even in the outer envelope the gas phase CO abundance reaches again its pre-outburst level.

Table 2. Main stellar and structure properties for the FU Orions like outburst model

Quantity	Value
Stellar mass	$0.5\ M_\odot$
Stellar effective temperature	4000 K
Stellar luminosity	$1.0, 100.0\ L_\odot$
FUV luminosity	1.1×10^{31} erg s^{-1}
X-ray luminosity	10^{30} erg s^{-1}
X-ray emission temperature	2×10^{7} K
Disk gas mass	$0.02\ M_\odot$
Gas/dust mass ratio	100
Inner disk radius	0.6 au
Outer disk radius	200 au
Column density power index	1.0
Scale height at $r = 100$ au	10 au
Flaring power index	1.15
Envelope gas mass	$0.02\ M_\odot$
Inner envelope radius	0.6 au
Outer envelope radius	3500 au
Cavity opening angle	30 deg

In Fig. 11 we show the CO gas phase abundance before the outburst, during the outburst, and 800 yr and 24 000 yr after the outburst. Due to the increased luminosity the disk is heated and CO sublimates almost everywhere in the disk. In the upper layers of the disk+envelope structure CO is also dissociated by the enhanced stellar UV field. However, due to the warmer conditions the warm CO layer extends further out into the envelope structure (up to r≈2000 aur≈2000 au). Closer to the midplane the envelope structure (r>200 aur>200 au) is shielded by the disk and the CO abundance does not change significantly during the outburst (top right plot in Fig. 11).

This is a very simplified approach for modeling an outburst scenario as it ignores the possibly significant impact on the disk structure during these violent phases. However, such models allow us to study the chemical evolution in more detail (e.g. impact on other species than CO). The treatment of detailed dust radiative transfer also allows us to study the impact of different dust properties (size distribution, opacities) that may, for example, change the freeze-out timescales. Although such models are not very realistic in terms of the dynamical evolution, they can identify important chemical processes that can subsequently be incorporated in more complex hydrodynamical models. Furthermore, such models can be used to study the impact on observables in particular molecular line emission.

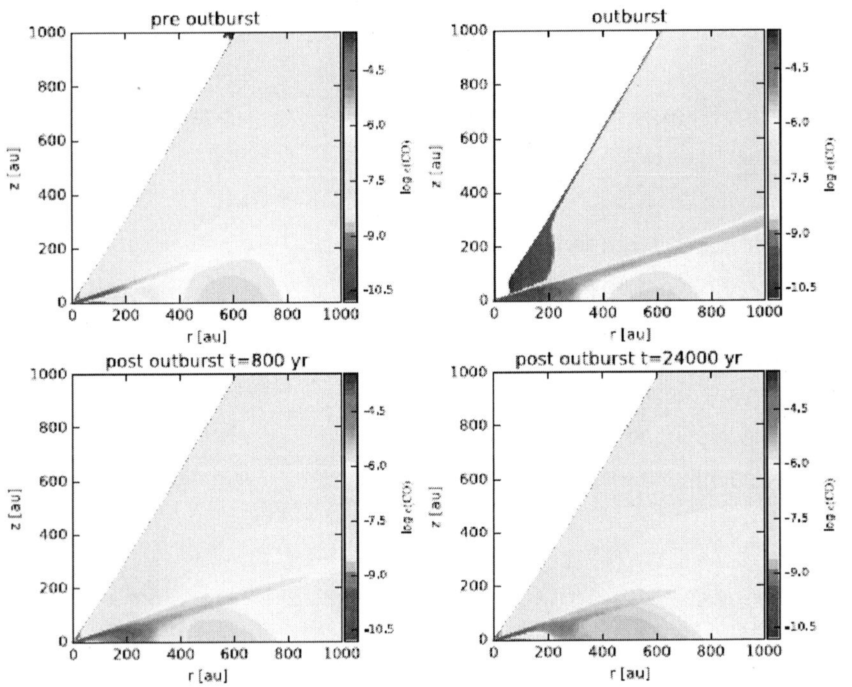

Figure 11. Time evolution of the CO gas phase abundance $\epsilon(CO)\epsilon(CO)$ for a FU Ori like outburst scenario. The structure includes a disk component and a remnant envelope (see Table 2for details). The *top left* plot shows $\epsilon(CO)\epsilon(CO)$ for the quiescent phase (stellar luminosity $L_*=1\ L_\odot$), the *top right* plot during the outburst phase ($L_*=100\ L_\odot$), the *bottom left and right* plots show the abundance 800 yr and 24 000 yr after the outburst, respectively

7 OBSERVATIONS OF THE GAS CONTENT OF DISKS

Gas represents the bulk of the mass content in disks. However, from an observational point of view the gas (in particular the gas mass) is difficult to trace. Most molecular gas lines originate within a thin layer, limited by the orption of molecules onto dust grains towards the disk midplane, and the influence of high energy radiation in the upper disk layers that dissociates molecules and ionizes atoms (see Sect. 4 and 6).

In the radial direction, the gas content in disks extends closer to the protostar than the dust component and the outer radius is influenced by the disk geometry and the radiation from the central source, although very recent observations suggest that desorption may occur in the outer disk due to non-thermal processes (Öberg et al. 2015). Still, radiation from the protostar is the main source of energy defining the gas excitation, especially in the inner disk, but also as the main catalyst driving diverse chemical pathways.

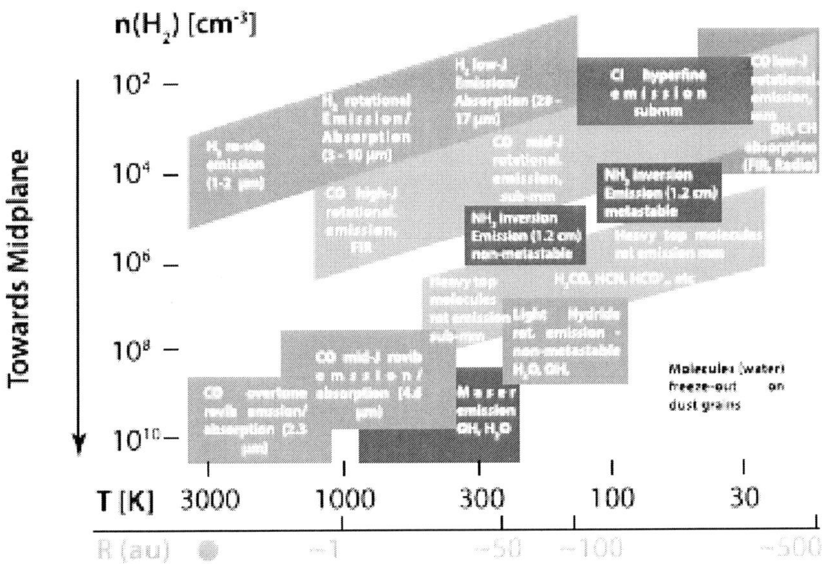

Figure 12. Commonly observed gas lines, represented as a function of the excitation conditions (temperature, density), corresponding roughly to the radial and vertical structure in a *T Tauri* disk. We note that the temperature scale does not refer to the disk midplane, but is just indicative of the (excitation) temperature as a function of distance from the star (from Dionatos 2015)

In this respect, higher excitation conditions corresponding to higher kinetic temperatures should occur closer to the protostar, while densities of different species shall probe the vertical structure of disk. In this simplified view the most common molecules detected in disks are represented in Fig. 12, as a function of radius and height in a disk. For more details on gas line diagnostics see also the reviews of Najita et al. (2007) and Carmona (2010) for the inner disk and Dutrey et al. (2014) for the outer disk.

The composition of the gas content in disks is subject to change, and chemical reactions play naturally a pivotal role, along with phase transitions (orption/desorption onto dust grains). It is therefore that chemistry in disks can trace different physical conditions but also plays an important role in controlling the disk dynamics. Consequently, understanding the chemical evolution of disks is central to decode the conditions for planet formation and habitability. The study of chemistry in disks in combination with the detailed study of the chemical compositions of solar system bodies is fundamental in understanding the conditions for life on Earth.

7.1 Emission-Line Profiles of Disks

Line broadening from gas following the Keplerian rotation of a disk results in a characteristic double-peaked profile (often coined as the Keplerian profile). This is a characteristic signature that can distinguish emission from gaseous disks from other sources of excited gas (envelope cavities for embedded protostars and jets for accreting disk sources).

The line-of-sight loci of gas moving at a constant velocity on a Keplerian rotating disk at an inclination θθ to the observer are given by the relation (Beckwith and Sargent 1993):

$$R = R_{\mathrm{out}} \left(\frac{u_{\mathrm{out}}}{u} \right)^2 \cos^2 \theta, \tag{11}$$

where uoutuout is the velocity at the terminating radius RoutRout (Fig. 13). The minimum of the line occurs when the velocity vector of the observer u→0u→0 and corresponds to the regions in the disk where the gas velocity is directed perpendicularly to the line-of-sight. As long as the velocity remains smaller than uoutuout, the iso-velocity contours on the disk surface are open and truncated at RoutRout. The maximum contour occurs for u=uoutu=uout that is the first closed (non-truncated) contour which contributes to the line peaks. Further increase in the velocity for u>uoutu>uout results in shorter, closed contours lying symmetrically on the horizontal disk mid-line that move towards to the center of the disk and contribute to the outer line wings. The contribution of the different curves of constant velocity to the line profile is illustrated in Fig. 13. The limiting uoutuout can be derived from the separation of the line-peaks and Eq. (1) once RoutRout and the inclination θθ are known.

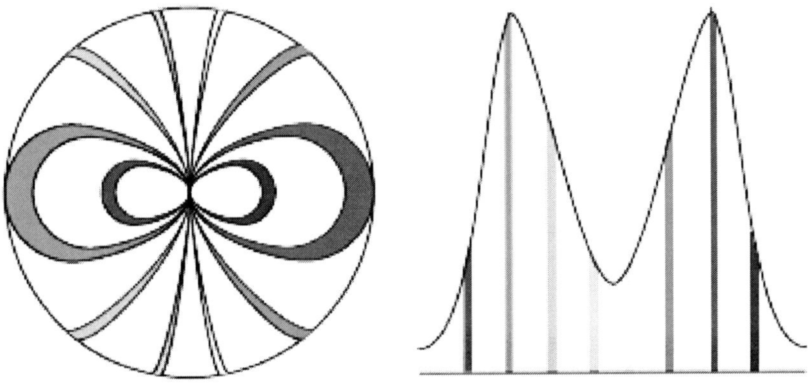

Figure 13. Contours of constant velocity on a Keplerian disk inclined at an angle θθ to the observer and their contributions in forming the characteristic Keplerian line-profile. Line minima come from the regions of the disk that is dissected by the observer's line-of-sight, while maxima are produced for the longest closed iso-velocity contours which occur at RoutRout(from Dionatos 2015)

7.2 New Input from ALMA

Keplerian profiles are commonly seen in disks of T Tauri and Herbig Ae/Be stars when observed with adequate angular and spectral resolutions, and indeed Keplerian rotation was already observed in such environments even before the advent of ALMA (e.g. Piétu et al. 2005, 2007; Dutrey et al. 2008). Careful analysis and modeling of spectroimaging observations can lead in constraining the physical conditions of specific layers in the disk traced by molecular lines (e.g. review by Dutrey et al. 2007). Most star formation theories predicted that such disks form early on, but it was only very recently that the first disks displaying Keplerian rotation were observed with ALMA around very young, embedded protostars (Tobin et al. 2012; Takakuwa et al. 2012; Murillo et al. 2013; Ohashi et al. 2014).

In disks around T Tauri and Herbig Ae/Be stars, observations of the CO snowline (Qi et al. 2013, see also Sect. 6.3) have for the first time detected the borderlines that define the conditions for the possible formation of rocky and gas-giant planets. High concentrations of dust observed in a number of transitional disks (e.g. van der Marel et al. 2015; Pérez et al. 2014b) may be the result of the interaction between planets or disk fragments (see Sect. 2.2) with the disk. In these cases, the gas appears to fill the dust cavities but at reduced levels (van der Marel et al. 2015), and in some cases show asymmetric distributions that may be related to the dust asymmetries (e.g. HCN and CS, van der Plas et al. 2014).

Large asymmetries and deviations from Keplerian rotation have also been observed in the distribution of the CO emission around the T Tauri star AS 205 (Salyk et al. 2014). However, in that case the influence of a close companion may drive tidal interactions, or a disk wind may be in action (see Sect. 5.3 and 8), or the combination of both.

An unexpected result comes from the observations of IM Lup in DCO^+ (Öberg et al. 2015), showing two co-centric rings at radii of ~90 and 300 au. While the inner ring can be explained in terms of thermal desorption of CO, the outer ring is interpreted in terms of non-thermal desorption of CO which reacts with H2D+H2D+ to form the observed DCO+DCO+.

8 OBSERVATIONAL CONSTRAINTS OF PHOTOEVAPORATIVE DISK WINDS

8.1 Atomic Tracers

Direct observations of photoevaporative disks winds are needed to test and constrain the different models and ultimately characterize them. In this section we will only cover the observational aspects of disk winds, and refer the reader to the Chap. 6 by Gorti et al. for a complete overview of photoevaporation in the context of disk dispersal and to the review by Alexander et al. (2014). A brief overview of the basic theory of photoevaporation is given in the present chapter (see Sect. 5.3).

Forbidden line emission often observed in T Tauri stars can potentially trace photoevaporative disk winds. Neon has high first and second ionization potentials (21.56 and 41.0 eV, respectively). Ionization can also occur by photoionization of the inner shell, requiring 0.87 and 0.88 keV for ejecting a K-shell electron for neutral and singly ionized neon. Only photons in the EUV range (or beyond) can ionize the outer electrons, while hard X-rays with energies in the range of keV are needed to eject a K-shell electron. Therefore neon can be used to constrain the high-energy stellar radiation reaching the disk. The mid-infrared forbidden transition of [Ne II] (12.81 μm) was first detected with the Spitzer Infrared Spectrograph (IRS, Pascucci et al. 2007, Lahuis et al. 2007). Although the Spitzer spectrograph lacks the spectral resolution needed to confirm or rule out a disk, wind, or jet origin of the line, interesting conclusions were extracted from studies based on statistically significant samples (see for example Lahuis et al. 2007, Flaccomio et al. 2009, Güdel et al. 2010, Baldovin-Saavedra et al. 2011). For example, the line luminosities of sources known to drive outflows and jets are brighter by 1–2 orders of magnitude than those of sources without outflows/jets. Correlations are observed between the line luminosity and different stellar and disk physical parameters, the most prominent ones being with the stellar X-ray luminosity LXLX and the disk mass accretion rate.

The [Ne III] line (15.55 μm) on the other hand is detected only in a handful of objects (e.g., Kruger et al. 2013, Espaillat et al. 2013). Espaillat et al. (2013) tested the possibility of the lines being emitted by an X-ray and EUV irradiated disk. The study shows no difference in the X-ray properties between stars with and without [Ne III] emission. In addition the authors used the ratio [Ne III]/[Ne II] to distinguish between X-ray and EUV production of the fine-structure emission. In fact, if the emission is dominated by X-ray production the ratio is expected to be ~0.1~0.1; in the case of EUV the line ratio can be much larger. These authors observed one object for which the ratio is ~1~1 and therefore consistent with EUV production. However, it is not possible to make general conclusions given the small number of detections.

A natural step in the study of [Ne II] emission was to follow-up with ground-based observations those stars that display [Ne II] emission in their Spitzer spectra. Thanks to the high-spectral resolution achievable with ground-based spectrographs in the mid-infrared (R~30000R~30000, or 10 km s^{-1} for VISIR at VLT), lines could be spectrally resolved, and the nature of the emission (disk, disk wind, or jet) for most of the objects could finally be unveiled. At the time of writing the neon line has been observed with ground-based instruments in ~50~50 objects (see Herczeg et al. 2007, Najita et al. 2009, Pascucci and Sterzik 2009, van Boekel et al. 2009, Pascucci et al. 2011, Sacco et al. 2012, Baldovin-Saavedra et al. 2012). About 50 % of the [Ne II] lines detected with Spitzer are confirmed with ground based observations. This hints at a high contribution from shocked material in the Spitzer spectra. The lines that are interpreted as emitted in jets present high-velocity blueshifts (30 km s^{-1} or more) and their full width at half maximum (FWHM) can be quite broad (FWHM up to ~140 kms−1~140 kms−1). Lines consistent with disk wind emission show small velocity blueshifts (~2--20 kms−1~2--20 kms−1) and are narrow (FWHM ~10--25 kms−1~10--25 kms−1). There is a handful of objects that show no shift in their line peak. However, they cannot be directly associated to disk emission; two of them (AA Tau, CoKu Tau 1) are high inclination objects known to drive bipolar jets. An interesting result from these studies is that the [Ne II] line shows only one velocity component, either a high-velocity jet, a disk wind, or a disk atmosphere. There is no evidence up to now of a neon line with two velocity components as is often observed in other atomic tracers such as [O I] in the optical.

Forbidden lines in the optical range are common in T Tauri stars and have been the subject of studies for quite some time (Jankovics et al. 1983, Edwards et al. 1987, Hamann 1994, Hartigan et al. 1995). Early studies reported the presence of blueshifted line emission and associated it with some sort of disk wind whose redshifted emission is obscured by the protoplanetary disk. Forbidden emission lines often display two-velocity components: a high-velocity component (HVC) blueshifted by hundred kilometers per second with respect to the stellar rest velocity, and a low-velocity component (LVC) blueshifted by a few kilometers per second. Kwan and Tademaru (1988) recognized that these two components are emitted in two different regions and attributed the low-velocity component to a slow-moving wind emitted from the disk, and the high-velocity component to a collimated jet near the star. In addition, Hartigan et al. (1995) observed a trend where the line velocity shifts are higher for lower critical densities, suggesting that the low-velocity component accelerates away from the surface of the disk.

The neutral atomic oxygen lines [O I] at 6300 Å and [O I] at 5577 Å are interesting because they are typically bright, in particular the 6300 Å line. Different authors have attempted to explain the observed properties of the [O I] line with thermal photoevaporative disk wind models considering different types of stellar radiation. The model presented in Font et al. (2004), where photoevaporation is driven by stellar EUV radiation, is difficult to reconcile with the observed [O I] line fluxes because it produces an almost fully ionized layer with a low fraction of neutral gas. Ercolano and Owen (2010) modeled photoevaporation driven by EUV and soft X-ray emission (0.1–10 keV) that can heat the gas to the high temperatures needed to reproduce the observed [O I] luminosity. Gorti et al. (2011) considered FUV and X-rays in their input stellar spectrum and found that the [O I] lines are emitted from dissociation of OH with only a minor contribution from thermal emission.

Recent observational studies have focused on the study of optical forbidden lines (Baldovin-Saavedra et al. 2012, Rigliaco et al. 2013, Natta et al. 2014, Manara et al. 2014). Baldovin-Saavedra et al. (2012) obtained UVES (VLT) high-resolution spectra for three objects with a [Ne II] line detected from the ground. In two of the objects the optical and the mid-infrared tracers display the same velocity shift and FWHM (see Fig. 14), and in one case the [Ne II] line traces clearly jet emission, while the [O I] lines show two velocity components. Rigliaco et al. (2013) performed a systematic study of the [O I] low-velocity component and compared it to the [Ne II] and CO lines. They find a trend between the line peak velocities, where the [Ne II] line is typically more blueshifted than [O I] which in turn is more blueshifted than CO. The authors obtained a small range in the line ratio of [O I] 6300/5577 Å which they claimed is difficult to reproduce in gas heated by thermal processes.

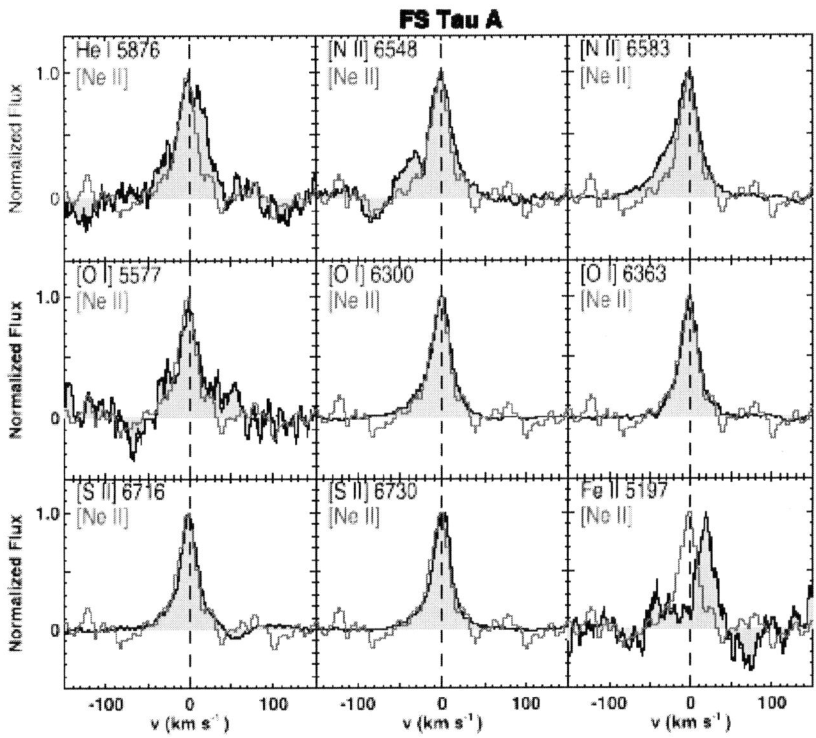

Figure 14. Comparison between the [Ne II] line (12.81 μm) detected with VISIR (VLT) and the optical lines detected with UVES (VLT) in the young system FS Tau A. In this case the line centroids of the mid-infrared tracer are in good agreement with the optical forbidden lines (Baldovin-Saavedra et al. 2012, ©ESO/A&A. Reproduced with permission.)

For two objects observed at very high spectral resolution the authors could decompose the line profile in two components: a broad and symmetric component attributed to disk emission and a narrower and blueshifted component attributed to unbound gas. Natta et al. (2014) found a correlation between the [O I] luminosity and both L⋆L⋆ and

LaccLacc, the latter being related to the stellar FUV emission. They found that, on average, the low-velocity component peaks at <20 kms−1<20 kms−1 and determined the physical conditions of the gas, finding high density (nH>108 cm−3nH>108 cm−3) and temperatures (T~5000--1000 KT~5000--1000 K). The authors warn that the absence of correlation between the luminosity of the [O I] line and LXLX, contradicts an X-ray driven disk wind.

8.2 Molecular Tracers

CO ro-vibrational emission at 4.7 μm is detected in many T Tauri stars. The line probes warm gas in the inner 1 au from the central star. Some of the observed lines show a singly peaked and narrow profile with a broad base that extends to >50 kms−1>50 kms−1; they show a small velocity shift of ~5 kms−1~5 kms−1 that cannot be explained by purely Keplerian rotation (e.g., Pontoppidan et al. 2011, Bast et al. 2011, Brown et al. 2013). Pontoppidan et al. (2011) performed spectroastrometry observations of the CO line. This technique allows to obtain spatial and kinematical information at very small scales and very high spectral resolution. In their study, they find that the single-peaked line profiles are consistent with a combination of gas in Keplerian rotation and a non-collimated disk wind. In addition, Brown et al. (2013) detected excess emission on the blue side of the line profile, giving further confirmation for the disk wind hypothesis.

Molecular hydrogen, H2H2, is the most abundant component in protoplanetary disks, but this homonuclear molecule is very difficult to observe since its rotational quadrupole transitions from low-energy levels are weak and lie in regions with poor atmospheric transmission. The ro-vibrational transitions observable in the near-infrared trace thermal emission of hot gas at about 1000 K, or gas excited by UV or X-rays (Carmona et al. 2007, 2011). Takami et al. (2004) reported the detection of a slow molecular wind in H2H2 for DG Tau. Baldovin-Saavedra et al. (in prep.) performed follow-up observations with CRIRES (VLT) of young systems with detections of [Ne II]. The program aimed at detecting the ro-vibrational transitions of H_2: 1–0 S(1)(1)(2.12 μm), 1–0 S(1)(1) (2.22 μm), and 2–1 S(1)(1) (2.25 μm). Emission from the three transitions is detected in three objects, two of them known to drive high-velocity jets. Two of the sources have line profiles consistent with emission from the disk atmosphere (no shift), and one shows a blueshift of ~−8 kms−1~−8 kms−1. The other four objects show detections from the 1–0 S(0)(0) transition only; two of them can be interpreted in terms of gas bound to the disk, and two present lines with small blueshifts (<−10 kms−1<−10 kms−1). Figure 15 (Baldovin-Saavedra et al. in prep.) shows a comparison between the H_2 1–0 S(1)(1) line and the [Ne II] line detected in the mid-infrared and published in Baldovin-Saavedra et al. (2012) for one of the stars in the sample. The spectra have been continuum subtracted and the lines normalized to their peak to allow for a better comparison. In this case the [Ne II] line is slightly blueshifted, vpeak=−2.9±0.7 kms−1vpeak=−2.9±0.7 kms−1 with a FWHM of 26.8 km s^{-1} and shows excess emission towards the blue side of the spectrum, while the H2H2 line peak is consistent with disk emission, vpeak=−0.1±0.2 kms−1vpeak=−0.1±0.2 kms−1 and FWHM=21.2±0.2 kms−1FWHM=21.2±0.2 kms−1. It is interesting to estimate the distance from the star needed to observe a line in Keplerian rotation that does not show a double peaked profile. Considering the inclination of the system, i=35∘i=35∘, the mass of the star M⋆=0.6 M⊙M⋆=0.6 M⊙, and the CRIRES spectral resolution of 3 km s^{-1}, the gas should be located at a distance R>77 auR>77 au.

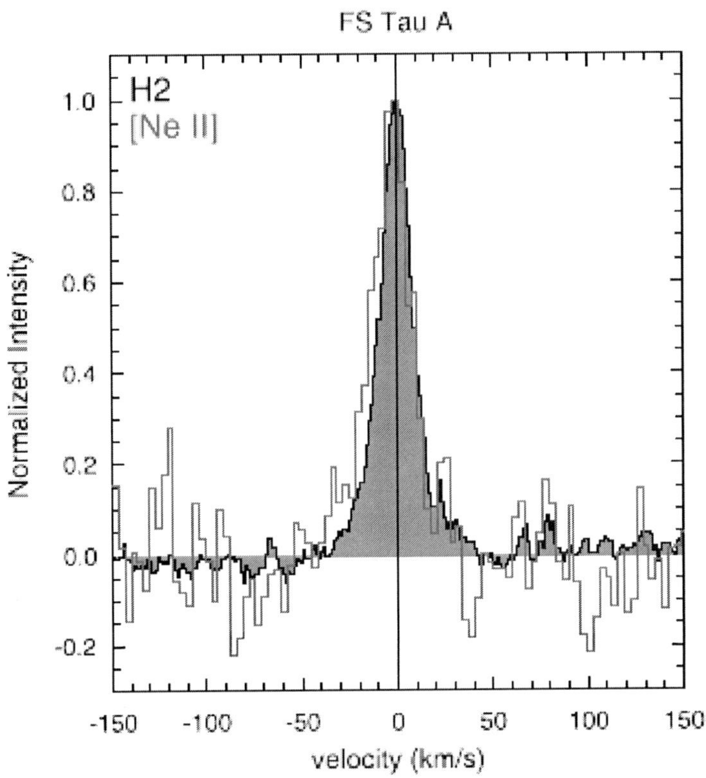

Figure 15. The *grey area* represents the H2H2 1–0 S(1) line at 2.12 μm detected with CRIRES and the *solid line* represents the [Ne II] line (12.81 μm) detected with VISIR in FS Tau A (Baldovin-Saavedra et al. in prep.)

The trend observed in H_2 is similar to what is observed in CO; the molecular lines tend to be less blueshifted than the [Ne II] line. However, at this stage it is difficult to establish a direct relation between the observed [Ne II] photoevaporative disk winds and their molecular counterpart. Furthermore, the nature of the molecular winds, i.e., magnetically driven or photoevaporation driven, remain uncertain.

9 SUMMARY

We discussed several physical processes which have significant impact on the evolution and the chemical composition of the gas disk. In early stages the disk evolution might be dominated by gravitational instabilities if the initial gas mass is high enough. At this stage most disks probably experience phases of highly increased accretion rates resulting in strong luminosity outbursts. In later phases the disk evolution is dominated by viscous accretion possibly driven by magneto-rotational instabilities. In addition disk winds may play a crucial role in the dispersal of the disk determining the time-scale for planet formation. One example are photo-evaporative winds which are driven by high energy stellar radiation of the star (see also Chap. 6 by Gorti et al.).

Chemical processes in the gas disks determine the chemical composition of planetary atmospheres but also have an impact on the solids due to freeze-out of molecules. Enhanced dust growth at the position of ice lines le to the formation of larger solid bodies which possibly grow further to planetesimals and planetary cores. The gas mass is another crucial parameter for planet formation. It is still challenging to measure directly the disk gas mass. Using the CO molecular line emission requires detailed chemical models (e.g. CO isotopologue chemistry and surface chemistry) and therefore the derived masses are still model dependent. The HD molecule offers a more direct method but is currently only detected in one disk. Upcoming missions like the James Webb Space Telescope will certainly improve this situation.

The high spatial resolution and sensitivity of ALMA provides detailed observations of the chemical composition of disks (e.g. detection of complex molecules, ice line locations). Future ALMA observations will certainly significantly improve our knowledge of disk chemistry. Interpretations of these observations are a formidable challenge for chemical models. Models consistently treating the dust and gas evolution are necessary. The incorporation of chemical processes in hydrodynamical models is required for a better understanding of the time evolution of the gaseous component of disks. Further detailed knowledge of the evolution of young stars is required as in particular high-energy stellar radiation is one of the main drivers of disk chemistry and evolution.

FOOTNOTES

[1] These data (and those provided in the subsequent sections) were obtained from numerical hydrodynamics simulations of collapsing pre-stellar cores using the method described in detail in Vorobyov and Basu (2015).

ACKNOWLEDGEMENTS

We want to thank the anonymous referee for a thorough and constructive report. We are grateful to the organizers of this International Space Science Institute (ISSI) workshop and the editors of this book. C. Rab and M. Güdel acknowledge funding by the Austrian Science Fund (FWF), project number P24790 and project S11601-N16 "Pathways to Habitability: From Disks to Active Stars, Planets and Life". C. Baldovin-Saavedra acknowledges funding by the Austrian Research Promotion Agency (FFG) under grant agreement FA 538022. E. Vorobyov acknowledges support from the RFBR grant 14-02-00719. The research leading to these results has received funding from the European Union Seventh Framework Programme FP7-2011 under grant agreement no 284405. This work was partly supported by the Austrian Science Fund (FWF) under research grant I2549-N27. This publication was partly supported from the FFG ASAP 12 project *JetPro** (FFG-854025). The computational results presented have been achieved in part using the Vienna Scientific Cluster (VSC).

REFERENCES

1. Y. Aikawa, T. Umebayashi, T. Nakano, S. Miyama, Faraday Discuss. **109**, 281 (1998)
2. Y. Aikawa, G.J. van Zadelhoff, E.F. van Dishoeck, E. Herbst, Astron. Astrophys. **386**, 622–632 (2002)
3. Y. Aikawa, K. Furuya, H. Nomura, C. Qi, Astrophys. J. **807**, 120 (2015)
4. V. Akimkin, S. Zhukovska, D. Wiebe, D. Semenov, Y. Pavlyuchenkov, A. Vasyunin, T. Birnstiel, T. Henning, Astrophys. J. **766**, 8 (2013)
5. R.D. Alexander, C.J. Clarke, J.E. Pringle, Mon. Not. R. Astron. Soc. **369**, 216–228 (2006a)
6. R.D. Alexander, C.J. Clarke, J.E. Pringle, Mon. Not. R. Astron. Soc. **369**, 229–239 (2006b)
7. R. Alexander, I. Pascucci, S. Andrews, P. Armitage, L. Cieza, in *Protostars and Planets VI*(2014), pp. 475–496

8. ALMA Partnership, C.L. Brogan, L.M. Pérez, T.R. Hunter, W.R.F. Dent, A.S. Hales, R.E. Hills, S. Corder, E.B. Fomalont, C. Vlahakis, Y. Asaki, D. Barkats, A. Hirota, J.A. Hodge, C.M.V. Impellizzeri, R. Kneissl, E. Liuzzo, R. Lucas, N. Marcelino, S. Matsushita, K. Nakanishi, N. Phillips, A.M.S. Richards, I. Toledo, R. Aladro, D. Broguiere, J.R. Cortes, P.C. Cortes, D. Espada, F. Galarza, D. Garcia-Appadoo, L. Guzman-Ramirez, E.M. Humphreys, T. Jung, S. Kameno, R.A. Laing, S. Leon, G. Marconi, A. Mignano, B. Nikolic, L.-A. Nyman, M. Radiszcz, A. Remijan, J.A. Rodón, T. Sawada, S. Takahashi, R.P.J. Tilanus, B. Vila Vilaro, L.C. Watson, T. Wiklind, E. Akiyama, E. Chapillon, I. de Gregorio-Monsalvo, J. Di Francesco, F. Gueth, A. Kawamura, C.-F. Lee, Q. Nguyen Luong, J. Mangum, V. Pietu, P. Sanhueza, K. Saigo, S. Takakuwa, C. Ubach, T. van Kempen, A. Wootten, A. Castro-Carrizo, H. Francke, J. Gallardo, J. Garcia, S. Gonzalez, T. Hill, T. Kaminski, Y. Kurono, H.-Y. Liu, C. Lopez, F. Morales, K. Plarre, G. Schieven, L. Testi, L. Videla, E. Villard, P. Andreani, J.E. Hibbard, K. Tatematsu, Astrophys. J. Lett. **808**, 3 (2015)
9. P. André, D. Ward-Thompson, J. Greaves, Science **337**, 69 (2012)
10. S.M. Andrews, J.P. Williams, Astrophys. J. **631**, 1134–1160 (2005)
11. S.M. Andrews, D.J. Wilner, C. Espaillat, A.M. Hughes, C.P. Dullemond, M.K. McClure, C. Qi, J.M. Brown, Astrophys. J. **732**, 42 (2011)
12. S.M. Andrews, D.J. Wilner, Z. Zhu, T. Birnstiel, J.M. Carpenter, L.M. Pérez, X.-N. Bai, K.I. Öberg, A.M. Hughes, A. Isella, L. Ricci, Astrophys. J. Lett. **820**, 40 (2016)
13. G. Aresu, I. Kamp, R. Meijerink, P. Woitke, W.-F. Thi, M. Spaans, Astron. Astrophys. **526**, 163 (2011)
14. G. Aresu, I. Kamp, R. Meijerink, M. Spaans, S. Vicente, L. Podio, P. Woitke, F. Menard, W.-F. Thi, M. Güdel, A. Liebhart, Astron. Astrophys. **566**, 14 (2014)
15. M. Audard, P. Ábrahám, M.M. Dunham, J.D. Green, N. Grosso, K. Hamaguchi, J.H. Kastner, Á. Kóspál, G. Lodato, M.M. Romanova, S.L. Skinner, E.I. Vorobyov, Z. Zhu, in *Protostars and Planets VI* (2014), pp. 387–410
16. S.A. Balbus, J.F. Hawley, Astrophys. J. **376**, 214–233 (1991)
17. C. Baldovin-Saavedra, M. Audard, M. Güdel, L.M. Rebull, D.L. Padgett, S.L. Skinner, A. Carmona, A.M. Glauser, S.B. Fajardo-Acosta, Astron. Astrophys. **528**, 22 (2011)
18. C. Baldovin-Saavedra, M. Audard, A. Carmona, M. Güdel, K. Briggs, L.M. Rebull, S.L. Skinner, B. Ercolano, Astron. Astrophys. **543**, 30 (2012)
19. J. Bally, E. Feigelson, B. Reipurth, Astrophys. J. **584**, 843–852 (2003)
20. C. Baruteau, A. Crida, S.-J. Paardekooper, F. Masset, J. Guilet, B. Bitsch, R. Nelson, W. Kley, J. Papaloizou, in *Protostars and Planets VI* (2014), pp. 667–689
21. J.E. Bast, J.M. Brown, G.J. Herczeg, E.F. van Dishoeck, K.M. Pontoppidan, Astron. Astrophys. **527**, 119 (2011)
22. S. Basu, E.I. Vorobyov, Astrophys. J. **750**, 30 (2012)
23. M.R. Bate, Mon. Not. R. Astron. Soc. **392**, 590–616 (2009)
24. S.V.W. Beckwith, A.I. Sargent, Astrophys. J. **402**, 280–291 (1993)
25. E.A. Bergin, Y. Aikawa, G.A. Blake, E.F. van Dishoeck, in *Protostars and Planets V* (2007), pp. 751–766.
26. E.A. Bergin, L.I. Cleeves, U. Gorti, K. Zhang, G.A. Blake, J.D. Green, S.M. Andrews, N.J. Evans II, T. Henning, K. Öberg, K. Pontoppidan, C. Qi, C. Salyk, E.F. van Dishoeck, Nature **493**, 644–646 (2013)
27. E.A. Bergin, L.I. Cleeves, N. Crockett, G.A. Blake, Faraday Discuss. **168**, 61–79 (2014). http://abs.harvard.edu/abs/2014FaDi..168...61B
28. B. Bitsch, A. Johansen, M. Lambrechts, A. Morbidelli, Astron. Astrophys. **575**, 28 (2015)
29. A.C. Boley, Astrophys. J. Lett. **695**, 53–57 (2009)
30. A.C. Boley, T. Hayfield, L. Mayer, R.H. Durisen, Icarus **207**, 509–516 (2010)
31. A.P. Boss, Science **276**, 1836–1839 (1997)
32. A.P. Boss, Astrophys. J. **731**, 74 (2011)
33. J.M. Brown, K.M. Pontoppidan, E.F. van Dishoeck, G.J. Herczeg, G.A. Blake, A. Smette, Astrophys. J. **770**, 94 (2013)
34. S. Bruderer, E.F. van Dishoeck, S.D. Doty, G.J. Herczeg, Astron. Astrophys. **541**, 91 (2012)
35. J.R. Burke, D.J. Hollenbach, Astrophys. J. **265**, 223–234 (1983)

36. S. Cabrit, constraints on accretion-ejection structures in young stars, in *EAS Publications Series*, ed. by J. Bouvier, J.-P. Zahn, EAS Publications Series, vol. 3, 2002, pp. 147–182
37. N. Calvet, E. Gullbring, Astrophys. J. **509**, 802–818 (1998) Carmona, Earth Moon Planets **106**, 71–95 (2010)
38. Carmona, M.E. van den Ancker, T. Henning, M. Goto, D. Fedele, B. Stecklum, Astron. Astrophys. **476**, 853 (2007).
39. Carmona, G. van der Plas, M.E. van den Ancker, M. Audard, L.B.F.M. Waters, D. Fedele, B. Acke, E. Pantin, Astron. Astrophys. **533**, 39 (2011)
40. Ceccarelli, C. Dominik, A. López-Sepulcre, M. Kama, M. Padovani, E. Caux, P. Caselli, Astrophys. J. Lett. **790**, 1 (2014)
41. G. Chaparro Molano, I. Kamp, Astron. Astrophys. **537**, 138 (2012a)
42. G. Chaparro Molano, I. Kamp, Astron. Astrophys. **547**, 7 (2012b)
43. E.I. Chiang, P. Goldreich, Astrophys. J. **490**, 368 (1997)
44. C.J. Clarke, A. Gendrin, M. Sotomayor, Mon. Not. R. Astron. Soc. **328**, 485–491 (2001)
45. L.I. Cleeves, F.C. Adams, E.A. Bergin, Astrophys. J. **772**, 5 (2013a)
46. L.I. Cleeves, F.C. Adams, E.A. Bergin, R. Visser, Astrophys. J. **777**, 28 (2013b)
47. L.I. Cleeves, E.A. Bergin, C. Qi, F.C. Adams, K.I. Öberg, Astrophys. J. **799**, 204 (2015)
48. O. Dionatos, Gas line observations of disks. Eur. Phys. J. Web Conf. **102**, 00008 (2015)
49. F. Du, E.A. Bergin, Astrophys. J. **792**, 2 (2014)
50. M.M. Dunham, E.I. Vorobyov, Astrophys. J. **747**, 52 (2012)
51. M.M. Dunham, E.I. Vorobyov, H.G. Arce, Mon. Not. R. Astron. Soc. **444**, 887–901 (2014)
52. M.M. Dunham, N.J. Evans II, S. Terebey, C.P. Dullemond, C.H. Young, Astrophys. J. **710**, 470–502 (2010)
53. Dutrey, S. Guilloteau, P. Ho, in *Protostars and Planets V* (2007), pp. 495–506
54. Dutrey, S. Guilloteau, V. Piétu, E. Chapillon, F. Gueth, T. Henning, R. Launhardt, Y. Pavlyuchenkov, K. Schreyer, D. Semenov, Astron. Astrophys. **490**, 15–18 (2008)
55. Dutrey, D. Semenov, E. Chapillon, U. Gorti, S. Guilloteau, F. Hersant, M. Hogerheijde, M. Hughes, G. Meeus, H. Nomura, V. Piétu, C. Qi, V. Wakelam, in *Protostars and Planets VI*(2014), pp. 317–338
56. S. Edwards, S. Cabrit, S.E. Strom, I. Heyer, K.M. Strom, E. Anderson, Astrophys. J. **321**, 473–495 (1987)
57. Ercolano, J.E. Owen, Mon. Not. R. Astron. Soc. **406**, 1553–1569 (2010)
58. Ercolano, G. Rosotti, Mon. Not. R. Astron. Soc. **450**, 3008–3014 (2015)
59. Ercolano, J.J. Drake, J.C. Raymond, C.C. Clarke, Astrophys. J. **688**, 398–407 (2008)
60. Espaillat, L. Ingleby, E. Furlan, M. McClure, A. Spatzier, J. Nieusma, N. Calvet, E. Bergin, L. Hartmann, J.M. Miller, J. Muzerolle, Astrophys. J. **762**, 62 (2013)
61. Espaillat, J. Muzerolle, J. Najita, S. Andrews, Z. Zhu, N. Calvet, S. Kraus, J. Hashimoto, A. Kraus, P. D'Alessio, in *Protostars and Planets VI* (2014), pp. 497–520
62. M.G. Evans, J.D. Ilee, A.C. Boley, P. Caselli, R.H. Durisen, T.W. Hartquist, J.M.C. Rawlings, Mon. Not. R. Astron. Soc. **453**, 1147–1163 (2015)
63. Favata, C.V.M. Fridlund, G. Micela, S. Sciortino, A.A. Kaas, Astron. Astrophys. **386**, 204–210 (2002)
64. Favata, G. Micela, B. Silva, S. Sciortino, M. Tsujimoto, Astron. Astrophys. **433**, 1047–1054 (2005)
65. Favre, L.I. Cleeves, E.A. Bergin, C. Qi, G.A. Blake, Astrophys. J. Lett. **776**, 38 (2013)
66. Fedele, M.E. van den Ancker, T. Henning, R. Jayawardhana, J.M. Oliveira, Astron. Astrophys. **510**, 72 (2010)
67. E.D. Feigelson, G.P. Garmire, S.H. Pravdo, Astrophys. J. **572**, 335–349 (2002)
68. Flaccomio, B. Stelzer, S. Sciortino, G. Micela, I. Pillitteri, L. Testi, Astron. Astrophys. **505**, 695–706 (2009)
69. A.S. Font, I.G. McCarthy, D. Johnstone, D.R. Ballantyne, Astrophys. J. **607**, 890–903 (2004)
70. Frank, T.P. Ray, S. Cabrit, P. Hartigan, H.G. Arce, F. Bacciotti, J. Bally, M. Benisty, J. Eislöffel, M. Güdel, S. Lebedev, B. Nisini, A. Raga, in *Protostars and Planets VI* (2014), pp. 451–474
71. S. Fromang, C. Terquem, S.A. Balbus, Mon. Not. R. Astron. Soc. **329**, 18–28 (2002)
72. K. Furuya, Y. Aikawa, Astrophys. J. **790**, 97 (2014)
73. C.F. Gammie, Astrophys. J. **457**, 355 (1996)

74. P.G. Gibbons, G.R. Mamatsashvili, W.K.M. Rice, Mon. Not. R. Astron. Soc. **453**, 4232–4243 (2015)
75. A.E. Glassgold, D. Galli, M. Padovani, Astrophys. J. **756**, 157 (2012)
76. A.E. Glassgold, J. Najita, J. Igea, Astrophys. J. **480**, 344–350 (1997)
77. A.E. Glassgold, J. Najita, J. Igea, Astrophys. J. **615**, 972–990 (2004)
78. U. Gorti, D. Hollenbach, Astrophys. J. **613**, 424–447 (2004)
79. U. Gorti, D. Hollenbach, Astrophys. J. **683**, 287–303 (2008)
80. U. Gorti, D. Hollenbach, J. Najita, I. Pascucci, Astrophys. J. **735**, 90 (2011)
81. R. Gredel, S. Lepp, A. Dalgarno, E. Herbst, Astrophys. J. **347**, 289–293 (1989)
82. M. Güdel, Annu. Rev. Astron. Astrophys. **12**, 71–237 (2004)
83. M. Güdel, A. Telleschi, Astron. Astrophys. **474**, 25–28 (2007)
84. M. Güdel, S.L. Skinner, M. Audard, K.R. Briggs, S. Cabrit, Astron. Astrophys. **478**, 797–807 (2008)
85. M. Güdel, F. Lahuis, K.R. Briggs, J. Carr, A.E. Glassgold, T. Henning, J.R. Najita, R. van Boekel, E.F. van Dishoeck, Astron. Astrophys. **519**, 113 (2010)
86. Hamann, Astrophys. J. Suppl. Ser. **93**, 485–518 (1994)
87. N. Harada, E. Herbst, V. Wakelam, Astrophys. J. **721**, 1570–1578 (2010)
88. P. Hartigan, S. Edwards, L. Ghandour, Astrophys. J. **452**, 736 (1995)
89. Helling, P. Woitke, P.B. Rimmer, I. Kamp, W.-F. Thi, R. Meijerink, Life **4**(2), 142–173 (2014)
90. T. Henning, D. Semenov, Chem. Rev. **113**, 9016–9042 (2013)
91. G.J. Herczeg, J.R. Najita, L.A. Hillenbrand, I. Pascucci, Astrophys. J. **670**, 509–515 (2007)
92. J. Hernández, L. Hartmann, N. Calvet, R.D. Jeffries, R. Gutermuth, J. Muzerolle, J. Stauffer, Astrophys. J. **686**, 1195–1208 (2008)
93. D.J. Hollenbach, A.G.G.M. Tielens, Rev. Mod. Phys. **71**, 173–230 (1999)
94. Hollenbach, D. Johnstone, S. Lizano, F. Shu, Astrophys. J. **428**, 654–669 (1994)
95. J. Igea, A.E. Glassgold, Astrophys. J. **518**, 848–858 (1999)
96. J.D. Ilee, A.C. Boley, P. Caselli, R.H. Durisen, T.W. Hartquist, J.M.C. Rawlings, Mon. Not. R. Astron. Soc. **417**, 2950–2961 (2011)
97. K. Imanishi, K. Koyama, Y. Tsuboi, Astrophys. J. **557**, 747–760 (2001)
98. Jankovics, I. Appenzeller, J. Krautter, Publ. Astron. Soc. Pac. **95**, 883–885 (1983)
99. J.K. Jørgensen, R. Visser, N. Sakai, E.A. Bergin, C. Brinch, D. Harsono, J.E. Lindberg, E.F. van Dishoeck, S. Yamamoto, S.E. Bisschop, M.V. Persson, Astrophys. J. Lett. **779**, 22 (2013)
100. Kamp, G.-J. van Zadelhoff, Astron. Astrophys. **373**, 641–656 (2001)
101. Kamp, I. Tilling, P. Woitke, W. Thi, M. Hogerheijde, Astron. Astrophys. **510**(26), 260000 (2010)
102. J.H. Kastner, D.P. Huenemoerder, N.S. Schulz, C.R. Canizares, D.A. Weintraub, Astrophys. J. **567**, 434–440 (2002)
103. S.J. Kenyon, L.W. Hartmann, K.M. Strom, S.E. Strom, Astron. J. **99**, 869–887 (1990)
104. H.J. Kim, N.J. Evans II, M.M. Dunham, J.-E. Lee, K.M. Pontoppidan, Astrophys. J. **758**, 38 (2012)
105. K.M. Kratter, R.A. Murray-Clay, Astrophys. J. **740**, 1 (2011)
106. K.M. Kratter, C.D. Matzner, M.R. Krumholz, Astrophys. J. **681**, 375–390 (2008)
107. A.J. Kruger, M.J. Richter, J.S. Carr, J.R. Najita, M.M. Moerchen, G.W. Doppmann, A. Seifahrt, Astrophys. J. **764**, 127 (2013)
108. Kwan, E. Tademaru, Astrophys. J. **332**, L41–L44 (1988).
109. Lahuis, E.F. van Dishoeck, G.A. Blake, N.J. Evans II, J.E. Kessler-Silacci, K.M. Pontoppidan, Astrophys. J. **665**, 492–511 (2007)
110. J.-E. Lee, J. Korean Astron. Soc. **40**, 83–89 (2007)
111. Z.-Y. Li, R. Banerjee, R.E. Pudritz, J.K. Jørgensen, H. Shang, R. Krasnopolsky, A. Maury, in *Protostars and Planets VI* (2014), pp. 173–194
112. M.N. Machida, S.-i. Inutsuka, T. Matsumoto, Astrophys. J. **724**, 1006–1020 (2010)
113. M.N. Machida, S.-i. Inutsuka, T. Matsumoto, Astrophys. J. **729**, 42 (2011)
114. E.E. Mamajek, Initial conditions of planet formation: lifetimes of primordial disks, in *American Institute of Physics Conference Series*, ed. by T. Usuda, M. Tamura, M. Ishii, American Institute of Physics Conference Series, vol. 1158 (2009), pp. 3–10

115. C.F. Manara, L. Testi, A. Natta, G. Rosotti, M. Benisty, B. Ercolano, L. Ricci, Astron. Astrophys. **568**, 18 (2014)
116. B.C. Matthews, A.V. Krivov, M.C. Wyatt, G. Bryden, C. Eiroa, in *Protostars and Planets VI*(2014), pp. 521–544
117. D. McElroy, C. Walsh, A.J. Markwick, M.A. Cordiner, K. Smith, T.J. Millar, Astron. Astrophys. **550**, 36 (2013)
118. R. Meijerink, G. Aresu, I. Kamp, M. Spaans, W.-F. Thi, P. Woitke, Astron. Astrophys. **547**, 68 (2012)
119. Meru, M.R. Bate, Mon. Not. R. Astron. Soc. **427**, 2022–2046 (2012)
120. Miotello, S. Bruderer, E.F. van Dishoeck, Astron. Astrophys. **572**, 96 (2014)
121. N.M. Murillo, S.-P. Lai, S. Bruderer, D. Harsono, E.F. van Dishoeck, Astron. Astrophys. **560**, 103 (2013)
122. J.R. Najita, J.S. Carr, A.E. Glassgold, J.A. Valenti, in *Protostars and Planets V* (2007), pp. 507–522
123. J.R. Najita, G.W. Doppmann, M.A. Bitner, M.J. Richter, J.H. Lacy, D.T. Jaffe, J.S. Carr, R. Meijerink, G.A. Blake, G.J. Herczeg, A.E. Glassgold, Astrophys. J. **697**, 957–963 (2009)
124. Natta, L. Testi, J.M. Alcalá, E. Rigliaco, E. Covino, B. Stelzer, V. D'Elia, Astron. Astrophys. **569**, 5 (2014)
125. S. Nayakshin, Mon. Not. R. Astron. Soc. **408**, 36–40 (2010)
126. S. Nayakshin, Mon. Not. R. Astron. Soc. **413**, 1462–1478 (2011)
127. S. Nayakshin, S.-H. Cha, J.C. Bridges, Mon. Not. R. Astron. Soc. **416**, 50–54 (2011)
128. Nomura, T.J. Millar, Astron. Astrophys. **438**, 923–938 (2005)
129. K.I. Öberg, C. Qi, J.K.J. Fogel, E.A. Bergin, S.M. Andrews, C. Espaillat, D.J. Wilner, I. Pascucci, J.H. Kastner, Astrophys. J. **734**, 98 (2011)
130. K.I. Öberg, K. Furuya, R. Loomis, Y. Aikawa, S.M. Andrews, C. Qi, E.F. van Dishoeck, D.J. Wilner, Astrophys. J. **810**, 112 (2015)
131. N. Ohashi, K. Saigo, Y. Aso, Y. Aikawa, S. Koyamatsu, M.N. Machida, M. Saito, S.Z. Takahashi, S. Takakuwa, K. Tomida, K. Tomisaka, H.-W. Yen, Astrophys. J. **796**, 131 (2014)
132. J.E. Owen, B. Ercolano, C.J. Clarke, Mon. Not. R. Astron. Soc. **412**, 13–25 (2011)
133. J.E. Owen, B. Ercolano, C.J. Clarke, R.D. Alexander, Mon. Not. R. Astron. Soc. **401**, 1415–1428 (2010)
134. Padovani, P. Hennebelle, A. Marcowith, K. Ferrière, Astron. Astrophys. **582**, 13 (2015)
135. Palau, L.A. Zapata, L.F. Rodríguez, H. Bouy, D. Barrado, M. Morales-Calderón, P.C. Myers, N. Chapman, C. Juárez, D. Li, Mon. Not. R. Astron. Soc. **444**, 833–845 (2014)
136. Pascucci, M. Sterzik, Astrophys. J. **702**, 724–732 (2009)
137. Pascucci, D. Hollenbach, J. Najita, J. Muzerolle, U. Gorti, G.J. Herczeg, L.A. Hillenbrand, J.S. Kim, J.M. Carpenter, M.R. Meyer, E.E. Mamajek, J. Bouwman, Astrophys. J. **663**, 383–393 (2007)
138. Pascucci, M. Sterzik, R.D. Alexander, S.H.P. Alencar, U. Gorti, D. Hollenbach, J. Owen, B. Ercolano, S. Edwards, Astrophys. J. **736**, 13 (2011)
139. Pascucci, L. Ricci, U. Gorti, D. Hollenbach, N.P. Hendler, K.J. Brooks, Y. Contreras, Astrophys. J. **795**, 1 (2014)
140. L.M. Pérez, A. Isella, J.M. Carpenter, C.J. Chandler, Astrophys. J. Lett. **783**, 13 (2014a)
141. L.M. Pérez, A. Isella, J.M. Carpenter, C.J. Chandler, Astrophys. J. Lett. **783**, 13 (2014b)
142. V. Piétu, A. Dutrey, S. Guilloteau, Astron. Astrophys. **467**, 163–178 (2007)
143. V. Piétu, S. Guilloteau, A. Dutrey, Astron. Astrophys. **443**, 945–954 (2005)
144. Pinte, W.R.F. Dent, F. Ménard, A. Hales, T. Hill, P. Cortes, I. de Gregorio-Monsalvo, Astrophys. J. **816**, 25 (2016)
145. K.M. Pontoppidan, S.M. Blevins, Faraday Discuss. **169**, 49–60 (2014)
146. K.M. Pontoppidan, G.A. Blake, A. Smette, Astrophys. J. **733**, 84 (2011)
147. S.H. Pravdo, E.D. Feigelson, G. Garmire, Y. Maeda, Y. Tsuboi, J. Bally, Nature **413**, 708–711 (2001)
148. Prisinzano, G. Micela, E. Flaccomio, J.R. Stauffer, T. Megeath, L. Rebull, M. Robberto, K. Smith, E.D. Feigelson, N. Grosso, S. Wolk, Astrophys. J. **677**, 401–424 (2008)
149. K.M. Punzi, P. Hily-Blant, J.H. Kastner, G.G. Sacco, T. Forveille, Astrophys. J. **805**, 147 (2015)

150. Qi, K.I. Öberg, D.J. Wilner, P. D'Alessio, E. Bergin, S.M. Andrews, G.A. Blake, M.R. Hogerheijde, E.F. van Dishoeck, Science **341**, 630–632 (2013)
151. Qi, K.I. Öberg, S.M. Andrews, D.J. Wilner, E.A. Bergin, A.M. Hughes, M. Hogherheijde, P. D'Alessio, Astrophys. J. **813**, 128 (2015)
152. R.R. Rafikov, Astrophys. J. Lett. **621**, 69–72 (2005)
153. S.N. Raymond, E. Kokubo, A. Morbidelli, R. Morishima, K.J. Walsh, in *Protostars and Planets VI* (2014), pp. 595–618
154. Reboussin, V. Wakelam, S. Guilloteau, F. Hersant, A. Dutrey, Astron. Astrophys. **579**, 82 (2015)
155. Z. Regály, A. Juhász, Z. Sándor, C.P. Dullemond, Mon. Not. R. Astron. Soc. **419**, 1701–1712 (2012)
156. W.K.M. Rice, G. Lodato, J.E. Pringle, P.J. Armitage, I.A. Bonnell, Mon. Not. R. Astron. Soc. **355**, 543–552 (2004)
157. Rigliaco, I. Pascucci, U. Gorti, S. Edwards, D. Hollenbach, Astrophys. J. **772**, 60 (2013)
158. P.D. Rogers, J. Wley, Mon. Not. R. Astron. Soc. **423**, 1896–1908 (2012)
159. G.G. Sacco, E. Flaccomio, I. Pascucci, F. Lahuis, B. Ercolano, J.H. Kastner, G. Micela, B. Stelzer, M. Sterzik, Astrophys. J. **747**, 142 (2012)
160. S. Sallum, K.B. Follette, J.A. Eisner, L.M. Close, P. Hinz, K. Kratter, J. Males, A. Skemer, B. Macintosh, P. Tuthill, V. Bailey, D. Defrère, K. Morzinski, T. Rodigas, E. Spalding, A. Vaz, A.J. Weinberger, Nature **527**, 342–344 (2015)
161. C. Salyk, K. Pontoppidan, S. Corder, D. Muñoz, K. Zhang, G.A. Blake, Astrophys. J. **792**, 68 (2014)
162. K.R. Schwarz, E.A. Bergin, L.I. Cleeves, G.A. Blake, K. Zhang, K.I. Öberg, E.F. van Dishoeck, C. Qi, Astrophys. J. **823**, 91 (2016)
163. D. Semenov, D. Wiebe, Astrophys. J. Suppl. Ser. **196**, 25 (2011)
164. D. Semenov, D. Wiebe, T. Henning, Astron. Astrophys. **417**, 93–106 (2004)
165. D. Semenov, D. Wiebe, T. Henning, Astrophys. J. Lett. **647**, 57–60 (2006)
166. F.H. Shu, Astrophys. J. **214**, 488–497 (1977)
167. Spitzer Jr., M.G. Tomasko, Astrophys. J. **152**, 971 (1968)
168. D. Stamatellos, A.P. Whitworth, Mon. Not. R. Astron. Soc. **392**, 413–427 (2009)
169. D. Stamatellos, A.P. Whitworth, D.A. Hubber, Astrophys. J. **730**, 32 (2011)
170. B. Stelzer, E. Flaccomio, K. Briggs, G. Micela, L. Scelsi, M. Audard, I. Pillitteri, M. Güdel, Astron. Astrophys. **468**, 463–475 (2007)
171. S. Takakuwa, M. Saito, J. Lim, K. Saigo, T.K. Sridharan, N.A. Patel, Astrophys. J. **754**, 52 (2012)
172. Takami, A. Chrysostomou, T.P. Ray, C. Davis, W.R.F. Dent, J. Bailey, M. Tamura, H. Terada, Astron. Astrophys. **416**, 213–219 (2004)
173. R. Teague, D. Semenov, S. Guilloteau, T. Henning, A. Dutrey, V. Wakelam, E. Chapillon, V. Pietu, Astron. Astrophys. **574**, 137 (2015)
174. Telleschi, M. Güdel, K.R. Briggs, M. Audard, F. Palla, Astron. Astrophys. **468**, 425–442 (2007)
175. Testi, T. Birnstiel, L. Ricci, S. Andrews, J. Blum, J. Carpenter, C. Dominik, A. Isella, A. Natta, J.P. Williams, D.J. Wilner, in *Protostars and Planets VI* (2014), pp. 339–361
176. W.-F. Thi, P. Woitke, I. Kamp, Mon. Not. R. Astron. Soc. **412**, 711–726 (2011)
177. Tilling, P. Woitke, G. Meeus, A. Mora, B. Montesinos, P. Riviere-Marichalar, C. Eiroa, W.-F. Thi, A. Isella, A. Roberge, C. Martin-Zaidi, I. Kamp, C. Pinte, G. Sandell, W.D. Vacca, F. Ménard, I. Mendigutía, G. Duchêne, W.R.F. Dent, G. Aresu, R. Meijerink, M. Spaans, Astron. Astrophys. **538**, 20 (2012)
178. J.J. Tobin, L. Hartmann, H.-F. Chiang, D.J. Wilner, L.W. Looney, L. Loinard, N. Calvet, P. D'Alessio, Nature **492**, 83–85 (2012)
179. J.J. Tobin, L. Hartmann, H.-F. Chiang, D.J. Wilner, L.W. Looney, L. Loinard, N. Calvet, P. D'Alessio, Astrophys. J. **771**, 48 (2013)
180. Tsujimoto, E.D. Feigelson, N. Grosso, G. Micela, Y. Tsuboi, F. Favata, H. Shang, J.H. Kastner, Astrophys. J. Suppl. Ser. **160**, 503–510 (2005)
181. N.J. Turner, S. Fromang, C. Gammie, H. Klahr, G. Lesur, M. Wardle, X.-N. Bai, in *Protostars and Planets VI* (2014), pp. 411–432

182. R. van Boekel, M. Güdel, T. Henning, F. Lahuis, E. Pantin, Astron. Astrophys. **497**, 137–144 (2009)
183. van der Marel, E.F. van Dishoeck, S. Bruderer, T. Birnstiel, P. Pinilla, C.P. Dullemond, T.A. van Kempen, M. Schmalzl, J.M. Brown, G.J. Herczeg, G.S. Mathews, V. Geers, Science **340**, 1199–1202 (2013)
184. van der Marel, P. Pinilla, J. Tobin, T. van Kempen, S. Andrews, L. Ricci, T. Birnstiel, Astrophys. J. Lett. **810**, 7 (2015)
185. van der Plas, S. Casassus, F. Ménard, S. Perez, W.F. Thi, C. Pinte, V. Christiaens, Astrophys. J. Lett. **792**, 25 (2014)
186. A.I. Vasyunin, D.S. Wiebe, T. Birnstiel, S. Zhukovska, T. Henning, C.P. Dullemond, Astrophys. J. **727**, 76 (2011)
187. R. Visser, E.A. Bergin, Astrophys. J. Lett. **754**, 18 (2012)
188. R. Visser, E.A. Bergin, J.K. Jørgensen, Astron. Astrophys. **577**, 102 (2015)
189. R. Visser, E.F. van Dishoeck, J.H. Black, Astron. Astrophys. **503**, 323–343 (2009)
190. E.I. Vorobyov, Astrophys. J. **723**, 1294–1307 (2010)
191. E.I. Vorobyov, Astrophys. J. Lett. **728**, 45 (2011a)
192. E.I. Vorobyov, Astrophys. J. **729**, 146 (2011b)
193. E.I. Vorobyov, Astron. Astrophys. **552**, 129 (2013)
194. E.I. Vorobyov, Astron. Astrophys. **590**, A115 (2016).
195. E.I. Vorobyov, S. Basu, Astrophys. J. **650**, 956–969 (2006)
196. E.I. Vorobyov, S. Basu, Astrophys. J. Lett. **714**, 133–137 (2010a)
197. E.I. Vorobyov, S. Basu, Astrophys. J. **719**, 1896–1911 (2010b)
198. E.I. Vorobyov, S. Basu, Astrophys. J. **805**, 115 (2015)
199. E.I. Vorobyov, I. Baraffe, T. Harries, G. Chabrier, Astron. Astrophys. **557**, 35 (2013)
200. V. Wakelam, J.-C. Loison, E. Herbst, B. Pavone, A. Bergeat, K. Béroff, M. Chabot, A. Faure, D. Galli, W.D. Geppert, D. Gerlich, P. Gratier, N. Harada, K.M. Hickson, P. Honvault, S.J. Klippenstein, S.D. Le Picard, G. Nyman, M. Ruaud, S. Schlemmer, I.R. Sims, D. Talbi, J. Tennyson, R. Wester, Astrophys. J. Suppl. Ser. **217**, 20 (2015)
201. C. Walsh, T.J. Millar, H. Nomura, Astrophys. J. **722**, 1607–1623 (2010)
202. C. Walsh, H. Nomura, T.J. Millar, Y. Aikawa, Astrophys. J. **747**, 114 (2012)
203. C. Walsh, E. Herbst, H. Nomura, T.J. Millar, S.W. Weaver, Faraday Discuss. **168**, 389 (2014)
204. C. Walsh, R.A. Loomis, K.I. Öberg, M. Kama, M.L.R. van 't Hoff, T.J. Millar, Y. Aikawa, E. Herbst, S.L.W. Weaver, H. Nomura, Astrophys. J. Lett. **823**(1), 10 (2016). http://stacks.iop.org/2041-8205/823/i=1/a=L10
205. J.P. Williams, W.M.J. Best, Astrophys. J. **788**, 59 (2014)
206. J.P. Williams, L.A. Cieza, Annu. Rev. Astron. Astrophys. **49**, 67–117 (2011)
207. Woitke, I. Kamp, W.-F. Thi, Astron. Astrophys. **501**, 383–406 (2009)
208. Woitke, M. Min, C. Pinte, W.-F. Thi, I. Kamp, C. Rab, F. Anthonioz, S. Antonellini, C. Baldovin-Saavedra, A. Carmona, C. Dominik, O. Dionatos, J. Greaves, M. Güdel, J.D. Ilee, A. Liebhart, F. Ménard, L. Rigon, L.B.F.M. Waters, G. Aresu, R. Meijerink, M. Spaans, Astron. Astrophys. **586**, 103 (2016)
209. Zhang, G.A. Blake, E.A. Bergin, Astrophys. J. Lett. **806**, 7 (2015)
210. Z. Zhu, L. Hartmann, N. Calvet, J. Hernandez, J. Muzerolle, A.-K. Tannirkulam, Astrophys. J. **669**, 483–492 (2007)
211. Z. Zhu, L. Hartmann, R.P. Nelson, C.F. Gammie, Astrophys. J. **746**, 110 (2012)

CHAPTER 4

Grain Surface Models and Data for Astrochemistry

H.M. Cuppen1, C. Walsh2,3, T. Lamberts1,4, D. Semenov5, R.T. Garrod6, E.M. Penteado1, S. Ioppolo7

1 Institute for Molecules and Materials, Radboud University Nijmegen, Heyendaalseweg 135, 6525 AJ Nijmegen, The Netherlands

2 Leiden Observatory, Leiden University, PO Box 9513, 2300 RA Leiden, The Netherlands

3 School of Physics and Astronomy, University of Leeds, Leeds LS2 9JT, UK

4 Computational Chemistry Group, Institute for Theoretical Chemistry, University of Stuttgart, Pfaffenwaldring 55, 70569 Stuttgart, Germany

5 Max Planck Institute for Astronomy, Königstuhl 17, 69117 Heidelberg, Germany

6 Depts. of Astronomy & Chemistry, University of Virginia, McCormick Road, PO Box 400319, Charlottesville, VA 22904, USA

7 Department of Physical Sciences, The Open University, Walton Hall, Milton Keynes MK7 6AA, UK

ABSTRACT

The cross-disciplinary field of astrochemistry exists to understand the formation, destruction, and survival of molecules in astrophysical environments. Molecules in space are synthesized via a large variety of gas-phase reactions, and reactions on dust-grain surfaces, where the surface acts as a catalyst. A broad consensus has been reached in the astrochemistry community on how to suitably treat gas-phase processes in models, and also on how to present the necessary reaction data in databases; however, no such consensus has yet been reached for grain-surface processes. A team of ~25~25 experts covering observational, laboratory and theoretical (astro)chemistry met in summer of 2014 at the

Lorentz Center in Leiden with the aim to provide solutions for this problem and to review the current state-of-the-art of grain surface models, both in terms of technical implementation into models as well as the most up-to-date information available from experiments and chemical computations. This review builds on the results of this workshop and gives an outlook for future directions.

Keywords: Surface reactions, Molecular ices, Accretion, Desorption, Photoprocesses, Diffusion

1 INTRODUCTION

The very presence of anything but atoms and obscuring minuscule dust grains in the interstellar medium (ISM) was inconceivable by astronomers merely a hundred years ago. Even the brightest minds of the time, such as Sir Arthur Eddington, were doubtful about the existence of molecules in the vast interstellar void. In his Bakerian lecture he pointed out that "...it is difficult to admit the existence of molecules in interstellar space because when once a molecule becomes dissociated there seems no chance of the atoms joining up again" (Eddington 1926).

However, around one decade later, absorption electronic transitions of the first interstellar molecular species, CN, CH, and CH+CH+, were identified (Swings and Rosenfeld 1937; McKellar 1940; Douglas and Herzberg 1941). The rapid development of radio and infrared detectors following World War II has since allowed the discovery of ~190~190 molecules in the ISM, as of March 2016 (see http://www.astro.uni-koeln.de/cdms/molecules). These interstellar species have a multitude of orbital electronic configurations and include stable molecules, radicals, open-shell molecules, cations, and anions.

Many interstellar molecules are recognizable from terrestrial and atmospheric chemistry. Among those are relatively stable species, e.g., water (H_2O), molecular hydrogen and nitrogen (H_2 and N_2), and carbon dioxide (CO_2), all of which consist of just a few atoms. More complex, hydrogen-rich saturated organic molecules are also present in space, e.g., formaldehyde (H_2CO), glycolaldehyde ($HCOCH_2OH$), methanol (CH_3OH), formic acid (HCOOH), and dimethyl ether (CH_3OCH_3) (Ehrenfreund and Charnley 2000; Herbst and van Dishoeck 2009). These species "bridge the gap" between the simple species listed previously and those considered of prebiotic and biological importance, e.g., amino acids. Other interstellar molecules are more exotic and unique to space. These include highly-unsaturated carbon chains and cages, e.g., $HC_{11}N$ (Bell et al. 1997), and the fullerenes, C_{60}, C60+C60+, and C_{70} (Cami et al. 2010; Berné et al. 2013; Campbell et al. 2015), the latter of which are also the largest molecular species discovered to date in the ISM. Even larger macromolecules, polyaromatic hydrocarbons (PAHs), consisting of between tens and hundreds of carbon atoms, are identifiable in space as a distinct class of species through their characteristic infrared bands (see the review by Tielens 2008). In summary, it is now known that the interstellar matter out of which stars and planets form has a substantial molecular component, which plays a pivotal role in the thermal balance of the ISM and its evolution (Tielens 2010).

The first theoretical models that successfully explained the presence and abundances of early observed molecular species were developed by Bates and Spitzer (1951), Watson and Salpeter (1972b), Herbst and Klemperer (1973), and Watson (1974), and thereafter significantly extended. The common perception in modern astrophysics is that many interstellar molecules, including complex unsaturated molecules, can be readily formed through purely gas-phase kinetics. Ion-molecule reactions and dissociative recombination reactions are of particular importance. Such processes typically do not have activation barriers and thus the rate coefficients are large, and possibly even enhanced, at low temperatures (~10–20 K, Adams et al. 1985; Herbst and Leung 1986). However, gas-phase chemistry alone cannot efficiently synthesize saturated organic species. The available reaction pathways typically require high temperatures and/or three-body reactions—conditions that are not usually met in the ISM.

Another efficient route towards increasing molecular complexity in the ISM is the chemical kinetics that occurs on dust-grain surfaces. Intriguingly, the most abundant molecule in space, molecular hydrogen, is formed almost exclusively via surface chemistry in the local universe (Gould and Salpeter 1963; Hollenbach and Salpeter 1971; Watson and Salpeter 1972a). The dust-grain surface has several roles. Firstly, the surface serves as a local "meeting point" for molecules or atoms that become bound to the dust grain via electrostatic or van der Waals forces, so-called physisorption, or by forming chemical bonds with its surface, so-called chemisorption. Secondly, the dust grain lattice can accommodate a portion of the excess energy usually generated during surface-mediated association reactions, stabilizing the product, and thus allowing large polyatomic species to be efficiently synthesized. Thirdly, in dense (≳104 cm−3≳104 cm−3) and cold (≪100 K≪100 K) interstellar environments, thick ice mantles can grow on dust grains (~100~100 monolayers). Molecular species trapped in the ice mantle under these conditions are protected from further processing by the FUV interstellar radiation field (ISRF), although they are exposed to the significantly lower strength ambient radiation field generated internally by the interaction of cosmic rays with molecular hydrogen. Eventually, over long timescales (≳104 yr≳104 yr) these pristine ices can be processed by the heating and enhanced irradiation associated with the star-formation process, potentially forming even more complex volatile and refractory organic compounds, including amino acids (Kvenvolden et al. 1970; Ehrenfreund and Charnley 2000; Elsila et al. 2009). These molecules may then be delivered to young protoplanets and planets via accretion early in the evolution of the planetary system, or at a later time via bombardment by pristine icy bodies (Anders 1989; Chyba et al. 1990; Cooper et al. 2001).

Most modern astrochemical models of ISM chemistry simulate dust-grain surface chemical kinetics processes with various degrees of complexity (Tielens and Hagen 1982; Hasegawa and Herbst 1993a; Garrod 2013). Models using large reaction networks typically adopt the rate-equation approach as is done for the gas-phase chemistry where the time evolution of surface species' abundances is described by a set of coupled ordinary differential equations, and the abundances considered "averaged" over the entire dust grain population, i.e., the mean-field approximation. One of the major challenges in these models, is the accurate treatment of the stochasticity of diffusive surface processes. This becomes critical when abundances of reactants on the dust-grain surface becomes very low, i.e., ≪1≪1 reactant per dust grain, and fluctuates with time, thus rendering the rate-equation approach unfeasible (Gillespie 1976; Green et al. 2001; Charnley 2001). This is the so-called "accretion-limited" case. A number of approximate or precise micro- and macroscopic Monte Carlo techniques have been proposed to overcome this issue (e.g., Biham et al. 2001; Charnley 2001; Lipshtat and Biham 2004; Stantcheva and Herbst 2004; Chang et al. 2005; Garrod 2008; Vasyunin and Herbst 2013). Another challenge is to account for the multilayered nature of dust-grain ice mantles, and to take all relevant processes into account in the modeling, e.g., inter-lattice diffusion, mobility/immobility of reactants, desorption, porosity trapping (see Cuppen and Herbst 2007; Chang et al. 2007; Kalvāns and Shmeld 2010; Wolff et al. 2011; Taquet et al. 2012; Vasyunin and Herbst 2013). A further obstacle in both approaches (rate equation and stochastic) is the lack of appropriate laboratory data on binding energies and desorption efficiencies of molecular ices of astrophysical interest, as well as the energy barriers and branching ratios for surface reactions. Large reaction networks can treat up to a few hundred different surface species; however, only a handful of reaction systems and molecules have been theoretically or experimentally studied. Moreover, the underlying molecular mechanism is not always fully understood, which makes it hard to scale up to astrophysically relevant timescales.

A wealth of evidence suggests that dust-grain surface processes are important over a wide range of interstellar conditions and star-formation environments, while models and observations are rapidly advancing to trace this chemical evolution through to at least the protoplanetary disk phase (Henning and Semenov 2013; Dutrey et al. 2014; Walsh et al. 2014). The new Atacama Large Millimeter Array (ALMA), with its orders-of-magnitude increase in sensitivity and resolving power, is expected to give us an unprecedented view of potentially pre-biotic and biologically-relevant molecules in various astrophysical environments over the coming years. The analysis of these new data will require much more elaborate, and more diverse, gas-grain astrochemical models than have been developed so far. Unfortunately, a major stumbling-block in our understanding of pre-biotic chemistry in the ISM is the lack of a standardized and comprehensive approach to simulate grain-surface chemistry. In the case of gas-phase chemistry, several publicly available databases with reactions and the corresponding rate data exist, of which the UMIST Database for Astrochemistry (UDfA) and the KInetic Database for Astrochemistry (KIDA) are the most widely used (McElroy et al. 2013; Wakelam et al. 2012). Consensus on how these data should be used, including how the rate coefficient is calculated and over which temperature ranges it is viable, has been reached, and the quality of the data in these databases is regularly reviewed. However, for grain-surface chemistry, this is not yet the case. Modelers often compile their own grain-surface reaction networks, and most are not publicly available, primarily due to the lack of an agreed and standardized approach.

Fortunately, many of the assumptions within the models can now be tested using surface science techniques with interstellar ice analogs. Over the past few years, substantial progress has been made on the understanding of various grain-surface reaction systems, including which processes are dominant and under what conditions, as well as the underlying mechanisms. In the summer of 2014, astronomers, experimentalists and theoretical chemists came together during a Lorentz Center Workshop ("Grain-Surface Networks and Data for Astrochemistry") to identify the needs of modelers for their models, the appropriate formalisms to use, and to identify how recent experimental techniques and results can help to test and improve the models. In this paper we summarize the key findings of this workshop and relay our recommendations for the treatment of grain-surface chemistry to the astrochemical community. First we describe the outline of a typical gas-grain model (Sect. 2) and in Sects. 3 to 8 we discuss, in turn, the various processes that need to be considered in astrochemical models of surface chemistry: accretion, desorption, surface reactions, diffusion (thermal diffusion in the surface and bulk and quantum tunneling), and photoprocesses. We also address the more technical aspects of writing and executing code such as numerical precision in chemical models (Sect. 9) and finally end with a discussion on a test case of CO hydrogenation to form complex molecules (Sect. 10) and a future outlook (Sect. 11).

2 OUTLINE OF A GENERIC GAS-GRAIN CODE

Gas-grain astrochemical models typically use the rate-equation approach to describe both the gas-phase and grain-surface chemistry using chemical kinetics. This generates a set of stiff ordinary differential equations that can be numerically solved using a multi-step integrator, e.g., via Runge-Kutta or Adams algorithms. In chemistry, rate equations are often applied to describe macroscopic experimental effects and account for many-body effects with a mean-field approach. As we mentioned above, this may not be the case on a dust-grain surface under particular conditions; hence, using the rate-equation approach to describe interstellar surface chemistry can lead to large errors when compared with results using more realistic stochastic techniques. The main reason why modelers persist with such a method is the convenience, stability, and the rather fast numerical performance of the pure chemical kinetics codes, even for reaction networks which consist of thousands of reactions involving hundreds of molecules (e.g., Dalgarno and Black 1976; Leung et al. 1984; D'Hendecourt et al. 1985; Brown and Charnley 1990; Hasegawa et al. 1992; Bergin et al. 1995; Millar et al. 1997; Aikawa et al. 1996; Willacy et al. 1998; Semenov et al. 2010; Wakelam et al. 2010; Agúndez and Wakelam 2013; Albertsson et al. 2013; McElroy et al. 2013; Grassi et al. 2014). As an indication, rate equations require CPU time of ~1--60~1--60 seconds. Typically, the addition of the modified rate approach to the rate equation model, which use the same numerical scheme, slows it by a factor of several due to the performance penalty for accounting for probabilities of reactants to be on the grain surface (Garrod et al. 2009). A multiphase model (Furuya et al. 2016) without bulk ice chemistry, but with swapping only, takes typically ~30–60 minutes per trajectory, using a full gas-ice network with deuterium chemistry. It is hence computationally feasible to use such a model to simulate a collapsing core model; tracing 35000 parcels from the prestellar core phase to the circumstellar disk phase results in ~35000 CPU hours or ~2~2 weeks on a ~100-core machine. When bulk ice chemistry is included, the CPU time increases by a factor of 10–100. Adding in bulk chemistry increases this to months and hence a multi-phase model with bulk ice chemistry coupled with 2-D/3-D physical models remains a computationally challenging problem. On the other hand, macroscopic Monte Carlo models like presented in Vasyunin et al. (2009) require much more CPU time, from hours to days, for a simulation of a TMC-1 type cloud. What is more important, Monte Carlo models usually have a rather limited range of physical conditions that can be considered due to their slow performance. Microscopic Monte Carlo models (Lamberts et al. 2014b; Cuppen et al. 2009; Chang and Herbst 2014, 2016) are restricted to an even smaller chemical network and require days to weeks. The method of choice is hence highly dependent on the available computer power, the problem that one would like to address which dictates the level of detail in the grain description required, and the heterogeneity and complexity of the astrophysical object that one is interested in. In recent times, efforts have been made to simulate both laboratory and astrophysical conditions with the same model, thus using the laboratory simulation as a benchmark (e.g., Lamberts et al. 2013). Especially for these cases, microscopic Monte Carlo methods are worth the extra computational effort since they allow to include more surface complexity that might be crucial to gain insight in the physical and chemical processes occurring in the experiments. At the same time, laboratory environments typically deal with well-constrained physical conditions and a limited chemical network.

Here we present a recipe for the construction of a chemical kinetics model based on the rate-equation approach (see also Semenov et al. 2010). The chemical system consists of two major phases: the gas-phase and the dust-grain surface ice mantle. If all the necessary kinetic data are provided (e.g., from a database) and the initial abundances are assumed (e.g., from observations) or generated (e.g., using a similar model), a chemical kinetics code numerically solves the equations of first- and second-order kinetics and returns time-dependent molecular concentrations. Under typical ISM conditions, i.e., low densities, three-body reactive collisions are usually irrelevant and hence ignored. Here, we focus solely on the grain-surface chemistry aspect of the code. The treatment of gas-phase chemistry has been described in a number of papers, including McElroy et al. (2013) and Wakelam et al. (2010).

As schematically shown in Fig. 1, species on a grain surface generally experience four types of processes: (i) accretion (or orption) onto the surface, (ii) desorption from the surface, (iii) diffusion across the surface or on/within the ice mantle, and (iv) reaction. When grain-surface ice mantles are still exposed to far-UV radiation, species contained within can also be photodissociated. This le to the following expression for the change in surface abundance:

$$\begin{aligned}\frac{\mathrm{d}}{\mathrm{d}t}n_{\mathrm{s}}(\mathrm{A},t) &= \sum_{i,j} f_{\mathrm{react}}(i+j \longrightarrow \mathrm{A}) - \sum_{i} f_{\mathrm{react}}(i+\mathrm{A} \longrightarrow) \\ &+ \sum_{i} f_{\mathrm{UV\ diss},i} - f_{\mathrm{UV\ diss,A}} \\ &+ f_{\mathrm{acc,A}} - [f_{\mathrm{evap,A}} + f_{\mathrm{nonth,A}}].\end{aligned} \tag{1}$$

The first four terms in this expression account for the gain and loss of species A due to grain-surface reactions or photodissociation reactions, respectively. The fifth term expresses the accretion of species A from the gas phase onto the grain, and the final term denotes the desorption of species A from the grain back into the gas phase. This latter process can occur via thermal desorption or by non-thermal processes, whereby desorption is trigged by the input of external energy in the form of far-UV or X-ray photons or high energy particles or by energy released during in-situ exothermic reactions. In the subsequent Sections we will discuss in detail the functional forms usually adopted for each of these chemical processes.

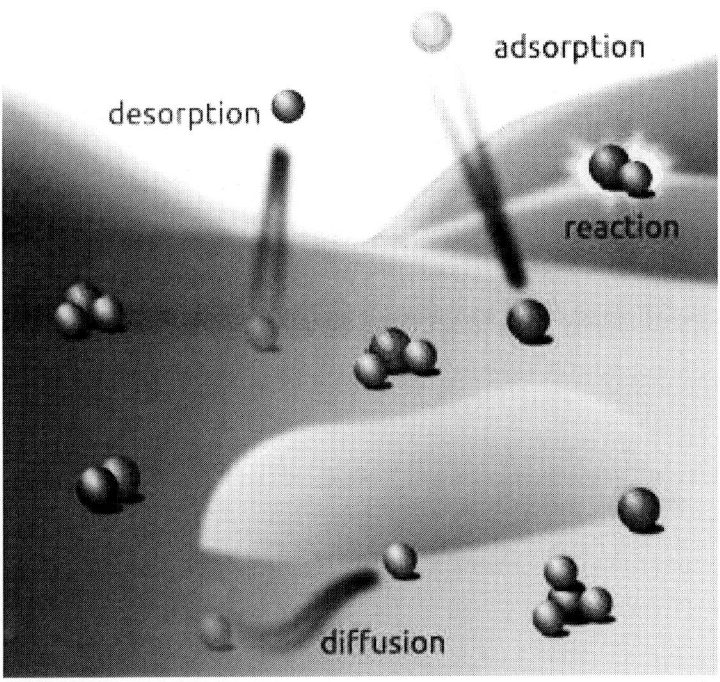

Figure 1. Overview of the most important grain surface processes that will be covered in this review

As previously mentioned, for low surface abundances, the mean-field assumption inherent in the rate-equation approach breaks down, and several stochastic methods have been developed to overcome this issue. Although the description of the chemistry is intrinsically more accurate, stochastic models are computationally much more demanding than rate equations, and for the purpose of this review we will limit ourselves to rate-equation models. Modifications to the rate-equation approach can be made to better treat the surface chemistry in the accretion-limited case. Caselli et al. (1998) were the first to propose such an adjustment. They applied a semi-empirical approach to scale down the reaction rates for those cases where the surface migration of atomic hydrogen is significantly faster than its accretion rate onto grains (Caselli et al. 1998). This method gave good agreement with stochastic methods for a number of cases; however, it was not clear how applicable the method was outside of the tested regime. More recently, a new modified-rate approach was suggested by Garrod (2008) which improves upon the original.

3 ACCRETION

The accretion term facc,Afacc,A in Eq. (1) accounts for the orption of gas-phase species onto the dust grains. It is determined by the collisional frequency of a gas-phase species with a grain, times a sticking efficiency, SASA:

$$f_{\mathrm{acc,A}} = S_{\mathrm{A}} v_{\mathrm{A}} n_{\mathrm{grain}} \pi r_{\mathrm{grain}}^{2} n_{\mathrm{g}}(\mathrm{A}), \tag{2}$$

where ng(A)ng(A) is the number density of species A in the gas phase, rgrainrgrain is the average radius of a dust grain (~0.1 μm~0.1 μm for ISM-like grains) with number density, ngrainngrain, and vAvA is the average gas-phase thermal velocity,

$$v_{\mathrm{A}} = \sqrt{\frac{8kT_{\mathrm{gas}}}{\pi m_{\mathrm{A}}}}. \tag{3}$$

This in turn depends on the gas temperature, TgasTgas, the mass of the species, mAmA, and Boltzmann's constant kk. The sticking efficiency or probability, SASA, of species A to the surface is determined by how well it can dissipate its kinetic energy. This depends on the dust-grain and gas temperature, on the relative masses of the substrate molecules and the incoming species, and on the presence of a barrier for sticking, typically restricted to chemisorption. For most species at low gas and grain temperatures, this results in a sticking fraction near unity, with the exception of hydrogen. Computationally, sticking fractions have been determined by Molecular Dynamics (Buch and Czerminski 1991; Al-Halabi et al. 2002, 2003, 2004; Batista et al. 2005; Veeraghattam et al. 2014), perturbation and effective Hamiltonian theories, close coupling wavepacket, and reduced density matrix approaches (Lepetit et al. 2011), or by the much-more-approximate soft-cube method (Logan and Keck 1968; Burke and Hollenbach 1983). These studies typically focus on the accretion of a single atom or molecule on an otherwise bare surface, whereas the sticking coefficient could be coverage dependent, especially for chemisorption where for high coverage there are simply fewer sites available for sticking. Experimentally it was found that the sticking coefficient of physisorbed H_2 increases linearly with the number of deuterium molecules already orbed on the surface (Amiaud et al. 2007; Chaabouni et al. 2012).

4 DESORPTION

The desorption term in Eq. (1) represents the desorption of the species from the grain surface back into the gas phase. Various desorption processes are possible and usually a particular distinction is made between thermal desorption and non-thermal desorption. For the latter process, a multitude of different mechanisms is possible, such as photodesorption (Westley et al. 1995b; Öberg et al. 2007, 2009a,b), sputtering by cosmic rays or grain heating by cosmic rays (Hasegawa and Herbst 1993a; Herbst and Cuppen 2006), and reactive or chemical desorption, where the excess heat generated upon reaction allows desorption of the products (Garrod et al. 2006, 2007; Dulieu et al. 2013). Here, thermal and reactive desorption are briefly discussed. Photodesorption is discussed in a separate section on photoprocesses since photodesorption and photodissociation are parallel processes which require a different treatment.

4.1 Thermal Desorption

The residence time of a species on the dust-grain surface is predominantly determined by its desorption rate. The thermal desorption rate, in turn, depends on the binding energy of the species to the surface, Ebind,AEbind,A,

$$k_{\mathrm{evap,A}} = \nu \exp\left(-\frac{E_{\mathrm{bind,A}}}{kT}\right), \tag{4}$$

where νν is a characteristic attempt frequency. Here νν and Ebind,AEbind,A are important input parameters for astrochemical models.

Usually the following equation for the characteristic frequency (Tielens and Allamandola 1987) is assumed,

$$\nu = \sqrt{\frac{2N_s E_{\mathrm{bind,A}}}{\pi^2 m_A}}, \tag{5}$$

where NsNs is the surface density of binding sites and mAmA is the mass of species AA. Tielens and Allamandola (1987) derived this expression assuming that the vibrational frequency perpendicular to the surface equals the vibrational frequency parallel to the surface and that the binding can be described by a harmonic potential, which might not be an accurate assumption for a physisorbed species. They also derived an expression including rotational degrees of freedom and one for the frequency of a free particle ν=kThν=kTh. Together this le to an estimation of ν=1012 s−1ν=1012 s−1, in accordance with all three approaches.

The binding energy, Ebind,AEbind,A, and the thermal desorption rate, kevap,Akevap,A, can be experimentally obtained using Temperature Programmed Desorption (TPD). These experiments are usually performed under ultra-high-vacuum conditions (base pressure better than ~10−9 mbar~10−9 mbar) coupled with a quadrupole mass spectrometer. The temperature of the substrate can be carefully controlled using a cryostat. A TPD experiment consists of two phases: (i) the substrate is brought to a constant low temperature and a known quantity of one or more species is deposited, and (ii) the temperature is linearly increased and the desorption monitored using the mass spectrometer.

Different types of analysis methods can be applied to obtain kinetic parameters, such as desorption energy, desorption order, and the so-called “prefactor”, which is analogous to (although not entirely equivalent to) the characteristic frequency, νν. Sometimes the latter two parameters are assumed and only the first is obtained from the analysis, other groups use for instance “leading edge fitting” to obtain all three simultaneously. Whichever method is applied, the three parameters are not completely independent and therefore a desorption energy derived from experiment should be used in combination with its corresponding prefactor. In most gas-grain codes, the computationally convenient description of the prefactor by Eq. (5) is adopted, although sometimes not physically accurate. It is recommended that experimentalists always quote the desorption energy derived in combination with the prefactor, and a fixed integer value for the desorption order, so that the binding energies are used in an appropriate manner in astrochemical models.

The desorption order is an important consideration worthy of further discussion. Zeroth-order desorption, i.e., a constant desorption rate, generally occurs when multiple layers of the same species are deposited. The number of surface species available for desorption (limited to the top few monolayers) remains the same; hence, the desorption rate is independent of the number of total species on the surface. In the sub-monolayer regime, first-order desorption is observed. Second-order desorption, i.e., a quadratic dependence of the desorption rate on the number of surface species, is also seen. This can occur in two cases: (i) when the surface exhibits a distribution of binding sites, and (ii) through chemical desorption of species that are formed via a second-order surface reaction.

In many astrochemical models, the first-order thermal desorption rate is assumed,

$$f_{\mathrm{thermal,A}} = k_{\mathrm{evap,A}} n_{\mathrm{s}}(\mathrm{A}) \tag{6}$$

where ns(A)ns(A) is the number density of species A orbed on the grain surface. As mentioned above, this only strictly occurs in the sub-monolayer regime. In two-phase gas-grain astrochemical models there is no positional information on the various species, so it is not known which species occupy the top layers of the ice mantle. However, it is possible to apply a fix to the thermal desorption rate to account for the fractional composition of the ice mantle, as well as treating thermal desorption as a zeroth-order process in the multilayer regime. This involves counting the number of monolayers present within the ice mantle,

$$N_{\mathrm{mono}} = \frac{\sum_i n_{\mathrm{s,i}}}{N_{\mathrm{s}} \sigma_g n_{\mathrm{grain}}}, \tag{7}$$

where the numerator is the total number density of surface species per unit volume, and the denominator is the total number of surface sites available per unit volume. For Nmono>NactNmono>Nact, where, NactNact is the assumed number of "active" monolayers (typically ~2--4~2--4), the thermal desorption rate is given by,

$$f_{\mathrm{thermal,A}} = k_{\mathrm{evap,A}} N_{\mathrm{act}} \chi_{\mathrm{A}} N_{\mathrm{s}} \sigma_g n_{\mathrm{grain}}, \tag{8}$$

where $\chi_{\mathrm{A}} = n_{\mathrm{s}}(\mathrm{A}) / \sum_i n_{\mathrm{s,i}}$,i is the fractional abundance of species A in the ice mantle. For $N_{\mathrm{mono}} \leq N_{\mathrm{act}}$, the rate switches to the first-order desorption rate.

Table 1. List of experimentally determined binding energies

Species	E_{bind} (K)	Prefactor ($cm^{-2}\,s^{-1}$)	Comment	Ref.
H_2	440		amorphous water ice	A
	480 ± 10		silicate	B
	555 ± 35		mixed $H_2O{:}CH_3OH{:}H_2$	C
H_2O	4800 ± 96	$10^{23\pm1}$	graphite	D
	4800		silicate	E
	4825 ± 5		unannealed water ice	F
	5027 ± 87		annealed water ice	F
	5600	10^{30}	amorphous water ice	G
	5770 ± 60	10^{30}	crystalline water ice	G
	5930 ± 240	10^{28}	amorphous silica	H
H_2CO	3260 ± 60	10^{28}	mixed $H_2CO{:}H_2O = 1{:}14$	I
	3730 ± 110	10^{28}	silicate	I
CH_3OH	4235 ± 15		pure	J
	4930 ± 98	$6 \times 10^{21\pm3}$	graphite/multilayer	D
	5770 ± 95	$9 \times 10^{9\pm3}$[a]	graphite/monolayer	D
CO	826 ± 24	$(7 \pm 2) \times 10^{26}$	pure	K
	828 ± 28	7.1×10^{26}	non-porous water ice	L
	830 ± 40	7.1×10^{26}	silicate	L
	850 ± 55	7.1×10^{26}	crystalline water ice	L
	858 ± 15	7.2×10^{26}	pure	M
	856 ± 15	7.2×10^{26}	layered CO–O_2	M
	865 ± 18	7.2×10^{26}	mixed $CO{:}O_2 = 1{:}1$/multilayer	M
	955 ± 18	7.6×10^{11}	mixed $CO{:}O_2 = 1{:}1$/monolayer	M
	855 ± 25		pure	N
	855 ± 25	$7 \times 10^{26\pm1}$	pure	O
	880 ± 36		layered	H
	955 ± 4		pure	F
	1180 ± 24	$(5 \pm 1) \times 10^{14}$[a]	amorphous water ice	K
CO_2	2235 ± 50		mixed $SO_2{:}CO_2 = 20{:}1$	J
	2270 ± 71	9.3×10^{26}	non-porous water ice	L
	2270 ± 80	9.3×10^{26}	silicate	L
	2360 ± 83	9.3×10^{26}	crystalline water ice	L
	2690 ± 150		pure	P
	2860 ± 150		mixed $H_2O{:}CO_2 = 20{:}1$	P
O	1100		amorphous silicate	E
	1660 ± 60		porous amorphous water ice	Q
	1765 ± 232		amorphous silicate/monolayer	R
	1850 ± 90		bare amorphous silicate	Q

O_2	912 ± 15	6.9×10^{26}	pure	M
	904 ± 15	6.9×10^{26}	layered CO–O_2	M
	896 ± 18	6.8×10^{26}	mixed O_2:CO = 1:1	M
	895 ± 36	6.9×10^{26}	silicate	L
	898 ± 30	6.9×10^{26}	non-porous water ice	L
	902 ± 24		layered	H
	904		amorphous silicate/monolayer	R
	925 ± 25		pure	O
	936 ± 40	7×10^{26}	water/crystalline	L
	1250		amorphous silicate	E
N_2	790 ± 25		pure	O
	830 ± 36		pure	H
NH_3	2790 ± 144	$(8 \pm 3) \times 10^{21}$	graphite	D
	3075 ± 25		pure	J

[a]In s^{-1}

A—Amiaud et al. (2007); B—Acharyya (2014); C—Sandford and Allamandola (1993b); D—Brown and Bolina (2007), Bolina et al. (2005b,a), Bolina and Brown (2005); E—Dulieu et al. (2013); F—Sandford and Allamandola (1988); G—Fraser et al. (2001); H—Collings et al. (2015); I—Noble et al. (2012a); J—Sandford and Allamandola (1993a); K—Collings et al. (2003); L—Noble et al. (2012b); M—Acharyya et al. (2007); N—Öberg et al. (2005); O—Fuchs et al. (2006); P—Sandford and Allamandola (1990); Q—He et al. (2015); R—He et al. (2014)

Table 1 lists the binding energies of a wide collection of stable species that have been determined using the TPD technique and are therefore relatively well constrained. The binding energies have been mostly determined for the desorption of pure ices from different substrates. The differences between the different substrates are rather small and become negligible in the multilayer regime (Green et al. 2009). The uncertainties on the binding energies quoted in Table 1 can have different origins: experimental errors, errors in the fit, or—especially for the amorphous silicate surfaces—they can represent a range of binding energies which is an intrinsic property of the substrate.

Desorption rates depend exponentially on binding energies and uncertainties in these binding energies can have a large effect, even at dark cloud conditions where the temperature is well below the desorption temperature of the vast majority of the surface species (Penteado et al. 2016). Since in most systems the diffusion barrier is calculated by taking a fixed ratio with the binding energy, changing binding energies not only affects the temperature at which species desorb, i.e., the temperature at which species cannot participate in the grain surface chemistry, but also the onset temperature at which species start to diffuse. A sensitivity analysis of grain surface chemistry under dark cloud conditions to binding energies of ice species showed that the model results appear particularly sensitive to the binding energy of H_2 (Penteado et al. 2016). The dust temperatures in molecular cloud cores are relatively well constrained to precision of ~1 K by a number of *Herschel* studies (e.g., Stutz et al. 2010; Launhardt et al. 2013; Lippok et al. 2016).

The experiments show that the molecules indeed desorb with a (close to) zeroth-order rate in the multilayer regime whereas they desorb with a (close to) first-order rate in the monolayer regime, as explained above (Fraser et al. 2001). It is difficult to experimentally obtain similar results for radical species due to their high reactivity (and correspondingly short lifetime). Binding energies for radicals can only be obtained in an indirect manner, usually involving the simulation of experimental data, and an exploration of the possible parameter space, using stochastic chemical models. However, there are recent experimental results reporting the experimental determination of the binding energy of atomic oxygen on a range of surfaces (Dulieu et al. 2013; He et al. 2014, 2015), showing that for some species, direct measurements are possible.

Most TPD experiments are performed using pure ices to allow an unambiguous interpretation of the results and to minimize the chance of contamination. Some studies on mixed and layered ice have been done to better mimic the composition of interstellar ice mantles. The introduction of more species in the ice immediately increases the complexity of the desorption process.

The binding energy of individual species will vary depending on its surrounding material, and the dominant ice-mantle species can prevent other species from desorbing. Collings et al. (2004) showed, for instance, that a fraction of molecules like CO and CO_2 can become trapped in an ice mantle which consists predominantly of water ice. The trapped CO and CO_2 are then released at the higher temperatures expected for water desorption. However, laboratory timescales are significantly shorter than those in the ISM; hence, trapped species may have sufficient time to escape the ice mantle since they will also have had sufficient time to segregate (Öberg et al. 2009d). This process depends on a large number of parameters including surface temperature, ice composition and mixing ratio.

Two-phase astrochemical models can include the effects of trapping in a somewhat empirical manner by allowing a fraction of volatile species such as CO to have a binding energy similar to the species within which they are trapped (e.g., CO_2 or H_2O, see, e.g., Viti et al. 2004). Three-phase and multilayer models can simulate the effects of trapping by allowing diffusion of surface species into the bulk ice mantle (and vice versa). We discuss the treatment of bulk diffusion in Sect. 7.

4.2 Reactive/Chemical Desorption

Chemical desorption is desorption of reaction products from the grain surface by excess reaction energy. This type of desorption is also referred to as reactive desorption. Garrod et al. (2006) were the first to suggest this mechanism to explain, e.g., the gas-phase detection of methanol in cold dark clouds. They based their initial model on the Rice-Ramsperger-Kessel (RRK) theory, which relates the excess energy and the binding energy of species to a desorption probability. They modified this theory by adding an unconstrained aa parameter which they chose to be 0.1. In a follow-up study, Garrod et al. (2007) showed that chemical desorption may play an important role in explaining the observed abundances of different gas-phase chemical species, particularly in dark molecular clouds. Later Cazaux et al. (2010) came to similar conclusions when they included this mechanism in their model for water formation on grains.

The first constraints on the probability of this mechanism were obtained using Molecular Dynamics simulations (Andersson et al. 2006; Arasa et al. 2010, 2011). In these studies, the fate of photoproducts of water ice photodissociation—OH and H—were monitored in time. In some cases, the photoproducts were found to recombine to form water which subsequently escaped from the ice mantle: this can loosely be described as reactive desorption driven by photodissociation. However, as will be discussed later, this can also be thought of as "photodesorption" (see Sect. 8). In these simulations, the desorption probability was highly dependent on the location of the dissociated molecule in the ice mantle. Recombination events in the fourth layer of the ice or further below almost exclusively resulted in trapping of the reformed water molecule. These results are limited to a water-rich environment and they may not be applicable to the formation of the first monolayers of the water ice mantle.

What remains to be quantified is the efficiency of reactive desorption which is not driven by photoprocessing. The first experimental study by Dulieu et al. (2013) measured the chemical desorption of reaction products through sequential O_2 hydrogenation experiments on an amorphous silicate or a graphite surface, where the amount of deposited O_2 remained in the (sub)monolayer regime. They find substantial desorption of the formed H_2O molecules, which is caused, at least in part, by the lack of binding opportunities with surrounding molecules. Moving to the multilayer regime, they find that the desorption probability for the O+OO+Orecombination reaction drops to negligible values (Minissale and Dulieu 2014). Expanding their studies to other reaction systems (e.g., CO+HCO+H, H2CO+HH2CO+H), they determined relatively low reactive desorption rates in the (close-to) sub-monolayer regime ($\lesssim$10 %, Minissale et al. 2016).

Despite the lack of conclusive experimental evidence for chemical desorption driven purely by exothermicity of reactions (and not by photoprocessing), especially in the multi-layer regime, astrochemical models typically still account for such a process by implementing the Garrod et al. (2007) prescription

$$P = \frac{a\Xi}{1 + a\Xi} \tag{9}$$

with

$$\Xi = \left[1 - \frac{E_{\text{bind}}}{E_{\text{exo}}}\right]^{s-1} \tag{10}$$

where EexoEexo is the exothermicity of the reaction, and ss is the number of vibrational modes in the molecule/surface-bond system. This number is s=2s=2 for diatomic species; for all others, s=3N−5s=3N−5, where NN is the number of atoms in the molecule, which is assumed to be non-linear and forms an extra "bond" to the surface. The efficiency parameter aa is not well constrained and is generally used as universal input parameter with a value between 0.01 and 0.1. Figure 2 shows the sensitivity of gas phase and ice abundances to this parameter in a laminar solar nebula model by changing aa from 0.05 to 0.01.

The figure shows that changes can locally be several orders of magnitude, but integrated over the height of the disk the changes are relatively small for several species. Ices in disks concentrate around the midplane and are nearly absent in upper layers due to thermal evaporation and photodesorption.

The changes in column densities are hence mainly determined by the changes in abundance around the midplane. Interestingly, the abundances of CH_3OH and CH_4 increase both in the gas phase and in the ice by lowering the reactive desorption. This is presumably since these species are formed in several steps and a lower reactive desorption efficiency keeps the intermediate species on the grain, enabling the full reaction route to proceed.

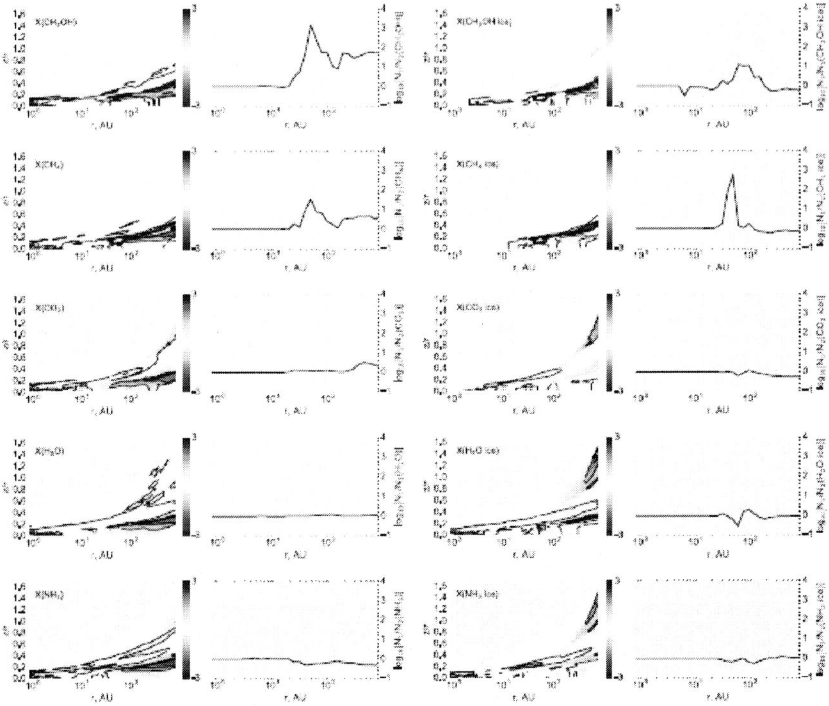

Figure 2. The change in molecular abundance of a selection of species in a laminar solar nebula at 1 Myr, when using a chemical desorption efficiency of 0.01 (N1N1) instead of 0.05 (N2N2). The log of relative ratios are given both as function of location (height zz and radius rr) and integrated over zz as a function of rr. Relative gas phase abundances are in the *left panels*, the corresponding ice abundances in the panels on the *right-hand side*

5 REACTIONS

Generally, surface reactions are thought to occur via one of three mechanisms: the diffusive *Langmuir-Hinshelwood* mechanism, where both species move over the surface and react upon meeting, the stick-and-hit *Eley-Rideal* mechanism where one (stationary) reactant is hit by another species from the gas phase, and the *hot-atom* mechanism (which is a combination of both) where non-thermalized species travel some distance over the surface before finding a fellow reactant. Photodissociation is usually treated separately and will be discussed in Sect. 8. Under astrophysical conditions where ices are abundant, the gas and dust grains typically have similar temperatures, and the chemical timescales tend to be significantly longer than the thermalization timescale; hence, the hot-atom mechanism is often considered not important.

The analytical expression to describe the Langmuir-Hinshelwood mechanism on a surface is

$$f_{\mathrm{react,LH}}(i + j \longrightarrow \mathrm{A}) = P_{\mathrm{react,LH},i,j}(k_{\mathrm{scan},i} + k_{\mathrm{scan},j})n_s(i,t)n_s(j,t) \quad (11)$$

where ns(i,t)ns(i,t) is the number of species ii present on the surface at time tt and kscan,ikscan,i the rate by which species ii scans the grain surface. The scanning rate is given by

$$k_{\mathrm{scan}} = \frac{k_{\mathrm{hop}}}{N_{\mathrm{sites}}} \quad (12)$$

where NsitesNsites is the number of binding sites per grain. For amorphous solid water with a density of 0.94 gcm−30.94 gcm−3, the site density is 1×10151×1015 molecules cm−2cm−2. Simulations of CO_2, which is a bulkier molecule, on top of water ice showed a site density of 6×10146×1014 molecules cm−2cm−2(Karssemeijer et al. 2014a). Both lead to approximately ≈106≈106 binding sites per grain for a standard grain of 0.1 μm.

The scanning rate determines the meeting frequency of the two particles ii and jj due to the mobility of one, or both, reactant(s). The Langmuir-Hinshelwood mechanism is dependent upon the abundances of both reactant and hence is a second-order process. The hopping rate, khopkhop, will be discuss in Sect. 6.

The rate coefficient, Preact,LH,i,jPreact,LH,i,j, accounts for the probability that a possible reaction barrier will be crossed during the encounter. This probability is assumed to be 1 for a reaction with zero activation energy, and 0.5 if the two reactants are the same species. For reactions with an activation barrier, EaEa, that occur on a dust-grain surface with a temperature, TT, this probability is

$$P_{\mathrm{react,LH},i,j} = \exp\left(-\frac{E_a}{kT}\right) \quad (13)$$

when the barrier is crossed through thermal activation as schematically depicted in Fig. 3 for the reaction H + H_2CO. In this case there is a clear transition state that determines the rate limiting energy barrier. Tunneling through the reaction barrier is also possible, greatly increasing the probability of reaction (see, e.g., Hasegawa et al. 1992). This occurs through delocalization of the transition state. As can be seen for the H--H2COH--H2CO complex, light species are much more delocalized and quantum-mechanical tunneling is hence of main importance for reactions where light species, e.g., H, D, are involved in the bond breaking or forming. Tunneling is discussed in more detail in Sect. 5.2.

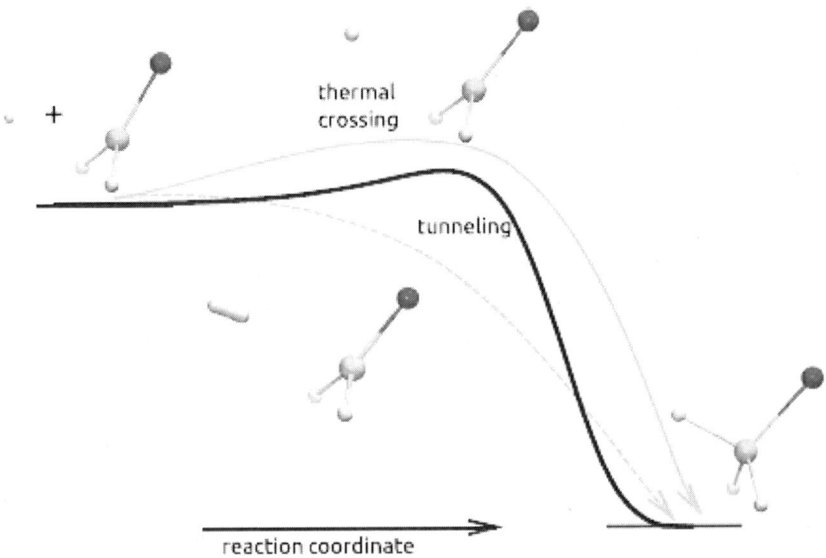

Figure 3. Schematic representation of crossing a reactive barrier either through tunneling or through thermal activation. The H+H2CO⟶ H3COH+H2CO⟶H3CO reaction is used as an example

Although conceptually simple, in reality the situation is more complex. First, a surface reaction may have several exit channels leading to a number of various products, similar to reactions in the gas phase. For most examples each of these channels will have its own transition state and corresponding (temperature-dependent) rate. The reaction constant does not need to include a special scaling to account for this effect, but the branching ratios αα are a natural outcome of the model

$$\alpha_l = \frac{k_l}{\sum_{j=1,m} k_j} \tag{14}$$

for the all possible mm reaction channels. In constructing a reaction network, one should be very careful when adding new product channels especially when the reaction rates come from very different sources (surface vs. gas phase experiments, computations) since some product channels might be heavily suppressed. For some reactions, only a destruction rate is known and branching ratios are determined separately. Individual product rate should in this case be adjusted accordingly.

Second, for a diffusive surface reaction to happen, the two molecules must remain orbed in close vicinity until they react, otherwise they can migrate away from each other or even desorb as schematically depicted in Fig. 4. Therefore, a surface reaction process is in competition with diffusion and desorption (Herbst and Millar 2008). Consequently, the reaction constant for product channel kk takes the following expression (see Equation 6 in Garrod and Pauly 2011),

$$P^{\mathrm{full},k}_{\mathrm{react,LH},i,j} = P^{k}_{\mathrm{react,LH},i,j}\frac{\nu\sum_{l=1,m}P^{l}_{\mathrm{react,LH},i,j}}{\nu\sum_{l=1,m}P^{l}_{\mathrm{react,LH},i,j}+k_{\mathrm{hop},i}+k_{\mathrm{hop},j}+k_{\mathrm{evap},i}+k_{\mathrm{evap},j}}, \quad (15)$$

where the khopkhop and kevapkevap are the thermal hopping (scanning) and evaporation rates for the reactants ii and jj, consequently. In the majority of astrophysically relevant situations the evaporation terms are small in magnitude compared with the hopping terms and can be safely neglected.

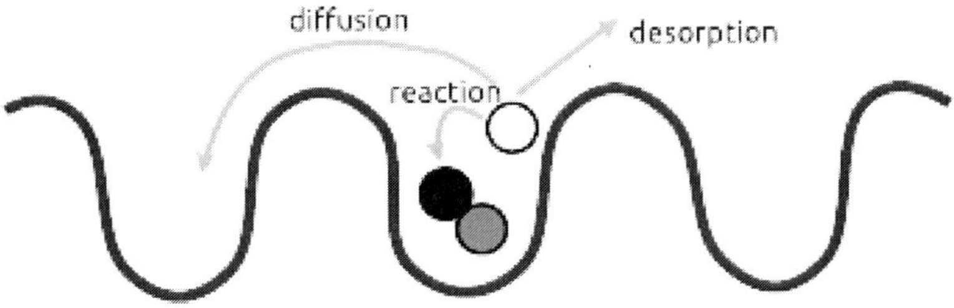

Figure 4. Surface reactions can be attempted as long as the reactions are each others vicinity. Hence, reaction competes with diffusion and desorption. Faster diffusion will lead to a higher meeting rate but also in a shorter reaction time

The Eley-Rideal mechanism is considered to be important only for high surface coverages or low surface mobility (Ruffle and Herbst 2001). An example where this can become important is during catastrophic freeze-out of CO in prestellar cores. During this phase, a layer of reactive CO ice forms on the grains (Pontoppidan 2006). Under these circumstances,

Eley-Rideal could be an important mechanism in the formation of methanol. It can be included in models by using the following expressions

$$f_{\mathrm{react,ER}}(i+j \longrightarrow \mathrm{A}) = P_{\mathrm{react,ER},i,j}\, f_{\mathrm{acc},i}\, n_{\mathrm{s}}(j,t) + P_{\mathrm{react,ER},i,j}\, f_{\mathrm{acc},j}\, n_{\mathrm{s}}(i,t). \qquad (16)$$

The reaction constant is different for the Eley-Rideal mechanism. Here the two reactants have only one attempt to cross the reaction barrier EaEa, so diffusion and desorption competition is of no importance. The corresponding rate coefficient is much simpler than in the LH-case (Eq. (15)):

$$P_{\mathrm{react,ER},i,j} = \alpha_l \exp\left(-\frac{E_a}{kT}\right). \qquad (17)$$

5.1 Surface Experiments

Surface reactions can be monitored in the laboratory using an ultra-high vacuum setup (better than ~10−9 mbar~10−9 mbar and H_2 as main gas residue in the chamber) and experiments are generally performed in two ways. Either the reactants are deposited in sequence, referred to as *pre-deposition* experiments, or in tandem in a so-called *co-deposition* experiment. The first gives a better control over the total dose and the predeposited amount can either be in the monolayer regime on top of an astrophysically relevant surface (e.g., a silicate or carbonaceous substrate), or a thicker ice if the reactant is a stable species, e.g., CO. For pre-deposition ice experiments, the final yield of the newly formed species for a selected radical fluence and ice temperature is largely limited by the penetration depth of the reactants in the ice. For the case of hydrogenation of CO ice, the maximal penetration depth of H atoms is four monolayers; therefore, only the upper layers of the ice are hydrogenated. Co-deposition experiments generally give a higher signal as they do not suffer from such penetration effects. Moreover, they are particularly useful in experiments involving radical species other than H atoms. Radicals are generally formed in a microwave discharge or RF (radio frequency) source in which the stable precursor is injected and subsequently dissociated. The dissociation products are then piped to the substrate, usually through a cooling pipe that thermalizes or cools the species to room temperature or even below. Generally, the dissociation is not 100 % efficient and recombinations of the radical species can occur in the cooling pipe. Depending on the initial stable precursor gas that is used, the desired radical might not be the only radical formed during the discharge which can lead to a beam of mixed composition. For hydrogen gas, this is H and H_2 where the latter sticks to the surface only in the monolayer regime. For oxygen gas, this is O and O_2 and in a predeposition experiment where first an ice is deposited which is then exposed to O atoms, O_2 has been seen to form a thick layer on top of the predeposited species depending on the surface temperature, which can inhibit any reaction.

Surface reactions are generally monitored by means of Fourier transform infrared spectroscopy and gas-phase molecules, generated by performing a post-experiment TPD, by mass spectrometry using a quadrupole mass spectrometer. With an *in-situ* technique like infrared spectroscopy, the amount of species formed can be obtained as a function of time if the band strength of the species is known. If the dose is also known, a formation rate in the experiment can then be determined. However, this depends on a combination of processes and their competition, i.e., diffusion, desorption, and reaction. Moreover, often several reactive species are present and a multitude of different reactions are possible at the same time. Thus, the reaction pathways and associated branching ratios can become nontrivial to disentangle as the systems studied increase in complexity. In addition, the characteristics of the substrate—its composition and surface structure—can affect reaction pathways and rates in the sub-monolayer regime. However, these data (rates, pathways, and branching ratios) are critical for advancing surface-chemical networks for use in astrochemistry. Simulating laboratory conditions can aid in extracting or constraining reaction data.

Although quantitative information on the reaction barrier or rate is not easily obtainable from surface experiments, experiments are extremely useful in detecting whether specific reactions can proceed under circumstances close to those present in the ISM and this is the aim of most experimental studies. It is also not trivial to *calculate* rates of surface reactions by quantum chemical methods. Surface reactions require many atoms in the calculation, which unequivocally increases the computational time. One of the approximations that can be made is the use of a model surface, e.g., a coronene molecule to model a carbonaceous grain surface (Adriaens et al. 2010), or a water cluster to mimic a thick ice layer (Rimola et al. 2014). Another promising method that can be used to include the effect of the surface on reaction rates is quantum mechanics/molecular mechanics (QM/MM, Goumans et al. 2009). Although such methods are robust for studying the effect of the surface on the reaction barrier height and shape, they cannot account for the orption of energy through phonon excitation. Therefore, they can result in different dynamical behavior of the products and thus, different rates.

Much more data is available for gas-phase reactions and a reasonable assumption for the construction of a surface reaction network is to lend from gas-phase data. As far as we are aware, there have been no reports of reactions that are efficient on the surface and not in the gas phase (or vice versa), except for association reactions of the form, A+B⟶ABA+B⟶AB. This type of reaction is very inefficient in the gas phase without the presence of a third body, since it has to proceed through radiative association. For surface reactions, the grain acts as a third body.

For reactions on clusters, the efficiency of association reactions likely lies somewhere in between. Examples of reactions that have a high barrier in the gas phase but nevertheless proceed in the solid phase at low temperatures are $CO + H$, $H_2CO + H$, $H_2 + OH$ and $H + H_2O_2$ (see Tables 2 and 3). This is thanks to the possibility of many crossing attempts, energy dissipation for reactions of the type A+B⟶ABA+B⟶AB, and tunneling; although the latter is relevant in the gas phase as well.

Table 2. Astrochemically relevant surface reactions for which tunneling has been confirmed experimentally

	Reaction	Comment		Ref.
1	$H + CO \longrightarrow HCO$	H-addition		A
2	$H + CH_2O \longrightarrow CH_3O/CH_2OH$	H-addition		B
3	$H + C_kH_l \longrightarrow C_kH_{l+1}$	H-addition	C_2H_2, C_2H_4, C_2H_6	C
		H-addition	C_6H_6	D[a]
4	$H + C_kH_lO_mN_n \longrightarrow H_2 + C_kH_{l-1}O_mN_n$	H-abstraction by H	CH_3OH[b]	E
			CH_3NH_2[c]	F[a]
			Glycine[b]	G
5	$H_2 + OH \longrightarrow H + H_2O$	H-abstraction by OH		H
6	$H + H_2O_2 \longrightarrow H_3O_2 \longrightarrow H_2O + OH$	H-addition		I[d]
7	$H + SiH_4 \longrightarrow SiH_3 + H_2$ $SiH_3 \longrightarrow$ Si–Si $\longrightarrow$ a–SiH[e]	H-abstraction		J[a]

[a]Based on high efficiency at low-T for high barrier process; no T-dependence or KIE studies

[b]Only on C-H sites

[c]On both C-H and N-H sites, but likely via a different mechanism

[d]See also additional comments in the Faraday Discussions

[e]Proposed mechanism

A—Hiraoka et al. (2002), Watanabe and Kouchi (2002), Hidaka et al. (2007); B—Watanabe and Kouchi (2002), Fuchs et al. (2009), Hidaka et al. (2009); C—Hiraoka et al. (1999, 2000); D—Hama et al. (2014); E—Nagaoka et al. (2007), Hidaka et al. (2009); F—Oba et al. (2014a); G—Oba et al. (2015); H—Oba et al. (2012); I—Miyauchi et al. (2008), Oba et al. (2014b); J—Hiraoka et al. (2001)

Table 3. Astrochemically relevant surface reactions for which tunneling has been studied theoretically

	Reaction	Comment	Method	$T_{\rm min}$	Ref.
1	$H + CO \longrightarrow HCO$	H-addition	HQTST	5 K	A
2	$CO + O \longrightarrow CO_2$	O-addition	HQTST	10 K	B
3	$CO + OH \longrightarrow CO_2 + H$		SCTST	200 K	C
4	$H + CH_3OH \longrightarrow CH_2OH + H_2$	H-abstraction by H	HQTST	30 K	D
5	$H + CH_4 \longrightarrow CH_3 + H_2$	H-abstraction by H	RPMD	200 K	E
6	$H + C_6H_6 \longrightarrow C_6H_7$	H-addition	HQTST	40 K	F
7	$O + CH_4 \longrightarrow OH + CH_3$	H-abstraction by O	RPMD, VTST/MT	200 K	G
8	$OH + H_2O \longrightarrow H_2O + OH$	H-abstraction by OH	Q-RPH	200 K	H[a]
9	$OH + CH_4 \longrightarrow H_2O + CH_3$	H-abstraction by OH	RPMD, VTST/MT	200 K	I
10	$OH + NH_3 \longrightarrow H_2O + NH_2$	H-abstraction by OH	CVT/μOMT	200 K	J
11	$OH + H_2 \longrightarrow H_2O + H$	H-abstraction by OH	SCTST	200 K	K
12	$OH + OH \longrightarrow H_2O + O$		SCTST	200 K	L

[a]Small water clusters, $(H_2O)_n$, $n = 1$–3

A—Andersson et al. (2011); B—Goumans and Andersson (2010); C—Nguyen et al. (2012), Weston et al. (2013); D—Goumans and Kästner (2011); E—Li et al. (2013); F—Goumans and Kästner (2010); G—Gonzalez-Lavado et al. (2014); H—Gonzalez et al. (2011); I—Suleimanov and Espinosa-Garcia (2015); J—Monge-Palacios et al. (2013a,b); K—Nguyen et al. (2011); L—Nguyen and Stanton (2013)

5.2 Tunneling

Experiments have shown that a number of surface reactions proceed through quantum-mechanical tunneling, a small summary of which is shown in Table 2. An extensive explanation can be found in Hama and Watanabe (2013). Briefly, both hydrogen-addition and hydrogen-abstraction reactions are important processes that can occur through tunneling.

When tunneling is involved in "crossing" the reaction barrier, the description of the rate coefficient changes. To accurately describe tunneling, a full quantum-mechanical calculation is preferred and there are numerous methods available to calculate tunneling rates (summarized recently by Kästner 2014). However, such methods typically lie beyond the scope of most astrochemical models. Instead, the most common way to account for tunneling in models is by using the Wentzel-Kramers-Brillouin approximation and the (crude) assumption of a rectangular barrier, which le to a rate constant,

$$k_{{\rm react},i,j} = \nu \exp\left(-\frac{2a}{h} \sqrt{2\mu E_{\rm a}} \right), \tag{18}$$

where νν is an attempt frequency (as described previously), aa is the width of the barrier, hh is Planck's constant, μμ is the effective mass of the system and EaEa is the reaction barrier. Temperature no longer plays a role, in contrast with the rate coefficient for thermally-activated reactions, and the reaction probability increases with decreasing reduced mass and barrier width.

For a particular reaction system, below the so-called cross-over temperature, the contribution of quantum-mechanical tunneling to the reaction rate dominates over the thermal contribution. Tunneling can only occur through the exothermic part of the barrier. For an endothermic reaction, the rate coefficient described by Eq. (18) should be scaled by a Boltzmann factor accounting for the difference in energy between the reactant and product states, following arguments of detailed balance

$$k_{\mathrm{react},j,i} = \nu \exp\left(-\frac{2a}{h}\sqrt{2\mu E_{\mathrm{a}}}\right)\exp\left(-\frac{\Delta E}{kT}\right) \tag{19}$$

where EaEa and ΔEΔE are defined in Fig. 5. This decreases the reaction rate considerably, and can potentially significantly alter the outcome of large astrochemical models (e.g., Lamberts et al. 2014a).

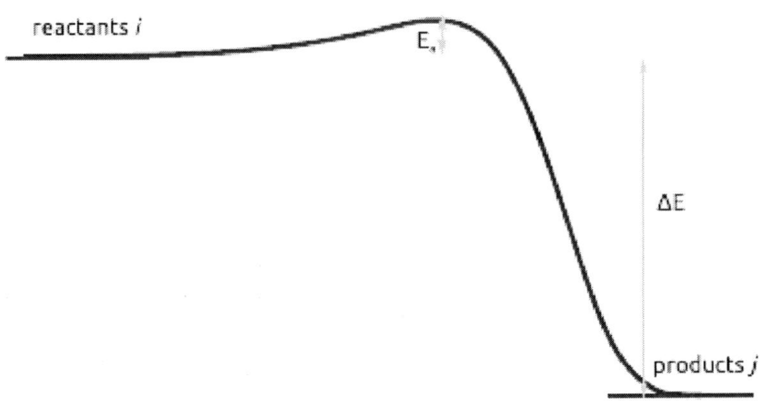

Figure 5. Energy profile of a reaction the forward reaction is exothermic; the backward reaction is endothermic and its rate should be described by Eq. (19)

First of all, the width of the barrier, aa, in the simple approximation mentioned above, is often not known and therefore commonly approximated as 1–2 Å (Garrod and Herbst 2006; Furuya et al. 2013;

Walsh et al. 2014), initiated by older literature such as Tielens and Hagen (1982) and Hasegawa et al. (1992). However, the shape of the barrier can also have a significant effect on the tunneling transmission coefficient. Eq. (18) assumes a rectangular barrier, i.e., with equal initial and final energies, which is not the case for most chemical reaction systems. A computationally cheap method to improve on this rectangular shape is using the Eckart model for the shape of the potential (Eckart 1930). However, this requires knowledge of the forward and backward reaction barriers, as well as the imaginary frequency of the transition state. Therefore, (quantum) calculations need to be available, which is typically the case only for the gas phase. Recently, Taquet et al. (2013) demonstrated the significance of using the Eckart model versus a rectangular barrier on the calculation of the rates for a set of surface reactions, using gas-phase theoretical data as input.

Secondly, let us consider the mass dependence, which can be intuitively understood by considering the uncertainty principle ΔxΔp>ℏ/2ΔxΔp>ℏ/2. A large mass will result in a large ΔpΔp, causing the accuracy of the position, ΔxΔx, to be better defined. Hence, a particle is more localized and is less likely to "leak" some of its probability density into or through a barrier. Eq. (18) requires the effective mass, μμ, along the reaction coordinate, which for a simple bond forming reaction can be approximated by,

$$\mu = \left(\frac{1}{m_1} + \frac{1}{m_2}\right)^{-1} = \frac{m_1 m_2}{m_1 + m_2}, \tag{20}$$

where m1m1 and m2m2 are the masses of the *individual atoms* forming the bond. However, in astrochemical models, generally the reduced mass of the *system* is used,

$$\mu = \left(\frac{1}{m_A} + \frac{1}{m_B}\right)^{-1} = \frac{m_A m_B}{m_B + m_A}, \tag{21}$$

where are A and B are the reacting species, regardless of the reaction mechanism. For abstraction reactions A+XB⟶AX+BA+XB⟶AX+B, the situation becomes more complex and depends on the incoming angle. The expression that can be applied to calculate the reduced mass for linear systems A–X–B is

$$\mu = \frac{m_A m_X (1+c)^2 + m_X (m_A + c^2 m_B)}{(m_A + m_X + m_B)(1 + c^2)}. \tag{22}$$

Here, cc is the ratio between the infinitesimal change in bond length of RAXRAX and RXBRXB. For reactions where species A and B have equal mass, or both have masses much larger than that of X, μμ is between 12mX12mX and mXmX. In general, however, this depends on the forces acting on the system. More extensive theoretical literature can be found in Bell (1980).

As an example, the reduced mass for abstraction reactions by OH, e.g.,

$$A\text{–}H + OH \longrightarrow A \cdot + H_2O, \tag{23}$$

should reflect that the tunneling species is the hydrogen atom and the effective mass should be μ≈mHμ≈mH. However, using Eq. (21) will yield a much higher mass and this has a large influence on the reaction rate. Consider, for example, a reaction with a barrier EaEa of 2000 K and a width of 1 Å. Using Eq. (18) with two different values for μ ($\mu_1 = 10$ and $\mu_2 = 1$ a.u.) yields $k_1/k_2 = 2 \times 10^{-3}$.

Typically, gas grain codes use information on barrier height and the masses of the products to calculate the rates. We recommend to include the tunneling reaction rate as obtained from high-level gas-phase calculations as an additional input parameter in an astrochemical model. When such rates are unavailable, the original approximate Eq. (18) can be reinstalled.

The importance of tunneling can be experimentally confirmed in two ways: either by studying the temperature dependence—to determine whether this confirms the predicted behavior of either Eq. (15), (17) or (18)—or by making use of the mass dependence of tunneling. In this latter case, an experiment is performed (at least) twice, using different isotopes of the same species between the two (or more) runs. Since isotopes are chemically equivalent, but have a different mass, this is a convenient way to detect the difference in tunneling efficiency. If the reaction products scale with the effective mass of the reaction, this is usually interpreted as a proof of tunneling and is called the Kinetic Isotope Effect (KIE). For instance, a reaction with a hydrogen atom, H+X⟶HXH+X⟶HX, can be compared to the same reaction with a deuterium atom, D+X⟶DXD+X⟶DX. In both cases, the experiments are performed at temperatures as low as possible, which is typically setup-dependent.

In order to study the influence of tunneling on a reaction with a relatively high precision from a computational perspective, it is necessary to use methods that encompass more than only an Eckart calculation of the barrier, or a Wigner correction (Wigner 1938) to the rate. For instance, a variety of Transition State Theories (TST) have been applied, such as Variational TST (or Canonical Variational Theory, CVT) combined with a Multidimensional Tunneling correction (VTST/MT), Semi-Classical TST (SCTST), and Harmonic Quantum TST (HQTST). Other methods, such as Quantum-Reaction Path Hamiltonian method (Q-RPH), Ring Polymer Molecular Dynamics (RPMD), and Free Energy Instanton Theory (FEIT) have also been employed. Several recent papers comment on the differences and accuracy of these various methods with respect to each other (Nyman 2014; Zhang et al. 2014; Hele et al. 2015). In Table 3, a number of gas-phase studies are listed that focus on reactions that may be important in surface astrochemistry. However, this is by no means an exhaustive list and focuses only on recent papers. Note that we have excluded the reaction H2+O⟶OH+HH2+O⟶OH+H from this table, based on its endothermicity (Lamberts et al. 2014a) and on the high barrier and, consequently, low reaction rate (Nguyen and Stanton 2014).

5.3 Temperature Dependence of Surface Reactions

Although many reaction rates are temperature independent, because they are either barrierless or occur through quantum tunneling, reactions still have a temperature window within which they are most effective. This is a consequence of the temperature dependence of the diffusion and desorption of reactants. Diffusion is required for reactants to meet in a Langmuir-Hinshelwood reaction and determines the lower bound of the temperature window; desorption, on the other hand, sets the upper bound. As a result, at low temperatures, surface chemistry will mostly be dominated by hydrogenation reactions,

$$X\cdot + H \longrightarrow X\text{–}H, \qquad (24)$$

effectively converting all radical species to more stable species such as H_2O, CH_4, NH_3, and CH_3OH. In parallel, hydrogen abstraction can recreate surface radicals,

$$X\text{–}H + Y\cdot \longrightarrow X\cdot + Y\text{–}H, \qquad (25)$$

but most likely with a much lower efficiency. Whereas hydrogenation reactions have been extensively studied, for hydrogen abstraction reactions, the reaction rates and the importance of tunneling remain unknown.

At higher temperatures, the residence time of hydrogen atoms on the surface becomes too short for hydrogenation reactions to dominate, and other surface reaction types increase in importance. For example, high-mass star-forming cores, that are in the so-called "hot core" phase, are found to be rich in gas-phase complex organic molecules, including alcohols, aldehydes, carboxylic acids, esters, and ethers (Bisschop et al. 2007b). While several of these classes of molecules may have viable gas-phase formation mechanisms, the larger species are thought to be formed primarily (or exclusively) on the surfaces of dust grains, or within/upon dust-grain ice mantles (Garrod and Herbst 2006; Garrod et al. 2008). The main reaction mechanism for the formation of complex organics on the grains is the creation of functional-group radicals on/within the ice mantle, such as CH_3 and CH_3O, both of which may be formed via the photodissociation of methanol (CH_3OH). As the temperature in the core increases to above 20 K20 K or so, these radicals become mobile, diffuse across or through the ice mantle, thereby allowing radical-radical association reactions to become competitive with hydrogenation of radicals by abundant H atoms. As temperatures rise yet further (≳100 K≳100 K), the grain-surface-formed molecules desorb into the gas phase, where they are detected typically via (sub)mm rotational spectroscopy.

Ubiquitous molecules, including methyl formate ($HCOOCH_3$) and dimethyl ether (CH_3OCH_3), appear to have efficient dust-grain surface formation routes. Astronomical observations and astrochemical modeling (e.g., Belloche et al. 2014) indicate that surface/ice-mantle formation mechanisms may also be efficient for very large molecules like propyl cyanide (C_3H_7CN); however, in such cases, the earlier formation of smaller homologues (e.g. ethyl cyanide, C_2H_5CN) is usually required. The removal of a hydrogen atom from these molecules produces the necessary large radicals, to which other functional groups may be added, further increasing chemical complexity.

The most abundant molecule in the ices is water itself, which may be photodissociated to form the highly reactive OH radical. At relatively low temperatures, most OH produced recombines with H, or with mobile, abundant radicals like CH_3 and HCO. However, models suggest that above around 60 K (Garrod 2013), OH becomes sufficiently mobile to find and react with large stable molecules, to abstract an H atom and produce a large radical. At this stage, H-abstraction by OH (along with NH_2, which is also formed by H-abstraction from ammonia) becomes the dominant formation mechanism for the molecular radicals that are the precursors to even larger species on the grains.

While relatively few of the H-abstraction reactions of OH invoked in this dust-grain chemistry have been directly measured, even in the gas phase, those for which rates have been determined display very small activation energy barriers to H abstraction by OH, typically less than 1000 K1000 K. Crucially, barriers of this size are comparable to the expected diffusion barrier for surface OH, meaning that for these reactions, the surface diffusion of OH is the rate-limiting step. Thus, as soon as temperatures are sufficiently high for OH diffusion, OH becomes the key instigator of radical production.

Unfortunately, few of these barriers to H-abstraction by OH from large, saturated molecules have been determined, and the calculation of rates in models is further complicated by the fact that the abstracted hydrogen atom may be able to tunnel through the barrier, meaning that information about the barrier shape is required. Alternatively, if tunneling is efficient, then in many cases the key quantity needed in astrochemical gas-grain calculations will be the diffusion barrier for OH. However, an accurate determination of each one of these quantities will be necessary for a full understanding of how complex organic molecules form in different temperature regimes in interstellar regions.

The rather elegant "warm-up" scenario for the origin of so-called "hot core" molecules has been muddied by recent observational and laboratory results. High sensitivity observations have shown that complex organic molecules are also present in the gas phase in cold environments (~ 10 K), albeit at a low level with respect to molecular hydrogen (Öberg et al. 2010, 2011; Cernicharo et al. 2012; Bacmann et al. 2012; Vastel et al. 2014). Bisschop et al. (2007a) investigated the laboratory hydrogenation of solid acetaldehyde, CH_3CHO, under interstellar relevant conditions, using both RAIRS and TPD to analyze the results. The experiments showed that the hydrogenation of CH_3CHO le to the formation of 20 % of ethanol, C_2H_5OH, showing for the first time that surface hydrogenation of unsaturated complex species can be responsible for the abundances of more complex saturated species detected in dense interstellar clouds. In addition, it has now been demonstrated experimentally that molecules such as glycolaldehyde ($HOCH_2CHO$) and ethylene glycol ($HOCH_2CH_2OH$) can form at 10 K *without* the need for irradiation (Fedoseev et al. 2015). Here, the reaction scheme also relies mainly on hydrogenation reactions which are known to be efficient at low temperatures; however, the scheme also involves dimerization of the HCO radical which allow the formation of the C–C backbone (see also Woods et al. 2013). These results suggest that complex molecules are already present in the ice mantles prior to warm up in the environs of young stars.

A few alternatives to the UV photo-induced surface chemistry hypotheses involve gas-phase reactions (Rawlings et al. 2013; Balucani et al. 2015). The so called "Rapid-Radical Association" mechanism proposes that COMs are formed by three-body gas-phase reactions between radicals in warm high density gas (Rawlings et al. 2013). This environment exists, for a very short period of time, following the sudden and total sublimation of dust-grain ice mantles driven by the catastrophic recombination of trapped hydrogen atoms, and other radicals, in the ice. More recently, Balucani et al. (2015) proposed a new gas-phase model to form some of the COMs, such as dimethyl ether and methyl formate, starting from methanol in the gas phase. In the proposed scheme, dimethyl ether is the precursor of methyl formate via an efficient reaction overlooked by previous models. Very recently, electronic structure and kinetic calculations showed that the gas-phase reaction NH2+H2CONH2+H2CO leading to formamide is barrierless. Hence, for some species, there is no need to invoke grain-surface chemistry provided that the necessary precursors are available in the gas phase (Barone et al. 2015). The main complicating factor in deciphering the chemistry of COMs is that their abundances and abundance ratios usually vary significantly from source to source. Moreover, not all the detected COMs and their abundances can be explained by gas-phase reactions alone. Therefore, it is likely that surface reactions on ice grains still play an important role in the dense regions of the ISM.

Figure 6 taken from Linnartz et al. (2015) summarizes the extensive laboratory work that investigated the surface formation of several simple and more complex species through atom-addition and radical-radical recombination reactions under dense cloud conditions. The figure shows how the surface reaction routes to H_2O, CO_2, HCOOH, CH_3OH, $(CH_2OH)_2$, HNCO, NH_3, and NH_2OH, are interlinked and are initiated by the H/O/N-atom addition to simpler precursor species, like CO molecules. The efficiency of these reaction routes, and their contribution to the molecular abundances observed in space, depends on a number of physico-chemical parameters.

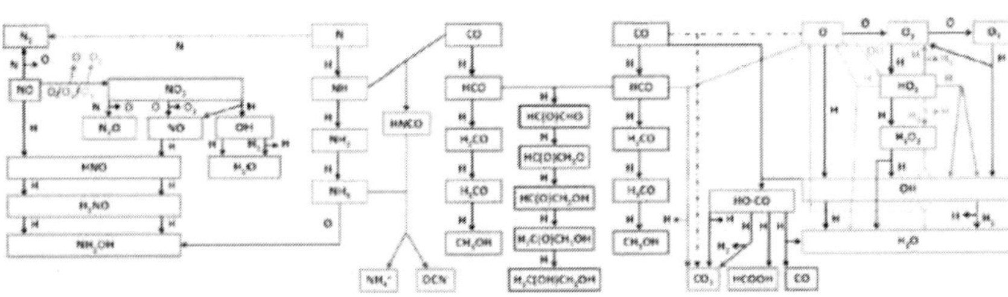

Figure 6. A summary of non-energetic surface chemistry through atom-addition and radical-radical recombination reactions based on experimental evidence. The *arrows* indicate possible pathways, but other (energetic) processes are at play as well. The figure clearly shows the complexity of non-energetic ice chemistry and the possibility for this type of chemistry to create complex molecules without additional energy input. Figure reproduced from Linnartz et al. (2015)

For instance, in atom addition/abstraction reactions, the height of an activation barrier determines whether or not a reaction can proceed at 10 K. Radical-radical recombination reactions are barrierless and therefore their inefficient thermal diffusion at 10 K is the limiting factor. Another important parameter is the molecular environment. In the case of exothermic reactions, polar (water-rich) ices can promote surface chemistry through the dissipation of extra energy in their H-bond network.

6 DIFFUSION

As discussed in the previous section, diffusion rates are key to determining the rate for the Langmuir-Hinshelwood reaction mechanism, because they regulate the meeting frequency between reactants. Models require as input diffusion barriers and binding energies for each of the surface species included in the reaction network to determine the hopping rate

$$k_{\mathrm{hop,th},i} = \nu \exp\left(-\frac{E_{\mathrm{diff}}}{kT}\right). \tag{26}$$

The scanning rate kscankscan is determined by either thermal hopping or quantum mechanical tunneling, depending on the mass of the diffusing species.

In early models, H atoms were assumed to diffuse via quantum tunneling. After experimental studies showed that the diffusion of H atoms is rather slow, thermal hopping became favored in models (Pirronello et al. 1997b,a, 1999; Katz et al. 1999). More recently, the discussion on the nature of the diffusion mechanism for atoms was reopened, with experiments suggesting H atoms can quantum tunnel under particular conditions (Watanabe et al. 2010; Kuwahata et al. 2015). It has also been postulated that atoms as heavy as oxygen can diffuse via quantum tunneling (Congiu et al. 2014). Heavier species are most likely to diffuse through thermal hopping.

Obtaining diffusion barriers experimentally remains challenging. Diffusion rates cannot be measured directly and have to be inferred from experiments using a model. Often diffusion barriers are measured through reaction, where a known dose of both mobile reactants and relatively immobile (stationary) reactants are deposited. If the reaction between species is diffusion limited, i.e., there exists no reaction barrier and in the regime of low surface coverage, the diffusion rate can be inferred from the disappearance of the reactants as a function of temperature, dose, and/or time. This method is limited to reactive species and, because one works within the submonolayer regime, sensitivity of detection is an issue. Usually products are measured using mass spectrometry during TPD. Another method is to deposit a layer of ASW (amorphous solid water) on top of an ice consisting of the species of interest. If the temperature is raised above the desorption temperature of this species, the rate limiting step for desorption is the diffusion of the species through the ASW layer. In the case of porous ASW, pore-wall diffusion is probed which mimics surface diffusion. For compact ASW, bulk diffusion is more likely studied. Here, the disappearance of the diffusing species is monitored, usually by IR spectroscopic techniques. This method is limited to stable species with a desorption temperature below that of ASW. Surface diffusion rates for CO molecules have been studied this way (Mispelaer et al. 2013; Karssemeijer et al. 2014b; Lauck et al. 2015).

For both methods of study, the model that is used to extract the diffusion rates is crucial. In the case of reaction, one has to ensure that no other reactions play a role. Radical beams are often not 100 % pure and other species will also be present on the surface. Moreover, species might not be instantaneously thermalized when deposited on the substrate, so that they are able to move some distance superthermally. One has to be sure that the second reactant is indeed stationary. Furthermore, the effect of the warm-up phase during TPD needs to be considered which can thermally enhance diffusion (and subsequent reaction). In the case of the two-layer experiments, care should be taken when considering to which type of diffusion the system is limited: bulk diffusion, pore wall diffusion, or surface diffusion upon a layer of the same species orbed on the ASW (see Fig. 7).

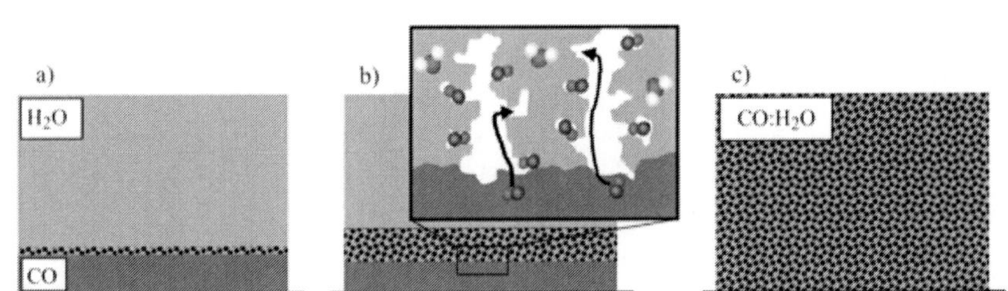

Figure 7. Schematic of the mixing process. (**a**) Layered system at t=0t=0. (**b**) Occurrence of mixing, with the inset showing the CO molecules diffusing along the micropore surfaces into the strong-binding nanopore sites. (**c**) After some period of time, the layer becomes fully mixed. Figure taken from Lauck et al. (2015)

The species that has been most studied to determine its surface diffusion is atomic hydrogen. Because it is a radical, it is studied through reaction, either with itself to form H_2, with D to form HD, or with O_2 to from H_2O_2, and ultimately H_2O. However, the results between different experimental groups are not in agreement (see Hama and Watanabe 2013, for a review). The latest results on an ASW surface show that very shallow-potential binding sites (Ediff≤18±2 meVEdiff≤18±2 meV) are dominant, while there are also middle- (Ediff=22±1 meVEdiff=22±1 meV) and deep- (Ediff=30 meVEdiff=30 meV) potential binding sites (Watanabe et al. 2010; Hama et al. 2012). These particular results were obtained with a new technique for the detection of product species. The remaining species are photodesorbed by a laser as a function of delay time between deposition and laser desorption and the detection occurs through REMPI (resonance-enhanced multi-photon ionization). This eliminates the effect of warm-up which can be an issue for TPD experiments. These experiments support the following picture (Watanabe et al. 2010; Hama et al. 2012; Kuwahata et al. 2015): H atoms land on the ASW surface, diffuse rapidly using the abundant shallow and middle sites, and are finally trapped in the deepest sites. Once a significant number of H atoms become trapped, the subsequent H atoms recombine with the trapped H atoms. Therefore, the effective diffusion rate becomes dependent upon coverage and is also affected by the number of deep binding sites present on the ice surface.

This picture is also observed at an atomistic level through adaptive kinetic Monte Carlo simulations of CO diffusion on an ASW surface (Karssemeijer et al. 2014b). Here, for a single CO molecule on an ASW surface, the diffusion is limited by diffusion out of the deep binding sites (ECOdiff=84--114 meVEdiffCO=84--114 meV), whereas when the deep sites are filled with CO molecules, the diffusion rate is increased (ECOdiff=48--79 meVEdiffCO=48--79 meV). In the simulations, a large spread in diffusion barriers between different surfaces is found, in addition to the large spread observed for each specific surface. However, given a specific surface and surface coverage, the temperature-dependent diffusion constant follows an Arrhenius-like behavior, i.e., it can be described by a single diffusion barrier, which is consistent the diffusion barrier for moving out of the deepest available binding site. This is good news for astrochemical gas-grain models because, for simplicity, they typically do not account for local chemical or structural differences on the grain. Instead, these models aim to provide a macroscopic view. For this purpose, a single diffusion barrier per species likely suffices. A collection of different barriers and binding energies obtained in this way is presented in Table 4.

Table 4. Desorption and diffusion energy barriers for CO and CO_2 on the proton-disordered and the ordered Fletcher phase of ice Ih

Substrate	Adsorbate	E_{bind}[a] (meV)	E_{diff} (meV)	f	Approach	Method	Ref.
Disordered	CO	143	49	0.34	comp.	AKMC	A
(frozen)	CO_2	299	127	0.42	comp.	AKMC	B
	H_2O	670	220	0.33	comp.	AKMC	C
Fletcher	CO	133	39	0.29	comp.	AKMC	B
(frozen)	CO_2	291	110	0.38	comp.	AKMC	B
Fletcher	CO	137	42	0.31	comp.	AKMC	B
(free)	CO_2	323	122	0.38	comp.	AKMC	B
ASW (frozen)[b]	CO (1 + 3)[c]	96	147	0.42	comp.	AKMC	D
ASW (frozen)[b]	CO (1 + 6)[c]	86	134	0.36	comp.	AKMC	D

[a]Binding energies are time-averaged over the kinetic Monte Carlo runs at temperatures of 25 and 70 K for CO and CO_2

[b]Amorphous solid water (ASW) values are calculated from simulations published in Karssemeijer et al. (2014b) on substrate S^c_2. Values are also at $T = 25$ K

[c]There is only one mobile CO molecule on the substrate. The remaining 3 or 6 CO molecules are immobilized in strong binding pore sites on the substrate

A—Karssemeijer et al. (2012); B—Karssemeijer and Cuppen (2014); C—Pedersen et al. (2015); D—Karssemeijer et al. (2014b)

Table 4 is far from complete, and to obtain the remaining diffusion barriers required for the astrochemical models, the diffusion energy of a species is assumed to be a universal, fixed fraction, ff, of the desorption energy: Ediff,i=fEbind,iEdiff,i=fEbind,i. There is no fundamental physical argument for such a universal ratio to exist and it is used solely due to the lack of data. This ratio is most likely dependent upon the species, the substrate material and structure, and on the surface coverage (which determines the likely binding "partner"). There already exist problems with the concept of using a single diffusion and a single desorption barrier even for a single species because they both vary strongly from site to site, especially for amorphous ices. As shown by Vasyunin and Herbst (2013) the value of ff influences the outcome of the models. Due to the lack of diffusion information, the existence and possible value of fraction ff is very poorly constrained and values between 0.3 and 0.8 are used by the modeling community and is often treated as a free parameter within these typical bounds (Hasegawa et al. 1992; Ruffle and Herbst 2000; Cuppen et al. 2009). An accurate value of ff is also important because it affects the conditions under which the accretion limit is reached and thereby whether or not modelers should turn to stochastic models. The data in Table 4 suggests that, at least for stable species, there is a more or less constant ratio and that this ratio, ff, is more likely to lie around 0.3--

0.40.3--0.4. It is not possible, currently, to give a definitive recommended value. We encourage the experimental and theoretical community to continue to work towards filling this large gap in the necessary input data for astrochemical models, especially for radical species, which, to date, have remained largely unstudied.

7 BULK PROCESSES

Although the chemistry on interstellar grains starts out with processes on bare carbonaceous or silicate grains, when the ice thickness increases to more than a few monolayers, it becomes important to differentiate between surface and bulk processes. The earliest models of grain-surface chemistry primarily considered the grain surface as a substrate for the formation of molecular hydrogen or other simple species, which could then rapidly desorb back into the gas phase upon recombination (e.g., Watson and Salpeter 1972a; Allen and Robinson 1977; Tielens and Hagen 1982). Hence, originally, ice chemistry was modeled using rate equations that consider only the averaged abundance of a species throughout the entire mantle (Hasegawa et al. 1992). This is a rather crude approximation because bulk species cannot diffuse as easily as those on the surface. "Bulk ice" implies that the species involved in "bulk" processes are fully surrounded by neighboring molecules and are therefore rather tightly bound, leading to the assumption that diffusion within the bulk ice at low temperature is inefficient and therefore, chemistry is also inhibited. The ice mantle can thus be conceptually divided into an ice surface and bulk ice. To this end, three-phase models have been introduced, where the distinction is made between gas-phase species, reactive surface species within the top (few) monolayer(s), and fully inert bulk species in the core of the ice mantle, with terms that allow surface material to be incorporated into the bulk (or vice versa), as the ice thickness increases or decreases (Hasegawa and Herbst 1993b; Garrod and Pauly 2011). Each phase is treated mathematically as an independent entity. More recent models, informed by the discovery of substantial ice mantles on interstellar dust grains via infrared absorption observations, have involved a more concerted effort to treat the build-up of surface ices (e.g., Hasegawa et al. 1992). It is in the extension of these models to the formation of multiple layers of ice that a significant conceptual error should be identified. The absolute reaction rates for surface processes are commonly formulated thus:

$$f_{\mathrm{react}}(A+B) = n_s(A)n_s(B)\left[k_{\mathrm{hop}}(A)+k_{\mathrm{hop}}(B)\right]/N_{\mathrm{sites}} = n_s(A)n_s(B)k_{\mathrm{scan}}(AB) \quad \text{CCC(27)}$$

where khopkhop is the site-to-site hopping rate for either species A or B, and NsitesNsites is the number of surface sites present on the grain surface. ns(A)ns(A) and ns(B)ns(B) are used here as a shorthand to represent the mean populations of each species, ⟨ns(A)⟩⟨ns(A)⟩ and ⟨ns(B)⟩⟨ns(B)⟩. This formulation is often further simplified, absorbing all coefficients into a single rate, kscan(AB)kscan(AB). The efficiency factor used for reactions involving activation energy barriers is omitted here for simplicity.

Equation (27) is valid in the case where the sum of all surface reactants, which we label nallnall, is smaller than NsitesNsites, i.e. that there is less than one layer of particles on the grain surface. However, if $n_{\text{all}} > N_{\text{sites}}$, the equation is necessarily invalid. This may be seen more clearly if Eq. (27) is re-arranged:

$$f_{\text{react}}(A+B) = n_s(A)k_{\text{hop}}(A)n_s(B)/N_{\text{sites}} + n_s(B)k_{\text{hop}}(B)n_s(A)/N_{\text{sites}} \qquad (28)$$

It may be seen that the reaction rate is composed of two analogous parts. The first represents the rate at which all surface species of type A may hop into an adjacent site, multiplied by the probability that the neighboring site is occupied by a species of type B. The second term may be described similarly. This latter probability is the simple ratio ns(B)/Nsitesns(B)/Nsites. Clearly, if ns(B)ns(B)exceeds NsitesNsites, such a probability is meaningless. More rigorously, this is also true where nall>Nsitesnall>Nsites.

A majority of two-phase models (i.e. gas-phase and grain surface) retain the usage of NsitesNsites in the reaction rates. In the case of an ice mantle with a thickness of, say, 100 monolayers, this could lead to a rate that is inaccurate by two to four orders of magnitude.

The simplest fix for this problem is to exchange NsitesNsites for nallnall, which may be calculated easily within the model code, for cases where nall>Nsitesnall>Nsites. This approach removes the immediate inconsistency in the probabilities; however, it retains another assumption implicit in the equations above, namely that not only are the species in all layers of the ice chemically active, but that all such reactants may diffuse within the bulk ice at the surface diffusion rate. Neither of these assumptions should necessarily be used, as explained elsewhere in this paper.

To remove the latter problem while retaining the structure of a two-phase model, one may adjust Eq. (28) further, by retaining NsitesNsites, but substituting ns(A)ns(A) for nmax(A)nmax(A), where:

$$n_{\max}(A) = n_s(A); \quad \text{for } n_{\text{all}} \leq N_{\text{sites}} \tag{29}$$

$$n_{\max}(A) = n_s(A)N_{\text{sites}}/n_{\text{all}}; \quad \text{for } n_{\text{all}} > N_{\text{sites}} \tag{30}$$

And similarly for nmax(B)nmax(B), giving

$$f_{\text{react}}(A+B) = n_{\max}(A)n_{\max}(B)\left[k_{\text{hop}}(A) + k_{\text{hop}}(B)\right]/N_{\text{sites}} \tag{31}$$

This solution removes the error in the probabilities, while limiting the reactive portion of the ice mantle to a single layer, on the assumption that a layer is composed of a total of NsitesNsitesparticles. Such an approach may be considered a quasi-three-phase model, although the chemically-active portion of the mantle, which is assumed to be the surface layer itself, nevertheless represents the composition of the entire mantle. Only a fully realized three-phase model, consisting of entirely separate phases for the gas, ice mantle, and ice-surface layer can resolve this point. However, the added complexity of the physical treatment, combined with technical challenges (see below), makes three-phase modeling of astrophysically complex objects such as protoplanetary disks including bulk chemistry a more distant prospect, as outlined in Sect. 2.

Another approximation that can be made is the multilayer approach by Taquet et al. (2012, 2013), where the abundances of each single layer are stored in the simulation and these compositions are used to determine the binding energy of the species in the overlying layer. Again only the top monolayer is considered chemically active. Furthermore, recently another multilayer model (using a Monte Carlo approach) has been reported where the chemically active surface is extended to the uppermost four monolayers (Vasyunin and Herbst 2013). Both models account for a correct description of thermal desorption, but still leaves the bulk reactively inert. In microscopic Kinetic Monte Carlo routines, on the other hand, bulk reactivity can be included, for instance through allowing two neighboring molecules to swap position (Öberg et al. 2009d) or incorporating interstitial positions between the molecules in the lattice (Lamberts et al. 2013; Chang and Herbst 2014).

A number of experimental and theoretical studies indicate that bulk processes can be important, both at low and high ice temperatures. Here we make the distinction between bulk diffusion and pore-wall diffusion, since the latter is essentially surface diffusion. Experimentally it is not always straightforward to distinguish the two effects. Bulk diffusion appears to play a role in the segregation of mixed H_2O:CO_2 and H_2O:CO ices which le to the occurrence of “islands” of both mixture components at temperatures of 30 K for CO and 60 K for CO_2 (Öberg et al. 2009d). Exposure of an O_2 ice to hydrogen atoms yields large quantities of H_2O_2 and H_2O, corresponding to tens of monolayers depending on the temperature (Ioppolo et al. 2010). For hydrogenation of CO, a maximum exposure of four monolayers was found experimentally (Fuchs et al. 2009). Furthermore, Molecular Dynamics simulations performed by Andersson et al. (2006) and Arasa et al. (2010) of photodissociation and photodesorption of water ice at low temperature (10–90 K) have shown that the desorption of species (H, OH, H_2O) depends strongly on the distance of the excited molecule to the top monolayer of the ice. The first 4–5 monolayers are, however, taking part in the desorption process, which indicates again that the assumption of a single reactive top monolayer cannot be valid. The desorption depth dependence has been observed for a range of excitation energies, ice temperatures and isotope compositions (Andersson et al. 2006; Andersson and van Dishoeck 2008; Arasa et al. 2010; Koning et al. 2013; Arasa et al. 2015). The studies further show that radical species which are created through photodissociation remain in the bulk of the ice, can use their excitation energy to move some short distance before they thermalize. This will most likely generally hold for radicals formed through an energetic process like photodissociation, exothermic reactions and through cosmic rays.

An analog for hydrogen diffusion at low temperature can also be brought about by reactions that have a net neutral effect, of the type H--A+A⟶A+H--AH--A+A⟶A+H--A as mentioned during the Lorentz Center Workshop by Fedoseev. This could be of interest, for instance, in the case of water ices where photodissociation can create OH radicals that may react with neighboring H_2O molecules, H2O+OH⟶OH+ H2OH2O+ OH⟶OH+H2O (Nguyen and Stanton 2013; Lamberts et al. 2016).

Recently, the discovery of a new class of thin films, *spontelectrics*, may also have interesting implications for diffusion and ordering in ISM ices (Field et al. 2013). Upon deposition onto a substrate at low temperatures, dipolar molecules were seen to spontaneously align, creating electric fields >108 eVm−2>108 eVm−2. The impact of such spontaneous ordering within ISM-like ices is yet to be quantified. An alternative mechanism that could work at high temperatures, is to allow the reactants to reach each other before the ice is close to evaporating, leaving the mantle molecules more freedom to diffuse.

Diffusion through the walls of macropores is less relevant for interstellar ices with respect to laboratory analogs. Interstellar water ice is mostly formed through reactions instead of deposition resulting in compact ice (Oba et al. 2009) by the exothermicity of surface reactions (Accolla et al. 2011). Moreover, laboratory experiments of deposited ices indicate that the macropores present in these ices collapse at least to a certain extent (Bossa et al. 2012; Mispelaer et al. 2013; Isokoski et al. 2014), making pores less relevant. Moreover, there is a class of reactions that are not (merely) diffusion limited, since one of the reactants is one of the main mantle species, but (mainly) thermally activated at higher temperatures. This is possible in particular for reactions between two neighboring species, for instance, in a water-rich environment. Examples of such reactions are H2O+D2O⟶2HDOH2O+D2O⟶2HDO (Lamberts et al. 2015), and H2O+HNCO⟶H3O++OCN−⟶HOCN+H2OH2O+HNCO⟶H3O++OCN−⟶HOCN+H2O (Theule et al. 2011). Other thermally activated reactions in an ice are possible, leading to the formation of salts (e.g., Bossa et al. 2009; Noble et al. 2014; Bergner et al. 2016). An example is the following network of equilibrium reactions

$$NH_3 + CO_2 \rightleftharpoons NH_3 + COO^- \tag{32}$$

$$NH_3 + NH_3 + COO^- \rightleftharpoons NH_4 + NH_2COO^- \tag{33}$$

$$NH_4 + NH_2COO^- \rightleftharpoons NH_3 + NH_2COOH \tag{34}$$

leading to the formation of carbamic acid in a protic environment, i.e. a NH3−NH3− or H_2O-dominated ice, where the *reaction* barrier is sufficiently low for proton transfer to occur thermally without any external source of processing. However, experimentally only mixtures with a NH_3:H_2O ratio greater than 1 were considered. In a more dilute environment, which is more representative for interstellar ices these processes become bulk diffusion limited and they require a suitable bulk diffusion mechanism in astrochemical gas-grain models.

Observations of interstellar ices show absorption bands attributed to various cations and anions (OCN^-, $HCOO^-$, NH+4NH4+, Boogert et al. 2015), and hence these processes should occur. Moreover, it is known that interstellar (and circumstellar) grains are likely charged (e.g., Draine 2003). Grains can become the main carrier of negative charge, for instance in dark and dense midplanes, where the gas is weakly ionized. As grain charge affects the chemistry, it is important to consider. Ionic chemistry in most grain surface models is currently limited to cation-grain recombination, where an ion recombines with a free electron and dissociates rather than sticking to the grain, applying gas-phase products and branching ratios analogous to those from dissociative recombination

$$XY^+ + GRAIN^- \longrightarrow X + Y + GRAIN. \quad (35)$$

The rate of this process is determined by the cation-grain collision rate and is analogous to the accretion rate (Eq. (2)) but can include an enhancement (e.g., Draine and Sutin 1987).

8 PHOTOPROCESSES

It has long been known experimentally that FUV irradiation of interstellar ice analogues induces complex chemistry (e.g., Hagen et al. 1979; D'Hendecourt et al. 1986; Allamandola et al. 1988; Grim et al. 1989; Gerakines et al. 1996; Öberg et al. 2009c). The interaction of FUV photons with ices has both primary and secondary chemical effects which influence both the ice and gas composition. Which of these effects dominates depends upon the FUV flux and shape of the FUV spectrum, the photoexcitation/photodissociation cross sections of the molecular components which make up the ice mixture, and the temperature and structure of the ice. One of these effects is termed "photodesorption", whereby absorption of FUV (or X-ray) photons by an ice le to desorption (or sublimation) of ice species into the gas phase. Photodesorption was originally thought to occur as a single-step process: the FUV photon excites the molecule into a state which is antibonding (or repulsive) with the surface and the species is ejected. As will be discussed later, it is now known that it can also occur as a multi-step process where the excited species decays back to the ground state and imparts kinetic energy to a surrounding molecule which is then "kicked out". Another effect is photodissociation, whereby the absorption of a FUV photon by a molecule induces dissociation, e.g., H2O+hν⟶OH+HH2O+hν⟶OH+H. It is this latter process which induces chemical changes in FUV-irradiated ices. As might be expected, photodesorption and photodissociation can occur in parallel. Photodissociation has primarily been induced to explain the origin of complex molecules in FUV-irradiated ices with an initial interstellar-like composition (e.g., CO, CO_2, H_2O, CH_4, N_2, NH_3, and CH_3OH).

Photodesorption is now considered an important process for releasing molecules into the gas phase in cold ($\lesssim$ 100 K) astrophysical regions where they would otherwise be frozen out onto icy grain mantles (and thus wholly depleted from the gas, e.g., Hartquist and Williams 1990). The origin of the FUV radiation depends upon the nature of the region. In cold and dark molecular clouds, FUV radiation is generated via the excitation of H_2 molecules by cosmic rays (Prasad and Tarafdar 1983), whereas in weak photodissociation regions (PDRs), the origin is the external, often enhanced, interstellar radiation field (ISRF, Hollenbach and Tielens 1997). In the envelopes surrounding forming stars, the innermost regions are sufficiently hot that thermal desorption is thought to dominate; however, molecules can be released back into the gas in the extremities of the envelope by photodesorption by the ISRF, similar to PDRs (e.g., van Dishoeck and Blake 1998).

Recent high-sensitivity observations have suggested that the origin of gas-phase molecules at offset positions from the location of the central forming star may be due to photodesorption by stellar FUV photons along the outflow cavity walls (e.g., Öberg et al. 2010). In protoplanetary disks around young stars, there are multiple origins, including FUV photons from the central star and the external interstellar radiation field, and those generated internally via the excitation of H_2 by cosmic rays or X-rays (e.g., Willacy and Langer 2000). Which source dominates the FUV radiation field depends on the radial distance away from the central star and the vertical distance away from the shielded disk midplane.

Given the potential importance of photodesorption when interpreting observations, this process has routinely been included in astrochemical models for some decades. However, it is only relatively recently that experiments have successfully quantified this process, and also that the underlying physical mechanisms at work have begun to be understood. Photoprocesses have only been included in models relatively recently once it was demonstrated that molecules, such as methanol (CH_3OH), were unable to form efficiently via gas-phase chemistry alone.

In this Section, we give a brief overview of the experimental studies specific to ice photodesorption and photodissociation (Sect. 8.1), and those theoretical calculations which have offered insight into the mechanisms at play (Sect. 8.2). We also provide a recommendation for the implementation of experimental and theoretical data into the computation of photodesorption and photodissociation rates in astrochemical models (Sect. 8.3).

8.1 Experimental Studies

The measured photodesorption yields for pure ices are listed in Table 5. Early experiments on the photodesorption of molecules of astrophysical interest resulted in yields spanning from ~10−6 photon−1~10−6 photon−1 (Greenberg 1973), to ~10−1 photon−1~10−1 photon−1 (Nishi et al. 1984; Hartquist and Williams 1990). The large range of values is likely due to the wide array of setups adopted and the challenges of direct detection of the photodesorbed molecules.

The former experiments were conducted in the 100–275 nm range at 77 K using a Hg-Xe high pressure arc lamp, whereas the latter were conducted using an ArF 193 nm laser at relatively high fluence between 90 and 135 K. It was not until the experiments by Westley et al. (1995a) and Westley et al. (1995b) that a robust quantification of water ice photodesorption at low temperatures (35--100 K35--100 K) was reported. They determined a yield of (3--8×10−3) photon−1(3--8×10−3) photon−1 at Lyman-αα(121.6 nm). Photodesorption was revisited much later by (Öberg et al. 2007, 2009a,b) who investigated the photodesorption yields of pure CO, N_2, CO_2, and H_2O ices under ultra-high vacuum conditions (base pressure better than 10−9 mbar10−9 mbar) at low temperatures using a broadband hydrogen microwave discharge lamp (120--170 nm120--170 nm). CO, CO_2, and H_2O were found to desorb from thick ices (≫10≫10 monolayers) at low temperatures (<20 K<20 K) with similar yields: (2.7±1.7× 10−32.7±1.7×10−3), (2.3±1.4×10−32.3±1.4×10−3), and (1.3±0.8×10−3) photon−1(1.3±0.8×10−3) photon−1. Pure N_2 ice, on the other hand, had a significantly lower yield (<2×10−4) photon−1(<2×10−4) photon−1. Öberg et al. (2009a) also provided evidence that photodesorption occurs primarily from the top 1–2 monolayers of the ice mantle. For a CO ice covered in a thin layer of N_2, the signal for CO photodesorption is decreased whilst that for N_2 is enhanced, and above that expected for pure N_2 ice.

This was the first evidence of *co-desorption* which we now term “kick out”. Öberg et al. (2009a,b) were also able to determine that photodesorption is a parallel process to photodissociation. For the pure H_2O ice experiments, they observe a product yield for H_2O:OH of 0.7:1.0 at 20 K to 1.7:1.0 at 100 K. The temperature dependence arises from the increased mobility of the photodissociation products at a higher temperature allowing more efficient reformation of the water ice.

For pure CO_2 ice, they find CO_2 desorbs as CO with a yield between 20–50 % that of CO_2. Again, at higher temperatures, the ratio of CO_2:CO increases due to increased mobility of the fragments. Hence, for both H_2O and CO_2, the primary process is photodissociation, followed by (i) desorption of the energetic fragments (OH or CO), (ii) recombination of the fragments which have excess energy to desorb, or (iii) “kick out’” of a neighboring molecule by an energetic fragment. Pure O_2 ice was also recently studied by Zhen and Linnartz (2014) under similar experimental conditions, and was found to have a desorption yield (6.0±2.0×10−4) photon−1 (6.0±2.0×10−4) photon−1. Ozone (O_3) was also detected (arising from photodissociation of O_2 and reaction between O_2 and O) and similar to CO_2, the yield had a weak temperature dependence where the yield increased at higher temperatures.

Table 5. Measured photodesorption yields for pure ice species and products (where determined)

Ice species	Temperature (K)	Irradiation source	Products	Yield (molecules photon^{-1})	Ref.
H_2O	35–100	H_2 lamp	–	$3–8 \times 10^{-3}$	A
	18–100	H_2 lamp	OH/H_2O	$1.3 \pm 0.4–4.5 \pm 1.2 \times 10^{-3}$	B[a]
CO	15–27	H_2 lamp	–	$2.7 \pm 1.3 \times 10^{-3}$	C, D
	7–15	H_2 lamp	–	$3.5 \pm 0.5–6.4 \pm 0.5 \times 10^{-2}$	E[b]
	14	H_2 lamp	–	$6.4 \pm 0.2–21.2 \pm 0.3 \times 10^{-2}$	F[c]
	18	Synchrotron	–	$4.1–16 \times 10^{-3}$	G[d], H
N_2	16	H_2 lamp	–	$\approx 2 \times 10^{-4}$	C,D
	14	Synchrotron	–	$1.5–5.3 \times 10^{-3}$	I[d], J
CO_2	16–60	H_2 lamp	CO/CO_2	$2.3 \pm 1.4 \times 10^{-3}$	C[e]
	10–40	Synchrotron	CO/CO_2	$1–3 \times 10^{-3}$	K
O_2	14–21	H_2 lamp	O_2/O_3	$6 \pm 2 \times 10^{-4}$	L
	14	Synchrotron	–	$2.1–3.3 \times 10^{-3}$	I[d]
CH_3OH	20–70	H_2 lamp	$C_2H_6/CH_3CHO/C_2H_5OH/CH_3OCH_3$ $HCOOCH_3/HCOCH_2OH/(CH_2OH)_2$	$2.1 \pm 1.0 \times 10^{-3}$	M[f]
	8–130	H_2 lamp	$CH_4/CO/HCOOH/CH_3CHO/X–CHO$	$< 3 \times 10^{-5}$	N[g]
	10	Synchrotron	$CO/CH_3/OH/H_2CO/CH_3O$	$< 10^{-5}$	O[h]

[a]The stated range reflects the temperature range, with higher yields at higher temperatures (see Öberg et al. 2009b, for more details)
[b]The stated range reflects the temperature range, with a slightly lower yield at the higher temperature (see Muñoz Caro et al. 2010, for more details)
[c]The stated range reflects the range of relative contributions from Lyman-α and H_2 molecular bands (see Chen et al. 2014, for more details)
[d]The stated range corresponds to the wavelength-dependent yield integrated over various radiation fields (see Fayolle et al. 2012, 2013, for more details)
[e]Includes the contribution from both CO and CO_2 photodesorption
[f]Products detected in situ via RAIRS and following TPD, and photodesorption yield determined indirectly by following loss of CH_3OH ice upon irradiation (see Öberg et al. 2009c, for details)
[g]CO and CH_4 detected in the gas-phase following irradiation with yields $\sim 2–60 \times 10^{-3}$ molecules photon^{-1} (see Cruz-Diaz et al. 2016, for details)
[h]Photofragments only detected in the gas-phase following irradiation with yields $\lesssim 10^{-4}$ molecules photon^{-1} (see Bertin et al. 2016, for details)
A—Westley et al. (1995b); B—Öberg et al. (2009b); C—Öberg et al. (2007) D—Öberg et al. (2009a); E—Muñoz Caro et al. (2010); F—Chen et al. (2014); G—Fayolle et al. (2011); H—Bertin et al. (2012); I—Fayolle et al. (2013); J—Bertin et al. (2013); K—Fillion et al. (2014); L—Zhen and Linnartz (2014); M—Öberg et al. (2009c); N—Cruz-Diaz et al. (2016); O—Bertin et al. (2016)

Later experimental efforts moved towards wavelength-dependent studies. These measurements are more challenging and are conducted at synchrotron facilities where the FUV flux is lower, requiring longer irradiation times to match the fluence typical of interstellar conditions. On the other hand, microwave discharge lamps have a significantly higher flux; however, independent measurements of CO photodesorption have shown that the derived yield is dependent on having a well-calibrated lamp spectrum. Muñoz Caro et al. (2010) determined a desorption yield for pure CO ice of (3.5±0.5×10−2) photon−1(3.5±0.5×10−2) photon−1 which is one order of magnitude higher than that from Öberg et al. (2009a). Recent investigations showed that the FUV spectrum, in particular the ratio of Lyman-αα (121.6 nm) to H_2 molecular emission bands (110–180 nm), is sensitive to the gas pressure, the type of lamp, the gas composition (pure H_2 versus a mixture of H_2 and He) and the window properties (Chen et al. 2014; Ligterink et al. 2015). Chen et al. (2014) determined CO desorption yields as high as (2.12±0.03×10−1) photon−1(2.12±0.03×10−1) photon−1 for a lamp setup with pure H_2 and a window giving a ratio of 20 % flux of Lyman-αα relative to the molecular emission bands. This showed that desorption of CO at Lyman-αα is less efficient than at wavelengths covering the H_2 emission bands and turned out to be the first step in quantifying the wavelength dependence of photodesorption. The same group have recently reported VUV (120–160 nm) absorption cross-sections for a plethora pure ices (Cruz-Diaz et al. 2014a,b). All data are publicly available for download.[1]

Fayolle et al. (2011) and Bertin et al. (2012) presented the first wavelength-dependent measurement of CO photodesorption (between 90 and 160 nm) conducted at the SOLEIL[2]synchrotron facility. The results showed a strong wavelength dependence corresponding to the vibronic transition spectrum of CO ice. CO was thus confirmed to desorb via the DIET mechanism: Desorption Induced by Electronic Transitions. The results also confirmed the low desorption yield at Lyman-αα wavelengths. Peak yields approaching ~5×10−2 photon−1~5×10−2 photon−1 were found to correspond with the A1Π−X1Σ+A1Π−X1Σ+ electronic transitions of CO (~130--160 nm~130--160 nm). Followup studies on pure N_2, O_2, and CO_2 ices showed interesting results (Fayolle et al. 2013; Fillion et al. 2014). The pure N_2 photodesorption yield was found to peak at significantly shorter wavelengths (≲100 nm≲100 nm) with a wavelength dependence corresponding to the b1Πu−X1Σ+gb1Πu−X1Σg+ electronic transition. A peak yield of 3.77±1.13×10−2 photon−13.77±1.13×10−2 photon−1 was found at 100.8 nm. Photodesorption yields at longer wavelengths were consistent with previous measurements using broad-band hydrogen lamps (Öberg et al. 2009a). Pure O_2 ice photodesorption was also studied and, in contrast to CO and N_2, was found to have a non-negligible contribution due to photodissociation (Fayolle et al. 2013), as also seen later in the broad-band photodesorption experiments (Zhen and Linnartz 2014). The peak photodesorption rates (~0.5×10−2 photon−1~0.5×10−2 photon−1) overlapped with the Schumann-Runge band for O_2 (X~3Σ−g−B~Σ−uX~3Σg−−B~Σu−) which le to dissociation into O(3P)O(3P) and O(1D)O(1D) below ~176 nm~176 nm(Parker 2000; Mason et al. 2006). Fillion et al. (2014) reported results from pure CO_2 ice. CO_2showed a more complicated behavior than the more volatile species discussed above. Both CO and CO_2 molecules were found to photodesorb at wavelengths below 118 nm only, with yields ranging from ~1--3×10−3 photon−1~1--3×10−3 photon−1. The steep increase in photodesorption signal at 118 nm is associated with the first dipole-allowed transition ((3pσu)1Σ+u−1Σ+g(3pσu)1Σu+−1Σg+) and thereafter follows the photoabsorption continuum. That both CO and CO_2 were seen with similar photodesorption yields suggests that the mechanism likely has multiple paths: (i) photoexcitation of CO_2 which "kick out" surface CO_2 (or CO) molecules (DIET mechanism), (ii) photodissociation of CO_2 generating energetic fragments (CO*+OCO*+O) which either "kick out" surface CO_2 (or CO) molecules (DIET mechanism), or desorb directly, or recombine to reform CO_2 which then desorbs. In all cases, the photodesorption signature of CO_2 and CO were similar and as described above. However, for pre-irradiated CO_2 ice, the photodesorption signatures of CO_2 and CO were found to match that of pure CO ice (Fillion et al. 2014). Here, the DIET mechanism is driven by photoexcitation of CO (as opposed to photoexcitation or photodissociation of CO_2).

Wavelength-dependent results from binary N_2:CO ice mixtures confirmed the desorption of surface species via sub-surface excitation (Bertin et al. 2013). The resulting photodesorption signature of surface molecules was found to correspond with the dominant component of the bulk ice composition, e.g., a thin surface layer of N_2 on top of CO ice photodesorbs with a signature of CO photodesorption whereas for a 1:1 mixture, both molecules photodesorb with the signature of both species. Also, Bertin et al. (2013) found that a thin layer (~3~3 monolayers) of N_2/CO on top of CO/N_2 bulk ice was sufficient to completely quench the CO/N_2photodesorption signal.

The quantification and understanding of the process of photodesorption has made leaps and bounds in the past few years and for species other than CO and N_2, it has been shown that irradiation also induces photodissociation, that is, photodesorption and photodissociation are parallel processes. For pure (and relatively) simple ices, the resultant products and reaction pathways are straightforward to ascertain, as described above. Many experiments have shown that FUV irradiation of mixed ices induce chemistry; however, these experiments often couple irradiation with thermal processing and it can be difficult to disentangle reaction pathways specific to either FUV or heat processing (e.g., Hagen et al. 1979; D'Hendecourt et al. 1986; Allamandola et al. 1988).

Several experiments have conducted in-situ monitoring of ice composition via infrared spectroscopy during irradiation at low temperatures. Gerakines et al. (1996) irradiated numerous pure ices composed of known interstellar molecules with a broad-band hydrogen lamp and monitored changes in the ice composition with a Transmission Fourier-transform infrared spectrometer. Gerakines et al. (1996) reported a multitude of identified products, even for relatively simple ices (e.g., H_2O, NH_3, and CH_4). They also studied more complex ices: formaldehyde (H_2CO) and methanol (CH_3OH). In the former experiment, products such as polyoxymethylene (POM, (--H2CO--)n(--H2CO--)n), CO, HCO, CO_2, CH_4, CH_3OH, and methyl formate ($HCOOCH_3$), were identified after 1 hour of irradiation. For pure CH_3OH irradiation, similar products were identified (excluding POM). Gerakines et al. (1996) were also the first to extract quantitative information from the experiments and provided photodissociation cross sections (averaged over the lamp spectrum) for those species for which photodissociation was deemed efficient, and proposed potential reaction pathways.

The irradiation of pure and mixed methanol ices was revisited by Öberg et al. (2009c). They also studied the influence of temperature and FUV fluence and monitored the composition of the ice in-situ via RAIRS (Reflection-Absorption Infrared Spectroscopy). In all experiments conducted, complex molecules were observed to form and included CH_3CHO (acetaldehyde), CH_3CH_2OH (ethanol), CH_3OCH_3 (dimethyl ether), $HCOOCH_3$ (methyl formate), $HOCH_2CHO$ (glycolaldehyde), and $(CH_2OH)_2$ (ethylene glycol). The exact products and relative ratios were found to have a weak dependence on temperature during irradiation and a stronger dependence on ice composition (pure CH_3OH versus CH_3OH mixed with CO or CH_4). For example, CO:CH_3OH ices were more enriched with --HCO--HCO molecules while CH_4:CH_3OH ice were more enriched in --CH3--CH3 molecules. Öberg et al. (2009c) also observed a large increase in the abundances of several complex molecules (around an order of magnitude) upon warming the irradiated sample from 20 to 70 K, confirming that radical mobility is temperature dependent and is important for building additional complexity. An analysis of the products allowed them to identify potential reaction pathways and formation cross sections. They also estimated the branching ratios for CH_3OH ice photodissociation at 20 K as 5:1:<15:1:<1 for CH2OH+HCH2OH+H, CH3O+HCH3O+H, and CH3+OHCH3+OH. The authors plan an extraction of fundamental data from the experimental results, for example, photodissociation rates, products, and branching ratios, diffusion rates, and binding energies of radical species. At the time of writing (March 2016), this challenging analysis was still ongoing.

As a final note in this Section, Öberg et al. (2009c) also report a photodesorption yield for CH_3OH of (2.1±1.0×10−3) photon−1(2.1±1.0×10−3) photon−1 which was determined by modelling the loss of methanol ice during irradiation as a combination of photodesorption (a zeroth-order process) and photodissociation (a first-order process). Very recently, several groups have revisited this system in light of recent revelations regarding the importance of robust lamp characterization for accurate determinations of photodesorption yields (Chen et al. 2014). The general consensus from several independent groups, is that pure methanol ice does not photodesorb intact. Photofragments only were observed, with a deduced upper limit to intact methanol photodesorption ∼10−6−10−5∼10−6−10−5 molecules per incident photon, i.e., at least two orders of magnitude less than that found for water ice (Cruz-Diaz et al. 2016; Bertin et al. 2016). This low photodesorption yield was found for both pure and mixed (with CO) methanol ice, showing also that the DIET mechanism appears not to work for large molecules such as methanol. Note that this low photodesorption yield for methanol now calls into question the origin of gas-phase methanol in both cold and irradiated environments (e.g., dark clouds, PDRs, and protoplanetary disks), providing an intriguing challenge for molecular astrophysicists and astrochemical modelers.

8.2 Theoretical Calculations

Photodesorption has been studied theoretically via classical molecular dynamics (MD) simulations. Most work to date has been conducted on water ice (Andersson et al. 2005, 2006; Andersson and van Dishoeck 2008; Arasa et al. 2010), including its deuterated analogues, HDO and D_2O (Koning et al. 2013). Andersson et al. (2005, 2006) studied the photodesorption mechanisms of H_2O in both crystalline and amorphous water ice at 10 K following photoexcitation of H_2O into the first excited state with photons with wavelengths between 130--150 nm130--150 nm. The MD calculations predicted a H_2O photodesorption efficiency between 0.5–1 % per absorbed photon in top few monolayers of an amorphous water ice surface (Andersson et al. 2005, 2006), which was in reasonable agreement with experimental data available at that time (e.g., Westley et al. 1995a). Andersson and van Dishoeck (2008) went a step further than earlier work and determined the relative contributions of all possible outcomes as function of depth into the ice. Upon photodissociation into H+OHH+OH, the following outcomes are possible: (i) H desorbs and OH is trapped, (ii) H is trapped and OH desorbs, (iii) H and OH are trapped, (iv) H and OH recombine and desorb as H_2O, and (v) H and OH recombine and are trapped as H_2O. There is also a parallel process whereby the excited H product had sufficient energy to "kick out" a neighboring water molecule.

Outcomes (i)–(iv) and the "kick out" processes all have the potential to affect the chemical balance of both the ice and gas. Andersson and van Dishoeck (2008) determined that outcome (i) has the highest probability in the surface monolayer (~90 %~90 %), and decreases in efficiency rapidly with depth into the ice reaching ~10 %~10 %efficiency by the sixth monolayer. On the other hand, trapping of either the fragments or the reformed molecule increased in probability with depth, becoming the most probable outcome by the forth monolayer. The MD calculations also provided insight into the behavior of the photodissociation fragments post dissociation. For outcomes in which no desorption occurred, H and OH had sufficient excess energy to travel up to 70 and 60 Å, respectively, on or within the uppermost monolayers of an amorphous ice surface (Andersson and van Dishoeck 2008). Thus photodissociation of ices also creates mobile radicals at 10 K which can take part in further surface reactions.

Arasa et al. (2010) extended the calculations to higher temperatures (up to 90 K). Similar results were found: photodesorption has the highest efficiency in the top few monolayers, and trapping increases in probability with depth into the ice. The efficiency of OH and H_2O photodesorption was found to have weak dependence on temperature with an increase in probability of around 30 %30 %. The probability of the parallel "kick out" mechanism was found to have a more complex behavior with temperature which the authors attribute to the amorphous nature of the ice. Koning et al. (2013) studied isotopic effects, and looked at the astrophysically relevant cases of HDO and D_2O photodissociation in an amorphous H_2O water ice. D atoms were found to photodesorb with a slightly lower efficiency than H atoms. On the other hand, the "kick out" probability of D atoms was found to be higher than that for H atoms due to more efficient energy transfer by the higher mass atom to the surrounding water molecules. Arasa et al. (2015) revisited these results to determine whether photodissociation of HDO (into either OH+D or OD+HOH+D or OD+H) could affect the deuterium fractionation of water in the gas phase. The conclusion was that the effects are likely minimal; however, the lower probability for D atom photodesorption may lead to fractionation effects within the ice. This latter prediction remains to be tested in astrochemical models. Probabilities for each outcome as a function of monolayer and temperature have now been made available for use in astrochemical models (Arasa et al. 2015). Also available are values averaged over the top few monolayers for use in those models which are unable to treat the ice chemistry in a stochastic manner.

Very recent MD calculations have also attempted to understand the underlying mechanism for CO photodesorption via the DIET process (van Hemert et al. 2015). van Hemert et al. (2015) excited selected CO molecules with 8.7--9.5 eV8.7--9.5 eV photons contained within both crystalline and amorphous CO clusters (containing up to 1200 CO molecules). They determined two potential mechanisms. In the first, the excited CO molecule decays to the ground state in an unfavorable configuration and subsequently ejected, and the second corresponds to the aforementioned “kick out” mechanism. Photodesorption yields of (4.0×10−3) photon−1(4.0×10−3) photon−1 agreed well with the lower end of the range of values currently constrained by experiment.

8.3 Implementation in Astrochemical Models

To date, the implementation of experimentally-determined photodesorption yields (YXYX, typically expressed in units of molecules photon−1photon−1) in astrochemical models has been fairly pragmatic. The photodesorption rate of a molecule is dependent upon the flux of FUV photons striking the grain surface, FUV photonscm−2s−1FUV photonscm−2s−1, the cross-sectional area of grain surface available per unit volume, σgng cm2cm−3σgng cm2cm−3, and the surface coverage fraction of the specific molecule in the top 1–2 monolayers of the ice mantle, θX=nX/nSθX=nX/nS. In the above expressions, σgσg is the geometrical cross section of a grain surface, ngng is the number density of grains, nXnX is the number density of molecules of species X in the surface layers of the ice mantle, and nSnS is the total number of species in the surface layers of the ice mantle. In two-phase models, there is no distinction between the surface coverage factor and the bulk coverage factor. However, these numbers will likely differ in three-phase and multilayer models.

The photodesorption rate is thus given by,

$$k_X^{pd} = \theta_X Y_X F_{UV} \sigma_g n_g \text{ cm}^{-3} \text{ s}^{-1}. \quad (36)$$

Derivation of the rate beginning instead with the cross-sectional area of an individual grain surface site le to an identical expression due to the relation between number density of surface sites, grain surface area, and total number of species per monolayer. Note that the experimentally determined yield implicitly includes the effects of ice opacity to FUV radiation (since the yield is given per incident photon) and the fact that photodesorption only occurs from the top few monolayers of the ice. This results in a zeroth-order rate coefficient: the photodesorption rate is independent of ice thickness (or number density of ice molecules) for thick ices. Within this framework, it is also possible to include branching ratios of the expected products, for those species for which photodissociation is the primary process (e.g., H_2O and CO_2). Several of the experimental groups also provide analytical fits to the photodesorption yields as a function of ice thickness (in monolayers) and temperature (Öberg et al. 2009a,c). This allows the inclusion of a more realistic treatment of the photodesorption rate at low surface coverage (important in models which consider evolution from diffuse to dense conditions).

To show the effect of photodesorption, Fig. 8 again shows a simulation of a laminar solar nebula where we have lowered the photodesorption yield from 10−210−2 to 10−510−5. Locally the abundance can change by several orders of magnitudes, but integrated over zz the changes are small, except for gas phase CO_2 which is strongly enhanced in the outer regions.

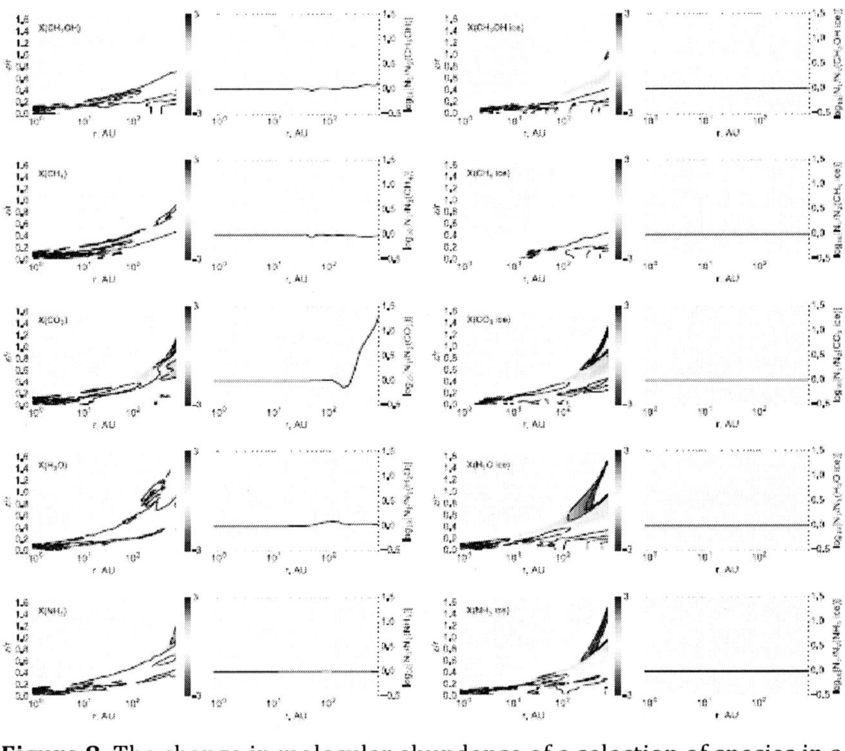

Figure 8. The change in molecular abundance of a selection of species in a laminar solar nebula at 1 Myr, when using a photodesorption yield of 1×10−51×10−5 instead of 1×10−21×10−2. Relative gas phase abundances are in the left panels, the corresponding ice abundances in the panels on the right-hand side

Now that wavelength-dependent data are available, astrochemical models for which the FUV spectrum is included explicitly (e.g., in protoplanetary disks or PDRs), the total photodesorption yield per species can be found by integrating over the experimentally determined yield and the FUV spectrum,

$$Y_X = \int_\lambda Y_X(\lambda) F_{UV}(\lambda) d\lambda. \quad (37)$$

These data are now available for pure CO, N_2, O_2, and CO_2 ices (Fayolle et al. 2011, 2013; Fillion et al. 2014). One caveat of this method is that one needs to take into account that the wavelength dependence of the desorption yield will depend upon the dominant molecular component of the ice mantle in the *subsurface* layers of the ice (Bertin et al. 2013). This is possible to track in three-phase and multilayer models, but non trivial in two-phase models. This is because additional assumptions are required, for instance, whether the ice is sequestered into layers and if so, with which layers (H_2O-dominated or CO-dominated) more minor species are most likely to be associated. In the case of H_2O-dominated ices, wavelength dependence has not been determined explicitly but is likely to follow the photodissociation (or absorption) cross section of amorphous solid water (see, e.g., Cruz-Diaz et al. 2014a). It is also worth bearing in mind that, in the regime where water ice is the dominant ice component, the ice temperature could be such that thermal desorption will dominate, except for those molecules which are comparable in volatility with water (e.g., CH_3OH).

In current gas-grain chemical models, photodissociation is treated as a parallel process to photodesorption. Because interstellar ices are optically thin to FUV radiation, photodissociation can occur throughout the ice mantle (in contrast to photodesorption which occurs only from the top few monolayers). Photodissociation rates for specific species are computed by integrating over the photodissociation cross section multiplied by the FUV spectrum, exactly as is done in the gas phase (e.g., van Dishoeck et al. 2006),

$$k_X^{ph} = \int_\lambda \sigma_X(\lambda) F_{UV}(\lambda) d\lambda, \quad (38)$$

with products and associated branching ratios usually adopted from the equivalent gas-phase measurements or estimates. Photoabsorption cross sections in the VUV range are now available for several molecules in the solid phase and are usually a good proxy for computing the photodissociation rates (e.g., Cruz-Diaz et al. 2014a,b). For other species, the gas-phase data is typically used and should be accurate to within a factor of a few. The derivation of photodissociation cross sections and branching ratios in the solid phase remains a complex process requiring careful modelling of ice experiments which explore the full range of multi-dimensional parameter space (Öberg et al. 2009c).

By treating photodissociation as a parallel process to photodesorption in full gas-grain models, models can implicitly account for the production of reactive radicals within and on the ice mantle. The models are also able to treat the fate of such radicals including loss to the gas phase via desorption (whether photo driven or thermally driven), or reaction with other radicals and atoms present in the ice, as discussed in the introduction to this Section. However, in models in which chemical desorption is also included, one runs into the issue in which photodesorption is accounted for twice: one explicitly as outlined above, and one implicitly, where radicals formed via photodissociation recombine to reform the molecule which then has an assumed efficiency for chemical (or reactive) desorption. This latter process is essentially direct photodesorption. Given the inherent uncertainties which exist in large astrochemical models due mainly to the lack of available experimental data, the effect of this "double counting" is likely minor. A potential solution to the problem is to treat photodissociation products as a separate (and likely) excited species in the ice mantle which have a timescale for thermalization guided by the MD calculations.

9 PRECISION CONSIDERATIONS IN RATE-BASED MODELING

A basic, two-phase model of interstellar gas and grain surface chemistry may be solved easily using the standard rate-equation approach that is commonly used for pure gas-phase chemistry. However, as explained earlier, this method can sometimes lead to unphysically high grain surface production rates, under conditions where two reactants on the surface have mean populations per dust grain that are less than unity. In a mathematically exact treatment, if the reaction between two such species is sufficiently rapid once both particles are present on the grain surface, then the surface populations of those species should become anti-correlated. The rate-equation method cannot reproduce such behavior, as it calculates only the mean, individual populations $\langle ns(A)\rangle\langle ns(A)\rangle$ and $\langle ns(B)\rangle\langle ns(B)\rangle$, rather than the mean of the product, $\langle ns(A)ns(B)\rangle$ $\langle ns(A)ns(B)\rangle$.

Garrod (2008) proposed a modification scheme (i.e. a so-called "modified rate" approach) to take account of the inaccuracies of the rate-equation method, which Garrod et al. (2009) showed to be effective over a wide range of physical conditions. However, the full implementation of this method involves the calculation of quantities such as (1−⟨ns(A)⟩)(1−⟨ns(A)⟩). In a numerical simulation, the value of this differential becomes highly imprecise as ⟨ns(A)⟩⟨ns(A)⟩approaches 1, due to the limited fractional precision available in the calculations (for double precision calculations, the fractional precision is on the order of 10−1610−16). This imprecision can induce the solver to use very small timesteps, significantly slowing the calculations. For this reason, the maximum available precision should be used for modified-rate calculations wherever possible. Several Fortran compilers now include quadruple precision as a standard feature. Converting a chemical code and differential equation solver to quadruple precision is a straight-forward task.

The use of a three-phase gas-grain model can also introduce precision problems. As explained by Hasegawa and Herbst (1993b), the three-phase treatment calculates a net deposition rate onto the dust grains that represents the difference between the total accretion rate of all gas-phase species onto the grains, and the total desorption rate of all surface species back into the gas (as well as any processes such as reaction that may also decrease the total number of surface particles). This net rate is used to determine the rate at which material in the surface layer is pushed down into the bulk ice (in the case of net deposition) or brought up from the bulk into the surface layer (net desorption), in order to maintain the same total population of particles in the surface layer. The net deposition rate therefore strongly couples the calculated abundances of all surface and mantle species with the abundances of whichever gas-phase or surface species have the greatest influence over that net deposition rate. Improving the overall precision of the calculations (i.e. moving to quadruple precision) allows the net rate to be more precisely determined, improving model run-times.

Unfortunately, in the case of three-phase models, simply increasing the overall precision of the calculations is often not enough to avoid a failure of the solver routine. A more careful consideration of the error associated with individual abundances is also required. We may illustrate this point with an example in which the net deposition rate onto dust grains is solely determined by the difference between (i) the rate of deposition of gas phase H_2 and (ii) the rate of desorption of grain surface H_2. If one rate strongly dominates over the other, then the *fractional error* on the net deposition rate is mainly dependent upon the solver-enforced fractional error (i.e. "EPS", error per step) on either the gas-phase or surface H_2 abundance. However, if the two rates are nearly equal, then the error on the net rate will depend upon the error on both abundances. Crucially, in this latter case, if the net rate is very much smaller than either the deposition or desorption rates, then the absolute error on the net rate may become much larger even than the rate itself. Because the calculated gas-phase and grain surface H_2abundances may fluctuate anywhere within their allowed error limits, this situation can produce wild fluctuations in the net deposition rate. This prompts a shift by the solver routine to smaller and smaller timesteps and ultimately a failure to integrate over manageable timescales.

To avoid this problem, we must enforce an error value that is more stringent than EPS for certain species in the model that affect the net deposition rate. The general EPS value used in a model may be, say, 10−410−4. The solver compares this EPS with the fractional error on each species resulting from the time-integration of the differential equations. If the EPS exceeds the computed error value for all chemical species, then convergence is assumed, and the timestep is complete. In order to impose stronger error handling on the net deposition rate, we may bias this comparison by some factor, according to the sensitivity of the net deposition rate to changes in the abundance of a particular species. If a small, fractional change in a chemical abundance induces a *larger* fractional change in the net deposition rate, then the ratio of these values is used to bias the error comparison to enforce smaller errors. The sensitivity of the net deposition rate to the abundance of each species can be determined at each Jacobian call by the solver. Typically only a few chemical abundances are important to the error dependence of the net deposition rate, such as gas-phase and grain surface H, H_2 and CO.

This more stringent error-handling approach is necessary for the use of three-phase models in all cases where the net deposition rate changes sign during the course of the integration, because such conditions induce extreme fractional errors as the net rate approaches zero.

10 TEST CASE: HYDROGENATION OF CO ICE LEADING TO COMPLEX MOLECULES

Catastrophic freeze-out of CO molecules from the gas phase le to the formation of an apolar (CO-rich) ice. The surface hydrogenation of CO in inter- and circumstellar ices drives the formation of more complex species. Here we will use the hydrogenation of CO as an example of how a combined experimental and modeling approach can result in an increased understanding of complex molecule formation. In 1992, Charnley et al. proposed that formaldehyde (H_2CO) and methanol (CH_3OH) form in the solid phase by the hydrogenation of CO ice. Formaldehyde is the first organic polyatomic molecule detected in the interstellar medium (ISM) (Snyder et al. 1969). Methanol has been observed abundantly in the ISM (Belloche et al. 2013) and is considered to be a resource for the formation of more complex molecules through surface reactions and after its sublimation in the gas phase (Balucani et al. 2015). Originally, gas-phase complex molecules were postulated to arise from dissociative recombination of protonated methanol, which itself was formed via radiative association of smaller ions and molecules (e.g., CH+3+H2O⟶CH3OH+2+hνCH3++H2O⟶CH3OH2++hν followed by CH3OH+2+e−⟶CH3OH+HCH3OH2++e−⟶CH3OH+H). Pioneering measurements with CRYRING, a low-temperature ion-storage ring, showed that dissociative recombination led predominantly to fragmentation of the protonated species, for instance, for the example listed above, only 3 % of dissociative recombinations where found to lead to methanol (Geppert et al. 2006). On the other hand, the surface formation of methanol has been long under debate, as the first sets of laboratory data gave conflicting results. In their study, Hiraoka et al. (2002) reported only the formation of formaldehyde, whereas Watanabe and Kouchi (2002) confirmed an efficient formation of formaldehyde and methanol ice. This controversy was then solved in 2009 by an independent investigation of the same reaction scheme performed for a wider range of H-atom fluxes and ice temperatures (Fuchs et al. 2009), which showed the different results from the two groups were due to the use of different H-atom fluxes and fluences.

The first (CO+HCO+H) and third (H2CO+HH2CO+H) reaction steps in this H-atom addition scheme presents barriers that have to be overcome, see also Fig. 3. The efficiency of the process therefore depends on the barrier height and how much excess energy is available to cross over. Figure 9 shows the time evolution of the surface abundance (in molecules cm−2cm−2) of CO, H_2CO and CH_3OH during H-atom bombardment of CO ice with an H-atom flux of 5×1013 cm−2s−15×1013 cm−2s−1at different temperatures. This figure is taken from Fuchs et al. (2009). At all the temperatures, the amount of CO decreases as the abundance of H_2CO increases. After an H-atom fluence of 1×1017 H1×1017 H atoms cm−2cm−2, the formation of methanol starts to increase at the expense of the H_2CO abundance. Similar trends are also reported in Watanabe et al. (2006). For temperatures higher than 15 K, a clear drop in the production rate of methanol is observed by Fuchs et al. (2009). This is mainly due to a lower lifetime of H atoms on the surface of the ice that causes the H-surface abundance to drop substantially and therefore reduce the probability of hydrogenation reactions occurring in the ice at temperatures higher than 15 K.

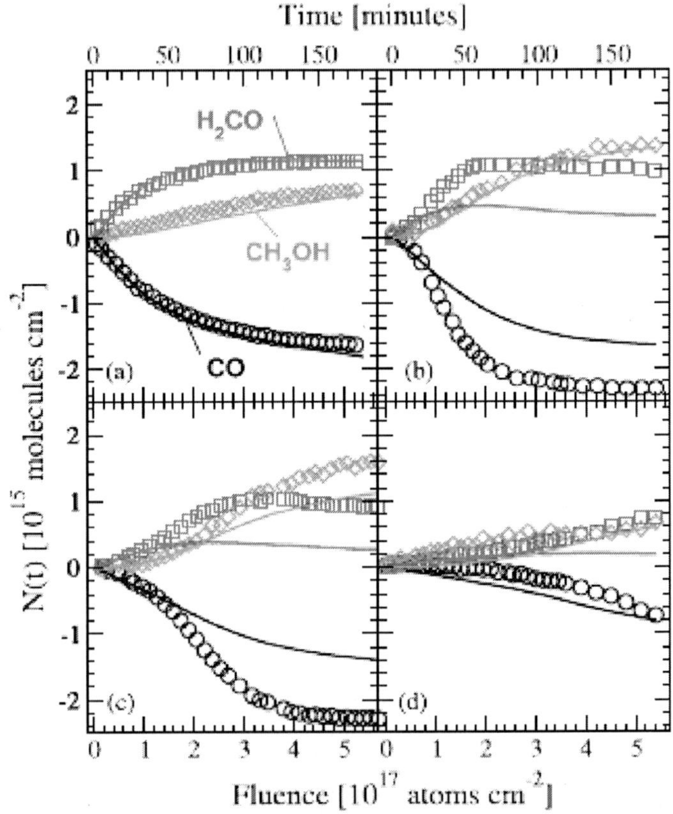

Figure 9. Time evolution of the surface abundance (in molecules cm−2cm−2) of CO, H_2CO and CH_3OH during H-atom bombardment of CO ice with a H-atom flux of 5×1013 cm−2s−15×1013 cm−2s−1 at surface temperatures of 12.0 K (**a**), 13.5 K (**b**), 15.0 K (**c**), and 16.5 K (**d**). Experimental data (*symbols*) and Monte Carlo simulation results (*solid lines*) are shown as well. Figure reproduced from Fuchs et al. (2009). Credit: G.W. Fuchs, H.M. Cuppen, S. Ioppolo, C. Romanzin, S.E. Bisschop, S. Andersson, E.F. van Dishoeck, and H. Linnartz: Astron. Astrophys. **505**, 629 (2009), reproduced with permission, copyright ESO

A detailed interpretation of laboratory measurements in terms of understanding processes like deposition of H atoms, diffusion, reaction and desorption is only possible if laboratory work is supported by simulations. Fuchs et al. (2009) used a microscopic kinetic Monte Carlo algorithm to fit effective reaction barriers for CO+HCO+H and H2CO+HH2CO+H to a large set of H_2CO and CH_3OH production and CO loss data as a function of H-atom fluence and temperature indicated by the solid lines in Fig. 9. The outcome of the model yielded typical values in the range of 400 to 500 K. In general, simulations and experimental data are in very good agreement at 12 K. The agreement for 13.5, 15, and 16.5 K is far less good, probably because of missing mechanisms that promote the penetration into the ice. The reaction rates for H+COH+CO and H+H2COH+H2CO have only a minor temperature dependence, which suggests that there is a clear tunneling component for the reaction at low temperature. Values from Fuchs et al. (2009) are in good absolute agreement with the barriers found by Awad et al. (2005), who also found a similar temperature behavior. Their values were obtained using a rate equation analysis for TT = 10, 15, and 20 K using the data from Watanabe et al. (2006).

Temperature-dependent values are important - within their experimental accuracies - for inclusion in astrochemical models that predict astronomical column densities. The Fuchs et al. Monte Carlo model is further extended to interstellar conditions to compare with observational H_2CO/CH_3OH data. The H_2CO and CH_3OH interstellar abundances obtained in the simulations do not scale directly with H-atom fluence because of the different relative importance of H_2 production and CO hydrogenation in space compared with the laboratory. However, laboratory experiments are required to derive the necessary rates that serve as input to the Monte Carlo program. The obtained H_2CO/CH_3OH ratios for the interstellar simulations are in closer agreement with observational limits than a direct translation of the experimental results. Cuppen et al. (2009) extended the Fuchs et al. (2009) model to simulate in detail the formation of both H_2CO and CH_3OH under interstellar conditions. Unlike the master equation (Stantcheva et al. 2002), rate equation (Ruffle and Herbst 2000), and macroscopic Monte Carlo (Ruffle and Herbst 2000) studies of methanol formation, the continuous-time, random-walk Monte Carlo technique accounts for the layering of the CO. This layering is crucial, because the constant addition of fresh CO tends to protect the underlying layers from hydrogenation and limits the time available for reaction.

The link between solid CO and CH_3OH was also recently confirmed observationally. Using VLT data on the CO band of a large collection of YSOs (Pontoppidan et al. 2003), peak width, peak position and upper limits on the 2152 cm−12152 cm−1 band were extracted and compared to laboratory spectra. This proved that the astronomical spectra are fully consistent with CO ice intimately mixed with CH_3OH, in agreement with a common chemical history as described above (Fraser et al. 2004; Cuppen et al. 2011). A more recent study expands this proposition by comparing literature laboratory spectra of different ice mixtures containing CO to high-resolution Keck/NIRSPEC spectra, of the massive YSO AFGL 7009S and of the low mass YSO L 1489 IRS (Penteado et al. 2015). Accurate band profile comparisons of both the 12CO12CO and 13CO13CO ice bands were carried out. Moreover, Penteado et al. gave evidence that a gradient of different ratios of CO:CH_3OH mixtures might exist in the grain mantles and that the spectrum of high-mass YSOs requires a higher content of methanol than the low-mass YSOs to reproduce the observed CO band profile.

In the past decade, to fully characterize the methanol formation reaction network, deuteration experiments were also performed on CO ices (Hidaka et al. 2007), which resulted in both deuterated formaldehyde and methanol. In this study, it was found that deuteration rate of CO is 12.5 times lower than the rate of CO hydrogenation under the same experimental conditions. A similar CO deuteration rate was also recently confirmed by Fedoseev et al. (2015). In addition to CO+DCO+D experiments, deuteration studies of H_2CO and CH_3OH ices have also been performed (Nagaoka et al. 2005; Hidaka et al. 2009; Watanabe and Kouchi 2008). Figure 8 of Hidaka et al. (2009) presents the complete surface reaction network as obtained when CO is exposed to H and D atoms based on several studies (Watanabe et al. 2006; Watanabe and Kouchi 2008; Hidaka et al. 2009). Reaction rate constants are shown relative to those of CO+HCO+H at 15 K (Hidaka et al. 2004). The general conclusion here is that although D-atom additions to CO/H_2CO proceed slower than H-atom additions to CO/D_2CO, D-atom induced substitution reactions in H_2CO and CH_2OH take place faster than the corresponding H-atom induced substitutions in D_2CO and CD_2OD.

Recently, Fedoseev et al. (2015) showed the first laboratory evidence that surface hydrogenation of CO ice le to the formation of molecules with more than one carbon atom along with the formation of H_2CO and CH_3OH. In their experiments, Fedoseev et al. confirmed the surface formation of two larger molecules of astrobiological importance, i.e., glycolaldehyde ($HC(O)CH_2OH$) and ethylene glycol ($H_2C(OH)CH_2OH$). The key step in this process is believed to be the recombination of two HCO radicals followed by the formation of a C-C bond as also indicated by modelling work (Woods et al. 2013). The dimerization of HCO is considered by Woods et al. (2013) as one of the formation routes of glycolaldehyde. Their astrophysical model matches the observed estimates in the hot molecular core G31.41+0.31G31.41+0.31 and low-mass binary protostar IRAS 1629-32422. Fedoseev et al. (2015) experimentally determined that the recombination reaction between two HCO radicals yields glyoxal (HC(O)CHO) at 13 K; sequential hydrogenation by 2 or 4 H-atoms turns HC(O)CHO into glycoladehyde and ethylene glycol, respectively. This is an important experimental finding, as so far surface hydrogenation reactions were mainly shown to be effective in the formation of relatively simple interstellar relevant species (e.g., H_2O, NH_3, and CH_3OH), while radical-radical recombination reactions were considered efficient in the ISM only at temperatures higher than 20–30 K. The latter assumption comes from the fact that radicals are commonly assumed to become mobile around 30 K (e.g., Garrod 2013; Butscher et al. 2015). Therefore, diffusion of radicals is assumed to be the limiting factor to the formation of COMs through radical-radical recombination reactions. Fedoseev et al. (2015) gave the first laboratory evidence that diffusion is not necessarily needed for radicals to find each other. In CO-rich ices, HCO radicals can be formed next to one another through the hydrogenation of CO molecules and react without diffusing in the ice.

To further strengthen the conclusions of Fedoseev et al. (2015), experimental results are implemented into the model used by Cuppen et al. (2009) based on the continuous-time, random walk Monte-Carlo method described before. This model allows for the simulation of microscopic grain-surface chemistry for the long timescales typical in interstellar space, including the layering of ice during the CO freeze out. The choice of microscopic simulations is because HCO radicals have to stay in close vicinity to each other in order for a reaction to occur. Therefore, a method that accounts for the position of the species in the ice lattice is required for the best representation of the system. In the simulations reported by Fedoseev et al. (2015), five new species, including glycolaldehyde and ethylene glycol, are introduced in the simulations (i.e., HC(O)CHO, $HC(O)CH_2O$/HC(O)CHOH, $HC(O)CH_2OH$, $H_2C(O)CH_2OH$/$HC(OH)CH_2OH$, and $H_2C(OH)CH_2OH$). The glycolaldehyde abundance was found to correlate with the formaldehyde abundance, while the abundance of ethylene glycol correlates with the abundance of methanol. This is not surprising since both CH_3OH and $H_2C(OH)CH_2OH$ are hydrogen saturated species while both H_2CO and $HC(O)CH_2OH$ are not. Finally, the correlation found between the abundances of $HC(O)CH_2OH$ and $H_2C(OH)CH_2O$ with those of formaldehyde and methanol is as expected typically in ratios of the order of a few percent and in good agreement with observations (Jørgensen et al. 2012; Schöier et al. 2002).

More recently, Chuang et al. (2016) presented new laboratory experiments on the low-temperature solid-state formation of three complex molecules, i.e., methyl formate (HC(O)OCH3), glycolaldehyde and ethylene glycol, through recombination of free radicals formed via H-atom addition and abstraction reactions that occur during the hydrogenation of CO ice at 15 K. The experiments include the co-deposition of H atoms and pure H_2CO, H atoms and pure CH_3OH as well as H atoms and mixtures of H_2CO with CO, H_2CO with CH_3OH, and CH_3OH with CO, in fact extending previous CO hydrogenation studies (e.g., Fuchs et al. 2009; Fedoseev et al. 2015). The aim of these experiments is to resemble the physical-chemical conditions typical of the CO freeze-out stage in dark molecular clouds, when H_2CO and CH_3OH form by the recombination of accreting CO molecules and H atoms on ice grains. The occurrence of H-atom abstraction reactions in ice mantles le to the formation of a higher number of intermediates (HCO, CH_3O, and CH_2OH) than previously thought, when assuming sequential H-atom addition reactions only. This enhances the probability to form COMs through radical-radical recombination without the need of UV photolysis or cosmic rays as external triggers.

Figure 10 shows the resulting COMs formation network based on the investigated reaction routes. From top to bottom, a sequence of H-atom addition and abstraction tunneling reactions is shown to lead to the formation of CH_3OH. The figure shows that H_2CO can undergo an abstraction reaction induced by H atoms to form HCO radicals, which can be further dehydrogenated to form CO. This pathway increases the total number of HCO formation events and its lifetime in the ice mantle. H_2CO also participates in addition reactions with H-atoms. Formation of CH_3O radicals is confirmed by the detection of solid methyl formate, a species formed most likely under the detection limit in Fedoseev et al. (2015). However, formation of CH_2OH radicals is not fully excluded by Chuang et al. (2016). In contrast, H-atom induced abstraction reaction involving CH_3OH likely yield CH_2OH while no proof for CH_3O formation is found by the authors. The barrier-less recombination of HCO intermediates yielding glyoxal followed by consequent hydrogenation (i.e., the mechanism investigated in Fedoseev et al. 2015) yields $HC(O)CH_2OH$ and $H_2C(OH)CH_2OH$ and is presented in the right panel of the reaction scheme in Fig. 10. Alternatively, the intermediate HCO radicals can directly recombine with CH_3O or CH_2OH to form $HC(O)OCH_3$ and $HC(O)CH_2OH$, respectively. According to the authors, the recombination of two CH_2OH radicals (dash arrow) seems to contribute less to the formation of $H_2C(OH)CH_2OH$.

Chuang et al. (2016) showed that both H-atom addition and abstraction reactions play an important role in effectively increasing the number of interaction events between reactants as well as the lifetime of radicals in the ice. This way, more COMs are formed through recombination of reactive intermediates than previously assumed at 10 K under interstellar conditions. It is important to note that Fuchs et al. (2009) did not include hydrogen abstraction reactions when calculating reaction rates. Clearly, the final products they observed in their experiments were the result of an equilibrium between addition and abstraction reactions. Therefore, the rates inferred by Fuchs et al. (2009) should be regarded as effective rates. The recent laboratory and modeling advances in our understanding of interstellar surface reactions discussed in this section give a strong indication that the COMs observed in cold dense molecular clouds have a solid phase non-energetic origin and are most likely linked to a combination of hydrogenation addition/abstraction reactions and radical-

radical recombination reactions at 10 K. CO-rich ices formed upon the catastrophic freeze-out of CO molecules onto interstellar grains seem to be the right environment for these surface reaction to occur efficiently in the ISM.

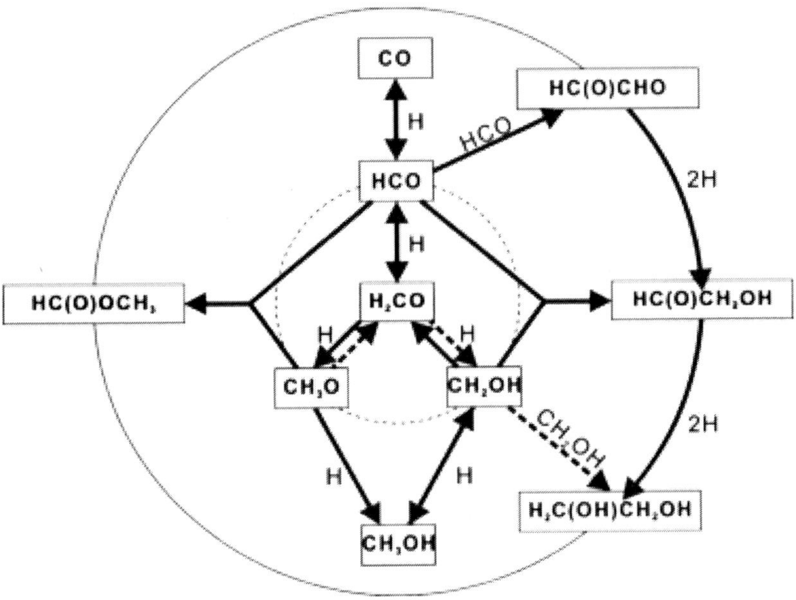

Figure 10. Extended COM formation network as obtained from the CO, H_2CO, and CH_3OH hydrogenation experiments by Fedoseev et al. (2015) and Chuang et al. (2016). *Solid arrows* indicate the reaction less efficient pathways. Figure reproduced from Chuang et al. (2016)

11 OUTLOOK

In this review, we have summarized the "state-of-the-art" in terms of the methods used and the data available for treating grain-surface chemistry under astrophysical conditions. Since the first conceptual models of interstellar grain-surface chemistry, now more than 40 years old (e.g., Watson and Salpeter 1972a; Allen and Robinson 1977; Tielens and Hagen 1982), the general structure of a gas-grain code has not significantly changed. What has come on leaps and bounds is the advances in experimental techniques that have enabled the first quantification of many of the parameters required for models including binding energies, reaction barriers, diffusion rates, and desorption yields. Such experiments remain challenging to perform, and as reaction systems become more complex, the various potential chemical pathways and branching ratios remain difficult to disentangle. As discussed here, using sophisticated stochastic astrochemical models to also simulate laboratory data can help extract this information from the experiments which is then input into the larger rate-equation astrochemical models for simulating astrophysical environments.

Despite vast advances in the quantification of interstellar grain-surface chemistry in the past decades, there remains much more to learn. Typical grain-surface networks consist of sometimes hundreds of grain-surface species, the vast majority of which are reactive radicals. Such species are transient by nature and remain difficult to isolate and thus study in the laboratory. We urge the experimental community to continue to push the boundaries of what is possible in the laboratory and develop innovative methods to quantify the necessary data for radicals. In astrochemical models used to simulate interstellar and circumstellar environments, the data adopted for radicals remains the most uncertain.

Thanks to the effort over the past decades, we have a reasonably clear picture of the different processes important for ice chemistry, summarized in each of the Sections in this review. However, one chemical process we have only touched upon is ion-molecule chemistry in ice mantles (e.g., Woon 2011), the gas-phase equivalent of which is arguably the most important for building molecular complexity in the gas phase in space.

A second process we have not covered in detail here, but debate remains on its relative importance, is the direct impact of cosmic-ray particles on ice-covered dust grains. Due the high energies attained by cosmic-ray particles (~ Mev–GeV), such a process is thought to have catastrophic effect on the composition of the ice due to the creation of an ionized particle track through the grain which can also destroy, ionize, and sublime molecules in an explosive manner from the ice mantle (e.g., D'Hendecourt et al. 1982; Leger et al. 1985). Such a process is also stochastic by nature, and remains challenging to account for in a "mean-field" approach, since each impact can have a different track on each individual grain, and not every dust grain will be impacted by a cosmic-ray particle. This process had fallen out of favor as it became increasingly evident that photodesorption was an efficient (and well understood) mechanism for releasing molecules from the ice mantle at low temperatures. Recent revisitations of this process suggest that the overall effect is of the same order of magnitude as other processes considered to influence the gas-grain balance in dark clouds such as photodesorption (e.g., Ivlev et al. 2015).

It is clear that much work remains in linking the observed gas-phase composition in dark clouds with the known ice composition, and astrochemical models remain critical for constraining the potential processes for determining the gas-grain balance. The outcome of this Lorentz Centre Workshop shows what can be achieved when scientists from various fields and disciplines (modelers, experimentalists, and observers) meet to intensively discuss a naturally interdisciplinary field, such as astrochemistry. We have presented the first detailed review of the data and methods used for interstellar/circumstellar grain-surface chemistry and made recommendations, based on a legacy of past and present experience and research, for the treatment of the reactions rates. We have also compiled in a systematic manner the ever-increasing pool of experimental and theoretical data currently available in the literature. As we move into a new era of observational astronomy, with ALMA and JWST both expected to have a transformational impact on the field of astrochemistry, we anticipate regular future meetings on this topic to review, compile, and update the current "state-of-the-art" for the treatment of grain-surface chemistry in astrochemical models.

FOOTNOTES

1 http://ghosst.osug.fr/.

2 http://www.synchrotron-soleil.fr/.

ACKNOWLEDGEMENTS

We would like to thank all other participants of the Lorentz Center Workshop "Grain-Surface Networks and Data for Astrochemistry": Yuri Aikawa, Wendy Brown, Francois Dulieu, Edith Fayolle, Gleb Fedoseev, Pierre Gratier, George Hassel, Eric Herbst, Leendertjan Karssemeijer, Gunnar Nyman, Laura Reboussin, Maxime Ruaud, Patrice Theulé, Matthijs van Spronsen, Anton Vasyunin, Vianney Taquet, and Ruud Visser for their input and open discussions. We further like to thank the Lorentz Center and the KIDA project for making this workshop possible. R. Garrod acknowledges the support of NASA through grants NNX11AC38G and NNX15AG07G. D. Semenov acknowledges support by the Deutsche Forschungsgemeinschaft through SPP 1385: "The first ten million years of the solar system: a planetary materials approach" (SE 1962/1-3). This research made use of the NASA Astrophysics Data System. C. Walsh acknowledges support from the Netherlands Organization for Scientific

Research (NWO, program number 639.041.335) H. Cuppen, T. Lamberts, and E. Penteado are financially supported by the European Research Council (ERC-2010-StG, Grant Agreement No. 259510-KISMOL). H. Cuppen further acknowledges the VIDI Research Program 700.10.427, which is financed by The Netherlands Organisation for Scientific Research (NWO). T. Lamberts is also financially supported by the European Union's Horizon 2020 research and innovation programme (grant agreement No. 646717, TUNNELCHEM). S. Ioppolo acknowledges the Royal Society for financial support.

REFERENCES

1. M. Accolla, E. Congiu, F. Dulieu, G. Manicò, H. Chaabouni, E. Matar, H. Mokrane, J.L. Lemaire, V. Pirronello, Changes in the morphology of interstellar ice analogues after hydrogen atom exposure. Phys. Chem. Chem. Phys. **13**, 8037 (2011).
2. K. Acharyya, Laboratory study of sticking and desorption of H2H2 and its significance in the chemical evolution of dense interstellar medium. Mon. Not. R. Astron. Soc. **443**, 1301–1309 (2014).
3. K. Acharyya, G.W. Fuchs, H.J. Fraser, E.F. van Dishoeck, H. Linnartz, Desorption of CO and O2O2 interstellar ice analogs. Astron. Astrophys. **466**, 1005–1012 (2007).
4. N.G. Adams, D. Smith, D.C. Clary, Rate coefficients of the reactions of ions with polar molecules at interstellar temperatures. Astrophys. J. **296**, 31 (1985).
5. D.A. Adriaens, T.P.M. Goumans, C.R.A. Catlow, W.A. Brown, Computational study of carbonyl sulphide formation on model interstellar dust grains. J. Phys. Chem. C **114**(4), 1892–1900 (2010).
6. M. Agúndez, V. Wakelam, Chemistry of dark clouds: databases, networks, and models. Chem. Rev. **113**, 8710 (2013).
7. Y. Aikawa, S.M. Miyama, T. Nakano, T. Umebayashi, Evolution of molecular abundance in gaseous disks around young stars: depletion of CO molecules. Astrophys. J. **467**, 684 (1996).
8. Al-Halabi, A.W. Kleyn, E.F. van Dishoeck, G.J. Kroes, Sticking of hydrogen atoms to crystalline ice surfaces: dependence on incidence energy and surface temperature. J. Phys. Chem. B **106**, 6515–6522 (2002)
9. Al-Halabi, A.W. Kleyn, E.F. van Dishoeck, M.C. van Hemert, G.J. Kroes, Sticking of hyperthermal CO to the (0001) face of crystalline ice. J. Phys. Chem. A **107**, 10615–10624 (2003)
10. Al-Halabi, H.J. Fraser, G.J. Kroes, E.F. van Dishoeck, orption of CO on amorphous water-ice surfaces. Astron. Astrophys. **422**, 777–791 (2004).
11. T. Albertsson, D.A. Semenov, A.I. Vasyunin, T. Henning, E. Herbst, New extended deuterium fractionation model: assessment at dense ISM conditions and sensitivity analysis. Astrophys. J. **207**, 27 (2013).
12. L. Allamandola, S. Sandford, G. Valero, Photochemical and thermal evolution of interstellar/precometary ice analogs. Icarus **76**, 225 (1988).
13. M. Allen, G.W. Robinson, The molecular composition of dense interstellar clouds. Astrophys. J. **212**, 396 (1977).
14. L. Amiaud, F. Dulieu, J.-H. Fillion, A. Momeni, J.L. Lemaire, Interaction of atomic and molecular deuterium with a nonporous amorphous water ice surface between 8 and 30 K. J. Chem. Phys. **127**(14), 144709 (2007).
15. E. Anders, Pre-biotic organic matter from comets and asteroids. Nature **342**, 255 (1989).
16. S. Andersson, E.F. van Dishoeck, Photodesorption of water ice. A molecular dynamics study. Astron. Astrophys. **491**, 907–916 (2008).

17. S. Andersson, G. Kroes, E.F. van Dishoeck, Photodissociation of water in crystalline ice: a molecular dynamics study. Chem. Phys. Lett. **408**, 415 (2005).
18. S. Andersson, A. Al-Halabi, G.-J. Kroes, E.F. van Dishoeck, Molecular-dynamics study of photodissociation of water in crystalline and amorphous ices. J. Chem. Phys. **124**(6), 064715 (2006).
19. S. Andersson, T.P.M. Goumans, A. Arnaldsson, Tunneling in hydrogen and deuterium atom addition to CO at low temperatures. Chem. Phys. Lett. **513**, 31–36 (2011).
20. Arasa, S. Andersson, H.M. Cuppen, E.F. van Dishoeck, G. Kroes, Molecular dynamics simulations of the ice temperature dependence of water ice photodesorption. J. Chem. Phys. **132**(18), 184510 (2010).
21. Arasa, S. Andersson, H.M. Cuppen, E.F. van Dishoeck, G.J. Kroes, Molecular dynamics simulations of D2OD2O ice photodesorption. J. Chem. Phys. **134**(16), 164503 (2011).
22. Arasa, J. Koning, G. Kroes, C. Walsh, E.F. van Dishoeck, Photodesorption of H2OH2O, HDO, and D2OD2O ice and its impact on fractionation. Astron. Astrophys. **575**, 121 (2015).
23. Z. Awad, T. Chigai, Y. Kimura, O.M. Shalabiea, T. Yamamoto, New rate constants of hydrogenation of CO on H2OCOH2OCO ice surfaces. Astrophys. J. **626**, 262 (2005).
24. Bacmann, V. Taquet, A. Faure, C. Kahane, C. Ceccarelli, Detection of complex organic molecules in a prestellar core: a new challenge for astrochemical models. Astron. Astrophys. **541**, 12 (2012).
25. N. Balucani, C. Ceccarelli, V. Taquet, Formation of complex organic molecules in cold objects: the role of gas-phase reactions. Mon. Not. R. Astron. Soc. **449**, 16 (2015).
26. V. Barone, C. Latouche, D. Skouteris, F. Vazart, N. Balucani, C. Ceccarelli, B. Lefloch, Gas-phase formation of the prebiotic molecule formamide: insights from new quantum computations. Mon. Not. R. Astron. Soc. **453**, 31–35 (2015).
27. D.R. Bates, L.J. Spitzer, The density of molecules in interstellar space. Astrophys. J. **113**, 441 (1951).
28. E.R. Batista, P. Ayotte, A. Bilić, B.D. Kay, H. Jónsson, What determines the sticking probability of water molecules on ice? Phys. Rev. Lett. **95**, 223201 (2005).
29. R.P. Bell, *The Tunnel Effect in Chemistry* (Chapman and Hall, London, 1980)
30. M.B. Bell, P.A. Feldman, M.J. Travers, M.C. McCarthy, C.A. Gottlieb, P. Thaddeus, Detection of HC[TINF]11[/TINF]N in the Cold Dust Cloud TMC-1. Astrophys. J. **483**, 61 (1997).
31. Belloche, H.S.P. Müller, K.M. Menten, P. Schilke, C. Comito, Complex organic molecules in the interstellar medium: IRAM 30 m line survey of Sagittarius B2(N) and (M). Astron. Astrophys. **559**, 47 (2013).
32. Belloche, R.T. Garrod, H.S.P. Muller, K.M. Menten, Detection of a branched alkyl molecule in the interstellar medium: iso-propyl cyanide. Science **345**, 1584 (2014).
33. E.A. Bergin, W.D. Langer, P.F. Goldsmith, Gas-phase chemistry in dense interstellar clouds including grain surface molecular depletion and desorption. Astrophys. J. **441**, 222 (1995).
34. J.B. Bergner, K.I. Öberg, M. Rajappan, E.C. Fayolle, Kinetics and mechanisms of the acid-base reaction between NH_3 and HCOOH in interstellar ice analogs. Astrophys. J. **829**, 85 (2016).
35. O. Berné, G. Mulas, C. Joblin, Interstellar C60+C60+. Astron. Astrophys. **550**, 4 (2013).
36. M. Bertin, E.C. Fayolle, C. Romanzin, K.I. Öberg, X. Michaut, A. Moudens, L. Philippe, P. Jeseck, H. Linnartz, J. Fillion, UV photodesorption of interstellar CO ice analogues: from subsurface excitation to surface desorption. Phys. Chem. Chem. Phys. **14**, 9929 (2012). doi
37. M. Bertin, E.C. Fayolle, C. Romanzin, H.A.M. Poderoso, X. Michaut, L. Philippe, P. Jeseck, K.I. Öberg, H. Linnartz, J. Fillion, Indirect ultraviolet photodesorption from CO:N_2 binary ices an efficient grain-gas process. Astrophys. J. **779**, 120 (2013).
38. M. Bertin, C. Romanzin, M. Doronin, L. Philippe, P. Jeseck, N. Ligterink, H. Linnartz, X. Michaut, J. Fillion, UV photodesorption of methanol in pure and co-rich ices: desorption rates of the intact molecule and of the photofragments. Astrophys. J. **817**, 12 (2016).
39. O. Biham, I. Furman, V. Pirronello, G. Vidali, Master equation for hydrogen recombination on grain surfaces. Astrophys. J. **553**, 595–603 (2001).

40. S.E. Bisschop, G.W. Fuchs, E.F. van Dishoeck, H. Linnartz, H-atom bombardment of CO_2, HCOOH, and CH_3CHO containing ices. Astron. Astrophys. **474**, 1061 (2007a).
41. S.E. Bisschop, J.K. Jørgensen, E.F. van Dishoeck, E.B.M. de Wachter, Testing grain-surface chemistry in massive hot-core regions. Astron. Astrophys. **465**, 913–929 (2007b).
42. A.S. Bolina, W.A. Brown, Studies of physisorbed ammonia overlayers orbed on graphite. Surf. Sci. **598**, 45–56 (2005).
43. A.S. Bolina, A.J. Wolff, W.A. Brown, Reflection absorption infrared spectroscopy and temperature programmed desorption investigations of the interaction of methanol with a graphite surface. J. Chem. Phys. **122**(4), 044713 (2005a).
44. A.S. Bolina, A.J. Wolff, W.A. Brown, Reflection absorption infrared spectroscopy and temperature-programmed desorption studies of the orption and desorption of amorphous and crystalline water on a graphite surface. J. Phys. Chem. B **109**, 16836–16845 (2005b).
45. A.A. Boogert, P.A. Gerakines, D.C. Whittet, Observations of the Icy Universe. Annu. Rev. Astron. Astrophys. **53**, 541 (2015).
46. J.B. Bossa, P. Theule, F. Duvernay, T. Chiavassa, NH_2CH_2OH thermal formation in interstellar ices contribution to the 5–8 μm region toward embedded protostars. Astrophys. J. **707**, 1524–1532 (2009).
47. J. Bossa, K. Isokoski, M.S. de Valois, H. Linnartz, Thermal collapse of porous interstellar ice. Astron. Astrophys. **545**, 82 (2012).
48. W.A. Brown, A.S. Bolina, Fundamental data on the desorption of pure interstellar ices. Mon. Not. R. Astron. Soc. **374**, 1006–1014 (2007).
49. P.D. Brown, S.B. Charnley, Chemical models of interstellar gas-grain processes. I. Modelling and the effect of accretion on gas abundances and mantle composition in dense clouds. Mon. Not. R. Astron. Soc. **244**, 432–443 (1990)
50. V. Buch, R. Czerminski, Eigenstates of a quantum-mechanical particle on a topologically disordered surface—H(D) atom physisorbed on an amorphous ice cluster $(H_2O)115$. J. Chem. Phys. **95**, 6026–6038 (1991).
51. J.R. Burke, D.J. Hollenbach, The gas-grain interaction in the interstellar medium—thermal accommodation and trapping. Astrophys. J. **265**, 223–234 (1983).
52. T. Butscher, F. Duvernay, P. Theule, G. Danger, Y. Carissan, D. Hagebaum-Reignier, T. Chiavassa, Formation mechanism of glycolaldehyde and ethylene glycol in astrophysical ices from HCO and CH_2OH recombination: an experimental study. Mon. Not. R. Astron. Soc. **453**, 1587 (2015).
53. J. Cami, J. Bernard-Salas, E. Peeters, S.E. Malek, Detection of C60C60 and C70C70 in a Young Planetary Nebula. Science **329**, 1180 (2010).
54. E.K. Campbell, M. Holz, D. Gerlich, J.P. Maier, Laboratory confirmation of C60+C60+ as the carrier of two diffuse interstellar bands. Nature **523**, 322 (2015).
55. P. Caselli, T.I. Hasegawa, E. Herbst, A proposed modification of the rate equations for reactions on grain surfaces. Astrophys. J. **495**, 309 (1998).
56. S. Cazaux, V. Cobut, M. Marseille, M. Spaans, P. Caselli, Water formation on bare grains: when the chemistry on dust impacts interstellar gas. Astron. Astrophys. **522**, 74 (2010).
57. J. Cernicharo, N. Marcelino, E. Roueff, M. Gerin, A. Jiménez-Escobar, G.M. Muñoz Caro, Discovery of the methoxy radical, CH_3O, toward B1: dust grain and gas-phase chemistry in cold dark clouds. Astrophys. J. **759**, 43 (2012).
58. H. Chaabouni, H. Bergeron, S. Baouche, F. Dulieu, E. Matar, E. Congiu, L. Gavilan, J.L. Lemaire, Sticking coefficient of hydrogen and deuterium on silicates under interstellar conditions. Astron. Astrophys. **538**, 128 (2012).
59. Q. Chang, E. Herbst, Interstellar simulations using a unified microscopic-macroscopic Monte Carlo model with a full gas-grain network including bulk diffusion in ice mantles. Astrophys. J. **787**, 135 (2014).
60. Q. Chang, E. Herbst, Unified microscopic–macroscopic Monte Carlo simulations of complex organic molecule chemistry in cold cores. Astrophys. J. **819**, 145 (2016).
61. Q. Chang, H.M. Cuppen, E. Herbst, Continuous-time random-walk simulation of H2H2 formation on interstellar grains. Astron. Astrophys. **434**, 599–611 (2005). doi

62. Q. Chang, H.M. Cuppen, E. Herbst, Gas-grain chemistry in cold interstellar cloud cores with a microscopic Monte Carlo approach to surface chemistry. Astron. Astrophys. **469**, 973–983 (2007).
63. S.B. Charnley, Stochastic theory of molecule formation on dust. Astrophys. J. Lett. **562**, 99–102 (2001).
64. S.B. Charnley, A.G.G.M. Tielens, T.J. Millar, On the molecular complexity of the hot cores in Orion A—grain surface chemistry as 'The last refuge of the scoundrel'. Astrophys. J. **399**, 71 (1992).
65. Y. Chen, K. Chuang, G.M. Muñoz Caro, M. Nuevo, C. Chu, T. Yih, W. Ip, C.R. Wu, Vacuum ultraviolet emission spectrum measurement of a microwave-discharge hydrogen-flow lamp in several configurations: application to photodesorption of CO ice. Astrophys. J. **781**, 15 (2014).
66. K.J. Chuang, G. Fedoseev, S. Ioppolo, E.F. van Dishoeck, H. Linnartz, H-atom addition and abstraction reactions in mixed CO, H_2CO and CH3OH ices—an extended view on complex organic molecule formation. Mon. Not. R. Astron. Soc. **455**, 1702 (2016).
67. Chyba, P. Thomas, L. Brookshaw, C. Sagan, Cometary delivery of organic molecules to the early Earth. Science **249**, 366 (1990).
68. M.P. Collings, J.W. Dever, H.J. Fraser, M.R.S. McCoustra, Laboratory studies of the interaction of carbon monoxide with water ice. Astrophys. Space Sci. **285**, 633–659 (2003).
69. M.P. Collings, M.A. Anderson, R. Chen, J.W. Dever, S. Viti, D.A. Williams, M.R.S. McCoustra, A laboratory survey of the thermal desorption of astrophysically relevant molecules. Mon. Not. R. Astron. Soc. **354**, 1133–1140 (2004).
70. M.P. Collings, V.L. Frankland, J. Lasne, D. Marchione, A. Rosu-Finsen, M.R.S. McCoustra, Probing model interstellar grain surfaces with small molecules. Mon. Not. R. Astron. Soc. **449**, 1826–1833 (2015).
71. Congiu, M. Minissale, S. Baouche, H. Chaabouni, A. Moudens, S. Cazaux, G. Manicò, V. Pirronello, F. Dulieu, Efficient diffusive mechanisms of O atoms at very low temperatures on surfaces of astrophysical interest. Faraday Discuss. **168**, 151 (2014).
72. Cooper, N. Kimmich, W. Belisle, J. Sarinana, K. Brabham, L. Garrel, Carbonaceous meteorites as a source of sugar-related organic compounds for the early Earth. Nature **414**, 879 (2001).
73. G.A. Cruz-Diaz, G.M. Muñoz Caro, Y. Chen, T. Yih, Vacuum-UV spectroscopy of interstellar ice analogs. Astron. Astrophys. **562**, 119 (2014a).
74. G.A. Cruz-Diaz, G.M. Muñoz Caro, Y. Chen, T. Yih, Vacuum-UV spectroscopy of interstellar ice analogs. Astron. Astrophys. **562**, 120 (2014b).
75. G.A. Cruz-Diaz, R. Martín-Doménech, G.M. Muñoz Caro, Y. Chen, Negligible photodesorption of methanol ice and active photon-induced desorption of its irradiation products. Astron. Astrophys. **592**, 68 (2016).
76. H.M. Cuppen, E. Herbst, Simulation of the formation and morphology of ice mantles on interstellar grains. Astrophys. J. **668**, 294–309 (2007).
77. H.M. Cuppen, E.F. van Dishoeck, E. Herbst, A.G.G.M. Tielens, Microscopic simulation of methanol and formaldehyde ice formation in cold dense cores. Astron. Astrophys. **508**, 275–287 (2009).
78. H.M. Cuppen, E.M. Penteado, K. Isokoski, N. van der Marel, H. Linnartz, CO ice mixed with CH_3OH: the answer to the non-detection of the 2152 cm−12152 cm−1 band? Mon. Not. R. Astron. Soc. **417**, 2809 (2011).
79. Dalgarno, J.H. Black, Molecule formation in the interstellar gas. Rep. Prog. Phys. **39**, 573 (1976).
80. L.B. D'Hendecourt, L.J. Allamandola, F. Baas, J.M. Greenberg, Interstellar grain explosions—molecule cycling between gas and dust. Astron. Astrophys. **109**, 12–14 (1982)
81. L.B. D'Hendecourt, L.J. Allamandola, J.M. Greenberg, Time dependent chemistry in dense molecular clouds. I. Grain surface reactions, gas/grain interactions and infrared spectroscopy. Astron. Astrophys. **152**, 130–150 (1985)

82. L.B. D'Hendecourt, L.J. Allamandola, R.J.A. Grim, J.M. Greenberg, Time-dependent chemistry in dense molecular clouds. II. Ultraviolet photoprocessing and infrared spectroscopy of grain mantles. Astron. Astrophys. **158**, 119–134 (1986)
83. A.E. Douglas, G. Herzberg, Note on CH+CH+ in Interstellar Space and in the Laboratory. Astrophys. J. **94**, 381 (1941).
84. B.T. Draine, Interstellar dust grains. Annu. Rev. Astron. Astrophys. **41**, 241–289 (2003).
85. B.T. Draine, B. Sutin, Collisional charging of interstellar grains. Astrophys. J. **320**, 803 (1987).
86. Dulieu, E. Congiu, J. Noble, S. Baouche, H. Chaabouni, A. Moudens, M. Minissale, S. Cazaux, How micron-sized dust particles determine the chemistry of our Universe. Sci. Rep. **3**, 1338 (2013).
87. Dutrey, D. Semenov, E. Chapillon, U. Gorti, S. Guilloteau, F. Hersant, M. Hogerheijde, M. Hughes, G. Meeus, H. Nomura, V. Piétu, C. Qi, V. Wakelam, Physical and chemical structure of planet-forming disks probed by millimeter observations and modeling, in *Protostars and Planets VI* (2014), pp. 317–338.
88. Eckart, The penetration of a potential barrier by electrons. Phys. Rev. **35**, 1303–1309 (1930).
89. A.S. Eddington, Bakerian lecture. Diffuse matter in interstellar space. Proc. R. Soc. A, Math. Phys. Eng. Sci. **111**, 424 (1926).
90. P. Ehrenfreund, S.B. Charnley, Organic molecules in the interstellar medium, comets, and meteorites: a voyage from dark clouds to the early Earth. Annu. Rev. Astron. Astrophys. **38**, 427 (2000).
91. J.E. Elsila, D.P. Glavin, J.P. Dworkin, Cometary glycine detected in samples returned by Stardust. Meteorit. Planet. Sci. **44**, 1323 (2009).
92. E.C. Fayolle, K.I. Öberg, H.M. Cuppen, R. Visser, H. Linnartz, Laboratory H_2O:CO_2 ice desorption data: entrapment dependencies and its parameterization with an extended three-phase model. Astron. Astrophys. **529**, 74 (2011).
93. E.C. Fayolle, K.I. Öberg, H.M. Cuppen, R. Visser, H. Linnartz, Laboratory H_2O:CO_2 ice desorption: entrapment and its parameterization with an extended three-phase model, in *European Conference on Laboratory Astrophysics*, ECLA. EAS Publications Series, vol. 58 (2012), pp. 327–331.
94. E.C. Fayolle, M. Bertin, C. Romanzin, H.A.M. Poderoso, L. Philippe, X. Michaut, P. Jeseck, H. Linnartz, K.I. Öberg, J. Fillion, Wavelength-dependent UV photodesorption of pure N_2 and O_2ices. Astron. Astrophys. **556**, 122 (2013).
95. Fedoseev, H.M. Cuppen, S. Ioppolo, T. Lamberts, H. Linnartz, Experimental evidence for glycolaldehyde and ethylene glycol formation by surface hydrogenation of CO molecules under dense molecular cloud conditions. Mon. Not. R. Astron. Soc. **448**, 1288–1297 (2015).
96. Field, O. Plekan, A. Cassidy, R. Balog, N. Jones, J. Dunger, Spontaneous electric fields in solid films: spontelectrics. Int. Rev. Phys. Chem. **32**, 345 (2013).
97. Fillion, E.C. Fayolle, X. Michaut, M. Doronin, L. Philippe, J. Rakovsky, C. Romanzin, N. Champion, K.I. Öberg, H. Linnartz, M. Bertin, Wavelength resolved UV photodesorption and photochemistry of CO_2 ice. Faraday Discuss. **168**, 533 (2014).
98. H.J. Fraser, M.P. Collings, M.R.S. McCoustra, D.A. Williams, Thermal desorption of water ice in the interstellar medium. Mon. Not. R. Astron. Soc. **327**, 1165–1172 (2001)
99. H.J. Fraser, M.P. Collings, J.W. Dever, M.R.S. McCoustra, Using laboratory studies of CO-H_2O ices to understand the non-detection of a 2152 cm−12152 cm−1 (4.647 μm) band in the spectra of interstellar ices. Mon. Not. R. Astron. Soc. **353**, 59–68 (2004).
100. G.W. Fuchs, K. Acharyya, S.E. Bisschop, K.I. Öberg, F.A. van Broekhuizen, H.J. Fraser, S. Schlemmer, E.F. van Dishoeck, H. Linnartz, Comparative studies of O_2 and N_2 in pure, mixed and layered CO ices. Faraday Discuss. **133**, 331 (2006).
101. G.W. Fuchs, H.M. Cuppen, S. Ioppolo, C. Romanzin, S.E. Bisschop, S. Andersson, E.F. van Dishoeck, H. Linnartz, Hydrogenation reactions in interstellar CO ice analogues. A combined experimental/theoretical approach. Astron. Astrophys. **505**, 629–639 (2009).
102. Furuya, Y. Aikawa, H. Nomura, F. Hersant, V. Wakelam, Water in protoplanetary disks: deuteration and turbulent mixing. Astrophys. J. **779**, 11 (2013).

103. Furuya, M.N. Drozdovskaya, R. Visser, E.F. van Dishoeck, C. Walsh, D. Harsono, U. Hincelin, V. Taquet, Water delivery from cores to disks: deuteration as a probe of the prestellar inheritance of H2OH2O. Astron. Astrophys. (2016, in press).
104. R.T. Garrod, A new modified-rate approach for gas-grain chemical simulations. Astron. Astrophys. **491**, 239–251 (2008).
105. R.T. Garrod, A three-phase chemical model of hot cores: the formation of glycine. Astrophys. J. **765**, 60 (2013).
106. R.T. Garrod, E. Herbst, Formation of methyl formate and other organic species in the warm-up phase of hot molecular cores. Astron. Astrophys. **457**, 927–936 (2006).
107. R.T. Garrod, T. Pauly, On the formation of CO_2 and other interstellar ices. Astrophys. J. **735**, 15 (2011).
108. R. Garrod, I.H. Park, P. Caselli, E. Herbst, Are gas-phase models of interstellar chemistry tenable? The case of methanol. Faraday Discuss. **133**, 51 (2006).
109. R.T. Garrod, V. Wakelam, E. Herbst, Non-thermal desorption from interstellar dust grains via exothermic surface reactions. Astron. Astrophys. **467**, 1103–1115 (2007).
110. R.T. Garrod, S.L.W. Weaver, E. Herbst, Complex chemistry in star-forming regions: an expanded gas-grain warm-up chemical model. Astrophys. J. **682**, 283–302 (2008).
111. R.T. Garrod, A.I. Vasyunin, D.A. Semenov, D.S. Wiebe, T. Henning, A new modified-rate approach for gas-grain chemistry: comparison with a unified large-scale Monte Carlo simulation. Astrophys. J. Lett. **700**, 43–46 (2009).
112. W.D. Geppert, M. Hamberg, R.D. Thomas, F. Österdahl, F. Hellberg, V. Zhaunerchyk, A. Ehlerding, T.J. Millar, H. Roberts, J. Semaniak, M. af Ugglas, A. Källberg, A. Simonsson, M. Kaminska, M. Larsson, Dissociative recombination of protonated methanol. Faraday Discuss. **133**, 177 (2006).
113. P.A. Gerakines, W.A. Schutte, P. Ehrenfreund, Ultraviolet processing of interstellar ice analogs. I. Pure ices. Astron. Astrophys. **312**, 289–305 (1996)
114. D.T. Gillespie, A general method for numerically simulating the stochastic time evolution of coupled chemical reactions. J. Comp. Physiol. **22**, 403 (1976).
115. J. Gonzalez, M. Caballero, A. Aguilar-Mogas, M. Torrent-Sucarrat, R. Crehuet, A. Solé, X. Giménez, S. Olivella, J.M. Bofill, J.M. Anglada, The reaction between HO and (H2O)n(H2O)n (n=1,3n=1,3) clusters: reaction mechanisms and tunneling effects. Theor. Chem. Acc. **128**(4–6), 579–592 (2011).
116. Gonzalez-Lavado, J.C. Corchado, Y.V. Suleimanov, W.H. Green, J. Espinosa-Garcia, Theoretical kinetics study of the o(3p)+ch4/cd4o(3p)+ch4/cd4 hydrogen abstraction reaction: the role of anharmonicity, recrossing effects, and quantum mechanical tunneling. J. Phys. Chem. A **118**(18), 3243–3252 (2014).
117. R.J. Gould, E.E. Salpeter, The interstellar abundance of the hydrogen molecule. I. Basic processes. Astrophys. J. **138**, 393 (1963).
118. T.P.M. Goumans, S. Andersson, Tunnelling in the O+COO+CO reaction. Mon. Not. R. Astron. Soc. **406**, 2213–2217 (2010).
119. T.P.M. Goumans, J. Kästner, Hydrogen-atom tunneling could contribute to h_2 formation in space. Angew. Chem. **49**(40), 7350–7352 (2010).
120. T.P.M. Goumans, J. Kästner, Deuterium enrichment of interstellar methanol explained by atom tunneling. J. Phys. Chem. A **115**(39), 10767–10774 (2011).
121. T.P.M. Goumans, C.R.A. Catlow, W.A. Brown, J. Kästner, P. Sherwood, An embedded cluster study of the formation of water on interstellar dust grains. Phys. Chem. Chem. Phys.. **11**, 5431 (2009).
122. T. Grassi, S. Bovino, D.R.G. Schleicher, J. Prieto, D. Seifried, E. Simoncini, F.A. Gianturco, KROME—a package to embed chemistry in astrophysical simulations. Mon. Not. R. Astron. Soc. **439**, 2386 (2014).
123. N.J.B. Green, T. Toniazzo, M.J. Pilling, D.P. Ruffle, N. Bell, T.W. Hartquist, A stochastic approach to grain surface chemical kinetics. Astron. Astrophys. **375**, 1111–1119 (2001).
124. S.D. Green, A.S. Bolina, R. Chen, M.P. Collings, W.A. Brown, M.R.S. McCoustra, Applying laboratory thermal desorption data in an interstellar context: sublimation of methanol thin films. Mon. Not. R. Astron. Soc. **398**, 357–367 (2009).

125. L.T. Greenberg, Photodesorption from interstellar dust grains, in *Interstellar Dust and Related Topics*, ed. by J.M. Greenberg, H.C. van de Hulst. IAU Symposium, vol. 52 (1973), p. 413
126. R.J.A. Grim, J.M. Greenberg, M.S. de Groot, F. Baas, W.A. Schutte, B. Schmitt, Infrared spectroscopy of astrophysical ices—new insights in the photochemistry. Astron. Astrophys. Suppl. Ser. **78**, 161–186 (1989)
127. W. Hagen, L.J. Allamandola, J.M. Greenberg, Interstellar molecule formation in grain mantles—the laboratory analog experiments, results and implications. Astrophys. Space Sci. **65**, 215–240 (1979).
128. T. Hama, N. Watanabe, Surface processes on interstellar amorphous solid water: orption, diffusion, tunneling reactions, and nuclear-spin conversion. Chem. Rev. **113**(12), 8783–8839 (2013).
129. T. Hama, K. Kuwahata, N. Watanabe, A. Kouchi, Y. Kimura, T. Chigai, V. Pirronello, The mechanism of surface diffusion of H and D atoms on amorphous solid water: existence of various potential sites. Astrophys. J. **757**, 185 (2012).
130. T. Hama, H. Ueta, A. Kouchi, N. Watanabe, H. Tachikawa, Quantum tunneling hydrogenation of solid benzene and its control via surface structure. J. Phys. Chem. Lett. **5**(21), 3843–3848 (2014).
131. T.W. Hartquist, D.A. Williams, Cosmic-ray induced desorption and high depletions in dense cores. Mon. Not. R. Astron. Soc. **247**, 343 (1990)
132. T.I. Hasegawa, E. Herbst, New gas-grain chemical models of quiescent dense interstellar clouds—the effects of H_2 tunneling reactions and cosmic ray induced desorption. Mon. Not. R. Astron. Soc. **261**, 83–102 (1993a)
133. T.I. Hasegawa, E. Herbst, Three-phase chemical models of dense interstellar clouds—gas dust particle mantles and dust particle surfaces. Mon. Not. R. Astron. Soc. **263**, 589 (1993b)
134. T.I. Hasegawa, E. Herbst, C.M. Leung, Models of gas-grain chemistry in dense interstellar clouds with complex organic molecules. Astrophys. J. Suppl. Ser. **82**, 167–195 (1992).
135. J. He, D. Jing, G. Vidali, Atomic oxygen diffusion on and desorption from amorphous silicate surfaces. Phys. Chem. Chem. Phys. **16**, 3493 (2014).
136. J. He, J. Shi, T. Hopkins, G. Vidali, M.J. Kaufman, A new determination of the binding energy of atomic oxygen on dust grain surfaces: experimental results and simulations. Astrophys. J. **801**, 120 (2015).
137. T.J.H. Hele, M.J. Willatt, A. Muolo, S.C. Althorpe, Communication: relation of centroid molecular dynamics and ring-polymer molecular dynamics to exact quantum dynamics. J. Chem. Phys. **142**(19), 191101 (2015).
138. T. Henning, D. Semenov, Chemistry in protoplanetary disks. Chem. Rev. **113**, 9016 (2013).
139. Herbst, H.M. Cuppen, Interstellar chemistry special feature: Monte Carlo studies of surface chemistry and nonthermal desorption involving interstellar grains. Proc. Natl. Acad. Sci. USA **103**, 12257–12262 (2006).
140. Herbst, W. Klemperer, The formation and depletion of molecules in dense interstellar clouds. Astrophys. J. **185**, 505–534 (1973).
141. E. Herbst, C.M. Leung, Effects of large rate coefficients for ion-polar neutral reactions on chemical models of dense interstellar clouds. Astrophys. J. **310**, 378 (1986).
142. E. Herbst, T.J. Millar, *Low Temperatures and Cold Molecules* (Imperial College Press, London, 2008)
143. E. Herbst, E.F. van Dishoeck, Complex organic interstellar molecules. Annu. Rev. Astron. Astrophys. **47**, 427–480 (2009).
144. Hidaka, N. Watanabe, T. Shiraki, A. Nagaoka, A. Kouchi, Conversion of H_2CO to CH_3OH by reactions of cold atomic hydrogen on ice surfaces below 20 K. Astrophys. J. **614**, 1124 (2004).
145. Hidaka, A. Kouchi, N. Watanabe, Temperature, composition, and hydrogen isotope effect in the hydrogenation of CO on amorphous ice surface at 10–20 K. J. Chem. Phys. **126**(20), 204707 (2007).

146. Hidaka, M. Watanabe, A. Kouchi, N. Watanabe, Reaction routes in the CO-H2CO-dn-CH3OH-dmCO-H2CO-dn-CH3OH-dm system clarified from H(D) exposure of solid formaldehyde at low temperatures. Astrophys. J. **702**, 291–300 (2009).
147. Hiraoka, K. Yamamoto, Y. Kihara, T. Takayama, T. Sato, Reaction of H atoms with solid C_2H_4 and C_2H_6 at 13 K. Astrophys. J. **514**, 524–528 (1999).
148. Hiraoka, T. Sato, T. Takayama, in *Astrochemistry: From Molecular Clouds to Planetary Systems*, ed. by Y.C. Minh, E.F. van Dishoeck (Astron. Soc. Pacific, Chelse, 2000), p. 283
149. Hiraoka, K. Sato, S. Sato, S. Hishiki, K. Suzuki, Y. Takahashi, T. Yokoyama, S. Kitagawa, Formation of amorphous silicon by the low-temperature tunneling reaction of H atoms with solid thin film of SiH_4 at 10 K. J. Phys. Chem. B **105**, 6950–6955 (2001).
150. Hiraoka, T. Sato, S. Sato, N. Sogoshi, T. Yokoyama, H. Takashima, S. Kitagawa, Astrophys. J. **577**, 265 (2002)
151. D. Hollenbach, E.E. Salpeter, Surface recombination of hydrogen molecules. Astrophys. J. **163**, 155–164 (1971)
152. D.J. Hollenbach, A.G.G.M. Tielens, Dense Photodissociation Regions (PDRs) 1. Annu. Rev. Astron. Astrophys. **35**, 179 (1997).
153. S. Ioppolo, H.M. Cuppen, C. Romanzin, E.F. van Dishoeck, H. Linnartz, Water formation at low temperatures by surface O_2 hydrogenation I: characterization of ice penetration. Phys. Chem. Chem. Phys. **12**, 12065 (2010).
154. K. Isokoski, J. Bossa, T. Triemstra, H. Linnartz, Porosity and thermal collapse measurements of H_2O, CH_3OH, CO_2, and H_2O:CO_2 ices. Phys. Chem. Chem. Phys. **16**, 3456 (2014).
155. A.V. Ivlev, T.B. Röcker, A. Vasyunin, P. Caselli, Impulsive spot heating and thermal explosion of interstellar grains revisited. Astrophys. J. **805**, 59 (2015).
156. J.K. Jørgensen, C. Favre, S.E. Bisschop, T.L. Bourke, E.F. van Dishoeck, M. Schmalzl, Detection of the simplest sugar, glycolaldehyde, in a solar-type protostar with ALMA. Astrophys. J. **757**, 4 (2012).
157. J. Kalvāns, I. Shmeld, Subsurface chemistry of mantles of interstellar dust grains in dark molecular cores. Astron. Astrophys. **521**, 37 (2010).
158. L.J. Karssemeijer, H.M. Cuppen, Diffusion-desorption ratio of orbed CO and CO_2 on water ice. Astron. Astrophys. **569**, 107 (2014).
159. L.J. Karssemeijer, A. Pedersen, H. Jónsson, H.M. Cuppen, Long-timescale simulations of diffusion in molecular solids. Phys. Chem. Chem. Phys. **14**, 10844–10852 (2012).
160. L.J. Karssemeijer, G.A. de Wijs, H.M. Cuppen, Interactions of orbed CO_2 on water ice at low temperatures. Phys. Chem. Chem. Phys. **16**, 15630–15639 (2014a).
161. L.J. Karssemeijer, S. Ioppolo, M.C. van Hemert, A. van der Avoird, M.A. Allodi, G.A. Blake, H.M. Cuppen, Dynamics of CO in amorphous water-ice environments. Astrophys. J. **781**, 16 (2014b).
162. J. Kästner, Theory and simulation of atom tunneling in chemical reactions. WIREs Comput. Mol. Sci. **4**(2), 158–168 (2014).
163. Katz, I. Furman, O. Biham, V. Pirronello, G. Vidali, Molecular hydrogen formation on astrophysically relevant surfaces. Astrophys. J. **522**, 305–312 (1999).
164. J. Koning, G.J. Kroes, C. Arasa, Isotope effects on the photodesorption processes of X_2O (X=H,DX=H,D) and HOD ice. J. Chem. Phys. **138**, 104701 (2013).
165. K. Kuwahata, T. Hama, A. Kouchi, N. Watanabe, Signatures of quantum-tunneling diffusion of hydrogen atoms on water ice at 10K. Phys. Rev. Lett. **115**, 133201 (2015).
166. K. Kvenvolden, J. Lawless, K. Pering, E. Peterson, J. Flores, C. Ponnamperuma, I.R. Kaplan, C. Moore, Evidence for extraterrestrial amino-acids and hydrocarbons in the murchison meteorite. Nature **228**, 923 (1970).
167. T. Lamberts, H.M. Cuppen, S. Ioppolo, H. Linnartz, Water formation at low temperatures by surface O_2 hydrogenation III: Monte Carlo simulation. Phys. Chem. Chem. Phys. **15**, 8287 (2013).
168. T. Lamberts, H.M. Cuppen, G. Fedoseev, S. Ioppolo, K.-J. Chuang, H. Linnartz, Relevance of the H2+OH2+O reaction pathway for the surface formation of interstellar water. Combined experimental and modeling study. Astron. Astrophys. **570**, 57 (2014a).

169. T. Lamberts, X. de Vries, H.M. Cuppen, The formation of ice mantles on interstellar grains revisited "the effect of exothermicity". Faraday Discuss. **168**, 327 (2014b).
170. T. Lamberts, S. Ioppolo, H.M. Cuppen, G. Fedoseev, H. Linnartz, Thermal H/D exchange in polar ice—deuteron scrambling in space. Mon. Not. R. Astron. Soc. **448**, 3820–3828 (2015).
171. T. Lamberts, G. Fedoseev, F. Puletti, S. Ioppolo, H.M. Cuppen, H. Linnartz, Low-temperature chemistry between water and hydroxyl radicals: H/D isotopic effects. Mon. Not. R. Astron. Soc. **455**, 634 (2016).
172. T. Lauck, L. Karssemeijer, K. Shulenberger, M. Rajappan, K.I. Öberg, H.M. Cuppen, CO diffusion into amorphous H_2O ices. Astrophys. J. **801**, 118 (2015).
173. R. Launhardt, A.M. Stutz, A. Schmiedeke, T. Henning, O. Krause, Z. Balog, H. Beuther, S. Birkmann, M. Hennemann, J. Kainulainen, T. Khanzadyan, H. Linz, N. Lippok, M. Nielbock, J. Pitann, S. Ragan, C. Risacher, M. Schmalzl, Y.L. Shirley, B. Stecklum, J. Steinacker, J. Tackenberg, The Earliest Phases of star Formation (EPoS): a Herschel key project. Astron. Astrophys. **551**, 98 (2013).
174. Leger, M. Jura, A. Omont, Desorption from interstellar grains. Astron. Astrophys. **144**, 147–160 (1985)
175. Lepetit, D. Lemoine, Z. Medina, B. Jackson, Sticking and desorption of hydrogen on graphite: a comparative study of different models. J. Chem. Phys. **134**(11), 114705 (2011).
176. C.M. Leung, E. Herbst, W.F. Huebner, Synthesis of complex molecules in dense interstellar clouds via gas-phase chemistry—a pseudo time-dependent calculation. Astrophys. J. **56**, 231 (1984).
177. Y. Li, Y.V. Suleimanov, J. Li, W.H. Green, H. Guo, Rate coefficients and kinetic isotope effects of the X+CH4→CH3+HX (X=H,D,Mu)X+CH4→CH3+HX (X=H,D,Mu) reactions from ring polymer molecular dynamics. J. Chem. Phys. **138**(9), 094307 (2013).
178. N.F.W. Ligterink, D.M. Paardekooper, K. Chuang, M.L. Both, G.A. Cruz-Diaz, J.H. van Helden, H. Linnartz, Controlling the emission profile of an H_2 discharge lamp to simulate interstellar radiation fields. Astron. Astrophys. **584**, 56 (2015).
179. H. Linnartz, S. Ioppolo, G. Fedoseev, Atom addition reactions in interstellar ice analogues. Int. Rev. Phys. Chem. **34**, 205 (2015).
180. Lippok, R. Launhardt, T. Henning, Z. Balog, H. Beuther, J. Kainulainen, O. Krause, H. Linz, M. Nielbock, S.E. Ragan, T.P. Robitaille, S.I. Sadavoy, A. Schmiedeke, Earliest phases of star formation (EPoS). Astron. Astrophys. **592**, 61 (2016).
181. Lipshtat, O. Biham, Efficient simulations of gas-grain chemistry in interstellar clouds. Phys. Rev. Lett. **93**(17), 170601 (2004).
182. R.M. Logan, J.C. Keck, Classical theory for the interaction of gas atoms with solid surfaces. J. Chem. Phys. **49**, 860–876 (1968).
183. N.J. Mason, A. Dawes, P.D. Holtom, R.J. Mukerji, M.P. Davis, B. Sivaraman, R.I. Kaiser, S.V. Hoffmann, D.A. Shaw, VUV spectroscopy and photo-processing of astrochemical ices: an experimental study. Faraday Discuss. **133**, 311 (2006).
184. D. McElroy, C. Walsh, A.J. Markwick, M.A. Cordiner, K. Smith, T.J. Millar, The UMIST database for astrochemistry 2012. Astron. Astrophys. **550**, 36 (2013).
185. McKellar, Evidence for the molecular origin of some Hitherto unidentified interstellar lines. Publ. Astron. Soc. Pac. **52**, 187 (1940).
186. T.J. Millar, P.R.A. Farquhar, K. Willacy, The UMIST database for astrochemistry 1995. Astron. Astrophys. **121**, 139 (1997).
187. Minissale, F. Dulieu, Influence of surface coverage on the chemical desorption process. J. Chem. Phys. **141**, 014304 (2014).
188. Minissale, F. Dulieu, S. Cazaux, S. Hocuk, Dust as interstellar catalyst I. Quantifying the chemical desorption process. Astron. Astrophys. **585**, 24 (2016).
189. F. Mispelaer, P. Theulé, H. Aouididi, J. Noble, F. Duvernay, G. Danger, P. Roubin, O. Morata, T. Hasegawa, T. Chiavassa, Diffusion measurements of CO, HNCO, H_2CO, and NH_3 in amorphous water ice. Astron. Astrophys. **555**, 13 (2013).

190. Miyauchi, H. Hidaka, T. Chigai, A. Nagaoka, N. Watanabe, A. Kouchi, Formation of hydrogen peroxide and water from the reaction of cold hydrogen atoms with solid oxygen at 10 K. Chem. Phys. Lett. **456**, 27–30 (2008).
191. M. Monge-Palacios, J.C. Corchado, J. Espinosa-Garcia, Dynamics study of the OH+NH3OH+NH3hydrogen abstraction reaction using QCT calculations based on an analytical potential energy surface. J. Chem. Phys. **138**(21), 214306 (2013a).
192. M. Monge-Palacios, C. Rangel, J. Espinosa-Garcia, Ab initio based potential energy surface and kinetics study of the OH+NH3OH+NH3 hydrogen abstraction reaction. J. Chem. Phys. **138**(8), 084305 (2013b).
193. G.M. Muñoz Caro, A. Jiménez-Escobar, J.Á. Martín-Gago, C. Rogero, C. Atienza, S. Puertas, J.M. Sobrado, J. Torres-Redondo, New results on thermal and photodesorption of CO ice using the novel InterStellar Astrochemistry Chamber (ISAC). Astron. Astrophys. **522**, 108 (2010).
194. Nagaoka, N. Watanabe, A. Kouchi, H-D substitution in interstellar solid methanol: a key route for D enrichment. Astrophys. J. Lett. **624**, 29–32 (2005).
195. Nagaoka, N. Watanabe, A. Kouchi, Effective rate constants for the surface reaction between solid methanol and deuterium atoms at 10 K. J. Phys. Chem. A **111**, 3016–3028 (2007).
196. T.L. Nguyen, J.F. Stanton, Ab initio thermal rate calculations of HO+HO= O(3P)+H2OHO+HO=O(3P)+H2Oreaction and isotopologues. J. Phys. Chem. A **117**(13), 2678–2686 (2013).
197. T.L. Nguyen, J.F. Stanton, Accurate ab initio thermal rate constants for reaction of o(3p) with h2 and isotopic analogues. J. Phys. Chem. A **118**(27), 4918–4928 (2014).
198. T.L. Nguyen, J.F. Stanton, J.R. Barker, Ab initio reaction rate constants computed using semiclassical transition-state theory: HO+H2→H2O+HHO+H2→H2O+H and isotopologues. J. Phys. Chem. A **115**(20), 5118–5126 (2011).
199. T.L. Nguyen, B.C. Xue, R.E. Weston, J.R. Barker, J.F. Stanton, Reaction of HO with CO: tunneling is indeed important. J. Phys. Chem. Lett. **3**(11), 1549–1553 (2012).
200. Nishi, H. Shinohara, T. Okuyama, Photodetachment, photodissociation, and photochemistry of surface molecules of icy solids containing NH3 and pure H_2O ices. J. Chem. Phys. **80**, 3898 (1984).
201. J.A. Noble, P. Theulé, F. Mispelaer, F. Duvernay, G. Danger, E. Congiu, F. Dulieu, T. Chiavassa, The desorption of H_2CO from interstellar grains analogues. Astron. Astrophys. **543**, 5 (2012a).
202. J.A. Noble, E. Congiu, F. Dulieu, H.J. Fraser, Thermal desorption characteristics of CO, O2O2 and CO_2 on non-porous water, crystalline water and silicate surfaces at submonolayer and multilayer coverages. Mon. Not. R. Astron. Soc. **421**, 768–779 (2012b).
203. J.A. Noble, P. Theule, F. Duvernay, G. Danger, T. Chiavassa, P. Ghesquiere, T. Mineva, D. Talbi, Kinetics of the NH_3 and CO_2 solid-state reaction at low temperature. Phys. Chem. Chem. Phys. **16**, 23604–23615 (2014).
204. G. Nyman, Computational methods of quantum reaction dynamics. Int. J. Quant. Chem. **114**(18), 1183–1198 (2014).
205. Y. Oba, N. Miyauchi, H. Hidaka, T. Chigai, N. Watanabe, A. Kouchi, Formation of compact amorphous H_2O ice by codeposition of hydrogen atoms with oxygen molecules on grain surfaces. Astrophys. J. **701**, 464–470 (2009).
206. Y. Oba, N. Watanabe, T. Hama, K. Kuwahata, H. Hidaka, A. Kouchi, Water formation through a quantum tunneling surface reaction, OH+H2OH+H2, at 10 K. Astrophys. J. **749**, 67 (2012).
207. Y. Oba, T. Chigai, Y. Osamura, N. Watanabe, A. Kouchi, Hydrogen isotopic substitution of solid methylamine through atomic surface reactions at low temperatures: a potential contribution to the D/H ratio of methylamine in molecular clouds. Meteorit. Planet. Sci. **49**, 117–132 (2014a).
208. Y. Oba, K. Osaka, N. Watanabe, T. Chigai, A. Kouchi, Reaction kinetics and isotope effect of water formation by the surface reaction of solid H2O2H2O2 with H atoms at low temperatures. Faraday Discuss. **168**, 185–204 (2014b).

209. Y. Oba, N. Watanabe, Y. Osamura, A. Kouchi, Chiral glycine formation on cold interstellar grains by quantum tunneling hydrogen–deuterium substitution reactions. Chem. Phys. Lett. **634**, 53–59 (2015).
210. K.I. Öberg, F. van Broekhuizen, H.J. Fraser, S.E. Bisschop, E.F. van Dishoeck, S. Schlemmer, Competition between CO and N_2 desorption from interstellar ices. Astrophys. J. Lett. **621**, 33–36 (2005).
211. K.I. Öberg, G.W. Fuchs, Z. Awad, H.J. Fraser, S. Schlemmer, E.F. van Dishoeck, H. Linnartz, Photodesorption of CO ice. Astrophys. J. Lett. **662**, 23–26 (2007).
212. K.I. Öberg, E.F. van Dishoeck, H. Linnartz, Photodesorption of ices I: CO, N_2, and CO_2. Astron. Astrophys. **496**, 281–293 (2009a).
213. K.I. Öberg, H. Linnartz, R. Visser, E.F. van Dishoeck, Photodesorption of ices. II. H_2O and D_2O. Astrophys. J. **693**, 1209–1218 (2009b).
214. K.I. Öberg, R.T. Garrod, E.F. van Dishoeck, H. Linnartz, Formation rates of complex organics in UV irradiated CH_3OH-rich ices. I. Experiments. Astron. Astrophys. **504**, 891–913 (2009c).
215. K.I. Öberg, E.C. Fayolle, H.M. Cuppen, E.F. van Dishoeck, H. Linnartz, Quantification of segregation dynamics in ice mixtures. Astron. Astrophys. **505**, 183–194 (2009d). doi
216. K.I. Öberg, S. Bottinelli, J.K. Jørgensen, E.F. van Dishoeck, A cold complex chemistry toward the low-mass protostar B1-b: evidence for complex molecule production in ices. Astrophys. J. **716**, 825–834 (2010).
217. K.I. Öberg, A.C.A. Boogert, K.M. Pontoppidan, S. van den Broek, E.F. van Dishoeck, S. Bottinelli, G.A. Blake, N.J. Evans II., The Spitzer ice legacy: ice evolution from cores to protostars. Astrophys. J. **740**, 109 (2011).
218. D.H. Parker, Laser photochemistry of molecular oxygen. Acc. Chem. Res. **33**, 563 (2000).
219. Pedersen, L. Karssemeijer, H.M. Cuppen, H. Jónsson, Long-timescale simulations of H2OH2Oadmolecule diffusion on Ice Ih(0001) surfaces. J. Phys. Chem. C **119**(29), 16528–16536 (2015).
220. E.M. Penteado, A.C.A. Boogert, K.M. Pontoppidan, S. Ioppolo, G.A. Blake, H.M. Cuppen, Spectroscopic constraints on CH_3OH formation: CO mixed with CH_3OH ices towards young stellar objects. Mon. Not. R. Astron. Soc. **454**, 531–540 (2015).
221. E.M. Penteado, C.W. Walsh, H.M. Cuppen, Sensitivity analysis of grain surface chemistry to binding energies of ice species (2016, submitted)
222. V. Pirronello, O. Biham, C. Liu, L. Shen, G. Vidali, Efficiency of molecular hydrogen formation on silicates. Astrophys. J. Lett. **483**, 131 (1997a).
223. V. Pirronello, C. Liu, L. Shen, G. Vidali, Laboratory synthesis of molecular hydrogen on surfaces of astrophysical interest. Astrophys. J. Lett. **475**, 69 (1997b).
224. V. Pirronello, C. Liu, J.E. Roser, G. Vidali, Measurements of molecular hydrogen formation on carbonaceous grains. Astron. Astrophys. **344**, 681–686 (1999)
225. K.M. Pontoppidan, Spatial mapping of ices in the Ophiuchus-F core. A direct measurement of CO depletion and the formation of CO_2. Astron. Astrophys. **453**, 47–50 (2006).
226. K.M. Pontoppidan, H.J. Fraser, E. Dartois, W. Thi, E.F. van Dishoeck, A.C.A. Boogert, L. d'Hendecourt, A.G.G.M. Tielens, S.E. Bisschop, A 3–5 μm VLT spectroscopic survey of embedded young low mass stars I. Structure of the CO ice. Astron. Astrophys. **408**, 981–1007 (2003).
227. S.S. Prasad, S.P. Tarafdar, UV radiation field inside dense clouds—its possible existence and chemical implications. Astrophys. J. **267**, 603 (1983).
228. J.M.C. Rawlings, D.A. Williams, S. Viti, C. Cecchi-Pestellini, W.W. Duley, Episodic explosions in interstellar ices. Mon. Not. R. Astron. Soc. **430**, 264 (2013).
229. Rimola, V. Taquet, P. Ugliengo, N. Balucani, C. Ceccarelli, Combined quantum chemical and modeling study of CO hydrogenation on water ice. Astron. Astrophys. **572**, 70 (2014).
230. D.P. Ruffle, E. Herbst, New models of interstellar gas-grain chemistry. I. Surface diffusion rates. Mon. Not. R. Astron. Soc. **319**, 837–850 (2000).
231. D.P. Ruffle, E. Herbst, New models of interstellar gas-grain chemistry. III. Solid CO_2. Mon. Not. R. Astron. Soc. **324**, 1054–1062 (2001).

232. S.A. Sandford, L.J. Allamandola, The condensation and vaporization behavior of H_2O:CO ices and implications for interstellar grains and cometary activity. Icarus **76**, 201–224 (1988).
233. S.A. Sandford, L.J. Allamandola, The volume- and surface-binding energies of ice systems containing CO, CO2, and H_2O. Icarus **87**, 188–192 (1990).
234. S.A. Sandford, L.J. Allamandola, Condensation and vaporization studies of CH3OH and NH3 ices: major implications for astrochemistry. Astrophys. J. **417**, 815–825 (1993a).
235. S.A. Sandford, L.J. Allamandola, H_2 in interstellar and extragalactic ices—infrared characteristics, ultraviolet production, and implications. Astrophys. J. Lett. **409**, 65–68 (1993b).
236. F.L. Schöier, J.K. Jrgensen, E.F. van Dishoeck, G.A. Blake, Does IRAS 16293-2422 have a hot core? Chemical inventory and abundance changes in its protostellar environment. Astron. Astrophys. **390**, 1001 (2002).
237. D. Semenov, F. Hersant, V. Wakelam, A. Dutrey, E. Chapillon, S. Guilloteau, T. Henning, R. Launhardt, V. Piétu, K. Schreyer, Chemistry in disks. Astron. Astrophys. **522**, 42 (2010).
238. L.E. Snyder, D. Buhl, B. Zuckerman, P. Palmer, Microwave detection of interstellar formaldehyde. Phys. Rev. Lett. **22**, 679 (1969).
239. T. Stantcheva, E. Herbst, Models of gas-grain chemistry in interstellar cloud cores with a stochastic approach to surface chemistry. Astron. Astrophys. **423**, 241–251 (2004).
240. T. Stantcheva, V.I. Shematovich, E. Herbst, On the master equation approach to diffusive grain-surface chemistry: The H, O, CO system. Astron. Astrophys. **391**, 1069–1080 (2002).
241. Stutz, R. Launhardt, H. Linz, O. Krause, T. Henning, J. Kainulainen, M. Nielbock, J. Steinacker, P. André, Dust-temperature of an isolated star-forming cloud: Herschel observations of the Bok globule CB244. Astron. Astrophys. **518**, 87 (2010).
242. Y.V. Suleimanov, J. Espinosa-Garcia, Recrossing and tunneling in the kinetics study of the OH+CH4→H2O+CH3OH+CH4→H2O+CH3 reaction. J. Phys. Chem. B (2015).
243. Swings, L. Rosenfeld, Considerations regarding interstellar molecules. Astrophys. J. **86**, 483 (1937). V. Taquet, C. Ceccarelli, C. Kahane, Multilayer modeling of porous grain surface chemistry. I. The GRAINOBLE model. Astron. Astrophys. **538**, 42 (2012).
244. V. Taquet, P.S. Peters, C. Kahane, C. Ceccarelli, A. López-Sepulcre, C. Toubin, D. Duflot, L. Wiesenfeld, Water ice deuteration: a tracer of the chemical history of protostars. Astron. Astrophys. **550**, 127 (2013).
245. Theule, F. Duvernay, A. Ilmane, T. Hasegawa, O. Morata, S. Coussan, G. Danger, T. Chiavassa, Kinetics of the OCN–OCN– and HOCN formation from the HNCO+ H2OHNCO+H2O thermal reaction in interstellar ice analogs. Astron. Astrophys. **530**, 96 (2011).
246. Tielens, Interstellar polycyclic aromatic hydrocarbon molecules. Annu. Rev. Astron. Astrophys. **46**, 289 (2008).
247. A.G.G.M. Tielens, *The Physics and Chemistry of the Interstellar Medium* (2010)
248. A.G.G.M. Tielens, L.J. Allamandola, Composition, structure, and chemistry of interstellar dust, in *Interstellar Processes*, ed. by D.J. Hollenbach, H.A. Thronson Jr. Astrophysics and Space Science Library, vol. 134 (1987), pp. 397–469
249. A.G.G.M. Tielens, W. Hagen, Model calculations of the molecular composition of interstellar grain mantles. Astron. Astrophys. **114**, 245–260 (1982)
250. E.F. van Dishoeck, G.A. Blake, Chemical evolution of star-forming regions. Annu. Rev. Astron. Astrophys. **36**, 317 (1998).
251. E.F. van Dishoeck, B. Jonkheid, M.C. van Hemert, Photoprocesses in protoplanetary disks. Faraday Discuss. **133**, 231 (2006).
252. M.C. van Hemert, J. Takahashi, E.F. van Dishoeck, Molecular dynamics study of the photodesorption of CO ice. J. Phys. Chem. A **119**, 6354 (2015).
253. Vastel, C. Ceccarelli, B. Lefloch, R. Bachiller, The origin of complex organic molecules in prestellar cores. Astrophys. J. **795**, 2 (2014).
254. A.I. Vasyunin, E. Herbst, A unified Monte Carlo treatment of gas-grain chemistry for large reaction networks. II. A multiphase gas-surface-layered bulk model. Astrophys. J. **762**, 86 (2013).

255. A.I. Vasyunin, D.A. Semenov, D.S. Wiebe, T. Henning, A unified Monte Carlo treatment of gas-grain chemistry for large reaction networks. I. Testing validity of rate equations in molecular clouds. Astrophys. J. **691**, 1459–1469 (2009).
256. V.K. Veeraghattam, K. Manrodt, S.P. Lewis, P.C. Stancil, The sticking of atomic hydrogen on amorphous water ice. Astrophys. J. **790**, 4 (2014).
257. Viti, M.P. Collings, J.W. Dever, M.R.S. McCoustra, D.A. Williams, Evaporation of ices near massive stars: models based on laboratory temperature programmed desorption data. Mon. Not. R. Astron. Soc. **354**, 1141–1145 (2004).
258. V. Wakelam, I.W.M. Smith, E. Herbst, J. Troe, W. Geppert, H. Linnartz, K. Öberg, E. Roueff, M. Agúndez, P. Pernot, H.M. Cuppen, J.C. Loison, D. Talbi, Reaction networks for interstellar chemical modelling: improvements and challenges. Space Sci. Rev. **156**, 13–72 (2010).
259. V. Wakelam, E. Herbst, J.-C. Loison, I.W.M. Smith, V. Chandrasekaran, B. Pavone, N.G. Adams, M.-C. Bacchus-Montabonel, A. Bergeat, K. Béroff, V.M. Bierbaum, M. Chabot, A. Dalgarno, E.F. van Dishoeck, A. Faure, W.D. Geppert, D. Gerlich, D. Galli, E. Hébrard, F. Hersant, K.M. Hickson, P. Honvault, S.J. Klippenstein, S. Le Picard, G. Nyman, P. Pernot, S. Schlemmer, F. Selsis, I.R. Sims, D. Talbi, J. Tennyson, J. Troe, R. Wester, L. Wiesenfeld, A KInetic Database for Astrochemistry (KIDA). Astrophys. J. Suppl. Ser. **199**, 21 (2012).
260. Walsh, T.J. Millar, H. Nomura, E. Herbst, S. Widicus Weaver, Y. Aikawa, J.C. Laas, A.I. Vasyunin, Complex organic molecules in protoplanetary disks. Astron. Astrophys. **563**, 33 (2014).
261. N. Watanabe, A. Kouchi, Efficient formation of formaldehyde and methanol by the addition of hydrogen atoms to CO in H_2O-CO ice at 10 K. Astrophys. J. Lett. **571**, 173–176 (2002).
262. N. Watanabe, A. Kouchi, Ice surface reactions: a key to chemical evolution in space. Prog. Surf. Sci. **83**, 439 (2008).
263. N. Watanabe, A. Nagaoka, H. Hidaka, T. Shiraki, T. Chigai, A. Kouchi, Dependence of the effective rate constants for the hydrogenation of CO on the temperature and composition of the surface. Planet. Space Sci. **54**, 1107–1114 (2006).
264. N. Watanabe, Y. Kimura, A. Kouchi, T. Chigai, T. Hama, V. Pirronello, Direct measurements of hydrogen atom diffusion and the spin temperature of nascent H2H2 molecule on amorphous solid water. Astrophys. J. Lett. **714**, 233–237 (2010).
265. W.D. Watson, Ion-molecule reactions, molecule formation, and hydrogen-isotope exchange in dense interstellar clouds. Astrophys. J. **188**, 35 (1974).
266. W.D. Watson, E.E. Salpeter, Molecule formation on interstellar grains. Astrophys. J. **174**, 321 (1972a).
267. W.D. Watson, E.E. Salpeter, On the abundances of interstellar molecules. Astrophys. J. **175**, 659 (1972b).
268. M.S. Westley, R.A. Baragiola, R.E. Johnson, G.A. Baratta, Photodesorption from low-temperature water ice in interstellar and circumsolar grains. Nature **373**, 405 (1995a).
269. M.S. Westley, R.A. Baragiola, R.E. Johnson, G.A. Baratta, Ultraviolet photodesorption from water ice. Planet. Space Sci. **43**, 1311–1315 (1995b).
270. R.E. Weston, T.L. Nguyen, J.F. Stanton, J.R. Barker, HO+COHO+CO reaction rates and H/D kinetic isotope effects: master equation models with ab initio SCTST rate constants. J. Phys. Chem. A **117**, 821–835 (2013).
271. Wigner, The transition state method. Trans. Faraday Soc. **34**, 29 (1938).
272. K. Willacy, W.D. Langer, The importance of photoprocessing in protoplanetary disks. Astrophys. J. **544**, 903 (2000).
273. K. Willacy, H.H. Klahr, T.J. Millar, T. Henning, Gas and grain chemistry in a protoplanetary disk. Astron. Astrophys. **338**, 995–1005 (1998)
274. Wolff, I. Lohmar, J. Krug, O. Biham, Diffusion-limited reactions on disordered surfaces with continuous distributions of binding energies. J. Stat. Mech. Theory Exp. **10**, 29 (2011).
275. P.M. Woods, B. Slater, Z. Raza, S. Viti, W.A. Brown, D.J. Burke, Glycolaldehyde formation via the dimerization of the formyl radical. Astrophys. J. **777**, 90 (2013).

276. D.E. Woon, Ion-ice astrochemistry: barrierless low-energy deposition pathways to HCOOH, CH_3OH, and CO_2 on icy grain mantles from precursor cations. Astrophys. J. **728**, 44 (2011).
277. Y. Zhang, T. Stecher, M.T. Cvitas, S.C. Althorpe, Which is better at predicting quantum-tunneling rates: quantum transition-state theory or free-energy instanton theory? J. Phys. Chem. Lett. **5**, 3976–3980 (2014).
278. J. Zhen, H. Linnartz, UV-induced photodesorption and photochemistry of O_2 ice. Mon. Not. R. Astron. Soc. **437**, 3190 (2014).

CHAPTER 5

Gamma-Ray Bursts and Fast Transients

R. Willingale1, P. Mészáros2

1.Department of Physics and AstronomyUniversity of LeicesterLeicesterUK

2.Pennsylvania State UniversityUniversity ParkUSA

ABSTRACT

The current status of observations and theoretical models of gamma-ray bursts and some other related transients, including ultra-long bursts and tidal disruption events, is reviewed. We consider the impact of multi-wavelength data on the formulation and development of theoretical models for the prompt and afterglow emission including the standard fireball model utilizing internal shocks and external shocks, photospheric emission, the role of the magnetic field and hadronic processes. In addition, we discuss some of the prospects for non-photonic multi-messenger detection and for future instrumentation, and comment on some of the outstanding issues in the field.

Keywords: Gamma-rays, bursts, Black holes, Neutron stars, Shocks, Neutrinos, Gravitational waves

1. INTRODUCTION

Gamma-ray bursts (GRBs) and other one-shot, fast, high energy transients are likely to produce UHE Cosmic rays, VHE neutrinos and Gravitational waves in addition to radiation across the entire electromagnetic spectrum from radio through to VHE gamma-rays. Multi-wavelength observations have been and will, without doubt, continue to be crucial in the study of GRBs and the detection of multi-messenger signals, along side the high energy photon prompt emission, will broaden our understanding of the extreme physical conditions and processes which give rise to such high energy phenomena.

Many of the major breakthroughs and results in GRB studies to date have arisen from multi-wavelength observations. The realisation that all GRBs are extra galactic was initiated by the first X-ray afterglow detection made by the *BeppoSAX* soft X-ray narrow field instruments for the GRB 970228 (Costa et al. 1997). This provided a position with a 90% confidence error radius of 50 arc seconds which lead to the identification of the optical emission (Van Paradijs et al. 1997) and subsequent deep observations with the Keck telescopes gave a redshift of $z=0.695z=0.695$ (Bloom et al. 2001). The exceptionally bright GRB 030329 in 2003 was detected by all three instruments on the High Energy Transient Explorer satellite (HETE) across a wide photon energy range: Fregate 7–400 keV; the WXM 2–25 keV; SXC 2–10 keV. Rapid communication of the $2'2'$ error box 73 minutes after the burst allowed the detection of a counterpart in optical and radio emission (Vanderspek et al. 2004). This multi-wavelength coverage lead to the measurement of a redshift, $z=0.1685z=0.1685$, the discovery of a supernova associated with the burst and established that long GRBs can be produced by the explosion of some extreme type Ic SNe.

The redshift distribution of GRBs is a moving target with new results being published all the time. A good up to date summary can be found at the web page maintained by Greiner.[1] The redshift measurements span the range 0.0085-9.4 with a mean of $\sim 2.5 \sim 2.5$. This distribution is the product of intense multi-wavelength observation activity and is subject to many selection biases and constraints, e.g. Coward et al. (2013). HST imaging of the error boxes of long GRBs reveals that they occur predominantly in UV bright star forming regions of their host galaxies, e.g. Fruchter et al. (2006).

The detection and accurate positioning of GRBs combined with a rapid alert system and fast slewing response to initiate near immediate follow-up observations was pioneered by the *Swift*satellite (Gehrels et al. 2004). This has provided a wealth of new and intriguing results including the first accurate localizations and redshifts for short GRBs (Gehrels et al. 2005), the identification of the highest redshift GRBs (Cucchiara et al. 2011), a population of low luminosity GRBs often associated with supernovae (Stanek et al. 2003; Zhang et al. 2012), the detection of the brightest optical GRB, visible with the naked eye (Bloom et al. 2009) and detailed multi-wavelength light curves and spectra from the prompt through to the afterglow emission. With the rapid development of new facilities like *LIGO*, *Virgo*, *IceCube* etc. there is now the possibility of the simultaneous detection of multi-messenger signals, including gravitational wave and neutrinos, from GRBs. The likelihood of such detections is, of course, dependant on a multitude of factors such as the local rate of mergers of compact objects and the beaming angle of GRB jets but it is clear that a rapid response and rapid alert enabling multi-wavelength and multi-messenger observations from the prompt through to late afterglow will be mandatory if we hope to gain a detailed understanding of GRBs and other fast, high energy transients.

2. THE FIREBALL MODEL THEN AND NOW

The theoretical paradigm underpinning most investigations of GRBs is the standard fireball shock model (see Mészáros 2002, and references therein). At early times, and close to the central engine, internal shocks within the relativistically expanding fireball produce the highly variable prompt hard X-ray and Gamma-ray emission we see as the GRB (Rees and Mészáros 1994) and as the fireball is decelerated by interaction with the low-density circumburst medium an external shock forward shock develops to produce the longer lasting afterglow emission seen initially from the soft X-ray band through to the IR (Rees and Mészáros 1992). The predominant emission mechanism is expected to be synchrotron radiation from relativistic electrons accelerated in the shocks, both internal and external, gyrating in the magnetic field wound up by compression and turbulence in the same shocks. In addition, inverse Compton up-scattering of low energy synchrotron photons by the relativistic electron population can produce a high energy component or tail to the expected synchrotron spectrum such that the broadband spectrum observed is described by a so-called Synchrotron-Self-Compton (SSC) model. When energy injection from the central engine ceases the forward shock emission enters a final power-law decay and produces a characteristic synchrotron radiation afterglow spectrum stretching from radio through to X-ray (Sari et al. 1998; Wijers and Galama 1999).

With the launch of *Fermi* late 2008, the detection of GRBs with both the Large Area Telescope (LAT, 20 MeV to >300 GeV) and the Gamma-ray Burst Monitor (GBM, 8 keV to 40 MeV) has given us exceptional broadband spectra of the prompt emission over more than seven decades of energy. An interesting and unexpected finding is that for many GRBs the GeV emission starts with a noticeable delay after the MeV emission. For example, the first pulse of GRB 080916C is visible only in the MeV while the second pulse, that appears ~4 seconds later, is seen strongly in the GeV band by the LAT, Fig. 1 (Abdo et al. 2009). For GRB 080916C there is no evidence for a second spectral component; the spectrum is fitted by a single component that evolves from soft to hard and back to soft again. Furthermore, for this and other LAT bursts the GeV emission persists into the afterglow phase >1000>1000 s after the initial trigger. However, for some bursts, for example GRB 090902B (Abdo et al. 2009) and GRB 090926A (Ackermann et al. 2011), a distinct second hard spectral component is observed. Explaining the GeV–MeV time delays and the second distinct hard spectral component is difficult in the context of a one-zone (internal shocks) SSC mechanism (Ackermann and the Fermi collab. 2011).

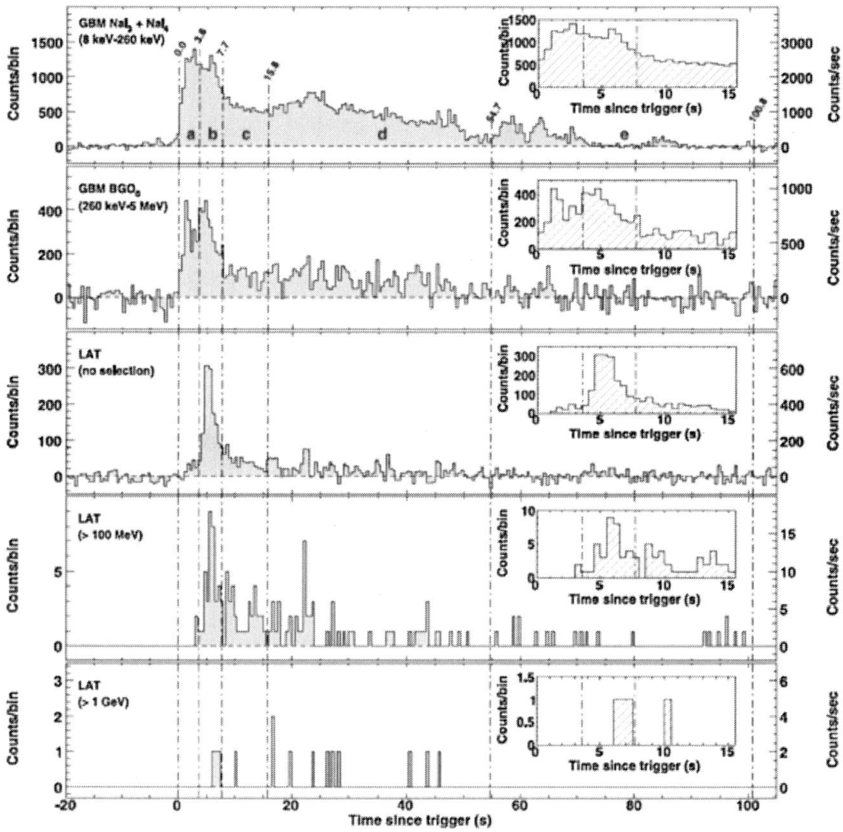

Figure 1. The Fermi GRM and LAT lightcurves for GRB 080916C (Abdo et al. 2009). The first pulse, 0.0–3.6 seconds is bright in the GBM but absent from the LAT

The initial GeV emission is obviously associated with the prompt pulse structure but the GeV emission extending into the afterglow, seen after the main prompt pulses, is unlikely to come from internal shocks and maybe SSC arising from the forward shock as the jet interacts with the surrounding circumburst medium.

Recent *Fermi* observations of the early prompt spectra of a few bursts have revealed a second mid-energy range spectral component in the emission again calling into question the validity of the one-zone internal shock/SSC radiation model for the prompt phase (Veres and Mészáros 2012). The new paradigm which may solve this problem evokes photospheric emission to explain at least part if not all of the prompt emission in the energy band 20 keV to 100 GeV.

3. THE PROMPT EMISSION

GRBs are primarily detected by virtue of the initial short burst of gamma-rays. A succession of instruments have provided GRB triggers, e.g. BATSE on *CGRO*, the Konus-Wind Experiment, FREGATE on *HETE-2*, IBIS on *INTEGRAL*, BAT on *Swift*, GRID on *AGILE* and the LAT and GRM on *Fermi*. The spectrum of the prompt emission is usually observed as a broken powerlaw characterised by the so-called Band function (Band et al. 1993), as shown in Fig. 2. The peak of the E.F(E)E.F(E) spectrum, EpeakEpeak, represents the characteristic photon energy for the GRB. In order to get a reliable estimate of EpeakEpeak broad energy band coverage is required.

There is a well known correlation, called the Amati Relation, between the EpeakEpeak energy expressed in the rest frame of the GRB and the isotropic energy, EisoEiso, estimated from the total fluence from the burst and the redshift (Amati et al. 2002; Amati 2006). This is illustrated in Fig. 3. The correlation is different for short (T90<2T90<2 seconds) and long (T90>2T90>2 seconds) with higher EpeakEpeak values for a given EisoEiso value for the short bursts. For both long and short bursts the correlation is not tight with a spread of ~ factor 10. There has been much discussion as to whether the correlation is an intrinsic property of the GRB emission or an observational or selection artifact (Butler et al. 2007). It may be that the relation is a signature of something fundamental about the GRB emission but better broad energy band spectral observations are required, coupled with more accurate determination of other burst properties, to reveal the true nature of the relationship.

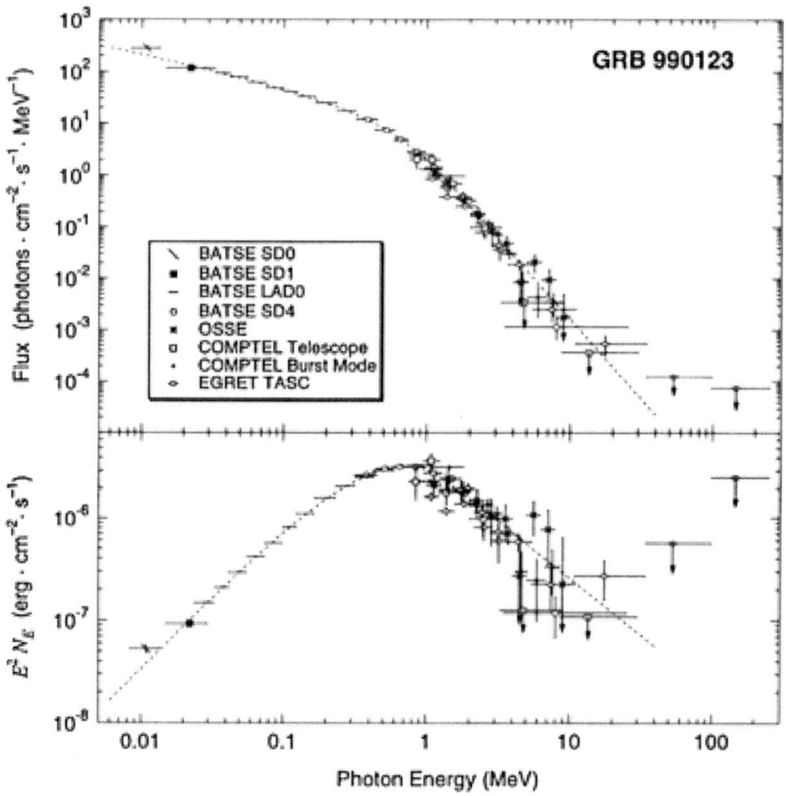

Figure 2. The Band function illustrated by spectrum of GRB 990123 with Epeak=720 keVEpeak=720 keV (Briggs et al. 1999)

The prompt emission of a few bright GRBs observed by *Fermi* have hard X-ray and Gamma-ray spectra which, as expected, are dominated by a Band function, usually attributed to non-thermal emission from internal shocks, but also include a highly significant component fitted by a thermal spectrum, e.g. GRB 100724B (Guiriec et al. 2011). Simultaneous observation across a broad energy band 10--107 keV10--107 keV is required to unambiguously confirm the presence of two components in the prompt spectra, see Fig. 4. Including the BB spectral component over the brightest part of the burst (T0−1.024 sT0−1.024 s to T0+83.969T0+83.969 seconds) shows a significant improvement in the fit compared to a Band function by itself however, the second component is weak compared to the dominant Band function and there is no guarantee that the additional component is thermal in origin and uniquely fitted by a Planck function. In GRB 100724B the two components seem to vary independently which favours the interpretation that the thermal component is of photospheric origin close to the central engine while the non-thermal component occurs at larger radii.

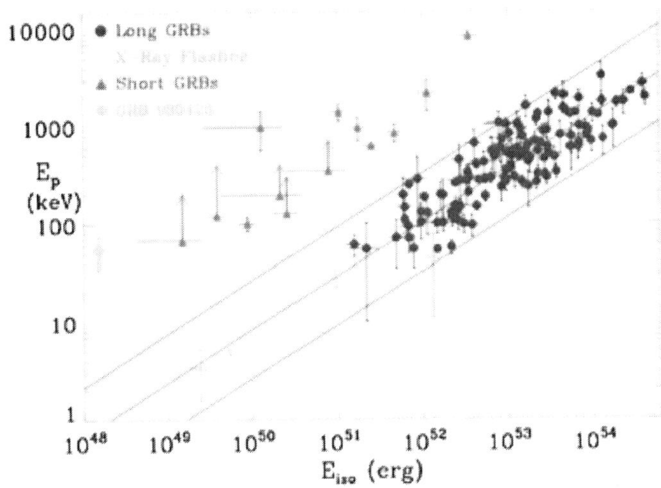

Figure 3. The Amati Relation—a correlation between the characteristic photon peak energy, EpEp, in the rest frame of the burst and the equivalent isotropic energy, EisoEiso. Long GRBs and X-ray Flashes lie in the correlation band which is ∼1∼1 dex wide in energy. All short GRBs are outliers and form a band of similar width but offset from the main correlation by ∼2∼2 dex in energy

The challenge to any theoretical model is to produce a Band-function-like broadband spectrum for the prompt emission with a characteristic peak energy in the 0.2–1 MeV range (Gehrels et al. 2009), with spectral evolution soft-to-hard-to-soft or more simply hard-to-soft and to include a (delayed) GeV tail component and possibly a lower energy Planck-function-like second component. Within the SSC model the peak is produced by a combination of the underlying relativistic electron spectrum, synchrotron cooling, synchrotron self absorption and up-scattering of the synchrotron photons by inverse Compton from the relativistic electrons. The observed energy peak is dependent on the bulk Lorentz factor of the outflow. Whether it is possible to produce the range of observed Band function parameters by SSC has been questioned by many authors (e.g. Ghisellini et al. 2000; Preece et al. 2000) and the radiation efficiency of internal shocks is low, 5%5%–10%10%, unless the colliding shells have widely differing Lorentz factors (Beloborodov 2000). An alternative is to associate the energy peak to a blackbody at the comoving pair combination temperature in the fireball's photosphere

(Eichler and Levinson 2000). The spectrum below the peak energy can then be attributed to the Rayleigh–Jeans part of the photospheric spectrum but the high energy power law tail and the GeV emission require an additional explanation (Meszaros and Rees 2000). If dissipation due to shocks or other processes occurs at or below the photosphere then a high efficiency is achievable and the thermal peak energies occur in the observed Band function range and Compton equilibrium of internal shock electrons or pairs with photospheric photons lead to spectra with a break at the right preferred energy and steep low energy slopes (Rees and Mészáros 2005; Pe'er et al. 2005, 2006). The photospheric model also provides a possible physical explanation for the Amati relation (or Ghirlanda relation, see below) (Thompson et al. 2007).

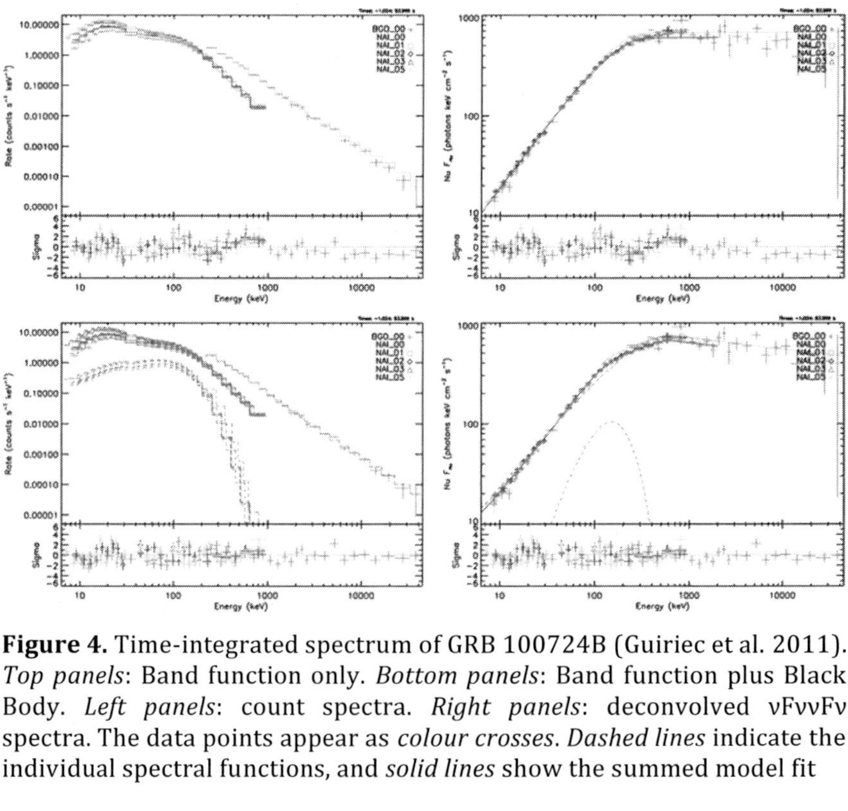

Figure 4. Time-integrated spectrum of GRB 100724B (Guiriec et al. 2011). *Top panels*: Band function only. *Bottom panels*: Band function plus Black Body. *Left panels*: count spectra. *Right panels*: deconvolved νFννFν spectra. The data points appear as *colour crosses*. *Dashed lines* indicate the individual spectral functions, and *solid lines* show the summed model fit

The magnetic field in the outflow plays a vital role. It is necessary for the production of synchrotron radiation and is certainly crucial in the acceleration of the relativistic charged particle distribution, by a Fermi mechanism, that ultimately produces the high energy tail of the spectrum via inverse Compton up-scattering. However, if the magnetic field is strong enough the magnetic stresses will be dominant and the whole dynamics of the jet will be driven by the magnetic field and must be modelled using magnetohydrodynamics (MHD). Magnetically dominated (or so-called Poynting dominated) GRB jets fall into two categories, baryon-free in which the baryons are absent or dynamically negligible (e.g. Drenkhahn 2002) and those with a non-negligible baryon load (e.g. Thompson 2006). In both cases the Alfvén (sound) speed in the ejecta is very high and therefore internal shocks maybe absent or weaker than in the standard hydrodynamics case. Similarly the reverse shock may be weaker although an external shock and associated forward blast wave will still be produced (Mészáros and Rees 1997). Any internal or reverse shocks that do form will have different radiation characteristics dependent on the magnetization parameter σσ. Along side or in place of internal shocks "internal dissipation regions" formed by magnetic reconnection will lead to particle acceleration via electric fields produced by magnetic reconnection, rather than a Fermi mechanism, and a high radiative efficiency is likely. In order to get the isotropically equivalent luminosities of Lγ≥1052 ergs−1Lγ≥1052 ergs−1 seen in GRBs the magnetic fields at the base of the jet must be B~1015 GB~1015 G. The baryon-free Poynting jet models are like pulsar wind models. In a similar way to the purely hydro baryon-loaded models the initially high magnetic stresses, involving pairs and photons, lead to acceleration and growth of the Lorentz factor, Γ∝rΓ∝r up to a pair annihilation photosphere which provides the first prompt emission peaking in the hard X-ray to MeV with up-scattering adding a high energy power law tail. In the absence of internal shocks the external shock must provide photons in the GeV–TeV range via inverse Compton scattering. On the other hand, baryon-loaded magnetically dominated jets behave differently with acceleration leading to Γ∝r1/3Γ∝r1/3 or similar intermediate power law dependencies (McKinney and Uzdensky 2012; Metzger et al. 2011). The spectral signatures expected from Poynting dominated jets have not been extensively studied. Calculations made using a one-zone steady state approximation for a simple magnetised outflow photosphere (Giannios and Spruit 2007) indicate that a Band-type spectrum can be produced. The time delays and spectral components of a simple baryon-loaded magnetic model have be calculated (Bošnjak and Kumar 2012) giving results in the range observed by *Fermi*.

In GRB jets with a significant baryonic content the charged baryons will be co-accelerated with the electrons in any shocks or magnetic reconnection zones, and hadronic processes may lead to secondary high energy photon emission and neutrinos. For example, the *Fermi* LAT observations of the short GRB 090510 were modelled by Asano et al.

(2009) as electron synchrotron for the MeV component and photohadronic cascade radiation for the distinct GeV power law component. Furthermore, since the acceleration and cascade development take time, a one-zone model including hadronic effects might explain the GeV–MeV photon delays seen by *Fermi* (Razzaque 2010). The hadronic interactions might also modify the low energy branch of the Band spectrum resulting in a GRB optical flash (Asano et al. 2010). The spectral signatures and GeV–MeV time delay can be produced by combining a magnetically dominated, baryon loaded model with a hadronic component involving nuclear collisions as calculated by Mészáros and Rees (2011). Murase et al. (2012) attempt to self consistently produce the GeV radiation, the MeV Band component and the low energy optical power law using a hadronic model. Protons accelerated by shocks or magnetic reconnection regions give hadronic cascades which in produce both the GeV and optical power laws. Cooled electron secondaries are then re-accelerated via a Fermi second order mechanism in the turbulent MHD waves produced by the same shocks, or reconnection regions, to yield a MeV Band spectrum.

More simultaneous, early, broad energy band/multi-wavelength observations with high signal to noise are required to reveal the detailed temporal behaviour of the components contributing to the prompt phase and to discriminate between the many theoretical possibilities.

4. THE CANONICAL GRB X-RAY AFTERGLOW

Before the arrival of the first *Swift XRT* observations of the X-ray afterglow of GRBs, starting only ~100~100 seconds after the prompt trigger from the Burst Alert Telescope (BAT), it was expected that the X-ray (and optical) afterglows would fall as a single simple power law. On the contrary, the vast majority of X-ray afterglows have the canonical form shown in Fig. 5 (Nousek et al. 2006; Evans et al. 2007). The afterglow seems to arise from two components. The fading emission from the tails of the prompt GRB pulses seen as the initial rapid decay and the plateau which turns over into a final powerlaw decay, typically 104104–105105 seconds after the prompt emission has died, which most likely arises from the external shock formed as jetted material sweeps up the circumburst medium (O'Brien et al. 2006; Willingale et al. 2007).

The slow decay over the plateau phase was unexpected and is difficult to explain. It requires some form of energy injection, probably from the central engine, although the initial fast decay would suggest that the central engine has already switched off. In some cases the energy to sustain the plateau may come from the spin-down of a magnetar formed during the initial collapse (Zhang and Mészáros 2001; Rowlinson et al. 2010), but in most instances it is probably continued low level activity of the central engine. For a few GRBs the afterglow plateau persists for >104>104 seconds in which case the required total late energy injection would have to be large and comparable to the prompt energy. Another unexpected feature in the X-ray afterglows is the appearance of late flares, seen in the *Swift* XRT but not in the BAT (Chincarini et al. 2007), see Fig. 6. These flares are probably indicative that the central engine can and does remain active after the prompt Gamma-ray and hard X-ray emission has ceased. Figure 6 shows the afterglow decay in two XRT energy bands, 0.3–1.5 keV and 1.5–10 keV. In all bursts the ratio of the afterglow fluxes in these bands, over the plateau and in the final decay, remains constant indicating that there is no spectral evolution and that the temporal break from the end of the plateau into the final decay is achromatic.

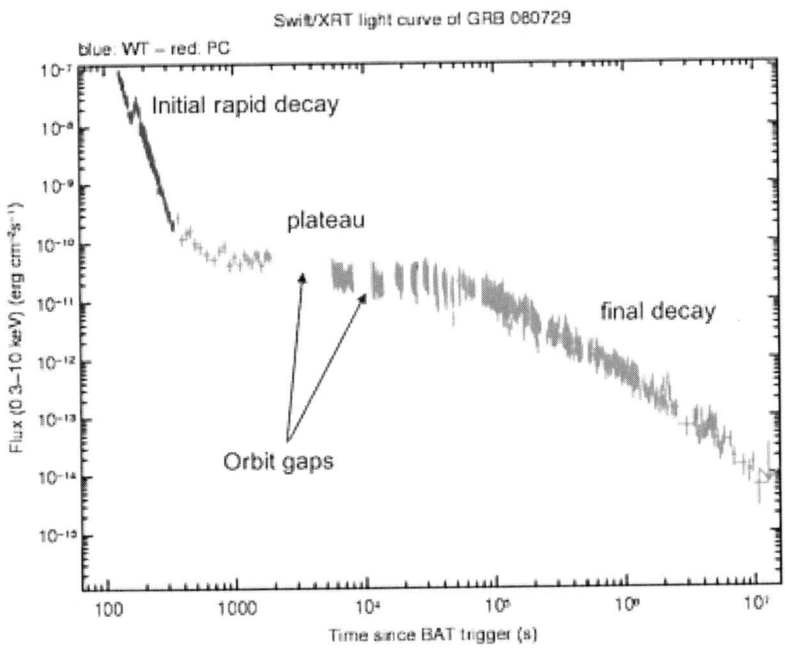

Figure 5. The canonical GRB X-ray afterglow. The prompt emission typically ceases after ~100~100seconds and ends in a rapid fall with temporal decay index >2>2. In the vast majority of bursts this rapid decay is halted by a late emission component which forms a plateau that decays slowly, temporal index ~0.5~0.5. Finally, the plateau turns down to give a final afterglow decay with index ~1.5~1.5. The duration and decay rate of each of the three phases vary considerably from burst to burst

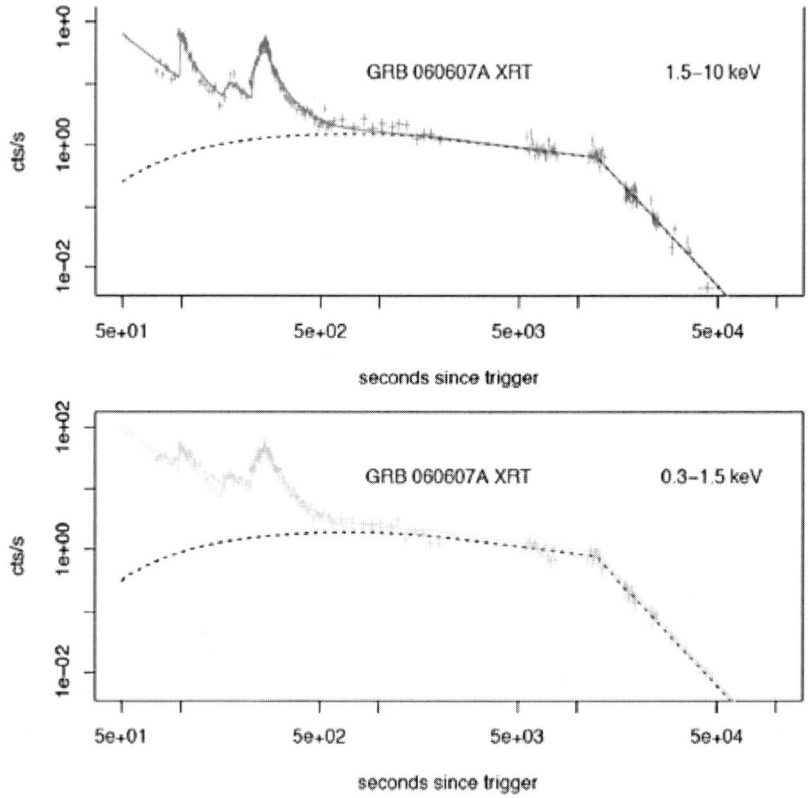

Figure 6. Late X-ray flares not seen in the *Swift* BAT. The lightcurves are shown in two XRT bands 0.3–1.5 keV and 1.5–10 keV. The ratio of the fluxes in these bands gives us a measure of the soft X-ray spectra of the features seen. The afterglow component is indicated by the *dotted line*

If the relativistic outflow is collimated and slowing down, as the circumburst medium is swept up, we expect an achromatic break in the afterglow decay when the edge of the jet becomes visible as the relativistic beaming angle opens up. The standard formulation for so-called jet-breaks is given by Sari et al. (1999) and reference therein. Temporal breaks identified as jet-breaks in the afterglow decay are often seen in the optical and IR, e.g. Tagliaferri et al. (2005), Fig. 7, and when combined with a redshift measurement and an estimate of the total equivalent isotropic energy released the break times can be used to estimate the collimation angle of the jet. However, although late breaks are sometimes seen in the X-ray afterglows they rarely coincide with a break at other wavelengths so the simple achromatic break scenario is probably complicated by other factors.

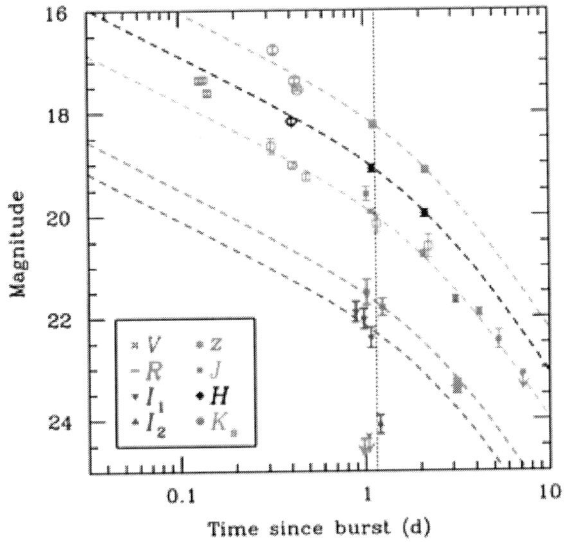

Figure 7. he optical/IR break seen in the afterglow of GRB 050904 (Tagliaferri et al. 2005). The profile of the break is the same in the zz, JJ, HH and KK bands indicating that the break is achromatic. When combined with the redshift the break time can be used to estimate the collimation angle of the jet

Using an estimate of the jet collimation angle, derived from the assumed jet-break time observed in the optical/IR, we can correct the isotropic energy EisoEiso to a true collimated energy value EγEγ. An original analysis performed by Ghirlanda et al. (2004) using 18 bursts showed that the correlation of EpeakEpeak with EγEγ (known as the Ghirlanda Relation) was significantly tighter than the Amati Relation for the same bursts. However, more recent analysis using *Swift* and *Fermi* bursts, illustrated in Fig. 8, are not so conclusive (Campana et al. 2007; McBreen et al. 2010). This again brings into question whether the optical late breaks are really jet-breaks and whether or not the EγEγ estimates are valid. More high quality multi-wavelength observations are required to tell us if the breaks are chromatic or achromatic to resolve this issue.

5. PROMPT TO AFTERGLOW

The prompt Gamma-ray and hard X-ray emission from GRBs is very variable, both spectrally and temporally, but it can be characterised as a stochastic sequence of pulses each with a distinctive shape and spectral signature including a fast rise, slower decay and spectral softening. Several authors have attempted to model these pulses, e.g. Norris et al. (2005) using an empirical mathematical model and Willingale et al. (2010) using a physically motivated formulation. It is supposed that each pulse is associated with internal shocks within a thin ultra-relativistically expanding shell. The time profile we observe is produced by the relativistic aberrations and the Equal Arrival Time Surfaces (EATS) arising from the large Lorentz factor of the relativistic jet pointing towards us and the spherical geometry of the shell. Figure 9 shows the typical model pulse profile and an example of a GRB multi-energyband light curve fitted as a sequence of pulses and an afterglow component. The fitting is only possible because of the multi-energy band coverage and nearly continuous temporal coverage from the trigger to very late in the afterglow phase. When combined with a redshift the model provides rest frame timing and pulse/afterglow duration estimates and luminosity estimates for all the components.

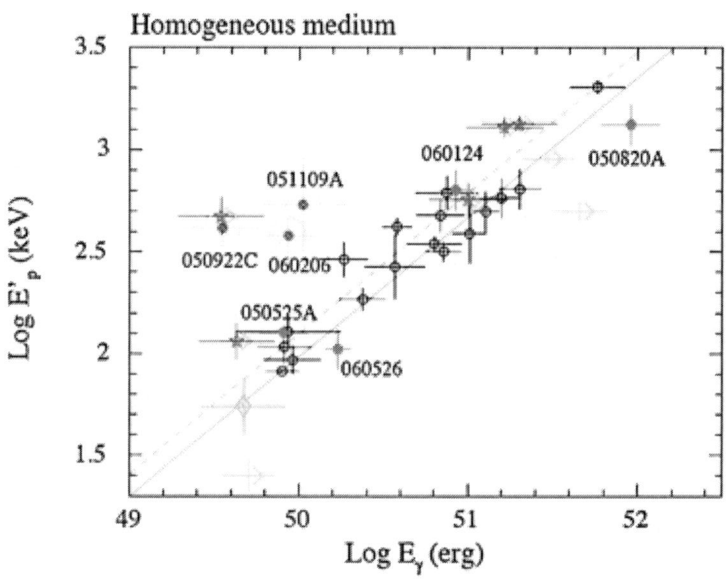

Figure 8. The Ghirlanda relation (Ghirlanda et al. 2004) between the photon energy EpeakEpeak and the collimated energy estimated from EisoEiso and the jet collimation angle (Campana et al. 2007; McBreen et al. 2010). The unlabelled *black points* are the original GRBs used by Ghirlanda. The other values are for *Swift* and *Fermi* bursts which don't all conform to the original tight correlation

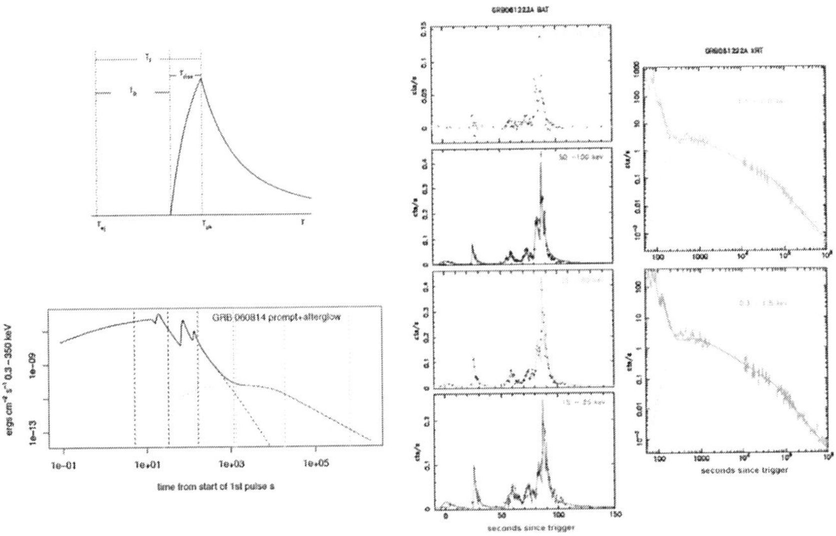

Figure 9. *Top left*: the model profile for prompt pulses. *Bottom left*: the combination of prompt pulses and an afterglow representing the entire light curve emission (in this case from GRB 060814). *Right*: the prompt and afterglow emission from GRB 061222A seen in a combination of the energy bands from the *Swift* BAT and XRT

On rare occasions the prompt emission is heralded by a precursor pulse which is bright enough to trigger the *Swift* BAT. In these cases the XRT and UVOT telescopes and alerted ground based instruments may be able to detect the latter stages of the prompt in which case we can get better multi-band coverage across the entire burst including the prompt phase as shown in Fig. 10. For GRB 080810 the burst also triggered the *Fermi* GBM so we also have simultaneous hard Gamma-ray coverage (Page et al. 2009) and for both GRB 061124 and GRB 080810 the afterglow flux was also detected in the radio as indicated on Fig. 10. Using data like these we can set up a temporally evolving Spectral Energy Distribution (SED) for both the prompt and afterglow emission. For GRB 061124 both the prompt and afterglow emission appear in the optical and the SED has a break between the optical and X-ray bands. The afterglow exhibits spectral evolution and decays too slowly compared with the standard synchrotron models. In addition there is evidence for a late break which might be a jet-break. For GRB 080810 the peak photon frequency in the SED, EpeakEpeak, evolves as a function of time and the afterglow optical-X-ray SED initially hardens but subsequently softens. The early rise and peak of the optical afterglow is also seen.

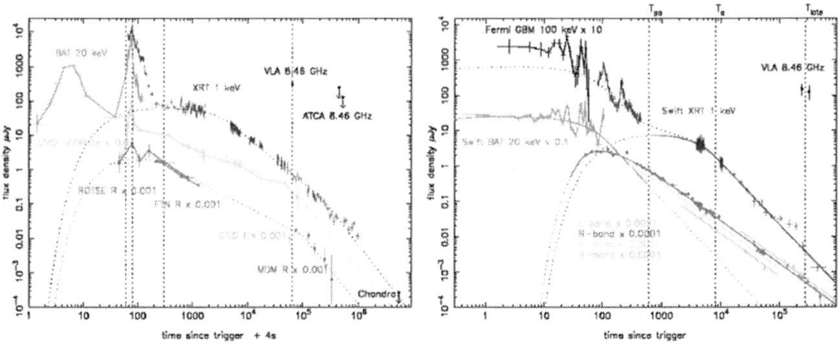

Figure 10. Multi-band observations of the prompt and afterglow for *left*: GRB 061124 (Page et al. 2007) and *right*: GRB 080810 (Page et al. 2009)

The rise and peak of the soft X-ray afterglow component is very rarely seen because it is hidden by the X-ray tail of the prompt hard X-ray and Gamma-ray emission. However, the early optical/IR afterglow peak can be observed (as in the case of GRB 080810 above). The optical emission most likely arises from the reverse shock which is expected to produce photons of lower energy than the forward shock which is responsible for the X-ray afterglow.

The optical/IR afterglow peak can be assumed to correspond to the deceleration radius of the expanding fireball. The Lorentz factor of the outflow at the time of the peak, tdtd, is half the initial value, Γ0Γ0 (Sari and Piran 1999). From observations of tdtd and using estimates of the isotropic energy released, EisoEiso, we can estimate the value of Γ0/2Γ0/2.

For both GRB 060418 and GRB 060607A the inferred deceleration radius is Rdec≈1017 cmRdec≈1017 cm and Γ0~400Γ0~400 giving us direct confirmation of the highly relativistic nature of the GRB fireballs (Molinari et al. 2007). Using values of the initial Lorentz factor derived in this way Liang et al. (2010) showed a correlation between Γ0Γ0 and EisoEiso, Fig. 11. More recently the same authors published a 3-parameter correlation between the isotropic luminosity, LisoLiso, EpeakEpeak and Γ0Γ0 (Liang et al. 2015). Clearly, multi-wavelength observations which cover the prompt emission and the early afterglow have a great potential to reveal the key parameters which are driving the burst fireball.

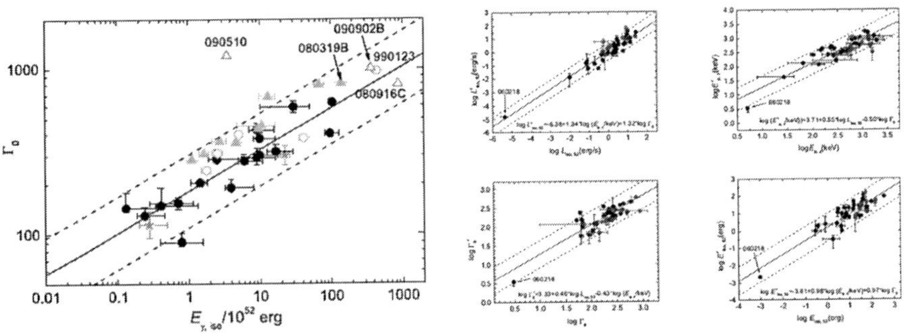

Figure 11. *Left*: the correlation between the initial Lorentz factor and the isotropic energy. *Right*: a 3-parameter correlation between the isotropic luminosity, EpeakEpeak and Γ0Γ0

6. ULTRA-LONG GRBS AND TIDAL DISRUPTION EVENTS

Recently it has been noted that a few GRBs have a prompt emission phase which lasts an extraordinarily long time, measured in kiloseconds (Levan et al. 2014). Figure 12 shows an example of such a burst monitored both by the *Swift* BAT and the *Konus-Wind* Experiment, GRB 130925A (Evans et al. 2014). There are currently just four bursts which have a prompt duration (T90T90) that is an order of magnitude longer than the main population of long GRBs (GRB 101225A, GRB 111209A, GRB 121027A, GRB 130925A). It may be that the progenitor and emission mechanism that gives rise to such persistent bursts is distinct from the majority but at the moment we don't have a detailed explanation for these ultra-long bursts. In addition there are one or two events detected by *Swift* which have repeated outbursts over a period of many days. For example, GRB 110328A was originally designated as a GRB but produced several BAT triggers and it soon became clear that the light curve didn't look like a conventional GRB, showing repeated flaring that eventually lasted for greater than 5×1065×106 seconds. It was established that it occurred in a host galaxy at a redshift of z=0.351z=0.351 and was most probably a Tidal Disruption Event (TDE) in which a star was ripped apart by tidal forces when it wandered too close to a massive galactic black hole (Cannizzo et al. 2011). This event was subsequently re-labelled as Sw 1644+57. We require continued operation of *Swift* and future satellites like *SVOM* (Cordier et al. 2015) coupled with extended multi-wavelength observations to find and characterise more of these oddball events which clearly don't match the paradigm of conventional GRBs.

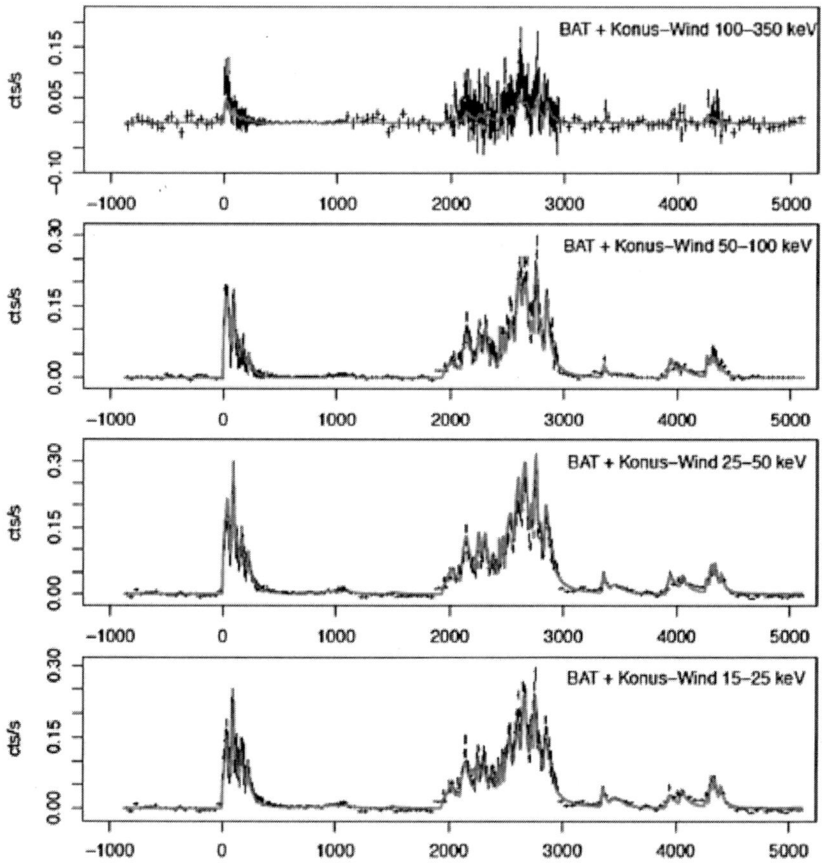

Figure 12 The combined *Swift* BAT and *Konus-Wind* prompt light curve of GRB 130925A with a prompt duration of ~5000~5000 seconds

7. FOLLOW-UP OF GRAVITATIONAL WAVE AND NEUTRINO TRIGGERS

A flux of neutrinos is expected from gamma-ray bursts (Asano and Mészáros 2014), but so far searching retrospectively through neutrino data from *IceCube* (IceCube Collaboration 2012; Aartsen et al. 2015), and *ANTARES* (Adrián-Martínez et al. 2013), looking for matches corresponding to the time and location of GRB triggers has not yielded any positive results. Alternatively, we can choose to follow-up neutrino triggers with an instrument like the *Swift*XRT (Evans et al. 2015). The XRT field of view is 24 arc minutes in diameter and the error radius of *IceCube* triggers is typically 1 degree at 50% confidence so this can be covered by a tiling of 7 XRT pointings as shown in Fig. 13. Each pointing has an exposure of 1–2 ksec depending on the satellite orbit conditions at the time of the trigger. The delay from neutrino trigger to the tiling of X-ray observations is typically 1–3 hours. The *Swift* XRT has a sensitivity of 5×10–13 ergcm–2s–15×10–13 ergcm–2s–1 in 1 ksec in the 0.3–10 keV band and can get source positions with a 90% confidence radius of 3.5 arcsec or better for brighter sources (Evans et al. 2009).

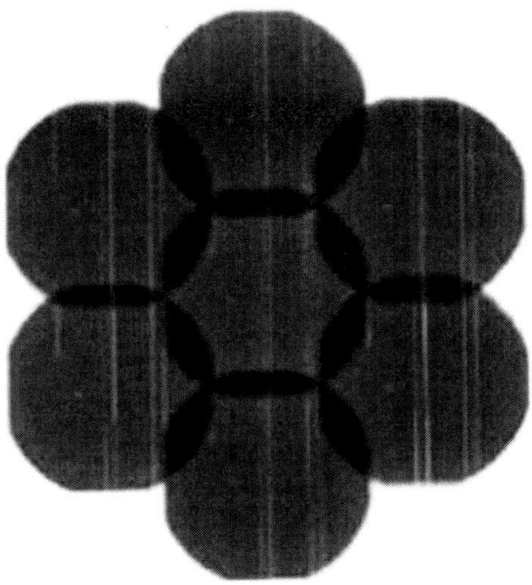

Figure 13. Exposure map of a 7-tile *Swift* XRT observation of an IceCube trigger. The *blank lines*and *dots* are bad columns and pixels on the CCD

Gravitational waves are expected from a range of transients, in particular the merger of two neutron stars which is also expected to produce a short GRB. Targeted searches for gravitational waves from short GRBs have so far failed to produce detections (Abadie et al. 2010). Similarly, *Swift* XRT follow-up of gravitational wave triggers has so far also drawn a blank (Evans et al. 2012), although this is to be expected because the error boxes associated with LIGO-Virgo triggers have to date been very large, many tens of square degrees.

The *Swift* XRT is probably the best instrument currently available for neutrino and gravitational wave follow-up, looking for the high energy transient counterparts, because of the reasonable sensitivity of the XRT in the 0.3–10 keV band and the rapid repointing response capability. However, it has a rather small field of view so tiling is required and the response time of a few hours means that any short transient may have faded significantly by the time the observations begin. A high sensitivity soft X-ray transient monitor with a wide field of many square degrees is required to increase the probability of finding any fast transients associated with these triggers.

8. THE LOBSTER EYE X-RAY TRANSIENT MONITOR

The wide field detectors used to find and position GRBs and similar transients, listed in Sect. 3, all use collimation and/or coded shadow mask techniques to detect and estimate the position of hard X-ray/Gamma-ray transients. Such devices have the required large fields of view but the sensitivity is limited by the multiplex nature of the positioning capability. If the X-ray flux could be focused in some way, so that multiplexing of the detector imaging elements could be reduced or eliminated, the sensitivity would be greatly improved. The lobster eye X-ray focusing geometry originally described by Angel (1979) provides just that. In a lobster eye the image is produced by grazing incidence reflections off the inside walls of microscopic square pores which are arranged over a spherical surface all pointing towards a common centre of curvature. The image is formed on a spherical focal surface with half the radius of curvature of the square pore optic. Unlike a fly's eye optic, which uses an array of refracting lenses, the lobster eye produces a single continuous image of the sky. The field of view is limited only by the size of the optic and size of the detector and if an optic (and detector) could be constructed across a complete sphere the whole sky would be captured in a single continuous image.

The idea of using the lobster eye principle to produce a soft X-ray all-sky monitor was first proposed by Priedhorsky et al. (1996) and a method of manufacturing the square pore array has been described by several groups (Chapman et al. 1991; Wilkins et al. 1989; Fraser et al. 1993). Recent proposals for soft X-ray wide field transient monitors using lobster eye optics include *ISS Lobster* submitted to NASA December 2014 and the *Theseus* SXT submitted the ESA M4 call at about the same time. Figure 14 shows a schematic view of a typical module and the flux sensitivity it can achieve. Each module has a field of view of ~1/6~1/6 steradian so an array of 6 modules can view 1 steradian. An angular resolution of ~5~5 arc minutes is currently achievable using the available square pore micro-channel plate technology. The sensitivity is ~2×10−10 ergcm−2s−1~2×10−10 ergcm−2s−1 in 100 sec in the 0.3–5 keV energy band. This is a factor of ~100~100higher than the *Swift* XRT but the lobster eye transient monitor will see transients when they happen rather than 1–2 hours later, in response to a trigger.

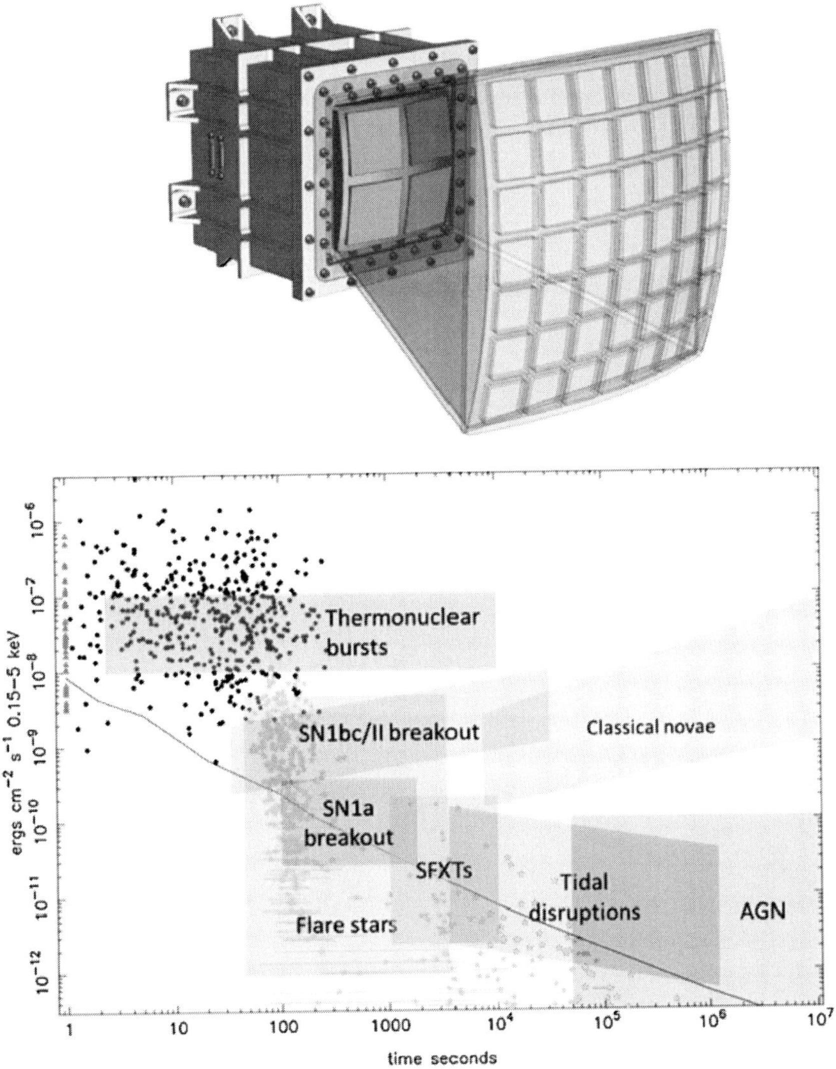

Figure 14. *Top*: a typical lobster eye module with a focal length of 300 mm and total mass of ~20 kg~20 kg. *Bottom*: the flux sensitivity of the module compared to typical flux ranges and timescales associated with types of transients. The *black dots* indicate the prompt flux from *Swift* GRBs detected by the BAT and the *green dots* are the afterglow flux levels seen in the afterglows by the XRT. The *black curve* is the module flux sensitivity in the energy band 0.15–5 keV as a function of integration time

It is predicted that such a monitor would see greater than 90% of the GRBs seen by the *Swift* BAT and provide positions <1<1minute for all transients. A lobster eye X-ray transient monitor is the ideal instrument to hunt for and position fast soft X-ray transients or look for the X-ray counterparts to neutrino and gravitational wave triggers.

9. DISCUSSION AND CONCLUSION

The study of GRBs is entering an exciting new phase. After great advances obtained through multi-wavelength studies made possible by *CGRO*, *Beppo-SAX*, *HETE-2* and *Swift*, GRB studies have been extended well into the GeV range by the *Fermi* mission (Bissaldi et al. 2015a), and higher energies reaching the TeV range are being probed by the High Altitude Water Cherenkov Observatory (HAWC) (Lennarz et al. 2015) and the High Energy Stereoscopic System (HESS https://www.mpi-hd.mpg.de/ hfm/ HESS/). The next instruments after that are expected to be, at X-rays to MeV energies, the SVOM satellite (Cordier et al. 2015), and in the TeV to multi-TeV range the ground-based (air) Cherenkov Telescope Array (CTA) (Bissaldi et al. 2015b).

Non-photonic messengers are also being probed, most notably high energy neutrinos, with *IceCube* setting strong constraints against classical (Swift-detected) GRBs providing more than 1%1% of the observed neutrino background (Aartsen et al. 2015). One possibility is that the gamma-rays arise at lower radii, e.g. in the photosphere, while cosmic rays are only accelerated in shocks further out, where the efficiency for photo-hadronic interaction producing neutrinos is low (Asano and Mészáros 2016). For a discussion of the physics and gamma-ray emission from GRB photospheres, see Beloborodov and Mészáros, 2016, elsewhere in this volume.

Following the recent announcement of the first detection of gravitational waves from the merger of stellar mass binary black holes by LIGO (Abbott et al. 2016) results from non-photonic GW searches are now eagerly awaited. Neutron star or neutron star-black hole mergers are also expected to produce gravitational waves detectable by LIGO and VIRGO, and their detection in coincidence with a short GRB gamma-ray signal is the smoking gun signal being looked for in order to confirm the progenitor's identity.

Gamma-ray bursts continue to surprise us, and it is clear that the array of major new facilities that are being brought to bear on these sources will play a major role in solving some of the many questions that they pose.

FOOTNOTES

1 http://www.mpe.mpg.de/~jcg/grbgen.html.

REFERENCES

1. M.G. Aartsen, M. Ackermann, J. Adams, J.A. Aguilar, M. Ahlers, M. Ahrens, D. Altmann, T. Anderson, C. Arguelles, T.C. Arlen et al., Search for prompt neutrino emission from gamma-ray bursts with IceCube. Astrophys. J. Lett. **805**, 5 (2015).
2. J. Abadie, B. Abbott, R. Abbott, T. Accadia, F. Acernese, R. Adhikari, P. Ajith, B. Allen, G. Allen, E.A. Ceron et al., Search for gravitational-wave inspiral signals associated with short gamma-ray bursts during ligo's fifth and virgo's first science run. Astrophys. J. **715**(2), 1453 (2010)
3. B.P. Abbott, R. Abbott, T.D. Abbott, M.R. Abernathy, F. Acernese, K. Ackley, C. Adams, T. Adams, P. Addesso, R.X. Adhikari et al., Observation of gravitational waves from a binary black hole merger. Phys. Rev. Lett. **116**(6), 061102 (2016).
4. A.A. Abdo, M. Ackermann, M. Arimoto, K. Asano, W.B. Atwood, M. Axelsson, L. Baldini, J. Ballet, D.L. Band, G. Barbiellini et al., Fermi observations of high-energy gamma-ray emission from GRB 080916C. Science **323**, 1688 (2009).
5. A.A. Abdo, M. Ackermann, M. Ajello, K. Asano, W.B. Atwood et al., Astrophys. J. Lett. **706**, 138 (2009)
6. M. Ackermann, the Fermi collab., Detection of a spectral break in the extra hard component of GRB 090926A. Astrophys. J. **729**, 114 (2011).
7. M. Ackermann, M. Ajello, K. Asano, M. Axelsson, L. Baldini, J. Ballet, G. Barbiellini, M. Baring, D. Bastieri, K. Bechtol et al., Detection of a spectral break in the extra hard component of GRB 090926a. Astrophys. J. **729**(2), 114 (2011)
8. S. Adrián-Martínez, A. Albert, I. Al Samarai, M. André, M. Anghinolfi, G. Anton, S. Anvar, M. Ardid, T. Astraatmadja, J.-J. Aubert et al., Search for muon neutrinos from gamma-ray bursts with the ANTARES neutrino telescope using 2008 to 2011 data. Astron. Astrophys. **559**, 9 (2013)
9. L. Amati, The Ep,i-Eiso correlation in gamma-ray bursts: updated observational status, re-analysis and main implications. Mon. Not. R. Astron. Soc. **372**, 233–245 (2006).
10. L. Amati, F. Frontera, M. Tavani, J.J.M. in't Zand, A. Antonelli, E. Costa, M. Feroci, C. Guidorzi, J. Heise, N. Masetti, E. Montanari, L. Nicastro, E. Palazzi, E. Pian, L. Piro, P. Soffitta, Intrinsic spectra and energetics of BeppoSAX gamma-ray bursts with known redshifts. Astron. Astrophys. **390**, 81–89 (2002).
11. J.R.P. Angel, Lobster eyes as X-ray telescopes. Astrophys. J. **233**, 364–373 (1979).
12. K. Asano, P. Mészáros, Ultrahigh-energy cosmic ray production by turbulence in gamma-ray burst jets and cosmogenic neutrinos. Phys. Rev. D **94**(2), 023005 (2016).
13. K. Asano, P. Mészáros, Neutrino and cosmic-ray release from gamma-ray bursts: time-dependent simulations. Astrophys. J. **785**(1), 54 (2014)
14. K. Asano, S. Guiriec, P. Mészáros, Hadronic models for the extra spectral component in the short GRB 090510. Astrophys. J. Lett. **705**(2), 191 (2009)
15. K. Asano, S. Inoue, P. Mészáros, Prompt x-ray and optical excess emission due to hadronic cascades in gamma-ray bursts. Astrophys. J. Lett. **725**(2), 121 (2010)
16. J.-L. Atteia, N. Kawai, R. Vanderspek, G. Pizzichini, G.R. Ricker, C. Barraud, M. Boer, J. Braga, N. Butler, T. Cline, G.B. Crew, J.-P. Dezalay, T.Q. Donaghy, J. Doty, E.E. Fenimore, M. Galassi, C. Graziani, K. Hurley, J.G. Jernigan, D.Q. Lamb, A. Levine, R. Manchanda, F. Martel, M. Matsuoka, E. Morgan, Y. Nakagawa, J.-F. Olive, G. Prigozhin, T. Sakamoto, R. Sato, Y.

Shirasaki, M. Suzuki, K. Takagishi, T. Tamagawa, K. Torii, J. Villasenor, S.E. Woosley, M. Yamauchi, A. Yoshida, HETE-2 observation of two gamma-ray bursts at z>3z>3. Astrophys. J. **626**, 292–297 (2005).

17. D. Band, J. Matteson, L. Ford, B. Schaefer, D. Palmer, B. Teegarden, T. Cline, M. Briggs, W. Paciesas, G. Pendleton, G. Fishman, C. Kouveliotou, C. Meegan, R. Wilson, P. Lestrade, BATSE observations of gamma-ray burst spectra. I—Spectral diversity. Astrophys. J. **413**, 281–292 (1993).
18. A.M. Beloborodov, On the efficiency of internal shocks in gamma-ray bursts. Astrophys. J. Lett. **539**(1), 25 (2000)
19. E. Bissaldi, F. Longo, N. Omodei, G. Vianello, A. von Kienlin, Gamma-ray burst observations with Fermi (2015a). ArXiv e-prints arXiv:1507.03478
20. E. Bissaldi, T. Di Girolamo, F. Longo, P. Vallania, C. Vigorito, Prospects for gamma-ray bursts detection by the Cherenkov Telescope Array (2015b). ArXiv e-prints
21. J.S. Bloom, S.G. Djorgovski, S.R. Kulkarni, The redshift and the ordinary host galaxy of GRB 970228. Astrophys. J. **554**(2), 678 (2001). http://stacks.iop.org/0004-637X/554/i=2/a=678
22. J.S. Bloom, D.A. Perley, W. Li, N.R. Butler, A.A. Miller, D. Kocevski, D.A. Kann, R.J. Foley, H.-W. Chen, A.V. Filippenko, D.L. Starr, B. Macomber, J.X. Prochaska, R. Chornock, D. Poznanski, S. Klose, M.F. Skrutskie, S. Lopez, P. Hall, K. Glazebrook, C.H. Blake, Observations of the naked-eye GRB 080319B: implications of nature's brightest explosion. Astrophys. J. 691, 723–737 (2009).
23. Ž. Bošnjak, P. Kumar, Magnetic jet model for GRBs and the delayed arrival of >100 MeV photons. Mon. Not. R. Astron. Soc. Lett. **421**(1), 39–43 (2012)
24. M. Briggs, D. Band, R. Kippen, R. Preece, C. Kouveliotou, J. Van Paradijs, G. Share, R. Murphy, S. Matz, A. Connors et al., Observations of GRB 990123 by the Compton gamma ray observatory. Astrophys. J. **524**(1), 82 (1999)
25. N.R. Butler, D. Kocevski, J.S. Bloom, J.L. Curtis, A complete catalog of Swift gamma-ray burst spectra and durations: demise of a physical origin for pre-Swift high-energy correlations. Astrophys. J. **671**, 656–677 (2007).
26. S. Campana, C. Guidorzi, G. Tagliaferri, G. Chincarini, A. Moretti, D. Rizzuto, P. Romano, Are Swift gamma-ray bursts consistent with the Ghirlanda relation? Astron. Astrophys. **472**, 395–401 (2007).
27. J.K. Cannizzo, E. Troja, G. Lodato, GRB 110328A/Swift J164449.3+573451: the tidal obliteration of a deeply plunging star? Astrophys. J. **742**, 32 (2011).
28. H. Chapman, K. Nugent, S. Wilkins, X-ray focusing using square channel-capillary arrays. Rev. Sci. Instrum. **62**(6), 1542–1561 (1991)
29. G. Chincarini, A. Moretti, P. Romano, A.D. Falcone, D. Morris, J. Racusin, S. Campana, S. Covino, C. Guidorzi, G. Tagliaferri, D.N. Burrows, C. Pagani, M. Stroh, D. Grupe, M. Capalbi, G. Cusumano, N. Gehrels, P. Giommi, V. La Parola, V. Mangano, T. Mineo, J.A. Nousek, P.T. O'Brien, K.L. Page, M. Perri, E. Troja, R. Willingale, B. Zhang, The first survey of X-ray flares from gamma-ray bursts observed by Swift: temporal properties and morphology. Astrophys. J. **671**, 1903–1920 (2007).
30. B. Cordier, J. Wei, J.-L. Atteia, S. Basa, A. Claret, F. Daigne, J. Deng, Y. Dong, O. Godet, A. Goldwurm, D. Götz, X. Han, A. Klotz, C. Lachaud, J. Osborne, Y. Qiu, S. Schanne, B. Wu, J. Wang, C. Wu, L. Xin, B. Zhang, S.-N. Zhang, The SVOM gamma-ray burst mission (2015). ArXiv e-prints
31. E. Costa, F. Frontera, J. Heise, M. Feroci, J.I. Zand, F. Fiore, M.N. Cinti, D.D. Fiume, L. Nicastro, M. Orlandini et al., Discovery of the x-ray afterglow of the gamma-ray burst of February 28 1997 (1997). ArXiv preprint
32. D.M. Coward, E.J. Howell, M. Branchesi, G. Stratta, D. Guetta, B. Gendre, D. Macpherson, The Swift gamma-ray burst redshift distribution: selection biases and optical brightness evolution at high z? Mon. Not. R. Astron. Soc. **432**(3), 2141–2149 (2013).
33. Cucchiara, A.J. Levan, D.B. Fox, N.R. Tanvir, T.N. Ukwatta, E. Berger, T. Krühler, A. Küpcü Yoldaş, X.F. Wu, K. Toma, J. Greiner, F.E. Olivares, A. Rowlinson, L. Amati, T. Sakamoto, K. Roth, A. Stephens, A. Fritz, J.P.U. Fynbo, J. Hjorth, D. Malesani, P. Jakobsson, K. Wiersema, P.T. O'Brien, A.M. Soderberg, R.J. Foley, A.S. Fruchter, J. Rho, R.E. Rutledge, B.P. Schmidt,

M.A. Dopita, P. Podsiadlowski, R. Willingale, C. Wolf, S.R. Kulkarni, P. D'Avanzo, A photometric redshift of z~9.4z~9.4 for GRB 090429B. Astrophys. J. **736**, 7 (2011).
34. G. Drenkhahn, Acceleration of GRB outflows by Poynting flux dissipation. Astron. Astrophys. **387**(2), 714–724 (2002)
35. D. Eichler, A. Levinson, A compact fireball model of gamma-ray bursts. Astrophys. J. **529**(1), 146 (2000)
36. P.A. Evans, A.P. Beardmore, K.L. Page, L.G. Tyler, J.P. Osborne, M.R. Goad, P.T. O'Brien, L. Vetere, J. Racusin, D. Morris, D.N. Burrows, M. Capalbi, M. Perri, N. Gehrels, P. Romano, An online repository of Swift/XRT light curves of γγ-ray bursts. Astron. Astrophys. **469**, 379–385 (2007).
37. P.A. Evans, R. Willingale, J.P. Osborne, P.T. O'Brien, N.R. Tanvir, D.D. Frederiks, V.D. Pal'shin, D.S. Svinkin, A. Lien, J. Cummings, S. Xiong, B.-B. Zhang, D. Götz, V. Savchenko, H. Negoro, S. Nakahira, K. Suzuki, K. Wiersema, R.L.C. Starling, A.J. Castro-Tirado, A.P. Beardmore, R. Sánchez-Ramírez, J. Gorosabel, S. Jeong, J.A. Kennea, D.N. Burrows, N. Gehrels, GRB 130925A: an ultralong gamma ray burst with a dust-echo afterglow, and implications for the origin of the ultralong GRBs. Mon. Not. R. Astron. Soc. **444**, 250–267 (2014).
38. P.A. Evans, J.P. Osborne, J.A. Kennea, M. Smith, D.M. Palmer, N. Gehrels, J.M. Gelbord, A. Homeier, M. Voge, N.L. Strotjohann, D.F. Cowen, S. Böser, M. Kowalski, A. Stasik, Swift follow-up of IceCube triggers, and implications for the advanced-LIGO era. Mon. Not. R. Astron. Soc. **448**, 2210–2223 (2015).
39. P. Evans, A. Beardmore, K. Page, J. Osborne, P. O'Brien, R. Willingale, R. Starling, D. Burrows, O. Godet, L. Vetere et al., Methods and results of an automatic analysis of a complete sample of Swift-XRT observations of GRBs. Mon. Not. R. Astron. Soc. **397**(3), 1177–1201 (2009)
40. P. Evans, J. Fridriksson, N. Gehrels, J. Homan, J. Osborne, M. Siegel, A. Beardmore, P. Handbauer, J. Gelbord, J. Kennea et al., Swift follow-up observations of candidate gravitational-wave transient events. Astrophys. J. Suppl. Ser. **203**(2), 28 (2012)
41. G. Fraser, A. Brunton, J. Lees, J. Pearson, W. Feller, X-ray focusing using square-pore microchannel plates first observation of cruxiform image structure. Nucl. Instrum. Methods Phys. Res., Sect. A, Accel. Spectrom. Detect. Assoc. Equip. **324**(1), 404–407 (1993)
42. A.S. Fruchter, A.J. Levan, L. Strolger, P.M. Vreeswijk, S.E. Thorsett, D. Bersier, I. Burud, J.M. Castro Cerón, A.J. Castro-Tirado, C. Conselice, T. Dahlen, H.C. Ferguson, J.P.U. Fynbo, P.M. Garnavich, R.A. Gibbons, J. Gorosabel, T.R. Gull, J. Hjorth, S.T. Holland, C. Kouveliotou, Z. Levay, M. Livio, M.R. Metzger, P.E. Nugent, L. Petro, E. Pian, J.E. Rho, A.G. Riess, K.C. Sahu, A. Smette, N.R. Tanvir, R.A.M.J. Wijers, S.E. Woosley, Long γγ-ray bursts and core-collapse supernovae have different environments. Nature **441**, 463–468 (2006).
43. N. Gehrels, E. Ramirez-Ruiz, D.B. Fox, Gamma-ray bursts in the Swift era. Annu. Rev. Astron. Astrophys. **47**, 567–617 (2009).
44. N. Gehrels, G. Chincarini, P. Giommi, K.O. Mason, J.A. Nousek, A.A. Wells, N.E. White, S.D. Barthelmy, D.N. Burrows, L.R. Cominsky, K.C. Hurley, F.E. Marshall, P. Mészáros, P.W.A. Roming, L. Angelini, L.M. Barbier, T. Belloni, S. Campana, P.A. Caraveo, M.M. Chester, O. Citterio, T.L. Cline, M.S. Cropper, J.R. Cummings, A.J. Dean, E.D. Feigelson, E.E. Fenimore, D.A. Frail, A.S. Fruchter, G.P. Garmire, K. Gendreau, G. Ghisellini, J. Greiner, J.E. Hill, S.D. Hunsberger, H.A. Krimm, S.R. Kulkarni, P. Kumar, F. Lebrun, N.M. Lloyd-Ronning, C.B. Markwardt, B.J. Mattson, R.F. Mushotzky, J.P. Norris, J. Osborne, B. Paczynski, D.M. Palmer, H.-S. Park, A.M. Parsons, J. Paul, M.J. Rees, C.S. Reynolds, J.E. Rho, T.P. Sasseen, B.E. Schaefer, A.T. Short, A.P. Smale, I.A. Smith, L. Stella, G. Tagliaferri, T. Takahashi, M. Tashiro, L.K. Townsley, J. Tueller, M.J.L. Turner, M. Vietri, W. Voges, M.J. Ward, R. Willingale, F.M. Zerbi, W.W. Zhang, The Swift gamma-ray burst mission. Astrophys. J. **611**, 1005–1020 (2004).
45. N. Gehrels, C.L. Sarazin, P.T. O'Brien, B. Zhang, L. Barbier, S.D. Barthelmy, A. Blustin, D.N. Burrows, J. Cannizzo, J.R. Cummings, M. Goad, S.T. Holland, C.P. Hurkett, J.A. Kennea, A. Levan, C.B. Markwardt, K.O. Mason, P. Meszaros, M. Page, D.M. Palmer, E. Rol, T. Sakamoto, R. Willingale, L. Angelini, A. Beardmore, P.T. Boyd, A. Breeveld, S. Campana,

M.M. Chester, G. Chincarini, L.R. Cominsky, G. Cusumano, M. de Pasquale, E.E. Fenimore, P. Giommi, C. Gronwall, D. Grupe, J.E. Hill, D. Hinshaw, J. Hjorth, D. Hullinger, K.C. Hurley, S. Klose, S. Kobayashi, C. Kouveliotou, H.A. Krimm, V. Mangano, F.E. Marshall, K. McGowan, A. Moretti, R.F. Mushotzky, K. Nakazawa, J.P. Norris, J.A. Nousek, J.P. Osborne, K. Page, A.M. Parsons, S. Patel, M. Perri, T. Poole, P. Romano, P.W.A. Roming, S. Rosen, G. Sato, P. Schady, A.P. Smale, J. Sollerman, R. Starling, M. Still, M. Suzuki, G. Tagliaferri, T. Takahashi, M. Tashiro, J. Tueller, A.A. Wells, N.E. White, R.A.M.J. Wijers, A short γγ-ray burst apparently associated with an elliptical galaxy at redshift z=0.225z=0.225. Nature **437**, 851–854 (2005).

46. G. Ghirlanda, G. Ghisellini, D. Lazzati, The collimation-corrected gamma-ray burst energies correlate with the peak energy of their νFννFν spectrum. Astrophys. J. **616**, 331–338 (2004).
47. G. Ghisellini, A. Celotti, D. Lazzati, Constraints on the emission mechanisms of gamma-ray bursts. Mon. Not. R. Astron. Soc. **313**, 1–5 (2000).
48. D. Giannios, H.C. Spruit, Spectral and timing properties of a dissipative gamma-ray burst photosphere. Astron. Astrophys. **469**(1), 1–9 (2007)
49. S. Guiriec, V. Connaughton, M.S. Briggs, M. Burgess, F. Ryde, F. Daigne, P. Mészáros, A. Goldstein, J. McEnery, N. Omodei, P.N. Bhat, E. Bissaldi, A. Camero-Arranz, V. Chaplin, R. Diehl, G. Fishman, S. Foley, M. Gibby, M.M. Giles, J. Greiner, D. Gruber, A. von Kienlin, M. Kippen, C. Kouveliotou, S. McBreen, C.A. Meegan, W. Paciesas, R. Preece, A. Rau, D. Tierney, A.J. van der Horst, C. Wilson-Hodge, Detection of a thermal spectral component in the prompt emission of GRB 100724b. Astrophys. J. **727**(2), 33 (2011).
50. IceCube Collaboration, An absence of neutrinos associated with cosmic-ray acceleration in [ggr]-ray bursts. Nature **484**(7394), 351–354 (2012)
51. D. Lennarz, I. Taboada, for the HAWC Collaboration, First results from HAWC on GRBs (2015). ArXiv e-prints arXiv:1508.07325
52. A.J. Levan, N.R. Tanvir, R.L.C. Starling, K. Wiersema, K.L. Page, D.A. Perley, S. Schulze, G.A. Wynn, R. Chornock, J. Hjorth, S.B. Cenko, A.S. Fruchter, P.T. O'Brien, G.C. Brown, R.L. Tunnicliffe, D. Malesani, P. Jakobsson, D. Watson, E. Berger, D. Bersier, B.E. Cobb, S. Covino, A. Cucchiara, A. de Ugarte Postigo, D.B. Fox, A. Gal-Yam, P. Goldoni, J. Gorosabel, L. Kaper, T. Krühler, R. Karjalainen, J.P. Osborne, E. Pian, R. Sánchez-Ramírez, B. Schmidt, I. Skillen, G. Tagliaferri, C. Thöne, O. Vaduvescu, R.A.M.J. Wijers, B.A. Zauderer, A new population of ultra-long duration gamma-ray bursts. Astrophys. J. **781**, 13 (2014).
53. E.-W. Liang, S.-X. Yi, J. Zhang, H.-J. Lü, B.-B. Zhang, B. Zhang, Constraining gamma-ray burst initial Lorentz factor with the afterglow onset feature and discovery of a tight Γ0−Eγ,isoΓ0−Eγ,isocorrelation. Astrophys. J. **725**, 2209–2224 (2010).
54. E.-W. Liang, T.-T. Lin, J. Lü, R.-J. Lu, J. Zhang, B. Zhang, A tight Liso−Ep,z−Γ0Liso−Ep,z−Γ0 correlation of gamma-ray bursts. Astrophys. J. **813**, 116 (2015).
55. S. McBreen, T. Krühler, A. Rau, J. Greiner, D.A. Kann, S. Savaglio, P. Afonso, C. Clemens, R. Filgas, S. Klose, A. Küpcü Yoldaş, F. Olivares E., A. Rossi, G.P. Szokoly, A. Updike, A. Yoldaş, Optical and near-infrared follow-up observations of four Fermi/LAT GRBs: redshifts, afterglows, energetics, and host galaxies. Astron. Astrophys. **516**, 71 (2010).
56. J.C. McKinney, D.A. Uzdensky, A reconnection switch to trigger gamma-ray burst jet dissipation. Mon. Not. R. Astron. Soc. **419**(1), 573–607 (2012)
57. P. Mészáros, Theories of gamma-ray bursts. Annu. Rev. Astron. Astrophys. **40**, 137–169 (2002).
58. P. Mészáros, M. Rees, Poynting jets from black holes and cosmological gamma-ray bursts. Astrophys. J. Lett. **482**(1), 29 (1997).
59. P. Mészáros, M. Rees, GeV emission from collisional magnetized gamma-ray bursts. Astrophys. J. Lett. **733**(2), 40 (2011)
60. P. Meszaros, M. Rees, Steep slopes and preferred breaks in gamma-ray burst spectra: the role of photospheres and comptonization. Astrophys. J. **530**(1), 292 (2000)
61. B. Metzger, D. Giannios, T. Thompson, N. Bucciantini, E. Quataert, The protomagnetar model for gamma-ray bursts. Mon. Not. R. Astron. Soc. **413**(3), 2031–2056 (2011)
62. E. Molinari, S.D. Vergani, D. Malesani, S. Covino, P. D'Avanzo, G. Chincarini, F.M. Zerbi, L.A. Antonelli, P. Conconi, V. Testa, G. Tosti, F. Vitali, F. D'Alessio, G. Malaspina, L. Nicastro, E.

Palazzi, D. Guetta, S. Campana, P. Goldoni, N. Masetti, E.J.A. Meurs, A. Monfardini, L. Norci, E. Pian, S. Piranomonte, D. Rizzuto, M. Stefanon, L. Stella, G. Tagliaferri, P.A. Ward, G. Ihle, L. Gonzalez, A. Pizarro, P. Sinclaire, J. Valenzuela, REM observations of GRB 060418 and GRB 060607A: the onset of the afterglow and the initial fireball Lorentz factor determination. Astron. Astrophys. **469**, 13–16 (2007).

63. K. Murase, K. Asano, T. Terasawa, P. Mészáros, The role of stochastic acceleration in the prompt emission of gamma-ray bursts: application to hadronic injection. Astrophys. J. **746**(2), 164 (2012)
64. J.P. Norris, J.T. Bonnell, D. Kazanas, J.D. Scargle, J. Hakkila, T.W. Giblin, Long-lag, wide-pulse gamma-ray bursts. Astrophys. J. **627**, 324–345 (2005).
65. J.A. Nousek, C. Kouveliotou, D. Grupe, K.L. Page, J. Granot, E. Ramirez-Ruiz, S.K. Patel, D.N. Burrows, V. Mangano, S. Barthelmy, A.P. Beardmore, S. Campana, M. Capalbi, G. Chincarini, G. Cusumano, A.D. Falcone, N. Gehrels, P. Giommi, M.R. Goad, O. Godet, C.P. Hurkett, J.A. Kennea, A. Moretti, P.T. O'Brien, J.P. Osborne, P. Romano, G. Tagliaferri, A.A. Wells, Evidence for a canonical gamma-ray burst afterglow light curve in the Swift XRT data. Astrophys. J. **642**, 389–400 (2006).
66. P.T. O'Brien, R. Willingale, J. Osborne, M.R. Goad, K.L. Page, S. Vaughan, E. Rol, A. Beardmore, O. Godet, C.P. Hurkett, A. Wells, B. Zhang, S. Kobayashi, D.N. Burrows, J.A. Nousek, J.A. Kennea, A. Falcone, D. Grupe, N. Gehrels, S. Barthelmy, J. Cannizzo, J. Cummings, J.E. Hill, H. Krimm, G. Chincarini, G. Tagliaferri, S. Campana, A. Moretti, P. Giommi, M. Perri, V. Mangano, V. LaParola, The early X-ray emission from GRBs. Astrophys. J. **647**, 1213–1237 (2006).
67. K.L. Page, R. Willingale, J.P. Osborne, J.P. Norris, P.T. O'Brien, V. Pal'shin, B. Zhang, O. Godet, F.E. Marshall, A. Melandri, E. Rol, P. Romano, R.L.C. Starling, P. Schady, S.A. Yost, S.D. Barthelmy, A.P. Beardmore, GRB 061121: broadband spectral evolution through the prompt and afterglow phases of a bright burst. Astrophys. J. **663**, 1125–1138 (2007).
68. K.L. Page, R. Willingale, E. Bissaldi, A.D.U. Postigo, S.T. Holland, S. McBreen, P.T. O'Brien, J.P. Osborne, J.X. Prochaska, E. Rol, E.S. Rykoff, R.L.C. Starling, N.R. Tanvir, A.J. van der Horst, K. Wiersema, B. Zhang, F.J. Aceituno, C. Akerlof, A.P. Beardmore, M.S. Briggs, D.N. Burrows, A.J. Castro-Tirado, V. Connaughton, P.A. Evans, J.P.U. Fynbo, N. Gehrels, C. Guidorzi, A.W. Howard, J.A. Kennea, C. Kouveliotou, C. Pagani, R. Preece, D. Perley, I.A. Steele, F. Yuan, Multiwavelength observations of the energetic GRB 080810: detailed mapping of the broad-band spectral evolution. Mon. Not. R. Astron. Soc. **400**, 134–146 (2009).
69. Pe'er, P. Mészáros, M.J. Rees, Peak energy clustering and efficiency in compact objects. Astrophys. J. **635**(1), 476 (2005)
70. Pe'er, P. Mészáros, M.J. Rees, The observable effects of a photospheric component on GRB and XRF prompt emission spectrum. Astrophys. J. **642**(2), 995 (2006)
71. R.D. Preece, M.S. Briggs, R.S. Mallozzi, G.N. Pendleton, W. Paciesas, D.L. Band, The BATSE gamma-ray burst spectral catalog. I. High time resolution spectroscopy of bright bursts using high energy resolution data. Astrophys. J. Suppl. Ser. **126**(1), 19 (2000)
72. W.C. Priedhorsky, A.G. Peele, K.A. Nugent, An X-ray all-sky monitor with extraordinary sensitivity. Mon. Not. R. Astron. Soc. **279**, 733–750 (1996)
73. S. Razzaque, A leptonic–hadronic model for the afterglow of gamma-ray burst 090510. Astrophys. J. Lett. **724**(1), 109 (2010)
74. M.J. Rees, P. Mészáros, Relativistic fireballs—energy conversion and time-scales. Mon. Not. R. Astron. Soc. **258**, 41–43 (1992).
75. M.J. Rees, P. Mészáros, Unsteady outflow models for cosmological gamma-ray bursts. Astrophys. J. **430**, 93–96 (1994).
76. M.J. Rees, P. Mészáros, Dissipative photosphere models of gamma-ray bursts and x-ray flashes. Astrophys. J. **628**(2), 847 (2005)
77. Rowlinson, P.T. O'Brien, N.R. Tanvir, B. Zhang, P.A. Evans, N. Lyons, A.J. Levan, R. Willingale, K.L. Page, O. Onal, D.N. Burrows, A.P. Beardmore, T.N. Ukwatta, E. Berger, J. Hjorth, A.S. Fruchter, R.L. Tunnicliffe, D.B. Fox, A. Cucchiara, The unusual X-ray emission of the short Swift GRB 090515: evidence for the formation of a magnetar? Mon. Not. R. Astron. Soc. **409**, 531–540 (2010).

78. R. Sari, T. Piran, Predictions for the very early afterglow and the optical flash. Astrophys. J. **520**, 641–649 (1999).
79. R. Sari, T. Piran, J.P. Halpern, Jets in gamma-ray bursts. Astrophys. J. **519**, 17–20 (1999).
80. R. Sari, T. Piran, R. Narayan, Spectra and light curves of gamma-ray burst afterglows. Astrophys. J. **497**, 17–20 (1998).
81. K.Z. Stanek, T. Matheson, P.M. Garnavich, P. Martini, P. Berlind, N. Caldwell, P. Challis, W.R. Brown, R. Schild, K. Krisciunas, M.L. Calkins, J.C. Lee, N. Hathi, R.A. Jansen, R. Windhorst, L. Echevarria, D.J. Eisenstein, B. Pindor, E.W. Olszewski, P. Harding, S.T. Holland, D. Bersier, Spectroscopic discovery of the supernova 2003dh associated with GRB 030329. Astrophys. J. **591**, 17–20 (2003).
82. G. Tagliaferri, L.A. Antonelli, G. Chincarini, A. Fernández-Soto, D. Malesani, M. Della Valle, P. D'Avanzo, A. Grazian, V. Testa, S. Campana, S. Covino, F. Fiore, L. Stella, A.J. Castro-Tirado, J. Gorosabel, D.N. Burrows, M. Capalbi, G. Cusumano, M.L. Conciatore, V. D'Elia, P. Filliatre, D. Fugazza, N. Gehrels, P. Goldoni, D. Guetta, S. Guziy, E.V. Held, K. Hurley, G.L. Israel, M. Jelínek, D. Lazzati, A. López-Echarri, A. Melandri, I.F. Mirabel, M. Moles, A. Moretti, K.O. Mason, J. Nousek, J. Osborne, L.J. Pellizza, R. Perna, S. Piranomonte, L. Piro, A. de Ugarte Postigo, P. Romano, GRB 050904 at redshift 6.3: observations of the oldest cosmic explosion after the Big Bang. Astron. Astrophys. **443**, 1–5 (2005).
83. Thompson, P. Mészáros, M. Rees, Thermalization in relativistic outflows and the correlation between spectral hardness and apparent luminosity in gamma-ray bursts. Astrophys. J. **666**(2), 1012 (2007)
84. Thompson, Deceleration of a relativistic, photon-rich shell: end of preacceleration, damping of magnetohydrodynamic turbulence, and the emission mechanism of gamma-ray bursts. Astrophys. J. **651**(1), 333 (2006)
85. J. Van Paradijs, P. Groot, T. Galama, C. Kouveliotou, R. Strom, J. Telting, R. Rutten, G. Fishman, C. Meegan, M. Pettini et al., Transient optical emission from the error box of the γγ-ray burst of 28 February 1997. Nature **386**(6626), 686–689 (1997)
86. R. Vanderspek, T. Sakamoto, C. Barraud, T. Tamagawa, C. Graziani, M. Suzuki, Y. Shirasaki, G. Prigozhin, J. Villasenor, J. Jernigan et al., Hete observations of the gamma-ray burst GRB 030329: evidence for an underlying soft x-ray component. Astrophys. J. **617**(2), 1251 (2004)
87. P. Veres, P. Mészáros, Single- and two-component gamma-ray burst spectra in the Fermi GBM-LAT energy range. Astrophys. J. **755**(1), 12 (2012).
88. R.A.M.J. Wijers, T.J. Galama, Physical parameters of GRB 970508 and GRB 971214 from their afterglow synchrotron emission. Astrophys. J. **523**, 177–186 (1999).
89. S.W. Wilkins, A.W. Stevenson, K.A. Nugent, H. Chapman, S. Steenstrup, On the concentration, focusing, and collimation of x-rays and neutrons using microchannel plates and configurations of holes. Rev. Sci. Instrum. **60**(6), 1026–1036 (1989)
90. R. Willingale, P.T. O'Brien, J.P. Osborne, O. Godet, K.L. Page, M.R. Goad, D.N. Burrows, B. Zhang, E. Rol, N. Gehrels, G. Chincarini, Testing the standard fireball model of gamma-ray bursts using late X-ray afterglows measured by Swift. Astrophys. J. **662**, 1093–1110 (2007).
91. R. Willingale, F. Genet, J. Granot, P.T. O'Brien, The spectral-temporal properties of the prompt pulses and rapid decay phase of gamma-ray bursts. Mon. Not. R. Astron. Soc. **403**, 1296–1316 (2010).
92. B. Zhang, P. Mészáros, Gamma-ray burst afterglow with continuous energy injection: signature of a highly magnetized millisecond pulsar. Astrophys. J. **552**, 35–38 (2001).
93. B.-B. Zhang, Y.-Z. Fan, R.-F. Shen, D. Xu, F.-W. Zhang, D.-M. Wei, D.N. Burrows, B. Zhang, N. Gehrels, GRB 120422A: a low-luminosity gamma-ray burst driven by a central engine. Astrophys. J. **756**, 190 (2012).

CHAPTER 6

Gamma-Ray Burst Progenitors

Andrew Levan1, Paul Crowther2, Richard de Grijs3,4, Norbert Langer5, Dong Xu6,7,8, Sung-Chul Yoon9

1.Department of PhysicsUniversity of Warwick Coventry, UK

2.Department of Physics and AstronomyUniversity of Sheffield Sheffield, UK

3.Kavli Institute for Astronomy & Astrophysics and Department of Astronomy Peking University Beijing, China

4.International Space Science Institute–BeijingBeijingChina

5.Argelander-Institut für Astronomie der Universität Bonn Bonn, Germany

6.Dark Cosmology Centre, Niels-Bohr-InstituteUniversity of Copenhagen Copenhagen, Denmark

7.National Astronomical ObservatoriesChinese Academy of Sciences Beijing, China

8.Key Laboratory of Space Astronomy and Technology, National Astronomical Observatories Chinese Academy of Sciences Beijing, China

9.Department of Physics and AstronomySeoul National University Seoul, Republic of Korea

ABSTRACT

We review our current understanding of the progenitors of both long and short duration gamma-ray bursts (GRBs). Constraints can be derived from multiple directions, and we use three distinct strands; (i) direct observations of GRBs and their host galaxies, (ii) parameters derived from modelling, both via population synthesis and direct numerical simulation and (iii) our understanding of plausible analog progenitor systems observed in the local Universe. From these joint constraints, we describe the likely routes that can drive massive stars to the creation of long GRBs, and our best estimates of the scenarios that can create compact object binaries which will ultimately form short GRBs, as well as the associated rates of both long and short GRBs. We further discuss how different the progenitors may be in the case of black hole engine or millisecond-magnetar models for the production of GRBs, and how central engines may provide a unifying theme between many classes of extremely luminous transient, from luminous and super-luminous supernovae to long and short GRBs.

Keywords: Gamma-ray burst, general Supernovae

1 INTRODUCTION

Since their discovery in the late 1960s (Klebesadel et al. 1973), unveiling the origin of gamma-ray bursts (GRBs) has been a central goal of contemporary astrophysics. While at one point the number of proposed models was only modestly smaller than the number of detected GRBs (e.g. Nemiroff 1994),[1] over the past ~20 years we have finally narrowed down this progenitor list. The step change in our ability to study GRBs arose from the discovery of afterglow emission in 1997—these panchromatic afterglows (Costa et al. 1997; Frail et al. 1997; van Paradijs et al. 1997) precisely pinpointed GRBs on the sky, enabling their cosmological origin to be secured via observations of both afterglows and host galaxies. It is now clear that GRBs are exceptionally luminous cosmological explosions, with energies (if considered isotropic) of up to 1054 erg1054 erg (e.g. Maselli et al. 2014; Perley et al. 2014), and redshifts ranging from z=0.0085z=0.0085 (35 Mpc) to z>8z>8 (Tanvir et al. 2009), possibly z>9z>9 (Cucchiara et al. 2011).

It became apparent in the early years of GRB observation that the distribution of durations was not a smooth single population, but consisted of at least two peaks (e.g. Mazets et al. 1981, 1982). This difference, secured by observations with BATSE in the 1990s (Kouveliotou et al. 1993), led to the identification of short and long duration GRBs, one with a typical duration (normally defined as t90t90, the time over which 90 % of the total energy release in γγ-rays is recorded) of around 1 s, and the other with a characteristic duration of about a minute. While further observations have identified additional possible sub-classes (see Fig. 1) at low luminosity (Soderberg et al. 2004, 2006; Liang et al. 2007a), or at intermediate (Mukherjee et al. 1998; de Ugarte Postigo et al. 2011) or ultra-long duration (Levan et al. 2011, 2014a), it is the short and long GRBs that make up the vast majority of the *observed* GRB population,[2] and whose origin has been most intensely sought.

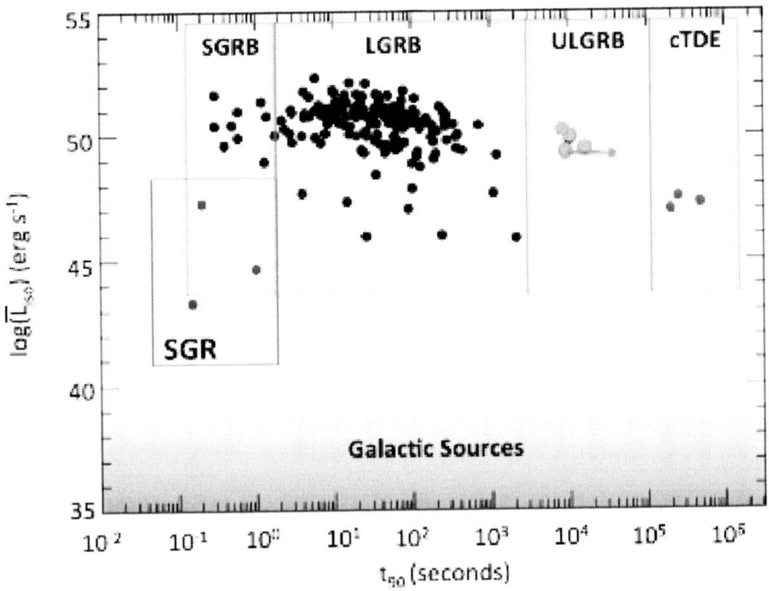

Figure 1. High energy phase space for gamma-ray bursts, adapted from Levan et al. (2014a). The duration of the bursts is shown, compared to their mean luminosity over that duration. This demonstrates that the majority of observed bursts arise from the long GRB (LGRB) or short GRB (SGRB) population, while outliers are clearly present both at extreme durations, visible as the ultra-long GRBs (ULGRBs) and candidate tidal disruption events (TDEs) and at low luminosities. Indeed, it is the low luminosity long GRBs that provide the best studied associated supernovae. A group of the short GRBs are also likely to arise from giant flares from soft gamma-repeaters (SGRs) in external galaxies (e.g. Palmer et al. 2005; Hurley et al. 2005; Tanvir et al. 2005; Levan et al. 2008), although these have yet to be firmly identified

In this review, we discuss progress towards GRB progenitors that can be made from three distinct strands. The first is direct observations of the GRBs and the host galaxies themselves. From these observations, we can determine the nature of any additional sources of energy in the GRB, be they supernovae (SNe) signatures in long GRBs (e.g. Hjorth et al. 2003), or radioactively powered kilonovae (KNe) created by nucleosynthesis in the neutron-rich disc or ejecta formed in short GRBs (Barnes and Kasen 2013;

Berger et al. 2013; Tanvir et al. 2013). These additional sources are frequently observed as late photometric bumps, interrupting the smooth decay of the afterglow light. Their study enables information about the energy and chemical make-up of the GRB explosion to be extracted, and has been very important in pinpointing GRBs progenitors. Studies of the afterglow light also provide the potential to measure beaming angles in GRBs, and hence to convert the observed rate of GRBs to the volumetric rate of GRB-like explosions. Finally, studies of the host galaxies themselves provide information about the stellar population from which the GRB is born, and the dynamics of the progenitors. Taken together these observations provide a significant, but still incomplete view of the stellar systems that create GRBs.

The second strand of our consideration comes from theoretical modelling of both GRB progenitors, and the pathways that lead to their creation. The observed GRB energetics, and the presence of photons from GRBs well above the pair production limit, directly implies relativistic outflows (e.g. Cavallo and Rees 1978), which must somehow pierce their progenitors. This presents immediate and significant constraints on the nature of GRB central engines that must be able to release a significant fraction of a solar rest-mass rapidly, and into a baryon-free environment (the presence of baryons would entrain any ejecta, and make achieving relativistic velocity extremely difficult e.g. Lei et al. 2013). In assessing the progenitors of GRBs from a theoretical perspective it is necessary to both model the details of the proposed progenitor (for example its rotation, the mass of the compact remnant formed, the baryon loading in its immediate environment etc.) and also understand the routes to obtaining these progenitors. Through this route, it is possible to determine both the types of star that might create GRBs, and whether these systems are born at a necessary rate to match the observed GRB population.

The third and final element of this work, and one which is often overlooked is to consider how local populations can inform our studies. Long GRBs are created from massive, and most likely rapidly rotating stars, and there is increasing evidence that these are drawn from low, but not excessively low metallicity (e.g. Fruchter et al. 2006; Graham and Fruchter 2013; Krühler et al. 2015). Indeed, most estimates have a rapid drop-off in the GRB rate somewhere between solar (Krühler et al. 2015; Perley et al. 2015) and 1/31/3 solar (Graham and Schady 2015b)—in other words, between the Milky Way (roughly solar) and the Small Magellanic Cloud (approximately 1/51/5th solar). In that sense, studies of the massive stellar populations within the Local Group could place strong constraints on evolutionary pathways that are viable based on star formation in differing environments. Similarly, for short GRBs, the local population of compact binaries, both in the field and clusters can be used as a route to informing the rate and pathways to their production, although again this is a challenging prospect. Such observations are likely incomplete, for example, most NS–NS binaries are found via the radio emission from a spun-up (recycled) pulsar, and there is a clear observational bias against binaries that merge very quickly after their initial formation (i.e. they have already merged and cannot be observed). However, despite these issues, local populations can provide unique diagnostics and constraints and may be able to directly identify plausible GRB progenitors.

By tackling the issue of GRB progenitors from these three routes it is possible to begin to constrain not only the basic properties of the progenitors, but also details about their likely rate, the necessary environmental conditions for their production, the possible presence of their remnants in the local Universe, or the likelihood that any stars identified today may ultimately be GRB progenitors. It is also vital if we wish to place GRBs among the full range of transients produced in the deaths of stars and to understand the links between them. Perhaps most import here are the links between GRBs and super-luminous supernovae (SLSNe, e.g. Gal-Yam 2012). These SNe peak a factor of 100 brighter than most SNe, but interestingly a rather similar set of models are invoked to explain their origin (Gal-Yam et al. 2009; Kasen and Bildsten 2010), and intriguing similarities are present in their environments (Lunnan et al. 2014, 2015; Leloudas et al. 2015; Angus et al. 2016). All of this work has substantial implications in its own right but is also a necessary first step if we hope to use GRBs as increasingly precise cosmological tools, for mapping the history of star formation, the build up of metals, or even as signatures of the collapse of the first stars.

2 WHAT DO OBSERVATIONS TELL US ABOUT LONG GRB PROGENITORS?

Progress towards understanding the nature of long-GRB progenitors has been one of the major success stories of the field and followed rapidly after the first precise locations became available. It is now clear that at least the majority of long GRBs arise from the core collapse of massive stars and are associated with hydrogen poor, high-velocity type Ic supernovae (e.g. Hjorth et al. 2003; Cano 2013). However, beyond this there remain central questions about the progenitors that have yet to be answered; just how massive are the stars creating GRBs, and are they classical Wolf–Rayet stars or something more exotic? Are binary channels important? What is the role of metallicity in creating GRB progenitors and what does this mean about their utility as cosmic probes? Exactly what central engines are created? Are the systems creating the long GRBs similar to those that are seen in the brightest supernovae?

2.1 Building the GRB-SNe Connection

The discovery of the extremely unusual, low luminosity GRB 980425 associated with a broad line type Ic supernova (SN 1998bw) marked the first strong hints as to the progenitors (Galama et al. 2000), although the total energy (if assumed isotropically released) of GRB 980425 was 4–6 orders of magnitude lower than seen in other GRBs, and it had little afterglow (AG) emission (Kouveliotou et al. 2004). Indeed, while GRB 980425 was amongst the first handful of events to be identified, it remains the closest known event of more than 1000 bursts with afterglows to date.

Despite these disparate observational properties, searches for similar supernova signatures in more distant GRBs gradually began, and possible examples were soon found as photometric bumps, causing a re-brightening of the optical counterpart on timescales of 20–30 days after the burst (e.g. Bloom et al. 1999). These photometric bumps apparently had broadly similar peak luminosities to SN 1998bw and importantly were a factor ten or more brighter than the mean peak luminosity of core-collapse SNe. A number of examples were found in the following years, predominantly in the GRB population observed at relatively low redshift $(z<0.5)(z<0.5)$ (e.g. Price et al. 2002, 2003a; Garnavich et al. 2003; Bersier et al. 2006), although in some cases with *HST* and 8-m telescopes at somewhat larger distances $(z<1z<1)$ (Masetti et al. 2003; Price et al. 2003b), and in a few cases inferred from bumps in the absence of a redshift (Bloom et al. 1999; Gorosabel et al. 2005; Levan et al. 2005). These observations gradually built a consensus that some form of supernova was present, in at least the majority of long GRBs. However, detailed spectroscopic study, necessary, for example, to identify if hydrogen was present, remained challenging, as the redshifts to the majority of GRBs detected at the time lay beyond the capabilities of available spectroscopy.

Unsurprisingly, the search was on for a low-redshift, but intrinsically energetic GRB that would enable a direct comparison with the local, low-energy GRB 980425. Such an opportunity was presented by GRB 030329, at $z=0.17z=0.17$ a redshift at which a supernova signature could be readily visible (SN2003dh). The light curve of GRB 030329 itself was extremely complex (Lipkin et al. 2004), consisting of many re-brightenings due to ongoing energy injection (e.g. Price et al. 2003c; Willingale et al. 2004). However, despite this, the optical light clearly transitioned from the featureless GRB-afterglow power-law into a spectrum with marked similarities to that of SN 1998bw (Hjorth et al. 2003; Stanek et al. 2003), cementing the association of GRBs with broad-lined SN Ic.

2.2 Distribution of Supernova Bulk Properties

A large sample of GRB-SNe or candidate GRB-SNe has now been accrued. The majority of these arise from photometric bumps in the late time light curve, but an increasing number also have direct spectroscopic detections of the SNe in question, although often through noisy spectra at a single epoch.[3] Inevitably this means that the quality of the spectra, and decisions of different authors about the merits of including a given burst vary. However, excepting GRB 111209A/SN 2011kl (Levan et al. 2014a; Greiner et al. 2015a) the spectra of all of these events bear marked similarities to that of SN 1998bw, showing broad lines consistent with high-velocity expansion >20,000 kms−1>20,000 kms−1, and no sign of hydrogen or helium emission features. All of these SNe are therefore spectroscopically classified as broad-lined type Ic supernovae (SN Ic-BL).

The task of obtaining broad-band spectral shapes and peak luminosities from light curves of these SNe is made challenging by their large luminosity distances, by the contribution of the host galaxy, and by the afterglow contribution. Indeed, at any given time the observed flux depends on the contribution of all three, e.g.

$$F_{\rm obs}(\nu,t) = F_{\rm AG}(\nu,t) + F_{\rm SN}(\nu,t) + F_{\rm host}(\nu). \tag{1}$$

FhostFhost can in principle be obtained from late-time observations and subtraction from early data, although in practice this is often complicated by slightly different filters or instrument combinations, as well as the difficulties introduced by matching ground-based seeing. A typical GRB host is of comparable (or perhaps slightly lower) luminosity than its SNe (Savaglio et al. 2009; Svensson et al. 2010; Perley et al. 2015) and so can create a significant uncertainty, especially for higher redshift bursts, and even more so when spectroscopy is used (since the spectroscopy is inevitably noisier than imaging). The afterglow can normally be treated as a power-law in both frequency (or wavelength) and time e.g.,

$$F_{\rm AG}(\nu,t) = t^{-\alpha}\nu^{-\beta}. \tag{2}$$

Measurements of the afterglow can be complicated by unseen temporal breaks (e.g. the jet-break due to lateral spreading of the GRB-jet, which typically steepens αα to ~2~2 (e.g. Rho 1999)) or spectral breaks (e.g. the cooling break, which imparts Δβ=0.5Δβ=0.5 (e.g. Sari et al. 1998)).

Given the spectral similarities to SN 1998bw, a common approach is to assume that the light curve of the SN evolves in time, tt, like that of SN 1998bw but modified by a stretch parameter ss, and luminosity scaling,[4]hh.

$$F_{\rm SN} = h F_{\rm 98bw}(t/s), \tag{3}$$

where F98bwF98bw is the specific flux of SN 1998bw as observed at the same rest-frame wavelength and time as that of the burst under consideration (i.e. taking into account a k-correction, and cosmological time dilation).

One can fit for these things simultaneously in well-sampled data (e.g. Zeh et al. 2004; Kann et al. 2010; Cano 2013; Cano et al. 2014), although degeneracies can exist (e.g. between hh and αα), and it is also common to attempt to extrapolate the afterglow contribution, either from earlier observations, or from data in the UV or IR where the SNe contribution is small (e.g. Levan et al. 2014b).

In practice, this is a relatively crude approach, since the SNe light curve may not be a good approximation to a stretched and scaled SN 1998bw. However, in many cases it appears to work well, at least given the data available. More recent attempts have moved beyond simple scaling factors and have attempted to infer bolometric properties, either from the scaled SN 1998bw light curves (e.g. Cano 2013) or using other SNe as templates, enabling single colours to be used for bolometric corrections (e.g. Lyman et al. 2014a, 2016a).

One thing which is clearly apparent in these observations is that the classical long duration, highly energetic GRBs that comprise the bulk of the *observed* GRB population are highly under-represented in the GRB-SNe sample (Fig. 2). In itself, this is not surprising, since SNe are most readily seen in low redshift examples, and when not outshone by the afterglow (whose luminosity broadly scales with that of the γγ-rays, (e.g. Nysewander et al. 2009a; Gehrels et al. 2008)). Indeed, many of the GRBs arise from a potentially distinct population of low-luminosity GRB (LLGRB), while even the Rosetta Stone of GRB 030329/SN 2003dh is of rather intermediate luminosity. Only GRB 130427A/SN 2013cq appears to arise from a highly luminous GRB (Xu et al. 2013; Levan et al. 2014b; Melandri et al. 2014).

While this may not be problematic, it is notable that in some cases these local, low luminosity bursts appear very different from their higher energy cousins. In particular, some are extremely long and dominated by thermal emission (Campana et al. 2006; Starling et al. 2011), features often not seen in other GRBs. It this sense it may be that they are not indicative of GRB-SNe in general, although it may also be that similar components in more distant, luminous GRBs escape detection, and some thermal components have been found in careful searches (Starling et al. 2012; Sparre and Starling 2012), perhaps suggesting similarities. Indeed, the strongest link of the similarities between these very different energies of high energy transient is actually the properties of their supernovae, which are sufficiently similar that it is likely they all arise from the same physical mechanism.

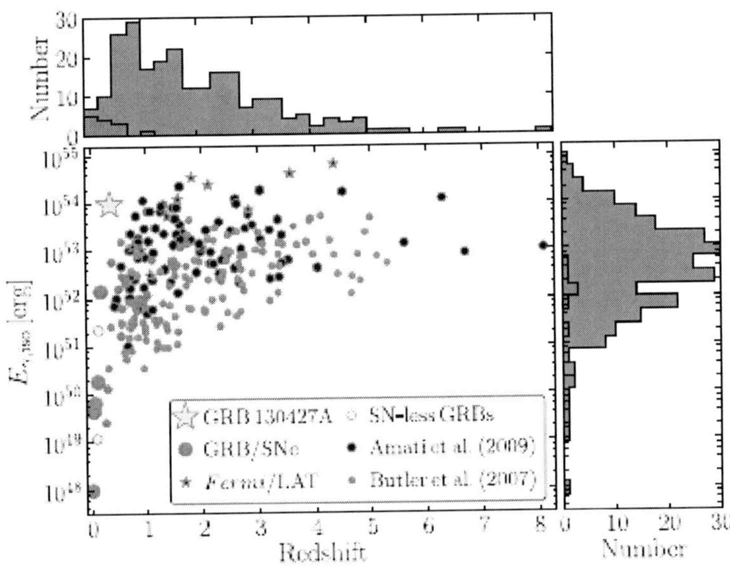

Figure 2. The distribution of isotropic equivalent energy with redshift for long GRBs from Xu et al. (2013), with the background points determined by Butler et al. (2007) and Amati et al. (2009). Those with spectroscopic detections of a supernova are marked in *red*. As can be seen these are typically low redshift events with a low luminosity (these bursts are likely intrinsically common, but only visible at low-z). Only a handful of events are typical of the more energetic GRBs that make up the bulk of the observed GRB population, with only GRB 130427A lying at the brightest end of the distribution

2.3 Test-Bed Examples

An alternative to the large scale samples accrued with data of variable quality is to attempt detailed modelling of well-studied examples, where time series spectra and excellent photometric coverage is available. In these cases it is possible to go beyond simple analytical fits to light curves, or snapshot velocities from spectroscopy, and build detailed models of the explosions of a range of stars that map the nickel releases, velocities, and kinetic energies, providing high-quality spectrophotometric predictions that can be compared with observations. The principle here is to conduct detailed spectral synthesis, in which energy injection from the core is coupled with radiation transport to predict the observed spectrum at given epochs as a function of the various input parameters. In principle, these models can also include detailed geometry, for example, ejecta which includes anisotropies in order to match both the rise/decay times of the SNe. This avoids the need to use scaling relations to obtain SNe properties and instead obtains them from ab initio approaches. It has been successful in several GRBs, including GRB 980425/SN 1998bw (e.g. Höflich et al. 1999; Woosley et al. 1999; Maeda et al. 2002), GRB 030329/SN 2003dh (e.g. Deng et al. 2005) and GRB 060218/SN 2006aj (Pian et al. 2006b; Mazzali et al. 2006). Results from such approaches are reassuringly often similar to those from light curves alone in terms of bulk properties (e.g. nickel mass, Cano 2013), although also provide evidence at times for anisotropies in the explosion, and stronger constraints on pre-explosion core mass and ejecta mass. However, the observational requirements for such work are frequently extreme, and so for the majority of GRBs such detailed information is either too expensive or impossible to obtain via current instrumentation.

2.4 Pushing the Boundaries: From no Supernovae to the Most Luminous Supernovae

While broad-lined type Ic supernovae have been discovered in the vast majority of cases where such a search was plausible there are a handful of cases where such searches have been unsuccessful. In particular, two local long GRBs, 060505 and 060614 (at z=0.09z=0.09 and z=0.125z=0.125) lie at redshifts where SNe similar to SN 1998bw should be readily identified, with peak apparent magnitudes in the range 18.5<mV<19.518.5<mV<19.5. However, deep searches failed to identify any such signatures to limits not only much fainter than GRB-SNe but also to other core-collapse events. Indeed, any SNe in these cases must have been a factor >100>100 fainter than SN 1998bw (Della Valle et al. 2006; Fynbo et al. 2006; Gal-Yam et al. 2006; Gehrels et al. 2006). It has been noted that GRB 060505 was of relatively short duration ~4 s~4 s, while GRB 060614 has subsequently been suggested to be an example of a short burst with extended, softer emission (Gehrels et al. 2006). This scenario seems likely, since further examples of SN-less, but unambiguously long GRBs have not been uncovered. However, other suggestions, for example, that these GRBs may arise in cases where no outward supernova shock is launched and the star collapses directly to a black hole (Fynbo et al. 2006; Fryer et al. 2007) have also been made. This may be particularly relevant given the apparent absence of very massive progenitors to local SN II-P (Smartt et al. 2009), which has led to renewed interest in the prospect of disappearing stars (Kochanek et al. 2008; Kochanek 2014), with recent surveys beginning to find candidate examples (Reynolds et al. 2015). Surprisingly, despite the decade since these discoveries, further examples have not been found, and so their nature remains mysterious.

At the other end of the scale, the recent discovery of an SN-GRB (SN 2011kl/GRB 111209A) which was a magnitude more luminous than SN 1998bw (Greiner et al. 2015a; Kann et al. 2016), challenges the picture of SN homogeneity that has been emerging from previous observations. This SN was found in an ultralong GRB with a duration of >10,000 s>10,000 s (Gendre et al. 2013; Levan et al. 2014a), and in addition to being more luminous than most GRB-SNe was also spectrally very different. In particular, while it also appears to belong to the SN Ic population, it was far bluer and more UV luminous than other GRB-SNe, and its spectrum bore a strong resemblance to the spectra of type I SLSNe, which also show weak absorption lines on an extremely blue continuum (e.g. Mazzali et al. 2016). Other ULGRBs do not show such strong apparent SNe, and so it remains unclear if this burst is a unique and unusual object, or if it in fact represents a broader range of SNe properties that should be considered, perhaps from cases where the central engine of the GRB begins to impact the supernova itself, as is likely the case for SLSNe (e.g. Cano et al. 2016).

2.5 Large Scale Environments

Additional constraints on the progenitors of long GRBs can be obtained from both their large scale and small scale environments. The metallicities of their host galaxies are of particular interest (in practice, the metallicities of the small scale environments are also of interest, but are not readily accessible with current technology, since one arcsecond seeing typically corresponds to several kiloparsecs at the redshift of a typical long GRB). Such constraints can either come from direct spectroscopic observations, or via inferences using the well-known correlation between mass (or luminosity) and metallicity. GRBs can provide exceptional diagnostics of metallicity at high redshift via studies of UV absorption lines in concert with the direct detection of Lyαα (e.g. Fynbo et al. 2009). However, this is predominantly for higher redshift bursts, since Lyαα does not enter the optical window until z>2z>2, and a measurement of the hydrogen column density NHNH along the line of sight is necessary to obtain the ratio of a given element to hydrogen (e.g. O/H, Fe/H). For this reason the metallicities of the host galaxies where the GRB-SNe connection has been determined are predominantly obtained either from emission line diagnostics such as, $R_{23} = ([\mathrm{O\,II}](3727\,\text{Å}) + [\mathrm{O\,III}](4959 + 5007\,\text{Å}))/\mathrm{H}\beta$ or $\mathrm{N2} = \log[\mathrm{N\,II}]/\mathrm{H}\alpha$, or indeed more complex approaches to line ratios that attempt to remove degeneracies that exist is the more simple examples (e.g. Dopita et al. 2016). This provides some insight into the chemical state of the gas-phase in these galaxies, although several important caveats should be noted in the use of these metallicities as a direct insight into GRB progenitors. Firstly, they are not a measurement of the metallicity of the progenitor star, and indeed, even if they are an accurate measurement of the oxygen metallicity of the star, this is still some way from the [Fe/H] ratios that are more commonly used in distinguishing different stellar evolution pathways. Secondly, significant metallicity gradients can exist within galaxies, and so the use of global proxies essentially derives central metallicities (i.e. those of the brightest regions), which may, or may not be indicative of the regions hosting a given transient event. Thirdly, in some cases, the GRB may be associated with a satellite galaxy of the presumed host. For example, Kelly et al. (2013) show that GRB 130702A (at z=0.145z=0.145), lies in a satellite galaxy at a significant offset (18 kpc). However, in other cases it is likely that more proximate galaxy/satellite pairs are not adequately resolved, at least by ground-based imaging, leading to the misidentification of the host. In extremum, there can even be chance alignment in which a background galaxy aligns with a foreground system leading not only to the misidentification of the precise host galaxy but the incorrect determination of the redshift. This has recently been demonstrated for an extremely well studied optically dark (i.e. no optical afterglow) GRB 080219B, which, rather than lying at z=0.41z=0.41 is in fact in a background galaxy at z=1.96z=1.96 (Perley et al. 2016). In these cases, the use of metallicity measurements inferred from the "apparent" host is clearly incorrect, and potentially problematic. For example, the case of GRB 080219B has been used to argue for the presence of GRBs in high metallicity environments, but in practice provides no such evidence since this metallicity is simply of the foreground system.

The range of metallicities seen in long GRB host galaxies (as well as short GRBs and luminous SNe) is shown in Fig. 3. It is clear that GRBs favour a lower metallicity than is typical in the local Universe, although given the redshift range considered this is not surprising. Interestingly, the metallicities of long GRB hosts at large appear systematically higher than those of the examples in which a spectroscopic signature of the SNe has been seen, with the median metallicity of the GRB-SNe sample being 0.2–0.3 dex lower. This may well be due to a bias against dusty systems, since obscured star formation is typically more metal rich, and it is clear that dusty GRBs arise in more luminous, and likely metal-rich host galaxies. In any case, it also interesting to note, that when considering only the GRBs in which spectroscopic SNe signatures are seen there is an apparent similarity between their metallicity distribution and those of SLSNe. The degree to which these differences arise due to observational selection effects, or alternatively due to genuine astrophysical differences is of significant interest for further study. It may reflect genuine difference between populations, or perhaps that even in the relatively low redshift regime there may be significant selection effects (e.g. Japelj et al. 2016).

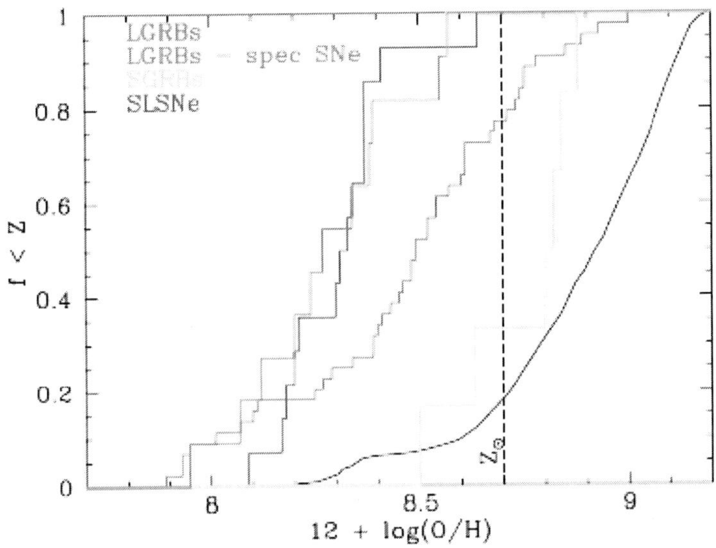

Figure 3 Host galaxy metallicities for long GRBs from Krühler et al. (2015), short GRBs (from Berger 2009) and SLSNe (from Leloudas et al. 2015). It is striking that the metallicities of LGRBs appear significantly lower than for the SGRBs, or indeed the general galaxy population (Graham and Fruchter 2013). However, those of the GRBs with apparent SNe appear to be somewhat lower still, and are very similar to those observed in SLSNe host galaxies

2.6 Small Scale Environments

The large scale environments can provide information about the chemical state of GRB hosts that is not currently available at smaller scales, due to both large luminosity distance, and the inability of large aperture telescopes to probe with sufficiently high resolution. However, star formation takes place on small spatial scales, from less than a parsec to tens of parsecs depending on the mode and intensity of star formation (e.g. Portegies Zwart et al. 2010). It is these small scale environments that in practice carry most information about the GRB progenitor itself, free from dilution from the light of the remaining galaxy. At a typical GRB redshift, the *HST* resolution (0.1 arcsec) still corresponds to physical scales of several hundred parsecs, and so while we are able to track the stellar populations in the regions of the GRBs, isolating the stellar population responsible for the GRB remains challenging. Two approaches can be considered when studying GRB locations. The first is simply to identify the GRB location relative to its host galaxy by measuring its offset, either relative to the galaxy host (e.g. Bloom et al. 2002), or perhaps to local regions of intense star formation, if such resolution is available (e.g. Hammer et al. 2006). The alternative is to try to study the population under the burst, despite the poor spatial resolution. This may be highly diagnostic for long GRBs, since the young massive stars are by far the most luminous in the host galaxy, especially at UV wavelengths (Fruchter et al. 2006). Indeed, at the typical redshifts of many GRBs rest frame UV light is redshifted into the optical window, making sensitive high-resolution imagery with *HST*an ideal route to characterising the immediate massive star populations surrounding GRBs. In addition, these hot massive stars also excite the gas phase of the interstellar medium, making narrow-band observations a sensitive probe of the young massive star population (James and Anderson 2006).

Such an approach has been taken by various groups studying both supernovae and GRBs. Fruchter et al. (2006) consider the total fraction of the galaxy light in pixels of lower surface brightness than the pixel containing the GRB or SNe, the so-called FlightFlight parameter. They show that GRBs are highly concentrated on the light of their host galaxy, while core collapse SNe broadly trace the distribution of rest-frame UV-light. In other words, the probability of a SN occurring in a given pixel is approximately proportional to the brightness of that pixel, in contrast, the probability of a GRB occurring in a given pixel is proportional to the brightness of that pixel squared.[5] Interestingly, further studies of local SNe show that a similar effect can be seen in the distribution of hydrogen-rich SNe II and hydrogen/helium poor SNe Ic (James and Anderson 2006; Kelly et al. 2008; Anderson et al. 2012). This is likely understood because of the strong correlation between stellar mass and luminosity. In the low mass range (10--20 M⊙10--20 M⊙) L~M3L~M3, while for higher mass stars (50--100 M⊙50--100 M⊙) L~M2L~M2 (Yusof et al. 2013) such that the most massive stars dominate the UV budget. In this sense, if core collapse SNe arise from essentially every star with an initial mass >8 M⊙>8 M⊙, then GRBs must arise from stars significantly more massive than this. Studies based on the expected distribution of stars in galaxies (Raskin et al. 2008), and on young clusters in nearby galaxies (Larsson et al. 2007) provide a consistent picture in which GRB progenitors have initial masses of >40 M⊙>40 M⊙, representing one of the few ways in which the masses of GRB progenitors can be ascertained.

2.7 Summary: LGRB Progenitors

The emerging picture for long GRB progenitors appears clear. A long GRB progenitor is a massive star, typically born at sub-solar metallicity, in which the hydrogen envelope is lost/burned prior to its explosion as a broad-lined type Ic supernova. There is some observational evidence that suggests these stars are very massive, for example, the large Nickel yields required to achieve absolute magnitudes a factor 10 brighter than SNe Ic (e.g. Mazzali et al. 2007a), or the location of the bursts on the brightest regions of their hosts (Fruchter et al. 2006; Svensson et al. 2010) could both be interpreted as signatures of very massive stars, perhaps with an initial mass at the zero-age main sequence (ZAMS) of MZAMS>40 M⊙MZAMS>40 M⊙. However, the locations of long GRBs are similar to those of the bulk SNe Ic population (Kelly et al. 2008), most of which can be explained by the explosion of initially far less massive stars (e.g. Mazzali et al. 2007b,c), and more complex models, involving the impact of chemically homogeneous evolution (e.g. Szécsi et al. 2015), or binaries (e.g. Stanway et al. 2016) on the observed environments may be necessary to provide tighter constraints on the progenitor mass, as well as on the progenitor rotation. These scenarios are not currently well constrained observationally, and theoretical progress is discussed in Sect. 5. None-the-less, the progress which has been made towards the progenitors of long GRBs has been remarkable, and while there remain many open questions about the details of long GRB progenitors, it is fair to say that barring a few special examples, we now know what forms the vast majority of the long GRB population.

3 OBSERVATIONAL CONSTRAINTS ON SHORT GRB PROGENITORS

3.1 Defining Short GRBs

Short GRBs are traditionally defined as those with durations of t90<2 st90<2 s. Observations with *Konus* and *BATSE* suggested they form a separate population from the long GRBs, and so may well also have different progenitors (Mazets et al. 1981, 1982; Kouveliotou et al. 1993). Following the identification of supernovae signatures in the afterglows of long GRBs, the favoured model for short bursts rapidly became the merger of two compact objects (neutron star—neutron star or neutron star—black hole), a model which had previously been popular for the long GRB population (e.g. Narayan et al. 1992). The reason for this was two-fold; both that the rapid time scale for the merger, combined with the clean environment (compared to e.g a collapsing massive star) meant that short durations were more naturally expected in merger scenarios. The second was the rather more speculative assertion that if mergers did not produce the long bursts, they must either produce the short bursts, or no high energy transient at all. Given this historical approach, it is unsurprising that the observational history of short GRBs has largely been benchmarked against this expectation, and hence the discussion of short GRBs has been largely framed as though their progenitors are known, even prior to the accrual of the significant observational data that now supports a merger origin.

Short bursts now represent approximately 10 % of the bursts observed by *Swift*, down somewhat from the 25 % observed in the BATSE observations. It seems likely that a significant element of this is due to the differing spectral responses of the different instruments, with BAT operating in a rather softer 15–150 keV band compared to BATSE (50–>300 keV). This has led to some discussion of whether the duration split between the two populations is correctly set at 2 seconds. Indeed, the duration distribution of *Swift* GRBs is far less clearly bimodal than that for BATSE, and so one cannot directly distinguish a "cut-off" based on these observations alone. Several alternative attempts in the early years of *Swift* addressed the potential problem by suggesting the use of additional diagnostics to distinguish between different progenitor types (e.g. Levan et al. 2007; Zhang et al. 2009). Indeed, Zhang et al. (2009) argued for a distinction between Type I (compact object merger) and Type II (collapsar) events based on a decision tree of multiple observational properties. These additional constraints often used rather more complex properties (for example the nature of the host galaxy, or the presence/absence of a supernova component) and so also risked significant confirmation bias, in finding bursts which met the expectations based on only a handful of systems and omitting other, potentially valuable systems. Using theoretical considerations, largely based on the break-out time for a collapsar, combined with the different spectral responses of different observatories Bromberg et al. (2013b) have suggested that very different values for T90T90 would be needed to identify the split between long and short GRBs for different missions. In particular, 2 s for BATSE and 0.7 s for *Swift*.[6] The essence of the argument is that hardness is a more important distinction than duration at durations where there is likely to be significant overlap between the two populations. This is shown graphically in Fig. 4, where the *Swift* short GRBs can be seen to be typically much softer than those detected by

BATSE. Applying a cut of this nature would remove some (although by no means all) of the best studied "short GRBs" from consideration, since they would then be more likely to be collapsars.

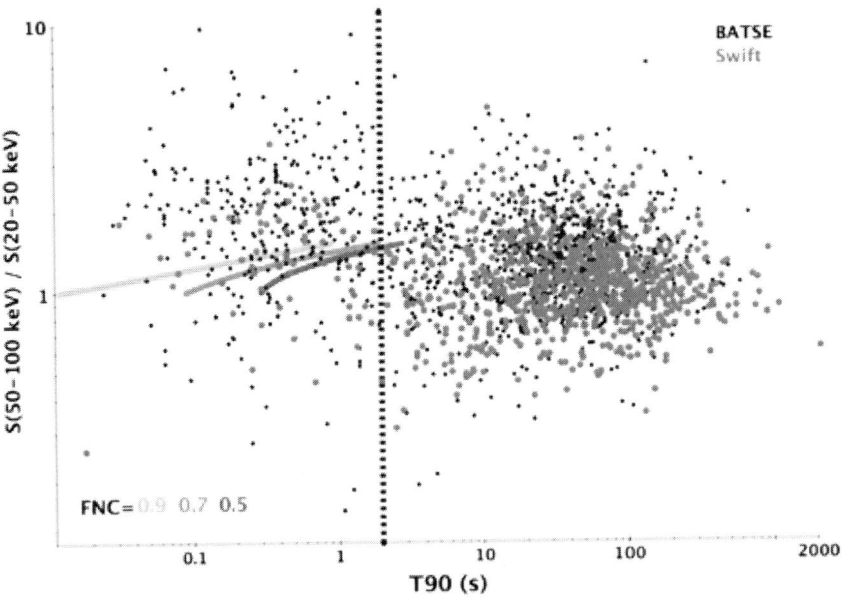

Figure 4. The hardness-duration distribution of GRBs detected by *Swift* (*red*) and BATSE (*black*). The *x-axis* shows the duration over which 90 % of the total fluence is recorded (T90T90, while the *y-axis* shows the γγ-ray hardness, defined as the ratio of flux between two adjacent channels (the channels chosen are those used by BATSE). The *Swift* data have been placed on the same scale by extrapolating the prompt spectral fits (either as a power-law or cut-off power-law) to the appropriate bands. The *dotted vertical line* shows the notional distinction between short and long GRBs at two seconds, but there is clearly significant overlap. It is apparent from the BATSE distribution that short GRBs are *on average* harder than long GRBs. However, for *Swift* the two distributions look very similar. Given this the contribution of the long GRB population is larger for *Swift* bursts at durations of <2 s than for BATSE. The lines suggested by Bromberg et al. (2013a) as distinguishing long and short GRBs at 50, 70 and 90 % confidence are also shown (where FNC is the fraction of non-collapsars), bursts below these lines are more likely to be collapsars from the short duration tail of the long GRB population. For *Swift*, only bursts above the 90 % *green line* are highly likely to be genuine members of the short burst population

3.2 Precise Locations from Afterglows

While progress towards the origin of long GRBs proceeded at pace from the discovery of the first afterglows to the identification of broad-lined SNe-Ic, pinpointing the progenitors of the short GRBs remained much more challenging. Firstly, short GRBs are typically significantly fainter, or more precisely exhibit markedly lower fluence than long GRBs despite similar peak fluxes (a parameter which more accurately describes their detectability to many $\gamma\gamma$-ray observatories). Since broad correlations exist between the fluence of the prompt emission and the brightness of either the X-ray or optical afterglow (Gehrels et al. 2008; Nysewander et al. 2009b) it is perhaps unsurprising that the afterglows of short GRBs escaped detection while the afterglow revolution was transforming our knowledge of the long GRB phenomena. Indeed, while a handful of short GRBs were detected and reasonably localised during the long GRB afterglow revolution, no successful afterglow campaigns were made prior to the launch of *Swift*. Its ability to re-point its X-ray and UV-optical telescopes rapidly provided the first X-ray afterglow detection for GRB 050509B (Gehrels et al. 2005; Hjorth et al. 2005a; Bloom et al. 2006). Interestingly, the first optical afterglow detection actually came from a HETE-2 burst, GRB 050709 (Hjorth et al. 2005b; Fox et al. 2005), which with the detection of another HETE-2 burst in early 2006 (de Ugarte Postigo et al. 2006; Levan et al. 2006), perhaps suggested that previous searches had been rather unlucky, although doubtless the ability of *Swift* to rapidly repoint and detect their X-ray afterglows was also an important component in the discovery of their optical counterparts.

As with long GRBs, the discovery of afterglows to short GRBs has revolutionised their study. Perhaps the most important diagnostic enabled by an afterglow is a precise location on the sky, and hence the ability to study the galaxy population hosting short GRBs. The first short burst with an afterglow was GRB 050509B, and its location was striking, being offset $\sim$30 kpc$\sim$30 kpc from an extremely massive galaxy in a merging cluster system at z=0.225z=0.225 (Gehrels et al. 2005; Bloom et al. 2006). Deep searches for star formation in this host galaxy revealed no sign, and immediately suggest that some short GRBs arise from ancient populations. However, this burst was localised only by its X-ray afterglow, and the large resulting error box contained additional background galaxies, many of which had blue colours consistent with star formation. Hence, while the probability of chance alignment was low, the association did not clinch the origin of at least some short GRBs in ancient populations. The discovery of the optical afterglow of GRB 050724 in a clearly elliptical host galaxy further strengthened this argument (Berger et al. 2005), and while bursts in low star formation rate host galaxies are clearly in a minority, perhaps 10–20 % of short GRBs overall do arise from elliptical hosts (Fong et al. 2013). Equally, while the remaining 80–90 % of the host galaxies do show signs of significant star formation, they remain distinct from the hosts of the long bursts in terms of their stellar masses and star formation rates. While long GRBs show a strong preference for star-forming dwarfs, with low metallicities and high specific star formation rates (e.g. Fruchter et al. 2006; Savaglio et al. 2009; Svensson et al. 2010; Graham and Fruchter 2013), the short bursts are in rather more typical galaxies, with a range of metallicities and higher stellar masses, therefore they appear to sample the entirety of the galaxy population (Leibler and Berger 2010; Fong et al. 2013).

Beyond the nature of the galaxies themselves, there is significant information contained within the distribution of the short GRBs on their hosts. Again, the differences between long and short GRBs become apparent rapidly in this distribution. While long GRBs are highly concentrated on their host light (Bloom et al. 2002; Fruchter et al. 2006; Svensson et al. 2010, see Sect. 2.6) the short GRBs are frequently scattered on their host light, and at large projected radii, in some cases there is apparently no underlying host galaxy, despite relatively bright galaxies nearby in the field, suggesting that these progenitors have been kicked from their birth sites (Berger 2010; Tunnicliffe et al. 2014), although it remains plausible that some lie at much higher redshifts in galaxies too faint for current observations (Levan et al. 2006; Berger et al. 2007). Beyond this much broader distribution, those short GRBs that do lie on galaxies tend to appear on much fainter pixels than those in the long GRBs (Fong et al. 2010; Fong and Berger 2013), and in particular show essentially no association with the blue light of their host galaxies. Taken together these properties demonstrate that short GRB progenitors are frequently old, and more importantly are also kicked from their birthplaces. These are all consistent with the expectations of binary mergers, in which the neutron stars receive substantial space velocities (kicks) from a combination of natal kicks and binary mass loss (e.g. Arzoumanian et al. 2002).

3.3 Kilonovae and Macronovae

The large scale observations are clearly broadly in favour of a model in which short GRBs arise from compact object mergers. However, they are in themselves not a smoking gun of the merger itself. This situation is reminiscent of early studies of long GRBs in which the star-forming host galaxies offered evidence of a link to massive stars and supernovae, but did not conclusively prove it. Since afterglows are generated at large radii and are essentially a product of the interaction of an outgoing relativistic shock with an external medium, they themselves are not uniquely diagnostic of the progenitor. Hence, some additional signature that could in principle only arise in compact object mergers was necessary. The most likely scenario is to identify a faint radioactively powered transient, created by nucleosynthesis in the neutron-rich material available in a compact object merger containing a neutron star. This material may reside either in the accretion disc, or be ejected into tidal tails. None-the-less it was recognised that for typical properties of a merger such a transient would yield a rapidly evolving, faint transient, significantly faster and fainter than normal core-collapse SNe (e.g. Li and Paczyński 1998). Like a supernova, these events would start faint and rise to peak on timescales of hours to days, and so would be seen as photometric bumps, interrupting the otherwise smooth decay of the afterglow. These events have had a range of names over the years, but those which are commonly used today are either kilonova (referring in essence to something about 1000 times brighter than a nova), or macronova. The precise properties of these transients were uncertain, and while some populations of faint, fast transients have been uncovered by recent synoptic sky surveys (e.g. Kasliwal et al. 2012; Foley et al. 2013), none have been interpreted as NS–NS or NS–BH mergers (although some have been suggested to arise from NS-white dwarf mergers, through channels similar to those creating NS–NS binaries (Metzger 2012; Lyman et al. 2014b, 2016b). Hence, there is little *observational* evidence as to the signatures that observers should search for. Indeed, early, deep observations of several short GRBs failed to uncover any sign of the moderate to late time “bumps” expected in this scenario (e.g. Hjorth et al. 2005a) and ruled out the most optimistic scenarios for these transients (Metzger and Berger 2012).

However, an important revelation came from detailed calculations conducted in 2013 (Barnes and Kasen 2013). It had long been recognised that NS–NS mergers were promising sites for r-process nucleosynthesis, and that NS–NS mergers (depending on the rates and individual r-process outputs) could be important, if not dominant sites for the creation of the heaviest elements (Rosswog et al. 1999, 2003). Indeed, a range of recent studies from the abundances of radioactive elements on the sea floor (Wallner et al. 2015) to in depth analysis of metal poor stars in the Milky Way (e.g. Macias and Ramirez-Ruiz 2016) suggest that the r-process abundances are not in an equilibrium that would be expected if they were regularly replenished with small amounts of additional material, as might be expected for core collapse supernovae. Instead, these analyses prefer a scenario in which rare events, with significantly more mass per event are dominant, favouring a merger origin. This nucleosynthesis should naturally power a luminous transient, but because of the dominance of r-process elements its evolution might be quite different to those previously assumed based on our studies of supernovae. Indeed, the opacities that were assumed in the earlier predictions were predominantly those of iron group elements which are dominant in normal SNe. The impact of including new opacities for heavy elements, in particular, lanthanides, was profound. The heavy opacity extinguishes essentially all the optical light for external observers and means that earlier observations were, in essence, unconstraining of the nucleosynthesis taking place. These observations instead suggested that IR observations were the most promising route to the identification of a kilonova (Barnes and Kasen 2013).

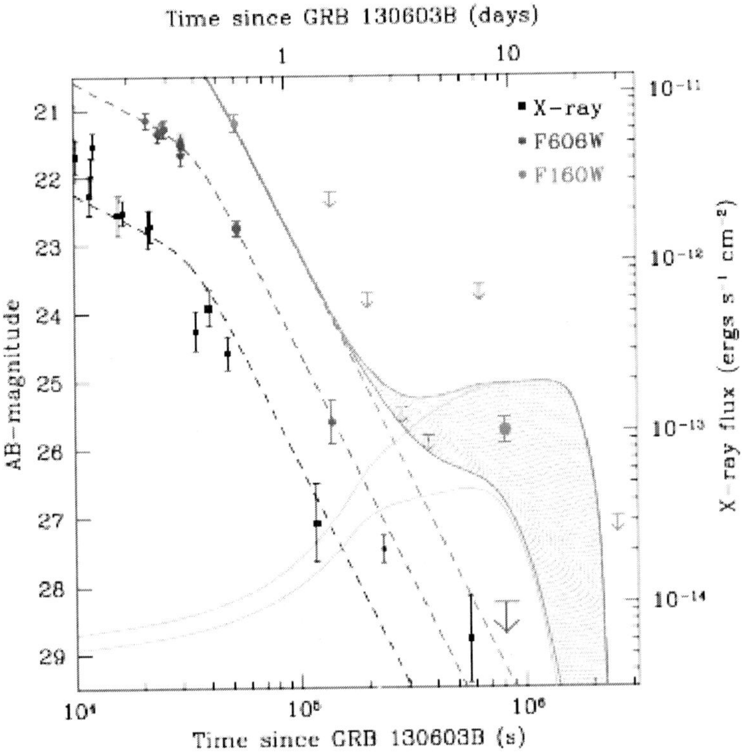

Figure 5. The *light curve* of the kilonova seen in GRB 130603B (updated from Tanvir et al. 2013). The points represent the X-ray (*black*) optical (*blue*) and IR (*red*) photometry of the afterglow, along with their expected decay. The *shaded region* is the expected kilonova range for 10–110–1 and 10–2 M⊙10–2 M⊙ of material from Barnes and Kasen (2013). While the *cyan line* shows the very faint expected optical emission. The re-brightening in the IR is strongly suggestive of a kilonova, although it is also relevant to note that late time X-ray observations by Fong et al. (2014) also imply that the X-ray lies above expectations

This supposition was observationally tested with *HST* observations of the short GRB 130603B in June 2013. These observations took place in two bands, one in the optical and one in the IR, for a burst at $z=0.35z=0.35$ (de Ugarte Postigo et al. 2014). Although only two epochs were obtained they successfully showed a fading IR source, while nothing was seen in the optical (Berger et al. 2013; Tanvir et al. 2013).

A comparison with the IR afterglow decay suggested a magnitude much brighter than expected, and hence a re-brightening from an associated kilonova (Fig. 5). The luminosity of the IR source was very similar to the prediction of kilonovae using the improved lanthanide opacities, and providing the first direct evidence for the origin of short GRBs in compact binary mergers.

Unfortunately, opportunities to further hone our understanding remain limited due to the rate of short GRBs at sufficiently low redshift, and so to date there are limited constraints on the true properties of this KNe. A detailed re-analysis of some archival short GRBs provides some evidence for similar components in their light curves (e.g. Jin et al. 2015), although other short GRBs show optical or X-ray bumps apparently not associated with the same physical process (Perley et al. 2009). Indeed, recent work, partly motivated by an apparent X-ray bump co-incident with the infrared kilonova in GRB 130603B (Fong et al. 2014) has focused on possible long-lived X-ray manifestations of mergers.

In particular, in the form of scattered X-ray's from the central engine (either millisecond magnetar or black hole) (Kisaka et al. 2015), the absorption and re-radiation of X-rays to provide IR and X-ray signals (Kisaka et al. 2016), or direct emission of X-rays from the engine itself (Sun et al. 2016), perhaps creating isotropic X-ray emission that could be of some value in searches for mergers without a GRB trigger (see below). However, given the paucity of observations to date we have still to distinguish between various possible suggestions for the emission processes at play, including the role of X-ray power, the true nuclear yields and the balance between material ejected into the accretion disc or into tidal tails around the merger.

3.4 Gravitational Wave Detection

The mergers of compact objects were long expected to be the first observed gravitational wave signatures, and indeed NS–NS and NS–BH were long considered prime suspects for this.

The recent discovery of a binary black hole merger with a particularly high total mass ($>60\ M_\odot>60\ M_\odot$) is therefore rather surprising at first sight (Abbott et al. 2016b), although it should be noted that the significantly higher masses here than normally considered result in a far larger horizon for massive BH–BH than for NS–NS systems (the measured strain scales approximately as M1M2M1M2), and so a significant astrophysical population of BH–BH binaries may make them preferentially detected (Abbott et al. 2016a,c). However, at the strain sensitivities now reached by the next generation gravitational wave detectors, and given the inferred rates of NS–NS and NS–BH mergers from both population synthesis and observed populations (see below and Abadie et al. (2010)) it is expected that mergers containing neutron stars could be found shortly.

The simultaneous detection of GW and GRB signals offers significant advantages. The GRB signal provides a precise time and location for a GW search. This in turn dramatically reduces the number of trials that must be run on the gravitational wave interferometer data stream, and means that the effective sensitivity increases substantially, perhaps by a factor ~2~2 or more (e.g. Dietz et al. 2013). However, these advantages also extend to the study of short GRB progenitors. For example, the chirp mass determined by the GW detection would immediately identify a system as either a NS–NS or NS–BH merger. Indeed, the combination of GRB, KNe, and GW detection provides a series of different routes to probing the compact binary population over very different distance scales (see Fig. 6). In particular, SGRBs can be detected out to z~1z~1 or beyond (e.g. Graham et al. 2009; Thöne et al. 2011b), but likely only illuminate a small fraction of the sky due to their relativistic jets.

This makes the probability of joint GRB–GW triggers relatively small, as the fraction of mergers in a volume limited sample where the GRB jet is aligned with the observer is low. Beaming is highly uncertain in short-GRBs, but beaming factors >10--100>10--100 seem likely (e.g. Chen and Holz 2013). This is somewhat offset since the relativistic jets are visible to observers face on to the merging binary, a geometry which also maximises the detectability of the gravitational waves (see Fig. 6 and Nissanke et al. 2011).

The most promising counterparts are therefore systems that may emit isotropically. Kilonova are visible to sensitive IR searches over distances of several hundred Mpc, well matched to the sensitivities of current gravitational wave detectors. Similarly, isotropic X-rays with modest luminosity (LX~1043 ergs−1LX~1043 ergs−1) should also be visible to *Swift* or other X-ray telescopes out to distances >100 Mpc. It is likely that both gravitational wave detections, and observations from the γγ-ray to IR, and possibly beyond (for example magnetars should provide long-lived radio emission, e.g. Gompertz et al. 2015; Fong et al. 2016) will be necessary to fully constrain compact object mergers, their rate, and their role in r-process production.

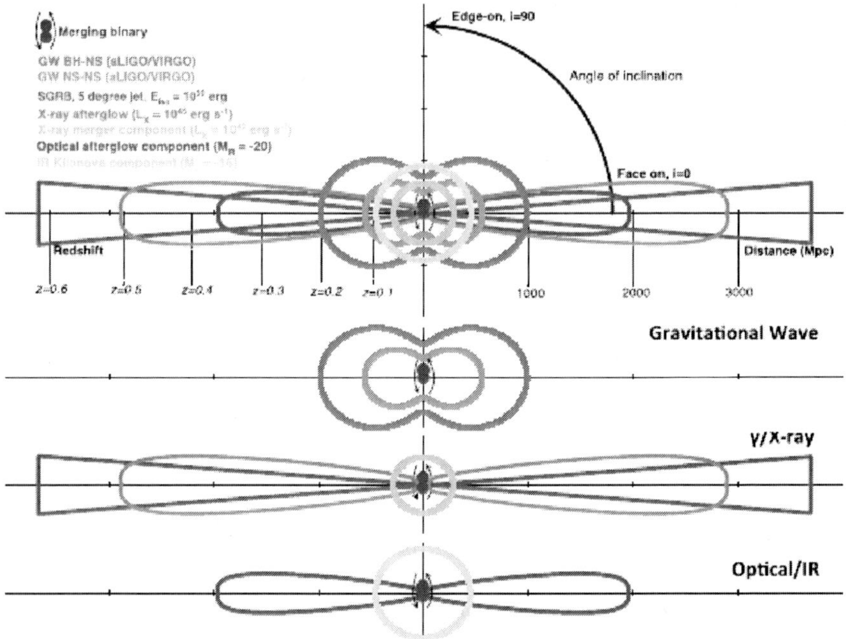

Figure 6. The characteristic visibility horizons for the detection of different observational manifestations of compact object mergers. The *top panel* shows all the likely signatures overplotted for comparison, while they are broken down by detection route in the *lower panels*. Mergers can be detected by the Advanced generation of LIGO and VIRGO detectors out to hundreds of Mpc for both NS–NS and NS–BH mergers (Nissanke et al. 2011), although there is a strong bias towards face-on events, which can be observed to distances factors of >2>2 larger than for edge on systems. This increases the probability of observing them in coincidence with short GRBs, which are also likely to be visible only to face on observers. However, such overlapping events will likely remain in a minority. Other wavelengths of emission are likely to require triggered observations after the detection of a gravitational wave or short-GRB trigger. The horizons given are based on typical depths likely to be achieved fairly short exposures, several hours–days after the trigger with the *Swift* BAT (Sγ=1×10−7 ergcm−2Sγ=1×10−7 ergcm−2, although the BAT in practice triggers on short GRBs based on their peak flux), *Swift* XRT (limiting FX∼1×10−12 ergs−1cm−2FX∼1×10−12 ergs−1cm−2 in ∼1000 s∼1000 s), ground based robotic optical telescopes (R=21R=21) and the VISTA survey telescope in the IR (J(AB)∼22J(AB)∼22). These horizons can obviously be scaled to different depths or time after trigger for a given assumption of temporal behaviour. In particular, observations of short-GRBs require much smaller areal coverage and so probe much deeper and farther. For example, *HST* can see kilonova signatures out to z∼0.4z∼0.4.

Furthermore, uncertainty as to the detailed properties of the emission means that any such horizons should be viewed as indicative rather than precise. In particular, we properties of off-axis GRB emission which enables bursts to be viewed at larger angles than a direct GRB remains to be directly tested, and the properties of isotropic optical/IR (Barnes and Kasen 2013) or X-ray kilonova/macronova (Kisaka et al. 2015; Sun et al. 2016) that might be powered either directly by radioactivity, or by the action of a central engine such as a magnetar, remain uncertain. For SGRBs beamed signals have been the dominant source of information to date, however, it seems likely that isotropic signatures will be more promising in the majority of gravitational wave transients

4 CENTRAL ENGINES: LINKING THE MOST LUMINOUS EXPLOSIONS

While progress towards the nature of the progenitor stars of long GRBs has been impressive there has been increasing focus on the nature of the compact object formed in the core of the collapsing star. These objects are the engines which drive the explosions and create the ultra-relativistic jets which ultimately pierce the star. Traditional models extract the GRB energy through some form of accretion onto a nascent black hole (Woosley 1993; Fryer et al. 1999). Accretion rates of up to several solar masses per second at typical efficiencies (~10 %~10 %) can create the necessary luminosities to explain both long and short GRBs (e.g. Oechslin and Janka 2006), provided that a sufficiently low density of baryons exists, which seems plausible along the rotation axis of rapidly rotating stars. Recently there has been a significant shift of focus to consider energy input from a millisecond-magnetar, a newly formed highly magnetic neutron star with B>1014 GB>1014 G, and spin periods of a few milliseconds (Metzger et al. 2011). Note that these, newly formed magnetars are quite different from the population of magnetars identified in the Milky Way as Soft Gamma Repeaters (SGRs) and Anomalous X-ray Pulsars. The Galactic systems have typically lower fields, and importantly much slower rotation rates (several seconds) than those powering GRBs (see e.g., Olausen and Kaspi 2014). Indeed, even if the spin-down evolution of the Galactic magnetars is reversed, it seems unlikely that their early properties were consistent with those postulated as GRB central engines (Rea et al. 2015). The strong magnetic fields result in rapid extraction of energy from the dipole field, and the total energy extracted can approach 1052 erg1052 erg, comparable to the typical isotropic energy releases of GRBs, and significantly above the beaming corrected values. Indeed an apparent upper limit on the kinetic energy releases of GRB-supernovae of around 1052 erg1052 erg provides additional support for such a model (Mazzali et al. 2014). These millisecond-magnetars would eventually spin down to much longer periods, at which point they would either become dormant or, depending on their mass may collapse to black holes at the point at which they cease to be centrifugally supported.

In any case, it is clear that in the majority of GRBs this central engine inputs a large fraction of a solar rest mass on a timescale of tens to hundreds of seconds, perhaps extending to several thousand seconds in some cases. Indeed, evidence for prolonged energy injection existed before the launch of *Swift* (Vaughan et al. 2004; Watson et al. 2006), but was only clearly identified following the direct detection of X-ray flares in the afterglows bursts early in the *Swift* era (Burrows et al. 2005; Nousek et al. 2006), and was surprisingly found in both long bursts and short bursts (Berger et al. 2005). While in long bursts this late activity might naturally arise from processes involving fallback accretion, such emission was not naturally expected in the case of short GRBs should they arise from neutron star mergers, since the merger itself should be over very rapidly, with a relatively clean environment. Hence, if a common origin for the long and short GRBs was in accretion onto the nascent black hole we might expect the clean environment to provide a significant observational distinction between the two scenarios. While short GRB afterglows are markedly fainter, the presence of long-lived emission does not offer such easy solutions. This may be due to instabilities that build up in the accretion discs around both long and short GRBs and deliver similar flares (Perna et al. 2006a), or because some fraction of short GRBs arise from black hole-neutron star mergers, in which the neutron star is tidal shredded over several pericenter passages (e.g. Davies et al. 2005).

However, the millisecond-magnetar model may be particularly appealing here. Millisecond-magnetars are likely created from the collapse of some fraction of massive stars (see below) while they could also be formed via the merger of either white dwarfs or neutron stars, provided the final mass is below the maximum mass of a neutron star (Usov 1992; Levan et al. 2006). These neutron stars with extreme magnetic fields can have rapid spin-down times, or may even be unstable and centrifugally supported, such that as magnetic braking slows their rotation from ~1 ms~1 ms at birth they ultimately collapse to form black holes. Importantly, millisecond-magnetars provide a route to providing energy input into the GRB afterglow on timescales of minutes to hours (or potentially even longer) after the initial burst. While the models have numerous free parameters, millisecond-magnetar models can explain many long and short GRB light curves (e.g. Rowlinson et al. 2013). However, this injection of energy is not without its problems, since it should in many cases yield a strong radio afterglow. The absence of such afterglows may be difficult to remedy with rapidly spinning magnetar models (e.g. Fong et al. 2016).

Interestingly, central engine models have become increasingly popular in explaining not only GRBs but also the most luminous SNe. In this case, the energy extracted from the millisecond-magnetar re-energises the outgoing supernova shock and creates the additional luminosity, boosting the original luminosity by a factor of 100 or more (Kasen and Bildsten 2010). The difference between GRB magnetars and those postulated in the SLSNe lies in the duration of the energy input. Millisecond-magnetars driving SNe must do much of their energy injection at late times when the supernova is large, otherwise, the energy may do work on the ejecta (increasing the ejecta velocity) but not create luminosity. In contrast, in most GRBs, the spin-down times of the millisecond-magnetars are short, such that most of their work is done more quickly. Broadly speaking GRB millisecond magnetars require high fields (B~1016 GB~1016 G) and spin down on timescales of hundreds of seconds, those powering SLSNe have more modest fields (B~1014--15 GB~1014--15 G) but spin down scales of days to weeks (e.g. Metzger et al. 2015). The ultra-long GRBs (Levan et al. 2014a) offer an interesting intermediate population in which millisecond-magnetars may be active for hours to days, but not longer. In this case the detection of a luminous (although not super luminous) supernova, SN2011kl associated with GRB 111209A (Greiner et al. 2015a) is of particular interest in providing evidence of a direct link between the progenitors of GRBs and those of SLSNe.

While central engine models are now increasingly used to explain a variety of exotic explosions, it is important to note that the engines themselves are not the same. Indeed, both black hole accretion and the magnetic extraction of rotational energy are considered as routes. The stars that create either black holes or rapidly spinning neutron stars may, in fact, be quite different immediately prior to the collapse, and so distinguishing between the different models is important, not only in understanding the explosions but in characterising their progenitors. Such a task is not trivial since the central engine itself is hidden at the core of the explosion. While the difference between asymmetric supernovae and ringing down magnetars is potentially distinguishable via gravitational wave observations (see e.g., Fryer et al. 2001; Davies et al. 2002; Rowlinson et al. 2013), such work likely beyond the capability of the current generation of detectors, that can only see such signatures for very local (e.g. local group) supernovae. However, the rate of energy deposition and the total energy budget are different between the two models. For example, in the millisecond-magnetar case rotational energy is released following Erot˙=IΩΩ˙Erot˙=IΩΩ˙, where ErotErot is the rotational energy, II the moment of inertia and ΩΩ the spin frequency. In this case the total energy budget is Erot=12IΩ2Erot=12IΩ2 (Lorimer and

Kramer 2004). For a 1.4 M⊙1.4 M⊙ neutron star, spinning at a period of a millisecond (approximately the maximum spin rate for most neutron star equations of state) the total energy is of order Erot~1052 ergErot~1052 erg. Alternatively, in the black hole case, the late time accretion is likely to follow the fall-back rate of t−5/3t−5/3, and the total energy is governed simply by the total mass accreted Eacc=ηmaccc2Eacc=ηmaccc2 or E˙acc=ηm˙accc2E˙acc=ηm˙accc2. For an efficiency of 10 % and a massive star with ~10 M⊙~10 M⊙of material in-falling (either directly, or via fallback) the total energy is a much larger Eacc~2×1054 ergEacc~2×1054 erg. Although the emission geometry can make it difficult to measure the true total energy of a given explosion, it is apparent that the most energetic GRBs and supernovae appear to exceed the limit for neutron star energy (e.g. Metzger et al. 2015), potentially posing a challenge to such models. However, there are some possibilities, such as very massive neutron stars (Metzger et al. 2015), that can provide a modest boost (factors of a few) to the total rotational energy, and so do enable such models to remain plausible, even at the high energy end of the distribution.

5 MASSIVE STAR PROGENITORS

5.1 The Role of Rotation

Observations make it clear that massive stars are now required in at least the vast majority, and probably all, long-duration GRBs. These massive stars must have somehow lost their hydrogen and helium envelope, create significant quantities of nickel, and have locations in galaxies consistent with the youngest and most massive stars. However, it is likely that not all stars with these conditions will launch a GRB since it would require very narrow beaming angles in GRBs for the GRB rate to match the massive star rate e.g. 1 degree for progenitors with MZAMS>40 M⊙MZAMS>40 M⊙ (Podsiadlowski et al. 2004). Indeed, the creation of a GRB is likely to require the specific conditions that can give rise to a central engine to power the burst. In the black hole engine model a centrifugally supported disc is formed outside a newly created black hole, the GRB jet can then be launched either through electrodynamic processes, or via neutrino—antineutrino annihilation off the disc (note that the T2T2 dependence of the weak interaction cross section means that at the extreme temperatures of these discs the cross section for neutrino interactions is significant). Alternatively, in millisecond-magnetar models, the magnetar must be created, and have sufficient field and spin to energise the explosion. In both of these scenarios a crucial factor in the creation of the GRB itself arises from *rotation*.

In particular, the critical rotation is that of the core immediately prior to core collapse. In order to create a centrifugally supported accretion disc at the innermost stable orbit of a black hole a minimum specific angular momentum (jj) is required (i.e. jj is the angular momentum (LL) per unit mass (MM), so j=L/Mj=L/M);

$$j > \frac{\sqrt{6}GM}{c}. \tag{4}$$

For typical core parameters this corresponds to j>1016 cm2s−1j>1016 cm2s−1. A newly formed millisecond-magnetar (radius ~10--20 km~10--20 km, and spin period 1 ms) has a rather similar specific angular momentum, and so the rotation properties of the core prior to core collapse are likely the same for the millisecond-magnetar or black hole scenarios.

In principle, such angular momentum should be easy to attain. As the core grows (and increases in density) during the main sequence its rotation should increase, potentially even to the point of breakup where the centrifugal force at the equator is equal to the gravitational force. However, in practice, angular momentum is effectively transported *outwards* in stars, both from the core to the envelope and subsequently from the envelope into the interstellar medium. The latter stage of this process is well understood, since mass loss from the star (at its surface) carries away angular momentum, resulting in a star with lower specific angular momentum. This process is strongly metallicity dependent, with the mass loss rate of iron group elements scaling broadly as metallicity, Z0.7--0.8Z0.7--0.8 (Vink et al. 2001; Vink and de Koter 2005). Angular momentum transport within the star is less well understood, although it seems likely that magnetic torques would be created by differential rotation of the core relative to the outer layers, and that these would ultimately create a rotation of the core that was tied to that of the envelope (Spruit 2002). Indeed, while such a model is far from universally accepted, its use does provide a reasonable match to the spins observed in neutron stars (Heger et al. 2005; Suijs et al. 2008). Since this mechanism couples the rotation of the core to that of the envelope, it therefore follows that the core is effectively braked by radiatively driven mass loss (Langer 1998), and so even at modest metallicity (e.g. somewhat less than solar) most massive stars will fall short of the critical specific angular momentum by an order of magnitude or more. Indeed, the apparent lack of angular momentum in most stars creates a signicant problem in understanding GRB creation. GRB-SNe are exclusively hydrogen poor events requiring no hydrogen envelope. However, standard pictures to create stars of this type remove the envelope via winds or binary interactions, through which significant angular momentum is also lost. At first sight, then the requirements of hydrogen deficiency and rapid rotation cannot both easily be met. However, there are solutions that may provide the necessary conditions for GRB creation.

5.2 Single Star Scenarios

We can consider solutions that involve both single and binary stars. In this sense we take single stars to mean stars that have either lived their entire lives (from zero age main sequence onwards) as single stars, or those in wide binaries in which there is no significant interaction of the two components.[7] "Normal" massive single stars evolve to have a strong central concentration, in which material in the core is used as the fuel for fusion of progressively heavier elements. On the main sequence, this core gradually builds from hydrogen to helium, and the outer layers of the star remain (chemically) as they were at the beginning of the main sequence. However, in rapidly rotating stars the mixing timescale can be shorter than the nuclear timescale, such that material synthesised in the core of the star is broadly mixed throughout the star, creating a star which undergoes so-called *chemically homogeneous evolution* (e.g. Yoon and Langer 2005; Woosley and Heger 2006; Yoon et al. 2012). This is possible because rotation creates hydrodynamic instabilities at the boundary between the convective core (where hydrogen fusion occurs) and the radiative envelope of the star. Of particular importance is the Eddington–Sweet circulation (Eddington 1926; Sweet 1950) which is driven by a thermal imbalance in rotating stars and may result in a short timescale of chemical mixing with a sufficiently high rotation rate. Because these stars evolve homogeneously they do not create the standard core-envelope structure, do not undergo giant branch phases, and do not experience the core-envelope breaking that dramatically slows rotation in the cores of slowly rotating massive stars. However, this phase must be initiated early in the life of the star before a significant chemical gradient at the boundary between the hydrogen-burning convective core and the radiative envelope is built up, as otherwise, rotationally-induced mixing becomes too inefficient to make the star undergo the chemically homogeneous evolution. This process of chemically homogeneous evolution further removes the hydrogen from the star (since it is burned) while not resulting in such significant mass loss, enabling relatively massive cores to be built from more modest initial masses (Yoon and Langer 2005).

However, chemically homogeneous evolution is not common. There is little evidence for it in massive stars in the Milky Way, although observations of massive stars in the Magellanic Clouds do show some that match the expectations of chemically homogeneous evolution (Martins et al. 2009). This suggests that there may well be some metallicity dependence on stars which undergo such an evolutionary pathway. Indeed, angular momentum loss due to stellar winds is important even early in the lives of stars, so that stars have a certain threshold rotational velocity which is a function of both their stellar mass and chemical composition.

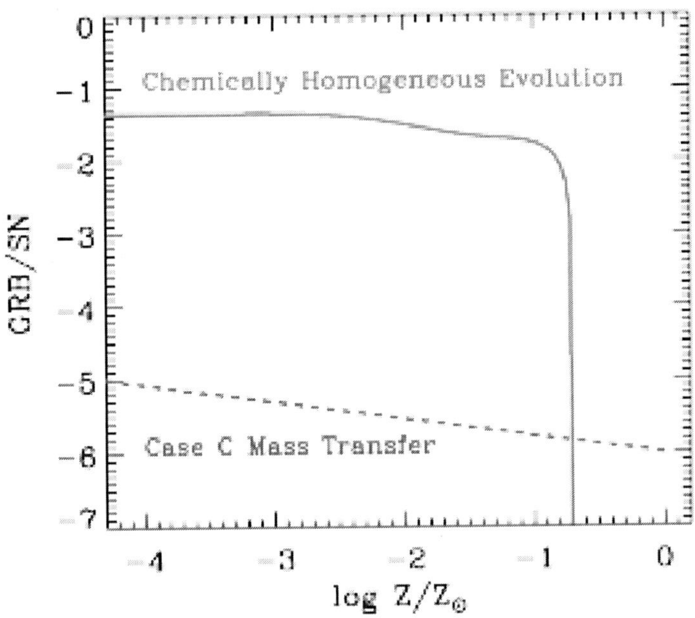

Figure 7. The expected relative rates for the creation of rapidly rotating massive stellar cores through late binary mass transfer after the main sequence (so called case C, *blue line*Podsiadlowski et al. 2010), and through chemically homogeneous evolution (*red line*Yoon et al. 2006). Both are plotted as a function of the metallicity of the star. It is clear that at low metallicities the effectively single star rate is much larger than the Case C mass transfer rate, and so will dominate over it as a route to GRB creation. In practice massive rapidly rotating stellar cores can be created through routes other than case C mass transfer (mergers etc.), but none-the-less it appears that a significant contribution from homogeneously evolving stars is likely at low metallicities. Any high metallicity events would necessarily arise through binary routes

The overall impact of this is that single stars can only evolve homogeneously, and hence create GRBs if they are of low metallicity. Indeed, the predicted rates of GRBs from chemically homogeneous stars drop rapidly at metallicity around 0.2 Z⊙0.2 Z⊙ (see Fig. 7). Interestingly, this is very similar to the critical metallicity of 0.3 Z⊙0.3 Z⊙ inferred from detailed observations of some host galaxies by Graham and Fruchter (2015a), but somewhat lower than that inferred from larger samples (typically at higher redshift) either via spectroscopy (Krühler et al. 2015) or photometric proxies (Perley et al. 2015). However, at these low metallicities, the contribution of these effectively single stars could be very important.

5.3 Binary Star Scenarios

Binary stars also provide a natural route of retaining or gaining angular momentum. In particular, stars may be spun up either via the accretion of material with high specific angular momentum, via tidal locking in tight binaries, or possibly even by direct mergers. It is pertinent to understand binary *population* properties in general, since these populations, in turn, provide direct constraints on the rate of much rarer events, such as those giving rise to both long and short GRBs. It is now well-established, at least in the general field of our Milky Way galaxy and in low-density star-forming regions, that of the most massive stars at least 60 %–80 % occur in main sequence–main sequence binary systems (e.g Mason et al. 1998; Kouwenhoven et al. 2005; Raghavan et al. 2010; Sana et al. 2010, 2011, 2012; Kobulnicky et al. 2014).

The highest binary fractions involving massive, and hence young, stars are found in populous compact star clusters (e.g. Elson et al. 1998; Hu et al. 2010, and references therein), interestingly, the places where the youngest most massive stars are likely to reside. This is intuitively exemplified by the much higher fraction of low- and high-mass X-ray binaries found in both globular clusters and actively star-forming regions (e.g. Coleiro and Chaty 2013) than among the general field star population. Indeed, interactions in the dense cores of these clusters tend to leave the most massive stars in binaries through exchange interactions. These simultaneously harden tight binaries, at the expense of widening, and potentially unbinding the wider binaries. This so-called hard/soft divide means that tight (hard) binaries get tighter, with their individual components more likely to interact and wider (soft) binaries become progressively wider, and less likely to interact (e.g. Heggie 1975).

Given the apparent preference that GRBs have for low(er) metallicity environments, studies in the Magellanic Clouds may be particularly insightful. For example, studies of the population in the 15–30 Myr-old compact Large Magellanic Cloud (LMC) star clusters NGC 1818 (Hu et al. 2010; de Grijs et al. 2013; Li et al. 2013), show that for binaries with an F-star primary and mass ratios, $q\equiv m2/m1>0.4q\equiv m2/m1>0.4$ (where m1m1 and m2m2 are the masses of the primary and secondary components, respectively), the cluster's binary fraction is $\sim0.35\sim0.35$. This suggests a total binary fraction for F stars of 0.55 to unity, depending on assumptions about the form of the mass-ratio distribution at low qq. A similarly high binary fraction (covering the clusters' full observable mass ranges) is obtained for the equivalently young, compact cluster NGC 1805 (Li et al. 2014), as well as for the intermediate-age ($\sim$ a few $\times10^8\times10^8$ yr-old), massive LMC clusters NGC 1831 and NGC 1868 (Li et al. 2014).

However, these binary fractions represent the entire population of binary systems. In a surprising development de Grijs et al. (2013), noticed that the binary fraction in the core region of NGC 1818 was significantly lower than that in the cluster's outer regions and even in the surrounding field. It seems likely that this is the effect of the dynamical disruption of soft binaries on short timescales in the dense inner core of the cluster where interaction times are low. Indeed, subsequent NN-body simulations confirm this scenario (Geller et al. 2013).

For a range of initial conditions, from smooth virialized density distributions to highly substructured and collapsing configurations it is possible to explain the structures of the cluster, although with a slight preference for structured initial conditions. These models produce the observed radial trend in binary frequency through disruption of soft binaries (with semi-major axes, a>3000 AUa>3000 AU), on approximately a crossing time, preferentially in the cluster core. Mass segregation subsequently causes the remaining binaries to sink towards the core (Geller et al. 2015). Thus, both a radial binary fraction distribution that falls towards the core (as observed for NGC 1818) and one that rises towards the core (as for dynamically older star clusters) can arise naturally from the same evolutionary sequence owing to binary disruption and mass segregation in rich star clusters. Indeed, the radial distribution of the binary fraction in another very young LMC cluster NGC 1805, showed an enhanced binary fraction in its core. This cluster is dynamically much older than NGC 1818, however, so that the effects of dynamical mass segregation had already become dominant (Geller et al. 2015). Such complex evolution within clusters means that studies of locations that attempt to derive the binarity of GRB progenitors (e.g. due to kicks to the binaries during a previous supernova explosion) are challenging since their initial locations might be widely varied.

Once binaries are formed, their evolution is dictated both by the evolution of the individual components, but crucially, also by the interaction between the two stars, which can have a marked impact on their evolutionary pathways. The introduction of binary channels greatly increases the diversity of stellar evolution pathways and their associated remnants. Indeed, it is visible on galaxy-wide scales where, for example, the rejuvenation of older stars via binary interactions can make them look much younger than they (or the bulk of the stellar population) really are (Eldridge and Stanway 2009; Stanway et al. 2016).

Because of the increased diversity of stellar products that can be created through binary evolution (e.g. unusual composition, rotation, mass etc), binary channels are popular routes to create exotic (and rare) stellar transients. Through binary evolution it is straightforward to remove the stellar envelope (and provide a stripped envelope core collapse supernova), while tidal locking in tight binaries can in principle provide the necessary angular momentum in the core to enable disc or millisecond-magnetar formation. However, the impacts of binary evolution go beyond the straightforward changes on the outer layers of the star due to the impact of centrifugal force and rotationally-induced chemical mixing. Hence, detailed studies require the construction of complex models that can simultaneously model angular momentum transport within individual stars and the binary components, as well as tracking mass loss and nuclear burning. Such calculations have been attempted by several teams (e.g. Yoon et al. 2010).

As with single stars, the crucial parameter to track through binary routes is the specific angular momentum of the stellar core immediately prior to core collapse. Stars can be spun up via mass transfer from a companion star, but are spun down if they themselves lose mass. This creates similar problems to those seen in massive single stars, in which it is very difficult to both lose the hydrogen envelope and retain both sufficient mass and rotation to obtain the necessary conditions for GRB creation. Several authors have considered late time mass transfer during a late stage of core helium burning or after core helium exhaustion (so-called Case C mass transfer) as a possible route to obtaining GRB-like conditions. This late time mass transfer can enable the formation of a massive stellar core, even at high metallicity (Brown and Lee 2004). Alternatively, depending on the properties of the binary this mass transfer may result in the formation of a common envelope, which will unbind the envelope of the star at the cost of the orbital energy (and hence separation). The resulting binary (after removal of the envelope) is a tight binary containing a carbon-oxygen core. Tidal locking within this binary may match the rotation period of the core to the period of the binary (typically a few hours) in which case the requirements on specific angular momentum can be met (e.g. Lee et al. 2002). Indeed, tidal locking through a variety of routes including neutron stars, post main sequence stars and low mass main sequence stars has been considered as a route of creating GRBs (Izzard et al. 2004; van den Heuvel and Yoon 2007), although detailed calculations disfavour this scenario (Detmers et al. 2008). Alternatively, evolution in binaries with relatively extreme mass ratios may also create the necessary conditions. Here mass transfer occurs from a massive star evolving away from the main sequence, and onto a low mass star on the main sequence. In this scenario, a further mass transfer from the low mass companion can occur on the CO core of the primary star. This, in turn, le to both the spin up of the core, and the explosive ejection of the common envelope (Podsiadlowski et al. 2010).

Finally, a route to obtaining the necessary angular momentum may be the direct merger of two stars. Such mergers do happen, for example, common envelope mergers have been used to explain unusual stellar outbursts of lower mass stars, such as V838 Mon, V1309 Sco or R136a1 (e.g. Tylenda and Soker 2006; Tylenda et al. 2011; Banerjee et al. 2012). More massive mergers may create the elusive Thorne–Zytkov objects. One of these has recently been claimed in the SMC (Levesque et al. 2014), although this is controversial, in particular, it seems unclear if it actually resides in the SMC (Worley et al. 2016), or is a foreground source (Maccarone and de Mink 2016).

Indeed, at higher masses the merger of either two Helium cores, or perhaps a black hole (or neutron star) with a Helium core has been proposed as a GRB mechanism, with the latter suggested as a possible origin for the Christmas-day burst, GRB 101225A (e.g. Thöne et al. 2011a). In these mergers the orbital angular momentum eventually is combined within the single merged object. In the case of He-star—He-star mergers this dramatically increases the total mass, and there is little time between the merger and the supernova. For black hole–He star mergers, the black hole accretion can essentially create the GRB immediately.

It is clear that there are a wide variety of binary channels that can give rise to the necessary conditions for GRB production. All of these scenarios are likely to occur in nature.

However, it is less clear if the most exotic events can create GRBs at the necessary rate to explain the observed populations of GRBs. In each of these scenarios, there is a limited parameter space over which the channel will work (in terms of mass, mass ratio, and initial separation), and so a given route is likely to provide only a modest rate of GRB-like events (see Fig. 7). Hence, while many stars are in binaries, the rates of creation of rapidly rotating massive cores through any one of these routes is significantly lower than the rate obtained through chemically homogeneous evolution for any star formation occurring at substantially sub-solar metallicity.

5.4 Predicted Long GRB Rates

Determining the rates of GRBs, and of their various progenitor channels is fraught with difficulty. Observationally, GRB detection is a sensitive function of both the properties of the detector (area, energy range etc.) and of the burst (hardness, peak flux, total fluence etc.). Furthermore, GRBs exhibit a broad range of luminosities, and so an extrapolation to the total rate of GRB-like events requires a significant correction factor. Indeed, it is apparent that there is a population of local, low-luminosity GRBs whose volumetric rate exceeds those of the more distant bursts by a factor of several hundred. If these are a separate population, or the faint end of a luminosity function remains uncertain (e.g. Chapman et al. 2007; Liang et al. 2007b), although the similarities in the supernovae in both the lowest and highest luminosity examples (e.g. Galama et al. 2000; Pian et al. 2006b; Xu et al. 2013; Levan et al. 2014b) suggests that similar physical mechanisms are at play. Finally, GRBs are relativistically beamed, and illuminate only a small fraction of the sky (Frail et al. 2001; Bloom et al. 2003). The beaming fraction for a given burst can be obtained from the so-called jet break when the relativistic jet expands laterally (Frail et al. 2001), but is a challenging and still highly uncertain parameter that significantly impacts the observed rates.

Similarly, when determining the likely rates of GRB production via the various routes considered above there are major uncertainties that enter. What is the relevant metallicity distribution for stars throughout the Universe, and how does this impact those that will undergo chemically homogeneous evolution (Langer and Norman 2006)? Are there any special environments in which the apparently universal top end of the initial mass function can become more top heavy (Bastian et al. 2010)? What is the binary fraction, and the range of initial separations? How is the mass ratio of binaries distributed, are these universal, and might this have any impact on the final products of stellar evolution (Li et al. 2013)? What is the efficiency of common envelope evolution (e.g. Nelemans and Tout 2005)? While these parameters can be studied from detailed simulations (or observations) (e.g. Yoon et al. 2006), or explored via rapid population synthesis (e.g. Izzard et al. 2004) there are still order of magnitude uncertainties in the true rate. Podsiadlowski et al. (2004) attempted to compare the rates of GRBs and massive stars by directly comparing the rates of GRBs (for some assumptions about beaming angles) with the rates of stars of a given mass within a typical galaxy. In Table 1 we provide an updated version of this table, including estimates of the low- and high- luminosity GRBs separately, as well as different types of SNe and massive stars. We also provide the rates of massive stars for a typical (i.e. Milky Way-like) galaxy, as well as the rates of massive stars below some threshold metallicity in the local Universe (extrapolated from Graham and Schady 2015b). We note that these scaled values assume a constant factor of 107107 between the volume averaged rate and the galactic rate, and so does not take into account chemical differences between galaxies of different masses.

Table 1. Approximate rates of various engine driven transients, and of massive stars at all metallicities, and below a set metallicity.

Object	Rate (galaxy^{-1} yr^{-1})	Rate (Gpc^{-3} yr^{-1})	References
Transients			
LGRB	8×10^{-8}	0.8	Sun et al. (2015)
−1°	6.6×10^{-4}	6600	
−5°	2.6×10^{-5}	260	
−20°	1.6×10^{-6}	16	
LLGRB	1.6×10^{-5}	160	Sun et al. (2015)
−1°	0.11	1.1×10^{6}	
−5°	4.2×10^{-3}	42 000	
−20°	2.6×10^{-4}	2600	
SGRB	2.0×10^{-7}	2.0	Sun et al. (2015)
−1°	1.3×10^{-3}	13 000	
−5°	5.3×10^{-5}	530	
−20°	3.0×10^{-6}	30	
ULGRB	1×10^{-8}	0.1	Gendre et al. (2013), Prajs et al. (2016)
−1°	6.6×10^{-5}	660	
−5°	2.5×10^{-6}	25	
−20°	2.0×10^{-7}	2.0	
rTDE	3.0×10^{-9}	0.03	Sun et al. (2015)
SLSNe*	3.0×10^{-6}	30	Quimby et al. (2013), Prajs et al. (2016)
Massive stars			
20 $M_\odot$ all Z	2×10^{-3}	20 000	Podsiadlowski et al. (2004)
20 $M_\odot$ $Z < 1/4\ Z_\odot$	2×10^{-4}	2000	Graham and Schady (2015b)
40 $M_\odot$ all Z	6×10^{-4}	6000	
40 $M_\odot$ $Z < 1/4\ Z_\odot$	6×10^{-5}	600	
80 $M_\odot$ all Z	2×10^{-4}	2000	
80 $M_\odot$ $Z < 1/4\ Z_\odot$	2×10^{-5}	200	
Compact binaries			
NS–NS	6×10^{-5}	600	Abadie et al. (2010)
NS–BH	2×10^{-7}	2	
BH–BH	5×10^{-7}	5	Abbott et al. (2016a)

Various attempts have been made to determine these rates, and the numbers given are rounded and approximate, rather than representing the full range of possibilities (which in some cases are quite large). A typical galaxy is assumed to have a B-band luminosity of 1010 L⊙1010 L⊙, and we assume a local B-band luminosity density of

108 Mpc−3108 Mpc−3 (e.g. Calura and Matteucci 2003), although clearly, this value evolves significantly over cosmic history. We note that the numbers per galaxy have been directly scaled from the volume averaged values with a fixed scaling of 107107. Since stars of a given chemical makeup are not the same "per galaxy" due to the mass-metallicity relation (Tremonti et al. 2004), the relative numbers of low metallicity stars should be viewed with caution since a given galaxy might contain stars entirely of low metallicity (if it was of low mass), or very few stars of low metallicity (if it was of high mass), i.e. if low metallicity is a requirement for GRB production then the rate per galaxy at low metallicity is boosted in low mass/metallicity galaxies, and depressed in high mass/metallicity galaxies. The massive star corrections to low metallicity (formally 12+log(O/H)<8.412+log(O/H)<8.4 following Graham and Schady (2015b)) are at z~0z~0 where 10 % of the star formation lies below this metallicity. This increases rapidly, and the correction likely becomes less than a factor 2 by z~2z~2.

The rate of classical, high luminosity long GRBs, for a beaming angle of a few degrees is remarkably similar to the rate of low metallicity >40 M⊙>40 M⊙ stars observed in the local Universe. Since both low metallicity (Graham and Fruchter 2015a; Perley et al. 2015) and high mass (Larsson et al. 2007; Raskin et al. 2008) have been observationally linked to GRBs this may suggest that a reasonable fraction of initially very massive stars create long GRBs. Although such interpretation is challenging because many high-luminosity long GRBs are at much higher redshift, where star formation rate densities are higher. None-the-less, this suggests that the rate of production of long GRBs is not much lower than the rate of production of massive (>40 M⊙>40 M⊙), low metallicity (<0.25 Z⊙<0.25 Z⊙) stars. In turn, this implies that they are probably not created only from stars which undergo very rare and unusual interactions. For example, for a typical 5-degree beaming angle, the LGRB rate is only a factor of 3 lower than the low metallicity massive star rate. While alternative (wider) beaming angles could lower the rate of GRBs by an order of magnitude or more, it still seems unlikely that the rarest channels, involving <1 %<1 % of stars are likely to be creating long GRBs. Indeed, if the low-luminosity GRBs are arising from a similar population, with similar progenitor masses and low metallicities then they would need to be born from a significant fraction of the very massive stars, even in the case of no beaming. Even with no beaming correction, the rate per galaxy of low-luminosity GRBs is ~2×10−5 galaxy−1yr−1~2×10−5 galaxy−1yr−1, again only a factor of a few lower than the low metallicity 40 M⊙40 M⊙stars. Indeed, since these low-luminosity GRBs are at low redshift there is less concern about evolution over cosmic history, and it seems likely that the low-luminosity GRBs are either lower-mass stars or nearly isotropic in emission in order to avoid low luminosity GRB rates that approach, or even exceed the massive star formation rate. It is also relevant to compare the rates to those of other transients, for example, the newly uncovered population of ULGRBs (Levan et al. 2014a) that may be linked to the SLSNe (Greiner et al. 2015a). The beaming corrections in this case are particularly uncertain, but for plausible beaming rates, comparable to those of long GRBs the inferred rate of ULGRBs is lower, of the same order of magnitude as the SLSNe rate. However, given the substantial uncertainties in both rate calculations, it is possible that this similarity is purely coincidental.

6 CONSTRAINTS FROM LOCAL STELLAR POPULATIONS

6.1 O Stars

It is apparent that long GRBs arise primarily from a subset of massive (MZAMS≥40 M⊙MZAMS≥40 M⊙), moderately metal-poor stars at cosmological redshifts. Although the nearest long GRBs lie at distances of tens to hundreds of megaparsec, we can study individual massive stars within appropriate environments in the Local Group.

The overwhelming majority of stars are believed to have their origins in dense star clusters, intermediate density OB associations or low-density star forming regions. If stars are randomly drawn from a Kroupa (2001) Initial Mass Function (IMF), cluster masses over ~100 M⊙~100 M⊙ are required to produce one star capable of ending its life as a core-collapse supernova. However, the most massive stars are preferentially found in the more massive clusters. In particular, stochastic sampling of the initial mass function means that the probability of a low mass cluster forming a very massive star is low (i.e. the massive star content of say 10, 100 M⊙100 M⊙ associations is less than that of a single 1000 M⊙1000 M⊙ cluster). Indeed, it has been suggested that the maximum mass of a star in a cluster (M∗,maxM∗,max) is related to the cluster total mass (MC)(MC), such that M∗,max≈0.39M2/3CM∗,max≈0.39MC2/3 (Weidner et al. 2010). In this case, for a star forming region to include at least one star with an initial mass of at least ~40 M⊙~40 M⊙, a total mass 103 M⊙103 M⊙ is required, with 10 such stars hosted by 104 M⊙104 M⊙ clusters.

The upper limit to star cluster mass is a sensitive function of how vigorously stars are being formed, so one would not expect long GRBs to occur in galaxies with low specific star formation rates. Therefore if progenitors of long GRBs arise from very massive stars, they will exclusively occur in galaxies with high specific star formation rates. Indeed, Kelly and Kirshner (2012) conclude that low zz broad-lined (BL) SNe Ic arise exclusively from hosts with the highest star-forming intensities, with Kelly et al. (2008) having earlier established that long GRBs and low zzSN Ic-BL originate from similar environments.

Since long GRBs strongly favour host galaxies with metallicities below 1/21/2 solar, the focus of our attention in the Local Group will be primarily the Large and Small Magellanic Clouds, with 1/21/2 and 1/51/5 solar metallicity, respectively. The most massive young star-forming region in the SMC, N66/NGC 346, hosts several tens of O stars, whereas many hundreds of O stars are known in the Tarantula Nebula region of the LMC.

The VLT FLAMES Tarantula Survey (VFTS Evans et al. 2011) has enabled properties of 800 OB stars in this region to be determined, revealing that 50 % of O stars are affected by binary evolution (Sana et al. 2013), with relatively low rotation rates, aside from a few very fast rotators presumably arising from binary interactions (Ramírez-Agudelo et al. 2015). Several star clusters are located within the Tarantula Nebula, most noticeably R136, the youngest (1.5 Myr) very high-mass cluster within the Local Group. Crowther et al. (2016) exploit HST/STIS spectroscopy to reveal that over three dozen stars more massive than 40 M$\odot$40 M$\odot$ are located within the central parsec.

Nearly 30 very massive stars (VMS, ≥100 M$\odot$≥100 M$\odot$) are located in the Tarantula Nebula, the majority within the R136 region, but others up to 100 parsecs away (Crowther et al. 2016). Some of the more remote VMS are plausible runaways from R136, but others likely were born in significantly low-density regions. In contrast, no VMS are located in SMC's N66 star-forming region, although it hosts a triple high mass system HD 5980.

In summary, large numbers of candidate long GRB progenitors exist in the LMC/SMC as far as their initial masses are concerned, yet there is no evidence that these are born with sufficiently high rotational rates for chemically homogeneous evolution. Rapid rotation (vsini>500 kms−1vsini>500 kms−1) is observed for small numbers of O stars in the Tarantula Nebula, presumably spun-up via binary evolution (de Mink et al. 2013). Similarly, no examples of rapid rotators are observed in N66 (Mokiem et al. 2006). Lamb et al. (2016) identify a large population of SMC field Oe stars, presumably arising from rapid rotation, although these too may result from close binary evolution.

This in itself raises interesting questions, since if a significant population of sufficiently massive and low metallicity stars can be identified in the local Universe, but essentially none of these appear as viable GRB progenitors then the fraction of massive low metallicity stars creating GRBs must be small. However, this begins to create tension with the necessary observed rates, especially when lower luminosity GRBs are considered (see Sect. 5.4).

6.2 Wolf–Rayet Stars

The immediate progenitors of long GRBs are massive stars whose hydrogen-rich envelopes have been stripped away to reveal compact cores, i.e. classical Wolf–Rayet (WR) stars (Crowther 2007). These stars possess dense, fast outflows with atmospheric compositions characteristic of hydrogen-burning (WN-type) or helium-burning (WC-type). In common with their O type progenitors, the strength of WR winds scales with metallicity (Vink and de Koter 2005), so metal-poor WR stars possess, lower density, slower outflows than their Milky Way counterparts (Hainich et al. 2015b).

The observed association between long GRBs and SN Ic-BL suggests the presence of very little helium in the progenitor star, favouring WC stars. The Magellanic Clouds host approximately 150 WR stars, of which only 15 % belong to the carbon sequence or rare oxygen sequence (WO). The binary frequency of Magellanic Cloud WR stars is approximately 40 %, similar to that observed in the Milky Way, with a lower binary fraction amongst WC and WO stars (Bartzakos et al. 2001).

Analysis of single WC and WO stars in the LMC (Crowther et al. 2002; Tramper et al. 2015) reveals He core masses of 10--20 M⊙10--20 M⊙. The sole SMC carbon/oxygen sequence WR star is in a close binary system, although Shenar et al. (2016) argue that the binary channel does not dominate the formation of WR stars at this metallicity.

The dense outflows from WR stars prevent direct rotational velocity measurements, but searches for non-spherical geometries can be done via linear spectropolarimetry. None of the four LMC WC stars observed by Vink (2007) exhibited the characteristic line depolarization with respect to the continuum, ruling out rapid rotation in these cases. Martins et al. (2009) have claimed that at least one of the hydrogen-rich WN stars in the SMC is the product of chemically homogeneous evolution. However, this single star might represent the product of spin-up arising from close binary evolution since such high initial rotation rates are not currently established amongst massive O stars in the SMC.

Finally, studies have been undertaken investigating the locations of WR stars relative to their host light, in principle directly comparable to the locations of transient phenomena (Leloudas et al. 2010; Bibby and Crowther 2012). These broadly show the locations of stripped envelope SNe match those of the WR population (Bibby and Crowther 2012), and that WC stars are more concentrated on their host light than WN system (Leloudas et al. 2010).

In summary, there are no robust long GRB progenitors amongst the Magellanic Cloud WR population at present, although at least one SMC WR star has been claimed to be the result of chemically homogeneous evolution, and their weak winds make measurements of their rotational rates via spectropolarimetry extremely challenging.

7 SUMMARY AND OPEN QUESTIONS

We are now approaching 20 years since the discovery of the first GRB afterglows, and to date over 1000 have been discovered. Intensive observations of the bursts themselves, their afterglows, associated supernovae and host galaxies have provided firm links to their progenitors in several cases, and in particular the link between long GRBs and type Ic supernovae appears secure. In turn, the link to young massive stars, coupled with their extreme luminosity makes GRBs a powerful probe of the distant Universe from the discovery of some of the most distant galaxies, to detailed work studying the buildup of stellar mass and metals across cosmic history. However, while this utility is clear, there remain many questions in addressing the nature of the progenitors themselves, and this, in turn, has a systematic impact on the cosmological utility of GRBs (e.g. in determining the correct factor to convert a GRB rate to a star formation rate at a given redshift/metallicity). In this review, we have highlighted much of the progress that has been made through a combination of intensive observation, large-scale numerical calculation and direct observation of local analogue populations. However, there are many questions that remain open, and will be the focus of research in the coming years, in particular;

- What fraction of massive O-stars are really required to create GRBs? What does this mean about routes to their progenitors?
- What is the true metallicity dependence for GRB production?
- How important is binary evolution in the creation of GRBs?
- Are any of the locally observed massive, rapidly rotating stars actually likely to create a GRB? Have any in the past that we can identify?
- What is the diversity of supernovae seen in long GRBs? Can they be standard candles? Is the presence of a luminous SN in one case a common or extremely rare occurrence?
- What is the range of kilonova/macronova properties seen in short GRBs? Can these create an important contribution to r-process enrichment, and explain the levels seen in the sea-floor on Earth?

- What is the role of NS–NS and NS–BH mergers in short GRB production? Can any BH–BH mergers make GRBs? What will simultaneous gravitational wave signatures tell us?

The answers to these questions are likely to come both from the application of the traditional techniques described above, with improved telescope aperture and response (e.g. JWST, E-ELT, GMRT etc.), or the every increasing computational power, and through new routes, such as direct multi-messenger observations (gravitational waves, neutrinos) that are now producing astrophysical detections. The answers to these questions are challenging, but it is likely that many will be uncovered with a degree of certainty in the coming decade.

FOOTNOTES

1 A time of launch of the Compton Gamma-Ray observatory (GCRO) and its BATSE instrument there were ~800~800 GRBs observed from a variety of missions, while approximately 110 models had been proposed. Interestingly, despite this progress, the now favoured collapsar model for long GRBs was not on that list.

2 The apparent distribution of GRB luminosities is an excellent example of Malmquist bias, where the brightest events are visible over a much larger volume. Hence, while the observed population is dominated by high luminosity events, the volumetric rates are dominated by much lower luminosity systems that generally escape detection.

3 The GRB/SNe pairs with some spectroscopic evidence include GRB 980425/SN 1998bw, GRB 021211/SN 2002lt, GRB 030329/SN2003dh, GRB031203/SN2003lw, GRB 060218/SN2006aj, GRB100316D/ SN2010bh, GRB 10129B/SN2010ma, GRB 111209A/SN 2011kl, GRB1204 22A/ SN2012bz, GRB 130427A/SN2013cq, GRB 130215A/ SN2013ez, GRB 130702A/SN 2013dx, GRB 140606B/iPTF14bfu.

4 Here we use hh to determine the luminosity scaling to avoid confusion with the k-correction, however in other works it is common to see the luminosity scaling expressed as kk.

5 The squared exponent here is approximate, and does not carry any specific meaning.

6 At 50 % probability. Since the standard assumption is to think of the duration distribution of long and short GRBs as Gaussians, there is inevitably overlap in the two populations.

7 Indeed, it has been suggested that these wide binaries are actually the best test-beds for single stars, since when observing a single star it is extremely difficult to tell if it has formerly been in a binary (Sana et al. 2012).

ACKNOWLEDGEMENTS

We thank the organisers and ISSI Beijing for a productive and well organised meeting in April 2015, and the referee for a very constructive and valuable report that has improved this manuscript. AJL acknowledges support from the Science and Technology Facilities Council under programme ST/L000733/1 and from the Philip Leverhulme Trust via a Leverhulme Prize. He is grateful for productive conversations with Nial Tanvir. RdG acknowledges financial support from the National Natural Science Foundation of China (grants 11373010, 11633005 and U1631102). He also acknowledges helpful insights into the evolution of binary systems provided by the participants of the 2014 conference on 'Binary systems, their evolution and environments' (particularly Philipp Podsiadlowski, Chris Belczyński and Dany Vanbeveren). SCY was supported by the Basic Science Research (2013R1A1A2061842) program through the National Research Foundation of Korea (NRF).

REFERENCES

1. J. Abadie, B.P. Abbott, R. Abbott, M. Abernathy, T. Accadia, F. Acernese, C. Adams, R. Adhikari, P. Ajith, B. Allen et al., Topical review: predictions for the rates of compact binary coalescences observable by ground-based gravitational-wave detectors. Class. Quantum Gravity **27**(17), 173001 (2010).
2. B.P. Abbott, R. Abbott, T.D. Abbott, M.R. Abernathy, F. Acernese, K. Ackley, C. Adams, T. Adams, P. Addesso, R.X. Adhikari et al., Astrophysical implications of the binary black hole merger GW150914. Astrophys. J. Lett. **818**, 22 (2016a).
3. B.P. Abbott, R. Abbott, T.D. Abbott, M.R. Abernathy, F. Acernese, K. Ackley, C. Adams, T. Adams, P. Addesso, R.X. Adhikari et al., Observation of gravitational waves from a binary black hole merger. Phys. Rev. Lett. **116**(6), 061102 (2016b).
4. B.P. Abbott, R. Abbott, T.D. Abbott, M.R. Abernathy, F. Acernese, K. Ackley, C. Adams, T. Adams, P. Addesso, R.X. Adhikari et al., The Rate of Binary Black Hole Mergers Inferred from Advanced LIGO Observations Surrounding GW150914. ArXiv e-prints (2016c)
5. L. Amati, F. Frontera, C. Guidorzi, Extremely energetic Fermi gamma-ray bursts obey spectral energy correlations. Astron. Astrophys. **508**, 173–180 (2009).
6. J.P. Anderson, S.M. Habergham, P.A. James, M. Hamuy, Progenitor mass constraints for core-collapse supernovae from correlations with host galaxy star formation. Mon. Not. R. Astron. Soc. **424**, 1372–1391 (2012).
7. C.R. Angus, A.J. Levan, D.A. Perley, N.R. Tanvir, J.D. Lyman, E.R. Stanway, A.S. Fruchter, A Hubble Space Telescope survey of the host galaxies of superluminous supernovae. Mon. Not. R. Astron. Soc. **458**, 84–104 (2016).

8. Z. Arzoumanian, D.F. Chernoff, J.M. Cordes, The velocity distribution of isolated radio pulsars. Astrophys. J. **568**, 289–301 (2002).
9. S. Banerjee, P. Kroupa, S. Oh, The emergence of super-canonical stars in R136-type starburst clusters. Mon. Not. R. Astron. Soc. **426**, 1416–1426 (2012).
10. J. Barnes, D. Kasen, Effect of a high opacity on the light curves of radioactively powered transients from compact object mergers. Astrophys. J. **775**, 18 (2013).
11. P. Bartzakos, A.F.J. Moffat, V.S. Niemela, Magellanic cloud WC/WO Wolf–Rayet stars—I. Binary frequency and Roche lobe overflow formation. Mon. Not. R. Astron. Soc. **324**, 18–32 (2001).
12. N. Bastian, K.R. Covey, M.R. Meyer, A universal stellar initial mass function? A critical look at variations. Annu. Rev. Astron. Astrophys. **48**, 339–389 (2010).
13. E. Berger, The host galaxies of short-duration gamma-ray bursts: luminosities, metallicities, and star-formation rates. Astrophys. J. **690**, 231–237 (2009).
14. E. Berger, A short gamma-ray burst "No-host" problem? Investigating large progenitor offsets for short GRBs with optical afterglows. Astrophys. J. **722**, 1946–1961 (2010).
15. E. Berger, W. Fong, R. Chornock, An r-process kilonova associated with the short-hard GRB 130603B. Astrophys. J. Lett. **774**, 23 (2013).
16. E. Berger, P.A. Price, S.B. Cenko, A. Gal-Yam, A.M. Soderberg, M. Kasliwal, D.C. Leonard, P.B. Cameron, D.A. Frail, S.R. Kulkarni, D.C. Murphy, W. Krzeminski, T. Piran, B.L. Lee, K.C. Roth, D.-S. Moon, D.B. Fox, F.A. Harrison, S.E. Persson, B.P. Schmidt, B.E. Penprase, J. Rich, B.A. Peterson, L.L. Cowie, The afterglow and elliptical host galaxy of the short γγ-ray burst GRB 050724. Nature **438**, 988–990 (2005).
17. E. Berger, D.B. Fox, P.A. Price, E. Nakar, A. Gal-Yam, D.E. Holz, B.P. Schmidt, A. Cucchiara, S.B. Cenko, S.R. Kulkarni, A.M. Soderberg, D.A. Frail, B.E. Penprase, A. Rau, E. Ofek, S.J.B. Burnell, P.B. Cameron, L.L. Cowie, M.A. Dopita, I. Hook, B.A. Peterson, P. Podsiadlowski, K.C. Roth, R.E. Rutledge, S.S. Sheppard, A. Songaila, A new population of high-redshift short-duration gamma-ray bursts. Astrophys. J. **664**, 1000–1010 (2007).
18. D. Bersier, A.S. Fruchter, L.-G. Strolger, J. Gorosabel, A. Levan, I. Burud, J.E. Rho, A.C. Becker, A. Cassan, R. Chornock, S. Covino, R.S. de Jong, D. Dominis, A.V. Filippenko, J. Hjorth, J. Holmberg, D. Malesani, B. Mobasher, K.A.G. Olsen, M. Stefanon, J.M. Castro Cerón, J.P.U. Fynbo, S.T. Holland, C. Kouveliotou, H. Pedersen, N.R. Tanvir, S.E. Woosley, Evidence for a supernova associated with the X-ray flash 020903. Astrophys. J. **643**, 284–291 (2006).
19. J.L. Bibby, P.A. Crowther, The Wolf–Rayet population of the nearby barred spiral galaxy NGC 5068 uncovered by the very large telescope and gemini. Mon. Not. R. Astron. Soc. **420**, 3091–3107 (2012).
20. J.S. Bloom, D.A. Frail, S.R. Kulkarni, Gamma-ray burst energetics and the gamma-ray burst Hubble diagram: promises and limitations. Astrophys. J. **594**, 674–683 (2003).
21. J.S. Bloom, S.R. Kulkarni, S.G. Djorgovski, The observed offset distribution of gamma-ray bursts from their host galaxies: a robust clue to the nature of the progenitors. Astron. J. **123**, 1111–1148 (2002).
22. J.S. Bloom, S.R. Kulkarni, S.G. Djorgovski, A.C. Eichelberger, P. Côté, J.P. Blakeslee, S.C. Odewahn, F.A. Harrison, D.A. Frail, A.V. Filippenko, D.C. Leonard, A.G. Riess, H. Spinrad, D. Stern, A. Bunker, A. Dey, B. Grossan, S. Perlmutter, R.A. Knop, I.M. Hook, M. Feroci, The unusual afterglow of the γγ-ray burst of 26 March 1998 as evidence for a supernova connection. Nature **401**, 453–456 (1999).
23. J.S. Bloom, J.X. Prochaska, D. Pooley, C.H. Blake, R.J. Foley, S. Jha, E. Ramirez-Ruiz, J. Granot, A.V. Filippenko, S. Sigurdsson, A.J. Barth, H.-W. Chen, M.C. Cooper, E.E. Falco, R.R. Gal, B.F. Gerke, M.D. Gladders, J.E. Greene, J. Hennanwi, L.C. Ho, K. Hurley, B.P. Koester, W. Li, L. Lubin, J. Newman, D.A. Perley, G.K. Squires, W.M. Wood-Vasey, Closing in on a short-hard burst progenitor: constraints from early-time optical imaging and spectroscopy of a possible host galaxy of GRB 050509b. Astrophys. J. **638**, 354–368 (2006).
24. O. Bromberg, E. Nakar, T. Piran, R. Sari, Short versus long and collapsars versus non-collapsars: a quantitative classification of gamma-ray bursts. Astrophys. J. **764**, 179 (2013a).

25. O. Bromberg, E. Nakar, T. Piran, R. Sari, Short versus long and collapsars versus non-collapsars: a quantitative classification of gamma-ray bursts. Astrophys. J. **764**, 179 (2013b).
26. G.E. Brown, C.-H. Lee, The case for case C mass transfer in the galactic evolution of black hole binaries. New Astron. **9**, 225–237 (2004).
27. D.N. Burrows, P. Romano, A. Falcone, S. Kobayashi, B. Zhang, A. Moretti, P.T. O'Brien, M.R. Goad, S. Campana, K.L. Page, L. Angelini, S. Barthelmy, A.P. Beardmore, M. Capalbi, G. Chincarini, J. Cummings, G. Cusumano, D. Fox, P. Giommi, J.E. Hill, J.A. Kennea, H. Krimm, V. Mangano, F. Marshall, P. Mészáros, D.C. Morris, J.A. Nousek, J.P. Osborne, C. Pagani, M. Perri, G. Tagliaferri, A.A. Wells, S. Woosley, N. Gehrels, Bright X-ray flares in gamma-ray burst afterglows. Science **309**, 1833–1835 (2005).
28. N.R. Butler, D. Kocevski, J.S. Bloom, J.L. Curtis, A complete catalog of Swift gamma-ray burst spectra and durations: demise of a physical origin for pre-Swift high-energy correlations. Astrophys. J. **671**, 656–677 (2007).
29. F. Calura, F. Matteucci, The cosmic evolution of the galaxy luminosity density. Astrophys. J. **596**, 734–747 (2003).
30. S. Campana, V. Mangano, A.J. Blustin, P. Brown, D.N. Burrows, G. Chincarini, J.R. Cummings, G. Cusumano, M. Della Valle, D. Malesani, P. Mészáros, J.A. Nousek, M. Page, T. Sakamoto, E. Waxman, B. Zhang, Z.G. Dai, N. Gehrels, S. Immler, F.E. Marshall, K.O. Mason, A. Moretti, P.T. O'Brien, J.P. Osborne, K.L. Page, P. Romano, P.W.A. Roming, G. Tagliaferri, L.R. Cominsky, P. Giommi, O. Godet, J.A. Kennea, H. Krimm, L. Angelini, S.D. Barthelmy, P.T. Boyd, D.M. Palmer, A.A. Wells, N.E. White, The association of GRB 060218 with a supernova and the evolution of the shock wave. Nature **442**, 1008–1010 (2006).
31. Z. Cano, A new method for estimating the bolometric properties of Ibc supernovae. Mon. Not. R. Astron. Soc. **434**, 1098–1116 (2013).
32. Z. Cano, K.G. Johansson Andreas, K. Maeda, A self-consistent analytical magnetar model: the luminosity of γγ-ray burst supernovae is powered by radioactivity. Mon. Not. R. Astron. Soc. **457**, 2761–2772 (2016).
33. Z. Cano, A. de Ugarte Postigo, A. Pozanenko, N. Butler, C.C. Thöne, C. Guidorzi, T. Krühler, J. Gorosabel, P. Jakobsson, G. Leloudas, D. Malesani, J. Hjorth, A. Melandri, C. Mundell, K. Wiersema, P. D'Avanzo, S. Schulze, A. Gomboc, A. Johansson, W. Zheng, D.A. Kann, F. Knust, K. Varela, C.W. Akerlof, J. Bloom, O. Burkhonov, E. Cooke, J.A. de Diego, G. Dhungana, C. Farina, F.V. Ferrante, H.A. Flewelling, O.D. Fox, J. Fynbo, N. Gehrels, L. Georgiev, J.J. González, J. Greiner, T. Güver, O. Hartoog, N. Hatch, M. Jelinek, R. Kehoe, S. Klose, E. Klunko, D. Kopač, A. Kutyrev, Y. Krugly, W.H. Lee, A. Levan, V. Linkov, A. Matkin, N. Minikulov, I. Molotov, J.X. Prochaska, M.G. Richer, C.G. Román-Zúñiga, V. Rumyantsev, R. Sánchez-Ramírez, I. Steele, N.R. Tanvir, A. Volnova, A.M. Watson, D. Xu, F. Yuan, A trio of gamma-ray burst supernovae: GRB 120729A, GRB 130215A/SN 2013ez, and GRB 130831A/SN 2013fu. Astron. Astrophys. **568**, 19 (2014).
34. G. Cavallo, M.J. Rees, A qualitative study of cosmic fireballs and gamma-ray bursts. Mon. Not. R. Astron. Soc. **183**, 359–365 (1978).
35. R. Chapman, N.R. Tanvir, R.S. Priddey, A.J. Levan, How common are long gamma-ray bursts in the local Universe? Mon. Not. R. Astron. Soc. **382**, 21–25 (2007).
36. H.-Y. Chen, D.E. Holz, Gamma-ray-burst beaming and gravitational-wave observations. Phys. Rev. Lett. **111**(18), 181101 (2013).
37. Coleiro, S. Chaty, Distribution of high-mass X-ray binaries in the Milky Way. Astrophys. J. **764**, 185 (2013).
38. E. Costa, F. Frontera, J. Heise, M. Feroci, J. in't Zand, F. Fiore, M.N. Cinti, D. Dal Fiume, L. Nicastro, M. Orlandini, E. Palazzi, M. Rapisarda#, G. Zavattini, R. Jager, A. Parmar, A. Owens, S. Molendi, G. Cusumano, M.C. Maccarone, S. Giarrusso, A. Coletta, L.A. Antonelli, P. Giommi, J.M. Muller, L. Piro, R.C. Butler, Discovery of an X-ray afterglow associated with the γγ-ray burst of 28 February 1997. Nature **387**, 783–785 (1997).
39. P.A. Crowther, Physical properties of Wolf–Rayet stars. Annu. Rev. Astron. Astrophys. **45**, 177–219 (2007).

40. P.A. Crowther, L. Dessart, D.J. Hillier, J.B. Abbott, A.W. Fullerton, Stellar and wind properties of LMC WC4 stars. A metallicity dependence for Wolf–Rayet mass-loss rates. Astron. Astrophys. **392**, 653–669 (2002).
41. P.A. Crowther, S.M. Caballero-Nieves, K.A. Bostroem, J. Maíz Apellániz, F.R.N. Schneider, N.R. Walborn, C.R. Angus, I. Brott, A. Bonanos, A. de Koter, S.E. de Mink, C.J. Evans, G. Gräfener, A. Herrero, I.D. Howarth, N. Langer, D.J. Lennon, J. Puls, H. Sana, J.S. Vink, The R136 star cluster dissected with Hubble Space Telescope/STIS. I. Far-ultraviolet spectroscopic census and the origin of He II λλ1640 in young star clusters. Mon. Not. R. Astron. Soc. **458**, 624–659 (2016).
42. Cucchiara, A.J. Levan, D.B. Fox, N.R. Tanvir, T.N. Ukwatta, E. Berger, T. Krühler, A. Küpcü Yoldaş, X.F. Wu, K. Toma, J. Greiner, F.E. Olivares, A. Rowlinson, L. Amati, T. Sakamoto, K. Roth, A. Stephens, A. Fritz, J.P.U. Fynbo, J. Hjorth, D. Malesani, P. Jakobsson, K. Wiersema, P.T. O'Brien, A.M. Soderberg, R.J. Foley, A.S. Fruchter, J. Rho, R.E. Rutledge, B.P. Schmidt, M.A. Dopita, P. Podsiadlowski, R. Willingale, C. Wolf, S.R. Kulkarni, P. D'Avanzo, A photometric redshift of z~9.4z~9.4 for GRB 090429B. Astrophys. J. **736**, 7 (2011).
43. M.B. Davies, A.J. Levan, A.R. King, The ultimate outcome of black hole-neutron star mergers. Mon. Not. R. Astron. Soc. **356**, 54–58 (2005).
44. M.B. Davies, A. King, S. Rosswog, G. Wynn, Gamma-ray bursts, supernova kicks, and gravitational radiation. Astrophys. J. Lett. **579**, 63–66 (2002).
45. R. de Grijs, C. Li, Y. Zheng, L. Deng, Y. Hu, M.B.N. Kouwenhoven, Gravitational conundrum? Dynamical mass segregation versus disruption of binary stars in dense stellar systems. Astrophys. J. **765**, 4 (2013).
46. S.E. de Mink, N. Langer, R.G. Izzard, H. Sana, A. de Koter, The rotation rates of massive stars: the role of binary interaction through tides, mass transfer, and mergers. Astrophys. J. **764**, 166 (2013).
47. de Ugarte Postigo, A.J. Castro-Tirado, S. Guziy, J. Gorosabel, G. Jóhannesson, M.A. Aloy, S. McBreen, D.Q. Lamb, N. Benitez, M. Jelínek, S.B. Pandey, D. Coe, M.D. Pérez-Ramírez, F.J. Aceituno, M. Alises, J.A. Acosta-Pulido, G. Gómez, R. López, T.Q. Donaghy, Y.E. Nakagawa, T. Sakamoto, G.R. Ricker, F.R. Hearty, M. Bayliss, G. Gyuk, GRB 060121: implications of a short-/intermediate-duration γγ-ray burst at high redshift. Astrophys. J. Lett. **648**, 83–87 (2006).
48. de Ugarte Postigo, I. Horváth, P. Veres, Z. Bagoly, D.A. Kann, C.C. Thöne, L.G. Balazs, P. D'Avanzo, M.A. Aloy, S. Foley, S. Campana, J. Mao, P. Jakobsson, S. Covino, J.P.U. Fynbo, J. Gorosabel, A.J. Castro-Tirado, L. Amati, M. Nardini, Searching for differences in Swift's intermediate GRBs. Astron. Astrophys. **525**, 109 (2011).
49. de Ugarte Postigo, C.C. Thöne, A. Rowlinson, R. García-Benito, A.J. Levan, J. Gorosabel, P. Goldoni, S. Schulze, T. Zafar, K. Wiersema, R. Sánchez-Ramírez, A. Melandri, P. D'Avanzo, S. Oates, V. D'Elia, M. De Pasquale, T. Krühler, A.J. van der Horst, D. Xu, D. Watson, S. Piranomonte, S.D. Vergani, B. Milvang-Jensen, L. Kaper, D. Malesani, J.P.U. Fynbo, Z. Cano, S. Covino, H. Flores, S. Greiss, F. Hammer, O.E. Hartoog, S. Hellmich, C. Heuser, J. Hjorth, P. Jakobsson, S. Mottola, M. Sparre, J. Sollerman, G. Tagliaferri, N.R. Tanvir, M. Vestergaard, R.A.M.J. Wijers, Spectroscopy of the short-hard GRB 130603B. The host galaxy and environment of a compact object merger. Astron. Astrophys. **563**, 62 (2014).
50. M. Della Valle, G. Chincarini, N. Panagia, G. Tagliaferri, D. Malesani, V. Testa, D. Fugazza, S. Campana, S. Covino, V. Mangano, L.A. Antonelli, P. D'Avanzo, K. Hurley, I.F. Mirabel, L.J. Pellizza, S. Piranomonte, L. Stella, An enigmatic long-lasting γγ-ray burst not accompanied by a bright supernova. Nature **444**, 1050–1052 (2006).
51. J. Deng, N. Tominaga, P.A. Mazzali, K. Maeda, K. Nomoto, On the light curve and spectrum of SN 2003dh separated from the optical afterglow of GRB 030329. Astrophys. J. **624**, 898–905 (2005).
52. R.G. Detmers, N. Langer, P. Podsiadlowski, R.G. Izzard, Gamma-ray bursts from tidally spun-up Wolf–Rayet stars? Astron. Astrophys. **484**, 831–839 (2008).
53. Dietz, N. Fotopoulos, L. Singer, C. Cutler, Outlook for detection of GW inspirals by GRB-triggered searches in the advanced detector era. Phys. Rev. D **87**(6), 064033 (2013).

54. M.A. Dopita, L.J. Kewley, R.S. Sutherland, D.C. Nicholls, Chemical abundances in high-redshift galaxies: a powerful new emission line diagnostic. Astrophys. Space Sci. **361**, 61 (2016).
55. A.S. Eddington, *The Internal Constitution of the Stars* (1926)
56. J.J. Eldridge, E.R. Stanway, Spectral population synthesis including massive binaries. Mon. Not. R. Astron. Soc. **400**, 1019–1028 (2009).
57. R.A.W. Elson, S. Sigurdsson, M. Davies, J. Hurley, G. Gilmore, The binary star population of the young cluster NGC 1818 in the Large Magellanic Cloud. Mon. Not. R. Astron. Soc. **300**, 857–862 (1998).
58. C.J. Evans, W.D. Taylor, V. Hénault-Brunet, H. Sana, A. de Koter, S. Simón-Díaz, G. Carraro, T. Bagnoli, N. Bastian, J.M. Bestenlehner, A.Z. Bonanos, E. Bressert, I. Brott, M.A. Campbell, M. Cantiello, J.S. Clark, E. Costa, P.A. Crowther, S.E. de Mink, E. Doran, P.L. Dufton, P.R. Dunstall, K. Friedrich, M. Garcia, M. Gieles, G. Gräfener, A. Herrero, I.D. Howarth, R.G. Izzard, N. Langer, D.J. Lennon, J. Maíz Apellániz, N. Markova, F. Najarro, J. Puls, O.H. Ramirez, C. Sabín-Sanjulián, S.J. Smartt, V.E. Stroud, J.T. van Loon, J.S. Vink, N.R. Walborn, The VLT-FLAMES Tarantula Survey. I. Introduction and observational overview. Astron. Astrophys. **530**, 108 (2011).
59. R.J. Foley, P.J. Challis, R. Chornock, M. Ganeshalingam, W. Li, G.H. Marion, N.I. Morrell, G. Pignata, M.D. Stritzinger, J.M. Silverman, X. Wang, J.P. Anderson, A.V. Filippenko, W.L. Freedman, M. Hamuy, S.W. Jha, R.P. Kirshner, C. McCully, S.E. Persson, M.M. Phillips, D.E. Reichart, A.M. Soderberg, Type Iax supernovae: a new class of stellar explosion. Astrophys. J. **767**, 57 (2013).
60. W. Fong, E. Berger, The locations of short gamma-ray bursts as evidence for compact object binary progenitors. Astrophys. J. **776**, 18 (2013).
61. W. Fong, E. Berger, D.B. Fox, Hubble space telescope observations of short gamma-ray burst host galaxies: morphologies, offsets, and local environments. Astrophys. J. **708**, 9–25 (2010).
62. W-f. Fong, B.D. Metzger, E. Berger, F. Ozel, Radio Constraints on Long-Lived Magnetar Remnants in Short Gamma-Ray Bursts. ArXiv e-prints (2016)
63. W. Fong, E. Berger, R. Chornock, R. Margutti, A.J. Levan, N.R. Tanvir, R.L. Tunnicliffe, I. Czekala, D.B. Fox, D.A. Perley, S.B. Cenko, B.A. Zauderer, T. Laskar, S.E. Persson, A.J. Monson, D.D. Kelson, C. Birk, D. Murphy, M. Servillat, G. Anglada, Demographics of the galaxies hosting short-duration gamma-ray bursts. Astrophys. J. **769**, 56 (2013).
64. W. Fong, E. Berger, B.D. Metzger, R. Margutti, R. Chornock, G. Migliori, R.J. Foley, B.A. Zauderer, R. Lunnan, T. Laskar, S.J. Desch, K.J. Meech, S. Sonnett, C. Dickey, A. Hedlund, P. Harding, Short GRB 130603B: discovery of a jet break in the optical and radio afterglows, and a mysterious late-time X-ray excess. Astrophys. J. **780**, 118 (2014).
65. D.B. Fox, D.A. Frail, P.A. Price, S.R. Kulkarni, E. Berger, T. Piran, A.M. Soderberg, S.B. Cenko, P.B. Cameron, A. Gal-Yam, M.M. Kasliwal, D.-S. Moon, F.A. Harrison, E. Nakar, B.P. Schmidt, B. Penprase, R.A. Chevalier, P. Kumar, K. Roth, D. Watson, B.L. Lee, S. Shectman, M.M. Phillips, M. Roth, P.J. McCarthy, M. Rauch, L. Cowie, B.A. Peterson, J. Rich, N. Kawai, K. Aoki, G. Kosugi, T. Totani, H.-S. Park, A. MacFadyen, K.C. Hurley, The afterglow of GRB 050709 and the nature of the short-hard $\gamma\gamma$-ray bursts. Nature **437**, 845–850 (2005).
66. D.A. Frail, S.R. Kulkarni, L. Nicastro, M. Feroci, G.B. Taylor, The radio afterglow from the $\gamma\gamma$-ray burst of 8 May 1997. Nature **389**, 261–263 (1997).
67. D.A. Frail, S.R. Kulkarni, R. Sari, S.G. Djorgovski, J.S. Bloom, T.J. Galama, D.E. Reichart, E. Berger, F.A. Harrison, P.A. Price, S.A. Yost, A. Diercks, R.W. Goodrich, F. Chaffee, Beaming in gamma-ray bursts: evidence for a standard energy reservoir. Astrophys. J. Lett. **562**, 55–58 (2001).
68. A.S. Fruchter, A.J. Levan, L. Strolger, P.M. Vreeswijk, S.E. Thorsett, D. Bersier, I. Burud, J.M. Castro Cerón, A.J. Castro-Tirado, C. Conselice, T. Dahlen, H.C. Ferguson, J.P.U. Fynbo, P.M. Garnavich, R.A. Gibbons, J. Gorosabel, T.R. Gull, J. Hjorth, S.T. Holland, C. Kouveliotou, Z. Levay, M. Livio, M.R. Metzger, P.E. Nugent, L. Petro, E. Pian, J.E. Rho, A.G. Riess, K.C. Sahu, A. Smette, N.R. Tanvir, R.A.M.J. Wijers, S.E. Woosley, Long $\gamma\gamma$-ray bursts and core-collapse supernovae have different environments. Nature **441**, 463–468 (2006).

69. C.L. Fryer, A.L. Hungerford, P.A. Young, Light-curve calculations of supernovae from fallback gamma-ray bursts. Astrophys. J. Lett. **662**, 55–58 (2007).
70. C.L. Fryer, S.E. Woosley, D.H. Hartmann, Formation rates of black hole accretion disk gamma-ray bursts. Astrophys. J. **526**, 152–177 (1999).
71. C.L. Fryer, S.E. Woosley, A. Heger, Pair-instability supernovae, gravity waves, and gamma-ray transients. Astrophys. J. **550**, 372–382 (2001).
72. J.P.U. Fynbo, D. Watson, C.C. Thöne, J. Sollerman, J.S. Bloom, T.M. Davis, J. Hjorth, P. Jakobsson, U.G. Jørgensen, J.F. Graham, A.S. Fruchter, D. Bersier, L. Kewley, A. Cassan, J.M. Castro Cerón, S. Foley, J. Gorosabel, T.C. Hinse, K.D. Horne, B.L. Jensen, S. Klose, D. Kocevski, J.-B. Marquette, D. Perley, E. Ramirez-Ruiz, M.D. Stritzinger, P.M. Vreeswijk, R.A.M. Wijers, K.G. Woller, D. Xu, M. Zub, No supernovae associated with two long-duration γγ-ray bursts. Nature **444**, 1047–1049 (2006).
73. J.P.U. Fynbo, P. Jakobsson, J.X. Prochaska, D. Malesani, C. Ledoux, A. de Ugarte Postigo, M. Nardini, P.M. Vreeswijk, K. Wiersema, J. Hjorth, J. Sollerman, H.-W. Chen, C.C. Thöne, G. Björnsson, J.S. Bloom, A.J. Castro-Tirado, L. Christensen, A. De Cia, A.S. Fruchter, J. Gorosabel, J.F. Graham, A.O. Jaunsen, B.L. Jensen, D.A. Kann, C. Kouveliotou, A.J. Levan, J. Maund, N. Masetti, B. Milvang-Jensen, E. Palazzi, D.A. Perley, E. Pian, E. Rol, P. Schady, R.L.C. Starling, N.R. Tanvir, D.J. Watson, D. Xu, T. Augusteijn, F. Grundahl, J. Telting, P.-O. Quirion, Low-resolution spectroscopy of gamma-ray burst optical afterglows: biases in the Swift sample and characterization of the absorbers. Astrophys. J. Suppl. Ser. **185**, 526–573 (2009).
74. Gal-Yam, Luminous supernovae. Science **337**, 927 (2012).
75. Gal-Yam, D.B. Fox, P.A. Price, E.O. Ofek, M.R. Davis, D.C. Leonard, A.M. Soderberg, B.P. Schmidt, K.M. Lewis, B.A. Peterson, S.R. Kulkarni, E. Berger, S.B. Cenko, R. Sari, K. Sharon, D. Frail, D.-S. Moon, P.J. Brown, A. Cucchiara, F. Harrison, T. Piran, S.E. Persson, P.J. McCarthy, B.E. Penprase, R.A. Chevalier, A.I. MacFadyen, A novel explosive process is required for the γγ-ray burst GRB 060614. Nature 444, 1053–1055 (2006).
76. Gal-Yam, P. Mazzali, E.O. Ofek, P.E. Nugent, S.R. Kulkarni, M.M. Kasliwal, R.M. Quimby, A.V. Filippenko, S.B. Cenko, R. Chornock, R. Waldman, D. Kasen, M. Sullivan, E.C. Beshore, A.J. Drake, R.C. Thomas, J.S. Bloom, D. Poznanski, A.A. Miller, R.J. Foley, J.M. Silverman, I. Arcavi, R.S. Ellis, J. Deng, Supernova 2007bi as a pair-instability explosion. Nature **462**, 624–627 (2009).
77. T.J. Galama, N. Tanvir, P.M. Vreeswijk, R.A.M.J. Wijers, P.J. Groot, E. Rol, J. van Paradijs, C. Kouveliotou, A.S. Fruchter, N. Masetti, H. Pedersen, B. Margon, E.W. Deutsch, M. Metzger, L. Armus, S. Klose, B. Stecklum, Evidence for a supernova in reanalyzed optical and near-infrared images of GRB 970228. Astrophys. J. **536**, 185–194 (2000).
78. P.M. Garnavich, K.Z. Stanek, L. Wyrzykowski, L. Infante, E. Bendek, D. Bersier, S.T. Holland, S. Jha, T. Matheson, R.P. Kirshner, K. Krisciunas, M.M. Phillips, R.G. Carlberg, Discovery of the low-redshift optical afterglow of GRB 011121 and its progenitor supernova SN 2001ke. Astrophys. J. **582**, 924–932 (2003).
79. N. Gehrels, C.L. Sarazin, P.T. O'Brien, B. Zhang, L. Barbier, S.D. Barthelmy, A. Blustin, D.N. Burrows, J. Cannizzo, J.R. Cummings, M. Goad, S.T. Holland, C.P. Hurkett, J.A. Kennea, A. Levan, C.B. Markwardt, K.O. Mason, P. Meszaros, M. Page, D.M. Palmer, E. Rol, T. Sakamoto, R. Willingale, L. Angelini, A. Beardmore, P.T. Boyd, A. Breeveld, S. Campana, M.M. Chester, G. Chincarini, L.R. Cominsky, G. Cusumano, M. de Pasquale, E.E. Fenimore, P. Giommi, C. Gronwall, D. Grupe, J.E. Hill, D. Hinshaw, J. Hjorth, D. Hullinger, K.C. Hurley, S. Klose, S. Kobayashi, C. Kouveliotou, H.A. Krimm, V. Mangano, F.E. Marshall, K. McGowan, A. Moretti, R.F. Mushotzky, K. Nakazawa, J.P. Norris, J.A. Nousek, J.P. Osborne, K. Page, A.M. Parsons, S. Patel, M. Perri, T. Poole, P. Romano, P.W.A. Roming, S. Rosen, G. Sato, P. Schady, A.P. Smale, J. Sollerman, R. Starling, M. Still, M. Suzuki, G. Tagliaferri, T. Takahashi, M. Tashiro, J. Tueller, A.A. Wells, N.E. White, R.A.M.J. Wijers, A short γγ-ray burst apparently associated with an elliptical galaxy at redshift z=0.225z=0.225. Nature **437**, 851–854 (2005).
80. N. Gehrels, J.P. Norris, S.D. Barthelmy, J. Granot, Y. Kaneko, C. Kouveliotou, C.B. Markwardt, P. Mészáros, E. Nakar, J.A. Nousek, P.T. O'Brien, M. Page, D.M. Palmer, A.M. Parsons, P.W.A. Roming, T. Sakamoto, C.L. Sarazin, P. Schady, M. Stamatikos, S.E. Woosley,

A new γγ-ray burst classification scheme from GRB060614. Nature **444**, 1044–1046 (2006).
81. N. Gehrels, S.D. Barthelmy, D.N. Burrows, J.K. Cannizzo, G. Chincarini, E. Fenimore, C. Kouveliotou, P. O'Brien, D.M. Palmer, J. Racusin, P.W.A. Roming, T. Sakamoto, J. Tueller, R.A.M.J. Wijers, B. Zhang, Correlations of prompt and afterglow emission in Swift long and short gamma-ray bursts. Astrophys. J. **689**, 1161–1172 (2008).
82. A.M. Geller, R. de Grijs, C. Li, J.R. Hurley, Consequences of dynamical disruption and mass segregation for the binary frequencies of star clusters. Astrophys. J. **779**, 30 (2013).
83. A.M. Geller, R. de Grijs, C. Li, J.R. Hurley, Different dynamical ages for the two young and coeval LMC star clusters, NGC 1805 and NGC 1818, imprinted on their binary populations. Astrophys. J. **805**, 11 (2015).
84. Gendre, G. Stratta, J.L. Atteia, S. Basa, M. Boër, D.M. Coward, S. Cutini, V. D'Elia, E.J. Howell, A. Klotz, L. Piro, The ultra-long gamma-ray burst 111209A: the collapse of a blue supergiant? Astrophys. J. **766**, 30 (2013).
85. B.P. Gompertz, A.J. van der Horst, P.T. O'Brien, G.A. Wynn, K. Wiersema, Broad-band modelling of short gamma-ray bursts with energy injection from magnetar spin-down and its implications for radio detectability. Mon. Not. R. Astron. Soc. **448**, 629–641 (2015).
86. J. Gorosabel, J.P.U. Fynbo, A. Fruchter, A. Levan, J. Hjorth, P. Nugent, A.J. Castro-Tirado, J.M. Castro Cerón, J. Rho, D. Bersier, I. Burud, A possible bright blue supernova in the afterglow of GRB 020305. Astron. Astrophys. **437**, 411–418 (2005).
87. J.F. Graham, A.S. Fruchter, The metal aversion of long-duration gamma-ray bursts. Astrophys. J. **774**, 119 (2013).
88. J.F. Graham, A.S. Fruchter, The Relative Rate of LGRB Formation as a Function of Metallicity. ArXiv e-prints (2015a)
89. J.F. Graham, P. Schady, The Absolute Rate of LGRB Formation. ArXiv e-prints (2015b)
90. J.F. Graham, A.S. Fruchter, A.J. Levan, A. Melandri, L.J. Kewley, E.M. Levesque, M. Nysewander, N.R. Tanvir, T. Dahlen, D. Bersier, K. Wiersema, D.G. Bonfield, A. Martinez-Sansigre, GRB 070714B: discovery of the highest spectroscopically confirmed short burst redshift. Astrophys. J. **698**, 1620–1629 (2009).
91. J. Greiner, P.A. Mazzali, D.A. Kann, T. Krühler, E. Pian, S. Prentice, F. Olivares E., A. Rossi, S. Klose, S. Taubenberger, F. Knust, P.M.J. Afonso, C. Ashall, J. Bolmer, C. Delvaux, R. Diehl, J. Elliott, R. Filgas, J.P.U. Fynbo, J.F. Graham, A.N. Guelbenzu, S. Kobayashi, G. Leloudas, S. Savaglio, P. Schady, S. Schmidl, T. Schweyer, V. Sudilovsky, M. Tanga, A.C. Updike, H. van Eerten, K. Varela, A very luminous magnetar-powered supernova associated with an ultra-long γγ-ray burst. Nature **523**, 189–192 (2015a).
92. . Hainich, D. Pasemann, H. Todt, T. Shenar, A. Sander, W.-R. Hamann, Wolf–Rayet stars in the Small Magellanic Cloud. I. Analysis of the single WN stars. Astron. Astrophys. **581**, 21 (2015b).
93. F. Hammer, H. Flores, D. Schaerer, M. Dessauges-Zavky, E. Le Floc'h, M. Puech, Detection of Wolf–Rayet stars in host galaxies of gamma-ray bursts (GRBs): are GRBs produced by runaway massive stars ejected from high stellar density regions? Astron. Astrophys. **454**, 103–111 (2006).
94. Heger, S.E. Woosley, H.C. Spruit, Presupernova evolution of differentially rotating massive stars including magnetic fields. Astrophys. J. **626**, 350–363 (2005).
95. D.C. Heggie, Binary evolution in stellar dynamics. Mon. Not. R. Astron. Soc. **173**, 729–787 (1975).
96. J. Hjorth, J. Sollerman, P. Møller, J.P.U. Fynbo, S.E. Woosley, C. Kouveliotou, N.R. Tanvir, J. Greiner, M.I. Andersen, A.J. Castro-Tirado, J.M. Castro Cerón, A.S. Fruchter, J. Gorosabel, P. Jakobsson, L. Kaper, S. Klose, N. Masetti, H. Pedersen, K. Pedersen, E. Pian, E. Palazzi, J.E. Rho, E. Rol, E.P.J. van den Heuvel, P.M. Vreeswijk, D. Watson, R.A.M.J. Wijers, A very energetic supernova associated with the γγ-ray burst of 29 March 2003. Nature **423**, 847–850 (2003).
97. J. Hjorth, J. Sollerman, J. Gorosabel, J. Granot, S. Klose, C. Kouveliotou, J. Melinder, E. Ramirez-Ruiz, R. Starling, B. Thomsen, M.I. Andersen, J.P.U. Fynbo, B.L. Jensen, P.M. Vreeswijk, J.M. Castro Cerón, P. Jakobsson, A. Levan, K. Pedersen, J.E. Rho, N.R. Tanvir, D.

Watson, R.A.M.J. Wijers, GRB 050509B: constraints on short gamma-ray burst models. Astrophys. J. Lett. **630**, 117–120 (2005a).

98. J. Hjorth, D. Watson, J.P.U. Fynbo, P.A. Price, B.L. Jensen, U.G. Jørgensen, D. Kubas, J. Gorosabel, P. Jakobsson, J. Sollerman, K. Pedersen, C. Kouveliotou, The optical afterglow of the short γγ-ray burst GRB 050709. Nature **437**, 859–861 (2005b).
99. P. Höflich, J.C. Wheeler, L. Wang, Aspherical explosion models for SN 1998BW/GRB-980425. Astrophys. J. **521**, 179–189 (1999).
100. Y. Hu, L. Deng, R. de Grijs, Q. Liu, S.P. Goodwin, The binary fraction of the young cluster NGC 1818 in the Large Magellanic Cloud. Astrophys. J. **724**, 649–656 (2010).
101. K. Hurley, S.E. Boggs, D.M. Smith, R.C. Duncan, R. Lin, A. Zoglauer, S. Krucker, G. Hurford, H. Hudson, C. Wigger, W. Hajdas, C. Thompson, I. Mitrofanov, A. Sanin, W. Boynton, C. Fellows, A. von Kienlin, G. Lichti, A. Rau, T. Cline, An exceptionally bright flare from SGR 1806-20 and the origins of short-duration γγ-ray bursts. Nature **434**, 1098–1103 (2005).
102. R.G. Izzard, E. Ramirez-Ruiz, C.A. Tout, Formation rates of core-collapse supernovae and gamma-ray bursts. Mon. Not. R. Astron. Soc. **348**, 1215–1228 (2004).
103. P.A. James, J.P. Anderson, The Hαα galaxy survey. III. Constraints on supernova progenitors from spatial correlations with Hαα emission. Astron. Astrophys. **453**, 57–65 (2006).
104. J. Japelj, S.D. Vergani, R. Salvaterra, L.K. Hunt, F. Mannucci, Taking stock of SLSN and LGRB host galaxy comparison using a complete sample of LGRBs. ArXiv e-prints (2016)
105. Z.-P. Jin, X. Li, Z. Cano, S. Covino, Y.-Z. Fan, D.-M. Wei, The light curve of the macronova associated with the long-short burst GRB 060614. Astrophys. J. Lett. **811**, 22 (2015).
106. D.A. Kann, S. Klose, B. Zhang, D. Malesani, E. Nakar, A. Pozanenko, A.C. Wilson, N.R. Butler, P. Jakobsson, S. Schulze, M. Andreev, L.A. Antonelli, I.F. Bikmaev, V. Biryukov, M. Böttcher, R.A. Burenin, J.M. Castro Cerón, A.J. Castro-Tirado, G. Chincarini, B.E. Cobb, S. Covino, P. D'Avanzo, V. D'Elia, M. Della Valle, A. de Ugarte Postigo, Y. Efimov, P. Ferrero, D. Fugazza, J.P.U. Fynbo, M. Gålfalk, F. Grundahl, J. Gorosabel, S. Gupta, S. Guziy, B. Hafizov, J. Hjorth, K. Holhjem, M. Ibrahimov, M. Im, G.L. Israel, M. Jelínek, B.L. Jensen, R. Karimov, I.M. Khamitov, Ü. Kiziloğlu, E. Klunko, P. Kubánek, A.S. Kutyrev, P. Laursen, A.J. Levan, F. Mannucci, C.M. Martin, A. Mescheryakov, N. Mirabal, J.P. Norris, J.-E. Ovaldsen, D. Paraficz, E. Pavlenko, S. Piranomonte, A. Rossi, V. Rumyantsev, R. Salinas, A. Sergeev, D. Sharapov, J. Sollerman, B. Stecklum, L. Stella, G. Tagliaferri, N.R. Tanvir, J. Telting, V. Testa, A.C. Updike, A. Volnova, D. Watson, K. Wiersema, D. Xu, The afterglows of Swift-era gamma-ray bursts. I. Comparing pre-Swift and Swift-era long/soft (type II) GRB optical afterglows. Astrophys. J. **720**, 1513–1558 (2010).
107. D.A. Kann, P. Schady, F. Olivares E, S. Klose, A. Rossi, D.A. Perley, T. Krühler, J. Greiner, A. Nicuesa Guelbenzu, J. Elliott, F. Knust, R. Filgas, E. Pian, P. Mazzali, J.P.U. Fynbo, G. Leloudas, P.M.J. Afonso, C. Delvaux, J.F. Graham, A. Rau, S. Schmidl, S. Schulze, M. Tanga, A.C. Updike, K. Varela, Highly Luminous Supernovae associated with Gamma-Ray Bursts I: GRB 111209A/SN 2011kl in the Context of Stripped-Envelope and Superluminous Supernovae. ArXiv e-prints (2016)
108. Kasen, L. Bildsten, Supernova light curves powered by young magnetars. Astrophys. J. **717**, 245–249 (2010).
109. M.M. Kasliwal, S.R. Kulkarni, A. Gal-Yam, P.E. Nugent, M. Sullivan, L. Bildsten, O. Yaron, H.B. Perets, I. Arcavi, S. Ben-Ami, V.B. Bhalerao, J.S. Bloom, S.B. Cenko, A.V. Filippenko, D.A. Frail, M. Ganeshalingam, A. Horesh, D.A. Howell, N.M. Law, D.C. Leonard, W. Li, E.O. Ofek, D. Polishook, D. Poznanski, R.M. Quimby, J.M. Silverman, A. Sternberg, D. Xu, Calcium-rich gap transients in the remote outskirts of galaxies. Astrophys. J. **755**, 161 (2012).
110. P.L. Kelly, R.P. Kirshner, Core-collapse supernovae and host galaxy stellar populations. Astrophys. J. **759**, 107 (2012).
111. P.L. Kelly, R.P. Kirshner, M. Pahre, Long γγ-ray bursts and type Ic core-collapse supernovae have similar locations in hosts. Astrophys. J. **687**, 1201–1207 (2008).
112. P.L. Kelly, A.V. Filippenko, O.D. Fox, W. Zheng, K.I. Clubb, Evidence that gamma-ray burst 130702A exploded in a dwarf satellite of a massive galaxy. Astrophys. J. Lett. **775**, 5 (2013).

113. S. Kisaka, K. Ioka, T. Nakamura, Isotropic detectable X-ray counterparts to gravitational waves from neutron star binary mergers. Astrophys. J. Lett. **809**, 8 (2015).
114. S. Kisaka, K. Ioka, E. Nakar, X-ray-powered macronovae. Astrophys. J. **818**, 104 (2016).
115. R.W. Klebesadel, I.B. Strong, R.A. Olson, Observations of gamma-ray bursts of cosmic origin. Astrophys. J. Lett. **182**, 85 (1973).
116. H.A. Kobulnicky, D.C. Kiminki, M.J. Lundquist, J. Burke, J. Chapman, E. Keller, K. Lester, E.K. Rolen, E. Topel, A. Bhattacharjee, R.A. Smullen, C.A. Vargas Álvarez, J.C. Runnoe, D.A. Dale, M.M. Brotherton, Toward complete statistics of massive binary stars: penultimate results from the Cygnus OB2 radial velocity survey. Astrophys. J. Suppl. Ser. **213**, 34 (2014).
117. C.S. Kochanek, Failed supernovae explain the compact remnant mass function. Astrophys. J. **785**, 28 (2014).
118. C.S. Kochanek, J.F. Beacom, M.D. Kistler, J.L. Prieto, K.Z. Stanek, T.A. Thompson, H. Yüksel, A survey about nothing: monitoring a million supergiants for failed supernovae. Astrophys. J. **684**, 1336–1342 (2008).
119. Kouveliotou, C.A. Meegan, G.J. Fishman, N.P. Bhat, M.S. Briggs, T.M. Koshut, W.S. Paciesas, G.N. Pendleton, Identification of two classes of gamma-ray bursts. Astrophys. J. Lett. **413**, 101–104 (1993).
120. Kouveliotou, S.E. Woosley, S.K. Patel, A. Levan, R. Blandford, E. Ramirez-Ruiz, R.A.M.J. Wijers, M.C. Weisskopf, A. Tennant, E. Pian, P. Giommi, Chandra observations of the X-ray environs of SN 1998bw/GRB 980425. Astrophys. J. **608**, 872–882 (2004).
121. M.B.N. Kouwenhoven, A.G.A. Brown, H. Zinnecker, L. Kaper, S.F. Portegies Zwart, The primordial binary population. I. A near-infrared adaptive optics search for close visual companions to A star members of Scorpius OB2. Astron. Astrophys. **430**, 137–154 (2005).
122. P. Kroupa, On the variation of the initial mass function. Mon. Not. R. Astron. Soc. **322**, 231–246 (2001).
123. T. Krühler, D. Malesani, J.P.U. Fynbo, O.E. Hartoog, J. Hjorth, P. Jakobsson, D.A. Perley, A. Rossi, P. Schady, S. Schulze, N.R. Tanvir, S.D. Vergani, K. Wiersema, P.M.J. Afonso, J. Bolmer, Z. Cano, S. Covino, V. D'Elia, A. de Ugarte Postigo, R. Filgas, M. Friis, J.F. Graham, J. Greiner, P. Goldoni, A. Gomboc, F. Hammer, J. Japelj, D.A. Kann, L. Kaper, S. Klose, A.J. Levan, G. Leloudas, B. Milvang-Jensen, A. Nicuesa Guelbenzu, E. Palazzi, E. Pian, S. Piranomonte, R. Sánchez-Ramírez, S. Savaglio, J. Selsing, G. Tagliaferri, P.M. Vreeswijk, D.J. Watson, D. Xu, GRB hosts through cosmic time. VLT/X-Shooter emission-line spectroscopy of 96 γγ-ray-burst-selected galaxies at 0.1 <z < 3.6. Astron. Astrophys. **581**, 125 (2015).
124. J.B. Lamb, M.S. Oey, D.M. Segura-Cox, A.S. Graus, D.C. Kiminki, J.B. Golden-Marx, J.W. Parker, The runaways and isolated O-type star spectroscopic survey of the SMC (RIOTS4). Astrophys. J. **817**, 113 (2016).
125. N. Langer, Coupled mass and angular momentum loss of massive main sequence stars. Astron. Astrophys. **329**, 551–558 (1998)
126. N. Langer, C.A. Norman, On the collapsar model of long gamma-ray bursts: constraints from cosmic metallicity evolution. Astrophys. J. Lett. **638**, 63–66 (2006).
127. J. Larsson, A.J. Levan, M.B. Davies, A.S. Fruchter, A new constraint for gamma-ray burst progenitor mass. Mon. Not. R. Astron. Soc. **376**, 1285–1290 (2007).
128. C.-H. Lee, G.E. Brown, R.A.M.J. Wijers, Discovery of a black hole mass-period correlation in soft X-ray transients and its implication for gamma-ray burst and hypernova mechanisms. Astrophys. J. **575**, 996–1006 (2002).
129. W.-H. Lei, B. Zhang, E.-W. Liang, Hyperaccreting black hole as gamma-ray burst central engine. I. Baryon loading in gamma-ray burst jets. Astrophys. J. **765**, 125 (2013).
130. C.N. Leibler, E. Berger, The stellar ages and masses of short gamma-ray burst host galaxies: investigating the progenitor delay time distribution and the role of mass and star formation in the short gamma-ray burst rate. Astrophys. J. **725**, 1202–1214 (2010).
131. G. Leloudas, J. Sollerman, A.J. Levan, J.P.U. Fynbo, D. Malesani, J.R. Maund, Do Wolf–Rayet stars have similar locations in hosts as type Ib/c supernovae and long gamma-ray bursts? Astron. Astrophys. **518**, 29 (2010).

132. G. Leloudas, S. Schulze, T. Krühler, J. Gorosabel, L. Christensen, A. Mehner, A. de Ugarte Postigo, R. Amorín, C.C. Thöne, J.P. Anderson, F.E. Bauer, A. Gallazzi, K.G. Hełminiak, J. Hjorth, E. Ibar, D. Malesani, N. Morell, J. Vinko, J.C. Wheeler, Spectroscopy of superluminous supernova host galaxies. A preference of hydrogen-poor events for extreme emission line galaxies. Mon. Not. R. Astron. Soc. **449**, 917–932 (2015).
133. A.J. Levan, N.R. Tanvir, A.S. Fruchter, E. Rol, J.P.U. Fynbo, J. Hjorth, G. Williams, E. Bergeron, D. Bersier, M. Bremer, T. Grav, P. Jakobsson, K. Nilsson, E. Olszewski, R.S. Priddey, D. Rafferty, J. Rho, The faint afterglow and host galaxy of the short-hard GRB 060121. Astrophys. J. Lett. **648**, 9–12 (2006).
134. A.J. Levan, P. Jakobsson, C. Hurkett, N.R. Tanvir, J. Gorosabel, P. Vreeswijk, E. Rol, R. Chapman, N. Gehrels, P.T. O'Brien, J.P. Osborne, R.S. Priddey, C. Kouveliotou, R. Starling, D. vanden Berk, K. Wiersema, A case of mistaken identity? GRB060912A and the nature of the long-short GRB divide. Mon. Not. R. Astron. Soc. **378**, 1439–1446 (2007).
135. A.J. Levan, N.R. Tanvir, P. Jakobsson, R. Chapman, J. Hjorth, R.S. Priddey, J.P.U. Fynbo, K. Hurley, B.L. Jensen, R. Johnson, J. Gorosabel, A.J. Castro-Tirado, M. Jarvis, D. Watson, K. Wiersema, On the nature of the short-duration GRB 050906. Mon. Not. R. Astron. Soc. **384**, 541–547 (2008).
136. A.J. Levan, N.R. Tanvir, S.B. Cenko, D.A. Perley, K. Wiersema, J.S. Bloom, A.S. Fruchter, A.d.U. Postigo, P.T. O'Brien, N. Butler, A.J. van der Horst, G. Leloudas, A.N. Morgan, K. Misra, G.C. Bower, J. Farihi, R.L. Tunnicliffe, M. Modjaz, J.M. Silverman, J. Hjorth, C. Thöne, A. Cucchiara, J.M.C. Cerón, A.J. Castro-Tirado, J.A. Arnold, M. Bremer, J.P. Brodie, T. Carroll, M.C. Cooper, P.A. Curran, R.M. Cutri, J. Ehle, D. Forbes, J. Fynbo, J. Gorosabel, J. Graham, D.I. Hoffman, S. Guziy, P. Jakobsson, A. Kamble, T. Kerr, M.M. Kasliwal, C. Kouveliotou, D. Kocevski, N.M. Law, P.E. Nugent, E.O. Ofek, D. Poznanski, R.M. Quimby, E. Rol, A.J. Romanowsky, R. Sánchez-Ramírez, S. Schulze, N. Singh, L. van Spaandonk, R.L.C. Starling, R.G. Strom, J.C. Tello, O. Vaduvescu, P.J. Wheatley, R.A.M.J. Wijers, J.M. Winters, D. Xu, An extremely luminous panchromatic outburst from the nucleus of a distant galaxy. Science **333**, 199 (2011).
137. A.J. Levan, N.R. Tanvir, R.L.C. Starling, K. Wiersema, K.L. Page, D.A. Perley, S. Schulze, G.A. Wynn, R. Chornock, J. Hjorth, S.B. Cenko, A.S. Fruchter, P.T. O'Brien, G.C. Brown, R.L. Tunnicliffe, D. Malesani, P. Jakobsson, D. Watson, E. Berger, D. Bersier, B.E. Cobb, S. Covino, A. Cucchiara, A. de Ugarte Postigo, D.B. Fox, A. Gal-Yam, P. Goldoni, J. Gorosabel, L. Kaper, T. Krühler, R. Karjalainen, J.P. Osborne, E. Pian, R. Sánchez-Ramírez, B. Schmidt, I. Skillen, G. Tagliaferri, C. Thöne, O. Vaduvescu, R.A.M.J. Wijers, B.A. Zauderer, A new population of ultra-long duration gamma-ray bursts. Astrophys. J. **781**, 13 (2014a).
138. A.J. Levan, N.R. Tanvir, A.S. Fruchter, J. Hjorth, E. Pian, P. Mazzali, R.A. Hounsell, D.A. Perley, Z. Cano, J. Graham, S.B. Cenko, J.P.U. Fynbo, C. Kouveliotou, A. Pe'er, K. Misra, K. Wiersema, Hubble space telescope observations of the afterglow, supernova, and host galaxy associated with the extremely bright GRB 130427A. Astrophys. J. **792**, 115 (2014b).
139. Levan, P. Nugent, A. Fruchter, I. Burud, D. Branch, J. Rho, A. Castro-Tirado, J. Gorosabel, J.M. Castro Cerón, S.E. Thorsett, C. Kouveliotou, S. Golenetskii, J. Fynbo, P. Garnavich, S. Holland, J. Hjorth, P. Møller, E. Pian, N. Tanvir, M. Ulanov, R. Wijers, S. Woosley, GRB 020410: a gamma-ray burst afterglow discovered by its supernova light. Astrophys. J. **624**, 880–888 (2005).
140. E.M. Levesque, P. Massey, A.N. Żytkow, N. Morrell, Discovery of a Thorne–Żytkow object candidate in the Small Magellanic Cloud. Mon. Not. R. Astron. Soc. **443**, 94–98 (2014).
141. Li, R. de Grijs, L. Deng, The binary fractions in the massive young Large Magellanic Cloud star clusters NGC 1805 and NGC 1818. Mon. Not. R. Astron. Soc. **436**, 1497–1512 (2013).
142. Li, R. de Grijs, L. Deng, Not-so-simple stellar populations in the intermediate-age Large Magellanic Cloud star clusters NGC 1831 and NGC 1868. Astrophys. J. **784**, 157 (2014).
143. Li, R. de Grijs, L. Deng, X. Liu, Blue straggler evolution caught in the act in the Large Magellanic Cloud globular cluster Hodge 11. Astrophys. J. Lett. **770**, 7 (2013).
144. L.-X. Li, B. Paczyński, Transient events from neutron star mergers. Astrophys. J. Lett. **507**, 59–62 (1998).

145. Liang, B. Zhang, F. Virgili, Z.G. Dai, Low-luminosity gamma-ray bursts as a unique population: luminosity function, local rate, and beaming factor. Astrophys. J. **662**, 1111–1118 (2007a).
146. Liang, B. Zhang, F. Virgili, Z.G. Dai, Low-luminosity gamma-ray bursts as a unique population: luminosity function, local rate, and beaming factor. Astrophys. J. **662**, 1111–1118 (2007b).
147. Y.M. Lipkin, E.O. Ofek, A. Gal-Yam, E.M. Leibowitz, D. Poznanski, S. Kaspi, D. Polishook, S.R. Kulkarni, D.W. Fox, E. Berger, N. Mirabal, J. Halpern, M. Bureau, K. Fathi, P.A. Price, B.A. Peterson, A. Frebel, B. Schmidt, J.A. Orosz, J.B. Fitzgerald, J.S. Bloom, P.G. van Dokkum, C.D. Bailyn, M.M. Buxton, M. Barsony, The detailed optical light curve of GRB 030329. Astrophys. J. **606**, 381–394 (2004).
148. D.R. Lorimer, M. Kramer, *Handbook of Pulsar Astronomy* (2004)
149. R. Lunnan, R. Chornock, E. Berger, T. Laskar, W. Fong, A. Rest, N.E. Sanders, P.M. Challis, M.R. Drout, R.J. Foley, M.E. Huber, R.P. Kirshner, C. Leibler, G.H. Marion, M. McCrum, D. Milisavljevic, G. Narayan, D. Scolnic, S.J. Smartt, K.W. Smith, A.M. Soderberg, J.L. Tonry, W.S. Burgett, K.C. Chambers, H. Flewelling, K.W. Hodapp, N. Kaiser, E.A. Magnier, P.A. Price, R.J. Wainscoat, Hydrogen-poor superluminous supernovae and long-duration gamma-ray bursts have similar host galaxies. Astrophys. J. **787**, 138 (2014).
150. R. Lunnan, R. Chornock, E. Berger, A. Rest, W. Fong, D. Scolnic, D.O. Jones, A.M. Soderberg, P.M. Challis, M.R. Drout, R.J. Foley, M.E. Huber, R.P. Kirshner, C. Leibler, G.H. Marion, M. McCrum, D. Milisavljevic, G. Narayan, N.E. Sanders, S.J. Smartt, K.W. Smith, J.L. Tonry, W.S. Burgett, K.C. Chambers, H. Flewelling, R.-P. Kudritzki, R.J. Wainscoat, C. Waters, Zooming in on the progenitors of superluminous supernovae with the HST. Astrophys. J. **804**, 90 (2015).
151. J.D. Lyman, D. Bersier, P.A. James, Bolometric corrections for optical light curves of core-collapse supernovae. Mon. Not. R. Astron. Soc. **437**, 3848–3862 (2014a).
152. J.D. Lyman, A.J. Levan, R.P. Church, M.B. Davies, N.R. Tanvir, The progenitors of calcium-rich transients are not formed in situ. Mon. Not. R. Astron. Soc. **444**, 2157–2166 (2014b).
153. J.D. Lyman, D. Bersier, P.A. James, P.A. Mazzali, J.J. Eldridge, M. Fraser, E. Pian, Bolometric light curves and explosion parameters of 38 stripped-envelope core-collapse supernovae. Mon. Not. R. Astron. Soc. **457**, 328–350 (2016a).
154. J.D. Lyman, A.J. Levan, P.A. James, C.R. Angus, R.P. Church, M.B. Davies, N.R. Tanvir, Hubble Space Telescope observations of the host galaxies and environments of calcium-rich supernovae. Mon. Not. R. Astron. Soc. **458**, 1768–1777 (2016b).
155. T.J. Maccarone, S.E. de Mink, Large proper motion of the Thorne-Żytkow object candidate HV 2112 reveals its likely nature as foreground Galactic S-star. Mon. Not. R. Astron. Soc. **458**, 1–5 (2016).
156. b.P. Macias, E. Ramirez-Ruiz, A Stringent Limit on the Mass Production Rate of rr-Process Elements in the Milky Way. ArXiv e-prints (2016)
157. K. Maeda, T. Nakamura, K. Nomoto, P.A. Mazzali, F. Patat, I. Hachisu, Explosive nucleosynthesis in aspherical hypernova explosions and late-time spectra of SN 1998bw. Astrophys. J. **565**, 405–412 (2002).
158. Martins, D.J. Hillier, J.C. Bouret, E. Depagne, C. Foellmi, S. Marchenko, A.F. Moffat, Properties of WNh stars in the Small Magellanic Cloud: evidence for homogeneous evolution. Astron. Astrophys. **495**, 257–270 (2009).
159. Maselli, A. Melandri, L. Nava, C.G. Mundell, N. Kawai, S. Campana, S. Covino, J.R. Cummings, G. Cusumano, P.A. Evans, G. Ghirlanda, G. Ghisellini, C. Guidorzi, S. Kobayashi, P. Kuin, V. La Parola, V. Mangano, S. Oates, T. Sakamoto, M. Serino, F. Virgili, B.-B. Zhang, S. Barthelmy, A. Beardmore, M.G. Bernardini, D. Bersier, D. Burrows, G. Calderone, M. Capalbi, J. Chiang, P. D'Avanzo, V. D'Elia, M. De Pasquale, D. Fugazza, N. Gehrels, A. Gomboc, R. Harrison, H. Hanayama, J. Japelj, J. Kennea, D. Kopac, C. Kouveliotou, D. Kuroda, A. Levan, D. Malesani, F. Marshall, J. Nousek, P. O'Brien, J.P. Osborne, C. Pagani, K.L. Page, M. Page, M. Perri, T. Pritchard, P. Romano, Y. Saito, B. Sbarufatti, R. Salvaterra, I. Steele, N. Tanvir, G. Vianello, B. Weigand, K. Wiersema, Y. Yatsu, T. Yoshii, G. Tagliaferri, GRB 130427A: a nearby ordinary monster. Science **343**, 48–51 (2014).

160. N. Masetti, E. Palazzi, E. Pian, A. Simoncelli, L.K. Hunt, E. Maiorano, A. Levan, L. Christensen, E. Rol, S. Savaglio, R. Falomo, A.J. Castro-Tirado, J. Hjorth, A. Delsanti, M. Pannella, V. Mohan, S.B. Pandey, R. Sagar, L. Amati, I. Burud, J.M. Castro Cerón, F. Frontera, A.S. Fruchter, J.P.U. Fynbo, J. Gorosabel, L. Kaper, S. Klose, C. Kouveliotou, L. Nicastro, H. Pedersen, J. Rho, I. Salamanca, N. Tanvir, P.M. Vreeswijk, R.A.M.J. Wijers, E.P.J. van den Heuvel, Optical and near-infrared observations of the GRB020405 afterglow. Astron. Astrophys. **404**, 465–481 (2003).
161. B.D. Mason, T.J. Henry, W.I. Hartkopf, T. ten Brummelaar, D.R. Soderblom, A multiplicity survey of chromospherically active and inactive stars. Astron. J. **116**, 2975–2983 (1998).
162. E.P. Mazets, S.V. Golenetskii, V.N. Ilinskii, V.N. Panov, R.L. Aptekar, I.A. Gurian, M.P. Proskura, I.A. Sokolov, Z.I. Sokolova, T.V. Kharitonova, Catalog of cosmic gamma-ray bursts from the KONUS experiment data. I. Astrophys. Space Sci. **80**, 3–83 (1981).
163. E.P. Mazets, S.V. Golenetskii, I.A. Gurian, V.N. Ilinskii, The 5 March 1979 event and the distinct class of short gamma bursts are they of the same origin. Astrophys. Space Sci. **84**, 173–189 (1982).
164. P.A. Mazzali, J. Deng, K. Nomoto, D.N. Sauer, E. Pian, N. Tominaga, M. Tanaka, K. Maeda, A.V. Filippenko, A neutron-star-driven X-ray flash associated with supernova SN 2006aj. Nature **442**, 1018–1020 (2006).
165. P.A. Mazzali, R.J. Foley, J. Deng, F. Patat, E. Pian, D. Baade, J.S. Bloom, A.V. Filippenko, D.A. Perley, S. Valenti, L. Wang, K. Kawabata, K. Maeda, K-i. Nomoto, Keck and European Southern Observatory very large telescope view of the symmetry of the ejecta of the XRF/SN 2006aj. Astrophys. J. **661**, 892–898 (2007a).
166. P.A. Mazzali, K.S. Kawabata, K. Maeda, R.J. Foley, K. Nomoto, J. Deng, T. Suzuki, M. Iye, N. Kashikawa, Y. Ohyama, A.V. Filippenko, Y. Qiu, J. Wei, The aspherical properties of the energetic type Ic SN 2002ap as inferred from its nebular spectra. Astrophys. J. **670**, 592–599 (2007b).
167. P.A. Mazzali, K.S. Kawabata, K. Maeda, R.J. Foley, K. Nomoto, J. Deng, T. Suzuki, M. Iye, N. Kashikawa, Y. Ohyama, A.V. Filippenko, Y. Qiu, J. Wei, The aspherical properties of the energetic type Ic SN 2002ap as inferred from its nebular spectra. Astrophys. J. **670**, 592–599 (2007c).
168. P.A. Mazzali, A.I. McFadyen, S.E. Woosley, E. Pian, M. Tanaka, An upper limit to the energy of gamma-ray bursts indicates that GRBs/SNe are powered by magnetars. Mon. Not. R. Astron. Soc. **443**, 67–71 (2014).
169. P.A. Mazzali, M. Sullivan, E. Pian, J. Greiner, D.A. Kann, Spectrum formation in superluminous supernovae (Type I). Mon. Not. R. Astron. Soc. **458**, 3455–3465 (2016).
170. Melandri, E. Pian, V. D'Elia, P. D'Avanzo, M. Della Valle, P.A. Mazzali, G. Tagliaferri, Z. Cano, A.J. Levan, Diversity of gamma-ray burst energetics vs. supernova homogeneity: SN 2013cq associated with GRB 130427A. Astron. Astrophys. **567**, 29 (2014).
171. B.D. Metzger, Nuclear-dominated accretion and subluminous supernovae from the merger of a white dwarf with a neutron star or black hole. Mon. Not. R. Astron. Soc. **419**, 827–840 (2012).
172. B.D. Metzger, E. Berger, What is the most promising electromagnetic counterpart of a neutron star binary merger? Astrophys. J. **746**, 48 (2012).
173. B.D. Metzger, D. Giannios, T.A. Thompson, N. Bucciantini, E. Quataert, The protomagnetar model for gamma-ray bursts. Mon. Not. R. Astron. Soc. **413**, 2031–2056 (2011).
174. B.D. Metzger, B. Margalit, D. Kasen, E. Quataert, The diversity of transients from magnetar birth in core collapse supernovae. Mon. Not. R. Astron. Soc. **454**, 3311–3316 (2015).
175. M.R. Mokiem, A. de Koter, C.J. Evans, J. Puls, S.J. Smartt, P.A. Crowther, A. Herrero, N. Langer, D.J. Lennon, F. Najarro, M.R. Villamariz, S.-C. Yoon, The VLT-FLAMES survey of massive stars: mass loss and rotation of early-type stars in the SMC. Astron. Astrophys. **456**, 1131–1151 (2006).
176. S. Mukherjee, E.D. Feigelson, G. Jogesh Babu, F. Murtagh, C. Fraley, A. Raftery, Three types of gamma-ray bursts. Astrophys. J. **508**, 314–327 (1998).
177. R. Narayan, B. Paczynski, T. Piran, Gamma-ray bursts as the death throes of massive binary stars. Astrophys. J. Lett. **395**, 83–86 (1992).

178. Nelemans, C.A. Tout, Reconstructing the evolution of white dwarf binaries: further evidence for an alternative algorithm for the outcome of the common-envelope phase in close binaries. Mon. Not. R. Astron. Soc. **356**, 753–764 (2005).
179. R.J. Nemiroff, A century of gamma ray burst models. Comments Astrophys. **17**, 189 (1994)
180. S. Nissanke, J. Sievers, N. Dalal, D. Holz, Localizing compact binary inspirals on the sky using ground-based gravitational wave interferometers. Astrophys. J. **739**, 99 (2011).
181. J.A. Nousek, C. Kouveliotou, D. Grupe, K.L. Page, J. Granot, E. Ramirez-Ruiz, S.K. Patel, D.N. Burrows, V. Mangano, S. Barthelmy, A.P. Beardmore, S. Campana, M. Capalbi, G. Chincarini, G. Cusumano, A.D. Falcone, N. Gehrels, P. Giommi, M.R. Goad, O. Godet, C.P. Hurkett, J.A. Kennea, A. Moretti, P.T. O'Brien, J.P. Osborne, P. Romano, G. Tagliaferri, A.A. Wells, Evidence for a canonical gamma-ray burst afterglow light curve in the Swift XRT data. Astrophys. J. **642**, 389–400 (2006).
182. M. Nysewander, A.S. Fruchter, A. Pe'er, A comparison of the afterglows of short- and long-duration gamma-ray bursts. Astrophys. J. **701**, 824–836 (2009a).
183. M. Nysewander, A.S. Fruchter, A. Pe'er, A comparison of the afterglows of short- and long-duration gamma-ray bursts. Astrophys. J. **701**, 824–836 (2009b).
184. R. Oechslin, H.-T. Janka, Torus formation in neutron star mergers and well-localized short gamma-ray bursts. Mon. Not. R. Astron. Soc. **368**, 1489–1499 (2006).
185. S.A. Olausen, V.M. Kaspi, The McGill magnetar catalog. Astrophys. J. Suppl. Ser. **212**, 6 (2014).
186. D.M. Palmer, S. Barthelmy, N. Gehrels, R.M. Kippen, T. Cayton, C. Kouveliotou, D. Eichler, R.A.M.J. Wijers, P.M. Woods, J. Granot, Y.E. Lyubarsky, E. Ramirez-Ruiz, L. Barbier, M. Chester, J. Cummings, E.E. Fenimore, M.H. Finger, B.M. Gaensler, D. Hullinger, H. Krimm, C.B. Markwardt, J.A. Nousek, A. Parsons, S. Patel, T. Sakamoto, G. Sato, M. Suzuki, J. Tueller, A giant $\gamma\gamma$-ray flare from the magnetar SGR 1806-20. Nature **434**, 1107–1109 (2005).
187. D.A. Perley, B.D. Metzger, J. Granot, N.R. Butler, T. Sakamoto, E. Ramirez-Ruiz, A.J. Levan, J.S. Bloom, A.A. Miller, A. Bunker, H.-W. Chen, A.V. Filippenko, N. Gehrels, K. Glazebrook, P.B. Hall, K.C. Hurley, D. Kocevski, W. Li, S. Lopez, J. Norris, A.L. Piro, D. Poznanski, J.X. Prochaska, E. Quataert, N. Tanvir, GRB 080503: implications of a naked short gamma-ray burst dominated by extended emission. Astrophys. J. **696**, 1871–1885 (2009).
188. D.A. Perley, S.B. Cenko, A. Corsi, N.R. Tanvir, A.J. Levan, D.A. Kann, E. Sonbas, K. Wiersema, W. Zheng, X.-H. Zhao, J.-M. Bai, M. Bremer, A.J. Castro-Tirado, L. Chang, K.I. Clubb, D. Frail, A. Fruchter, E. Göğüş, J. Greiner, T. Güver, A. Horesh, A.V. Filippenko, S. Klose, J. Mao, A.N. Morgan, A.S. Pozanenko, S. Schmidl, B. Stecklum, M. Tanga, A.A. Volnova, A.E. Volvach, J.-G. Wang, J.-M. Winters, Y.-X. Xin, The afterglow of GRB 130427A from 1 to 1016 GHz1016 GHz. Astrophys. J. **781**, 37 (2014).
189. D.A. Perley, N.R. Tanvir, J. Hjorth, T. Laskar, E. Berger, R. Chary, A. de Ugarte Postigo, J.P.U. Fynbo, T. Krühler, A.J. Levan, M.J. Michałowski, S. Schulze, The Swift GRB Host Galaxy Legacy Survey—II. Rest-Frame NIR Luminosity Distribution and Evidence for a Near-Solar Metallicity Threshold. ArXiv e-prints (2015)
190. D.A. Perley, T. Krühler, P. Schady, M.J. Michałowski, C.C. Thöne, D. Petry, J.F. Graham, J. Greiner, S. Schulze, S. Kim, A Revised Host Galaxy Association for GRB 020819B: A High-Redshift Dusty Starburst, Not a Low-Redshift Gas-Poor Spiral. ArXiv e-prints (2016)
191. R. Perna, P.J. Armitage, B. Zhang, Flares in long and short gamma-ray bursts: a common origin in a hyperaccreting accretion disk. Astrophys. J. Lett. **636**, 29–32 (2006a).
192. Pian, P.A. Mazzali, N. Masetti, P. Ferrero, S. Klose, E. Palazzi, E. Ramirez-Ruiz, S.E. Woosley, C. Kouveliotou, J. Deng, A.V. Filippenko, R.J. Foley, J.P.U. Fynbo, D.A. Kann, W. Li, J. Hjorth, K. Nomoto, F. Patat, D.N. Sauer, J. Sollerman, P.M. Vreeswijk, E.W. Guenther, A. Levan, P. O'Brien, N.R. Tanvir, R.A.M.J. Wijers, C. Dumas, O. Hainaut, D.S. Wong, D. Baade, L. Wang, L. Amati, E. Cappellaro, A.J. Castro-Tirado, S. Ellison, F. Frontera, A.S. Fruchter, J. Greiner, K. Kawabata, C. Ledoux, K. Maeda, P. Møller, L. Nicastro, E. Rol, R. Starling, An optical supernova associated with the X-ray flash XRF 060218. Nature **442**, 1011–1013 (2006b).
193. P. Podsiadlowski, P.A. Mazzali, K. Nomoto, D. Lazzati, E. Cappellaro, The rates of hypernovae and gamma-ray bursts: implications for their progenitors. Astrophys. J. Lett. **607**, 17–20 (2004).

194. P. Podsiadlowski, N. Ivanova, S. Justham, S. Rappaport, Explosive common-envelope ejection: implications for gamma-ray bursts and low-mass black-hole binaries. Mon. Not. R. Astron. Soc. **406**, 840–847 (2010).
195. S.F. Portegies Zwart, S.L.W. McMillan, M. Gieles, Young massive star clusters. Annu. Rev. Astron. Astrophys. **48**, 431–493 (2010).
196. S. Prajs, M. Sullivan, M. Smith, A. Levan, N.V. Karpenka, T.D.P. Edwards, C.R. Walker, W.M. Wolf, C. Balland, R. Carlberg, A. Howell, C. Lidman, R. Pain, C. Pritchet, V. Ruhlmann-Kleider, The Volumetric Rate of Superluminous Supernovae at z~1z~1. ArXiv e-prints (2016)
197. P.A. Price, E. Berger, D.E. Reichart, S.R. Kulkarni, S.A. Yost, R. Subrahmanyan, R.M. Wark, M.H. Wieringa, D.A. Frail, J. Bailey, B. Boyle, E. Corbett, K. Gunn, S.D. Ryder, N. Seymour, K. Koviak, P. McCarthy, M. Phillips, T.S. Axelrod, J.S. Bloom, S.G. Djorgovski, D.W. Fox, T.J. Galama, F.A. Harrison, K. Hurley, R. Sari, B.P. Schmidt, M.J.I. Brown, T. Cline, F. Frontera, C. Guidorzi, E. Montanari, GRB 011121: a massive star progenitor. Astrophys. J. Lett. **572**, 51–55 (2002).
198. P.A. Price, S.R. Kulkarni, E. Berger, D.W. Fox, J.S. Bloom, S.G. Djorgovski, D.A. Frail, T.J. Galama, F.A. Harrison, P. McCarthy, D.E. Reichart, R. Sari, S.A. Yost, H. Jerjen, K. Flint, A. Phillips, B.E. Warren, T.S. Axelrod, R.A. Chevalier, J. Holtzman, R.A. Kimble, B.P. Schmidt, J.C. Wheeler, F. Frontera, E. Costa, L. Piro, K. Hurley, T. Cline, C. Guidorzi, E. Montanari, E. Mazets, S. Golenetskii, I. Mitrofanov, D. Anfimov, A. Kozyrev, M. Litvak, A. Sanin, W. Boynton, C. Fellows, K. Harshman, C. Shinohara, A. Gal-Yam, E. Ofek, Y. Lipkin, Discovery of GRB 020405 and its late red bump. Astrophys. J. **589**, 838–843 (2003a).
199. P.A. Price, S.R. Kulkarni, B.P. Schmidt, T.J. Galama, J.S. Bloom, E. Berger, D.A. Frail, S.G. Djorgovski, D.W. Fox, A.A. Henden, S. Klose, F.A. Harrison, D.E. Reichart, R. Sari, S.A. Yost, T.S. Axelrod, P. McCarthy, J. Holtzman, J.P. Halpern, R.A. Kimble, J.C. Wheeler, R.A. Chevalier, K. Hurley, G.R. Ricker, E. Costa, F. Frontera, L. Piro, GRB 010921: strong limits on an underlying supernova from the Hubble Space Telescope. Astrophys. J. **584**, 931–936 (2003b).
200. P.A. Price, D.W. Fox, S.R. Kulkarni, B.A. Peterson, B.P. Schmidt, A.M. Soderberg, S.A. Yost, E. Berger, S.G. Djorgovski, D.A. Frail, F.A. Harrison, R. Sari, A.W. Blain, S.C. Chapman, The bright optical afterglow of the nearby γγ-ray burst of 29 March 2003. Nature **423**, 844–847 (2003c).
201. R.M. Quimby, F. Yuan, C. Akerlof, J.C. Wheeler, Rates of superluminous supernovae at z=0.2z=0.2. Mon. Not. R. Astron. Soc. **431**, 912–922 (2013).
202. Raghavan, H.A. McAlister, T.J. Henry, D.W. Latham, G.W. Marcy, B.D. Mason, D.R. Gies, R.J. White, T.A. ten Brummelaar, A survey of stellar families: multiplicity of solar-type stars. Astrophys. J. Suppl. Ser. **190**, 1–42 (2010).
203. O.H. Ramírez-Agudelo, H. Sana, S.E. de Mink, V. Hénault-Brunet, A. de Koter, N. Langer, F. Tramper, G. Gräfener, C.J. Evans, J.S. Vink, P.L. Dufton, W.D. Taylor, The VLT-FLAMES Tarantula Survey. XXI. Stellar spin rates of O-type spectroscopic binaries. Astron. Astrophys. **580**, 92 (2015).
204. Raskin, E. Scannapieco, J. Rho, M. Della Valle, Using spatial distributions to constrain progenitors of supernovae and gamma-ray bursts. Astrophys. J. **689**, 358–370 (2008).
205. N. Rea, M. Gullón, J.A. Pons, R. Perna, M.G. Dainotti, J.A. Miralles, D.F. Torres, Constraining the GRB-magnetar model by means of the galactic pulsar population. Astrophys. J. **813**, 92 (2015).
206. T.M. Reynolds, M. Fraser, G. Gilmore, Gone without a bang: an archival HST survey for disappearing massive stars. Mon. Not. R. Astron. Soc. **453**(3), 2889–2900 (2015)
207. J.E. Rho, The dynamics and light curves of beamed gamma-ray burst afterglows. Astrophys. J. **525**, 737–749 (1999).
208. S. Rosswog, E. Ramirez-Ruiz, M.B. Davies, High-resolution calculations of merging neutron stars—III. Gamma-ray bursts. Mon. Not. R. Astron. Soc. **345**, 1077–1090 (2003).
209. S. Rosswog, M. Liebendörfer, F.-K. Thielemann, M.B. Davies, W. Benz, T. Piran, Mass ejection in neutron star mergers. Astron. Astrophys. **341**, 499–526 (1999).

210. Rowlinson, P.T. O'Brien, B.D. Metzger, N.R. Tanvir, A.J. Levan, Signatures of magnetar central engines in short GRB light curves. Mon. Not. R. Astron. Soc. **430**, 1061–1087 (2013).
211. Sana, G. James, E. Gosset, The massive star binary fraction in young open clusters—III. IC 2944 and the Cen OB2 association. Mon. Not. R. Astron. Soc. **416**, 817 (2011)
212. H. Sana, Y. Momany, M. Gieles, G. Carraro, Y. Beletsky, V.D. Ivanov, G. de Silva, G. James, A MAD view of Trumpler 14. Astron. Astrophys. **515**, 26 (2010).
213. H. Sana, S.E. de Mink, A. de Koter, N. Langer, C.J. Evans, M. Gieles, E. Gosset, R.G. Izzard, J.-B. Le Bouquin, F.R.N. Schneider, Binary interaction dominates the evolution of massive stars. Science **337**, 444 (2012).
214. H. Sana, A. de Koter, S.E. de Mink, P.R. Dunstall, C.J. Evans, V. Hénault-Brunet, J. Maíz Apellániz, O.H. Ramírez-Agudelo, W.D. Taylor, N.R. Walborn, J.S. Clark, P.A. Crowther, A. Herrero, M. Gieles, N. Langer, D.J. Lennon, J.S. Vink, The VLT-FLAMES Tarantula Survey. VIII. Multiplicity properties of the O-type star population. Astron. Astrophys. **550**, 107 (2013).
215. R. Sari, T. Piran, R. Narayan, Spectra and light curves of gamma-ray burst afterglows. Astrophys. J. Lett. **497**, 17–20 (1998).
216. S. Savaglio, K. Glazebrook, D. Le Borgne, The galaxy population hosting gamma-ray bursts. Astrophys. J. **691**, 182–211 (2009).
217. b.T. Shenar, R. Hainich, H. Todt, A. Sander, W.-R. Hamann, A.F.J. Moffat, J.J. Eldridge, H. Pablo, L.M. Oskinova, N.D. Richardson, Wolf-Rayet stars in the Small Magellanic Cloud: II. Analysis of the binaries. ArXiv e-prints (2016)
218. S.J. Smartt, J.J. Eldridge, R.M. Crockett, J.R. Maund, The death of massive stars—I. Observational constraints on the progenitors of Type II-P supernovae. Mon. Not. R. Astron. Soc. **395**, 1409–1437 (2009).
219. A.M. Soderberg, S.R. Kulkarni, E. Berger, D.W. Fox, M. Sako, D.A. Frail, A. Gal-Yam, D.S. Moon, S.B. Cenko, S.A. Yost, M.M. Phillips, S.E. Persson, W.L. Freedman, P. Wyatt, R. Jayawardhana, D. Paulson, The sub-energetic $\gamma\gamma$-ray burst GRB 031203 as a cosmic analogue to the nearby GRB 980425. Nature **430**, 648–650 (2004).
220. A.M. Soderberg, S.R. Kulkarni, E. Nakar, E. Berger, P.B. Cameron, D.B. Fox, D. Frail, A. Gal-Yam, R. Sari, S.B. Cenko, M. Kasliwal, R.A. Chevalier, T. Piran, P.A. Price, B.P. Schmidt, G. Pooley, D.-S. Moon, B.E. Penprase, E. Ofek, A. Rau, N. Gehrels, J.A. Nousek, D.N. Burrows, S.E. Persson, P.J. McCarthy, Relativistic ejecta from X-ray flash XRF 060218 and the rate of cosmic explosions. Nature **442**, 1014–1017 (2006).
221. M. Sparre, R.L.C. Starling, A search for thermal X-ray signatures in gamma-ray bursts—II. The Swift sample. Mon. Not. R. Astron. Soc. **427**, 2965–2974 (2012).
222. H.C. Spruit, Dynamo action by differential rotation in a stably stratified stellar interior. Astron. Astrophys. **381**, 923–932 (2002).
223. K.Z. Stanek, T. Matheson, P.M. Garnavich, P. Martini, P. Berlind, N. Caldwell, P. Challis, W.R. Brown, R. Schild, K. Krisciunas, M.L. Calkins, J.C. Lee, N. Hathi, R.A. Jansen, R. Windhorst, L. Echevarria, D.J. Eisenstein, B. Pindor, E.W. Olszewski, P. Harding, S.T. Holland, D. Bersier, Spectroscopic discovery of the supernova 2003dh associated with GRB 030329. Astrophys. J. Lett. **591**, 17–20 (2003).
224. E.R. Stanway, J.J. Eldridge, G.D. Becker, Stellar population effects on the inferred photon density at reionization. Mon. Not. R. Astron. Soc. **456**, 485–499 (2016).
225. R.L.C. Starling, K. Wiersema, A.J. Levan, T. Sakamoto, D. Bersier, P. Goldoni, S.R. Oates, A. Rowlinson, S. Campana, J. Sollerman, N.R. Tanvir, D. Malesani, J.P.U. Fynbo, S. Covino, P. D'Avanzo, P.T. O'Brien, K.L. Page, J.P. Osborne, S.D. Vergani, S. Barthelmy, D.N. Burrows, Z. Cano, P.A. Curran, M. de Pasquale, V. D'Elia, P.A. Evans, H. Flores, A.S. Fruchter, P. Garnavich, N. Gehrels, J. Gorosabel, J. Hjorth, S.T. Holland, A.J. van der Horst, C.P. Hurkett, P. Jakobsson, A.P. Kamble, C. Kouveliotou, N.P.M. Kuin, L. Kaper, P.A. Mazzali, P.E. Nugent, E. Pian, M. Stamatikos, C.C. Thöne, S.E. Woosley, Discovery of the nearby long, soft GRB 100316D with an associated supernova. Mon. Not. R. Astron. Soc. **411**, 2792–2803 (2011).

226. R.L.C. Starling, K.L. Page, A. Pe'Er, A.P. Beardmore, J.P. Osborne, A search for thermal X-ray signatures in gamma-ray bursts—I. Swift bursts with optical supernovae. Mon. Not. R. Astron. Soc. **427**, 2950–2964 (2012).
227. M.P.L. Suijs, N. Langer, A.-J. Poelarends, S.-C. Yoon, A. Heger, F. Herwig, White dwarf spins from low-mass stellar evolution models. Astron. Astrophys. **481**, 87–90 (2008).
228. b.H. Sun, B. Zhang, H. Gao, X-ray counterpart of gravitational waves due to binary neutron star mergers: light curves, luminosity functions, and event rate densities. ArXiv e-prints (2016)
229. H. Sun, B. Zhang, Z. Li, Extragalactic high-energy transients: event rate densities and luminosity functions. Astrophys. J. **812**, 33 (2015).
230. K.M. Svensson, A.J. Levan, N.R. Tanvir, A.S. Fruchter, L.-G. Strolger, The host galaxies of core-collapse supernovae and gamma-ray bursts. Mon. Not. R. Astron. Soc. **405**, 57–76 (2010).
231. P.A. Sweet, The importance of rotation in stellar evolution. Mon. Not. R. Astron. Soc. **110**, 548 (1950)..
232. Szécsi, N. Langer, S.-C. Yoon, D. Sanyal, S. de Mink, C.J. Evans, T. Dermine, Low-metallicity massive single stars with rotation. Evolutionary models applicable to I Zwicky 18. Astron. Astrophys. **581**, 15 (2015).
233. N.R. Tanvir, R. Chapman, A.J. Levan, R.S. Priddey, An origin in the local Universe for some short γγ-ray bursts. Nature **438**, 991–993 (2005).
234. N.R. Tanvir, D.B. Fox, A.J. Levan, E. Berger, K. Wiersema, J.P.U. Fynbo, A. Cucchiara, T. Krühler, N. Gehrels, J.S. Bloom, J. Greiner, P.A. Evans, E. Rol, F. Olivares, J. Hjorth, P. Jakobsson, J. Farihi, R. Willingale, R.L.C. Starling, S.B. Cenko, D. Perley, J.R. Maund, J. Duke, R.A.M.J. Wijers, A.J. Adamson, A. Allan, M.N. Bremer, D.N. Burrows, A.J. Castro-Tirado, B. Cavanagh, A. de Ugarte Postigo, M.A. Dopita, T.A. Fatkhullin, A.S. Fruchter, R.J. Foley, J. Gorosabel, J. Kennea, T. Kerr, S. Klose, H.A. Krimm, V.N. Komarova, S.R. Kulkarni, A.S. Moskvitin, C.G. Mundell, T. Naylor, K. Page, B.E. Penprase, M. Perri, P. Podsiadlowski, K. Roth, R.E. Rutledge, T. Sakamoto, P. Schady, B.P. Schmidt, A.M. Soderberg, J. Sollerman, A.W. Stephens, G. Stratta, T.N. Ukwatta, D. Watson, E. Westra, T. Wold, C. Wolf, A γγ-ray burst at a redshift of z~8.2z~8.2. Nature **461**, 1254–1257 (2009).
235. N.R. Tanvir, A.J. Levan, A.S. Fruchter, J. Hjorth, R.A. Hounsell, K. Wiersema, R.L. Tunnicliffe, A 'kilonova' associated with the short-duration γγ-ray burst GRB 130603B. Nature **500**, 547–549 (2013).
236. C.C. Thöne, A. de Ugarte Postigo, C.L. Fryer, K.L. Page, J. Gorosabel, M.A. Aloy, D.A. Perley, C. Kouveliotou, H.T. Janka, P. Mimica, J.L. Racusin, H. Krimm, J. Cummings, S.R. Oates, S.T. Holland, M.H. Siegel, M. de Pasquale, E. Sonbas, M. Im, W.-K. Park, D.A. Kann, S. Guziy, L.H. García, A. Llorente, K. Bundy, C. Choi, H. Jeong, H. Korhonen, P. Kubànek, J. Lim, A. Moskvitin, T. Muñoz-Darias, S. Pak, I. Parrish, The unusual γγ-ray burst GRB 101225A from a helium star/neutron star merger at redshift 0.33. Nature **480**, 72–74 (2011a).
237. C.C. Thöne, S. Campana, D. Lazzati, A. de Ugarte Postigo, J.P.U. Fynbo, L. Christensen, A.J. Levan, M.A. Aloy, J. Hjorth, P. Jakobsson, E.M. Levesque, D. Malesani, B. Milvang-Jensen, P.W.A. Roming, N.R. Tanvir, K. Wiersema, M. Gladders, E. Wuyts, H. Dahle, Variable Lyααsheds light on the environment surrounding GRB 090426. Mon. Not. R. Astron. Soc. **414**, 479–488 (2011b).
238. Tramper, S.M. Straal, D. Sanyal, H. Sana, A. de Koter, G. Gräfener, N. Langer, J.S. Vink, S.E. de Mink, L. Kaper, Massive stars on the verge of exploding: the properties of oxygen sequence Wolf–Rayet stars. Astron. Astrophys. **581**, 110 (2015).
239. C.A. Tremonti, T.M. Heckman, G. Kauffmann, J. Brinchmann, S. Charlot, S.D.M. White, M. Seibert, E.W. Peng, D.J. Schlegel, A. Uomoto, M. Fukugita, J. Brinkmann, The origin of the mass-metallicity relation: insights from 53,000 star-forming galaxies in the Sloan Digital Sky Survey. Astrophys. J. **613**, 898–913 (2004).
240. R.L. Tunnicliffe, A.J. Levan, N.R. Tanvir, A. Rowlinson, D.A. Perley, J.S. Bloom, S.B. Cenko, P.T. O'Brien, B.E. Cobb, K. Wiersema, D. Malesani, A. de Ugarte Postigo, J. Hjorth, J.P.U. Fynbo, P. Jakobsson, On the nature of the 'hostless' short GRBs. Mon. Not. R. Astron. Soc. **437**, 1495–1510 (2014).

241. R. Tylenda, N. Soker, Eruptions of the V838 Mon type: stellar merger versus nuclear outburst models. Astron. Astrophys. **451**, 223–236 (2006).
242. R. Tylenda, M. Hajduk, T. Kamiński, A. Udalski, I. Soszyński, M.K. Szymański, M. Kubiak, G. Pietrzyński, R. Poleski, Ł. Wyrzykowski, K. Ulaczyk, V1309 Scorpii: merger of a contact binary. Astron. Astrophys. **528**, 114 (2011).
243. V.V. Usov, Millisecond pulsars with extremely strong magnetic fields as a cosmological source of gamma-ray bursts. Nature **357**, 472–474 (1992).
244. E.P.J. van den Heuvel, S.-C. Yoon, Long gamma-ray burst progenitors: boundary conditions and binary models. Astrophys. Space Sci. **311**, 177–183 (2007).
245. van Paradijs, P.J. Groot, T. Galama, C. Kouveliotou, R.G. Strom, J. Telting, R.G.M. Rutten, G.J. Fishman, C.A. Meegan, M. Pettini, N. Tanvir, J. Bloom, H. Pedersen, H.U. Nørdgaard-Nielsen, M. Linden-Vørnle, J. Melnick, G. van der Steene, M. Bremer, R. Naber, J. Heise, J. in't Zand, E. Costa, M. Feroci, L. Piro, F. Frontera, G. Zavattini, L. Nicastro, E. Palazzi, K. Bennett, L. Hanlon, A. Parmar, Transient optical emission from the error box of the γγ-ray burst of 28 February 1997. Nature **386**, 686–689 (1997).
246. S. Vaughan, R. Willingale, P.T. O'Brien, J.P. Osborne, J.N. Reeves, A.J. Levan, M.G. Watson, J.A. Tedds, D. Watson, M. Santos-Lleó, P.M. Rodríguez-Pascual, N. Schartel, The discovery of an evolving dust-scattered X-ray halo around GRB 031203. Astrophys. J. Lett. **603**, 5–8 (2004).
247. J.S. Vink, Constraining GRB progenitor models by probing Wolf–Rayet wind geometries in the Large Magellanic Cloud. Astron. Astrophys. **469**, 707–711 (2007).
248. J.S. Vink, A. de Koter, On the metallicity dependence of Wolf–Rayet winds. Astron. Astrophys. **442**, 587–596 (2005).
249. J.S. Vink, A. de Koter, H.J.G.L.M. Lamers, Mass-loss predictions for O and B stars as a function of metallicity. Astron. Astrophys. **369**, 574–588 (2001).
250. Wallner, T. Faestermann, J. Feige, C. Feldstein, K. Knie, G. Korschinek, W. Kutschera, A. Ofan, M. Paul, F. Quinto, G. Rugel, P. Steier, Abundance of live 244Pu244Pu in deep-sea reservoirs on Earth points to rarity of actinide nucleosynthesis. Nat. Commun. **6**, 5956 (2015).
251. D. Watson, S.A. Vaughan, R. Willingale, J. Hjorth, S. Foley, J.P.U. Fynbo, P. Jakobsson, A. Levan, P.T. O'Brien, J.P. Osborne, K. Pedersen, J.N. Reeves, J.A. Tedds, M.G. Watson, The soft X-ray blast in the apparently subluminous GRB 031203. Astrophys. J. **636**, 967–970 (2006).
252. Weidner, P. Kroupa, I.A.D. Bonnell, The relation between the most-massive star and its parental star cluster mass. Mon. Not. R. Astron. Soc. **401**, 275–293 (2010).
253. R. Willingale, J.P. Osborne, P.T. O'Brien, M.J. Ward, A. Levan, K.L. Page, The X-ray, optical and radio evolution of the GRB 030329 afterglow and the associated SN2003dh. Mon. Not. R. Astron. Soc. **349**, 31–38 (2004).
254. S.E. Woosley, Gamma-ray bursts from stellar mass accretion disks around black holes. Astrophys. J. **405**, 273–277 (1993).
255. S.E. Woosley, A. Heger, The progenitor stars of gamma-ray bursts. Astrophys. J. **637**, 914–921 (2006).
256. S.E. Woosley, R.G. Eastman, B.P. Schmidt, Gamma-ray bursts and Type IC supernova SN 1998BW. Astrophys. J. **516**, 788–796 (1999).
257. C.C. Worley, M.J. Irwin, C.A. Tout, A.N. Żytkow, M. Fraser, R.G. Izzard, The proper motion of HV2112: a TŻO candidate in the SMC. Mon. Not. R. Astron. Soc. **459**, 31–35 (2016).
258. Xu, A. de Ugarte Postigo, G. Leloudas, T. Krühler, Z. Cano, J. Hjorth, D. Malesani, J.P.U. Fynbo, C.C. Thöne, R. Sánchez-Ramírez, S. Schulze, P. Jakobsson, L. Kaper, J. Sollerman, D.J. Watson, A. Cabrera-Lavers, C. Cao, S. Covino, H. Flores, S. Geier, J. Gorosabel, S.M. Hu, B. Milvang-Jensen, M. Sparre, L.P. Xin, T.M. Zhang, W.K. Zheng, Y.C. Zou, Discovery of the broad-lined Type Ic SN 2013cq associated with the very energetic GRB 130427A. Astrophys. J. **776**, 98 (2013).
259. S.-C. Yoon, N. Langer, Evolution of rapidly rotating metal-poor massive stars towards gamma-ray bursts. Astron. Astrophys. **443**, 643–648 (2005).
260. S.-C. Yoon, A. Dierks, N. Langer, Evolution of massive population III stars with rotation and magnetic fields. Astron. Astrophys. **542**, 113 (2012).

261. S.-C. Yoon, N. Langer, C. Norman, Single star progenitors of long gamma-ray bursts. I. Model grids and redshift dependent GRB rate. Astron. Astrophys. **460**, 199–208 (2006).
262. S.-C. Yoon, S.E. Woosley, N. Langer, Type Ib/c supernovae in binary systems. I. Evolution and properties of the progenitor stars. Astrophys. J. **725**, 940–954 (2010).
263. N. Yusof, R. Hirschi, G. Meynet, P.A. Crowther, S. Ekström, U. Frischknecht, C. Georgy, H. Abu Kassim, O. Schnurr, Evolution and fate of very massive stars. Mon. Not. R. Astron. Soc. **433**, 1114–1132 (2013).
264. Zeh, S. Klose, D.H. Hartmann, A systematic analysis of supernova light in gamma-ray burst afterglows. Astrophys. J. **609**, 952–961 (2004).
265. Zhang, B.-B. Zhang, F.J. Virgili, E.-W. Liang, D.A. Kann, X.-F. Wu, D. Proga, H.-J. Lv, K. Toma, P. Mészáros, D.N. Burrows, P.W.A. Roming, N. Gehrels, Discerning the physical origins of cosmological gamma-ray bursts based on multiple observational criteria: the cases of z=6.7z=6.7 GRB 080913, z=8.2z=8.2 GRB 090423, and some short/hard GRBs. Astrophys. J. **703**, 1696–1724 (2009).

CHAPTER 7

Molecular Dications in Planetary Atmospheric Escape

Stefano Falcinelli [1,*], Fernando Pirani [2], Michele Alagia [3], Luca Schio [3], Robert Richter [4], Stefano Stranges [3,5], Nadia Balucani [2] and Franco Vecchiocattivi [1]

[1]*Department of Civil and Environmental Engineering, University of Perugia, Via G. Duranti 93, 06125 Perugia, Italy*

[2]*Department of Chemistry, Biology and Biotechnologies, University of Perugia, Via Elce di Sotto, 06123 Perugia, Italy*

[3]*IOM-CNR Tasc, Km 163.5, Area Science Park, Basovizza, 34149 Trieste, Italy*

[4]*Elettra-Sincrotrone Trieste, Area Science Park, Basovizza, 34149 Trieste, Italy*

[5]*Department of Chemistry and Drug Technology, University of Rome "La Sapienza", 00185 Rome, Italy*

ABSTRACT

Fundamental properties of multiply charged molecular ions, such as energetics, structure, stability, lifetime and fragmentation dynamics, are relevant to understand and model the behavior of gaseous plasmas as well as ionosphere and astrophysical environments. Experimental determinations of the Kinetic Energy Released (KER) for ions originating from dissociations reactions, induced by Coulomb explosion of doubly charged molecular ions (molecular dications) produced by double photoionization of CO_2, N_2O and C_2H_2 molecules of interest in planetary atmospheres, are reported. The KER measurement as a function of the ultraviolet (UV) photon energy in the range of 28–65 eV was extracted from the electron-ion-ion coincidence spectra obtained by using tunable synchrotron radiation coupled with ion imaging techniques at the ELETTRA Synchrotron Light Laboratory Trieste, Italy. These experiments, coupled with a computational analysis based on a Monte Carlo trajectory simulation, allow assessing the probability of escape for simple ionic species in the upper atmosphere of Mars, Venus and Titan. The measured KER in the case of H^+, C^+, CH^+, CH_2^+, N^+, O^+, CO^+, N_2^+ and NO^+ fragment ions range between 1.0 and 5.5 eV, being large enough to allow these ionic species to participate in the atmospheric escape from such planets into space. In the case of Mars, we suggest a possible explanation for the observed behavior of the O^+ and CO_2^{2+} ion density profiles.

Keywords: molecular dications; astrochemistry; planetary atmospheres; ionospheres; synchrotron radiation; ions escape; mass spectrometry; coincidence technique; ion imaging

1. INTRODUCTION

Gas phase ionization processes induced by energetic photons, electrons or by excited metastable neutral species (also called collisional autoionization reactions [1,2]) play an important role in several phenomena occurring in low energy ionized plasmas and electric discharges [3,4]. Furthermore, ionic species are extremely important in the upper atmosphere of planets, where they govern the chemistry of ionospheres [5,6]. In particular, the ionosphere chemistry of Titan has recently been revealed to be extremely active by the instruments on board Cassini [7,8]. Finally, molecular ions have also been detected in comet tails [9].

In space, ions are formed in various ways, the importance of which depends on the specific conditions of the extraterrestrial environment considered [10,11,12]. The interaction of neutral molecules with cosmic rays, UV photons, X-rays and other phenomena such as shock waves are all important processes for their production. In particular, the absorption of UV photons with an energy content higher than the ionization potential of the absorbing species can induce ionization with the formation of both singly and doubly charged ions. In the latter case, we have the so-called molecular dications. The reader can refer to some interesting recent review papers by Alagia et al. [6] and by Sabzyan et al. [13] in order to have an overview of the main characteristics and the importance of such doubly charged species. These ionic species can be produced by different techniques, such as mass spectrometry [14], ion-molecule reactions [15], and double photoionization processes [16,17,18,19]. In such a field, the importance of a number of experimental and theoretical works conducted by Eland and coworkers [20,21,22] and by Hochlaf et al. [23,24] concerning the molecules of interest in the present paper must be noted. The molecular dications can be stable or metastable [14,25] and can be used, in principle, as energy storage at a molecular level [14,15,26]. Multiply charged ionic species can also be produced by cosmic rays, which are significant since they are ubiquitous and carry a high energy content (up to 100 GeV). They consist of protons, alpha particles, electrons, γ-rays and (to a small extent) also heavier nuclei (such as C^{6+}). Furthermore, single or double ionization can also occur by absorption of X-rays. In this case, the ejection of a core electron followed by the Auger emission of another electron produces molecular dications, which have been suggested to play a role in the envelope of young stellar objects [27] and upper planetary atmospheres [28,29,30,31]. Therefore, molecular dications have recently attracted the attention of the scientific community in order to investigate their possible role in the upper planetary atmospheres. Moreover, molecular N_2^{2+} and CO_2^{2+} dications have been predicted to exist with significant densities in Mars, Venus and Titan (the largest moon of Saturn) ionospheres by Alcaraz and coworkers [32,33,34]. In fact, molecular dications can be produced in very stable electronic states that allow them to survive several seconds, a lifetime compatible with collision events in such rarefied environments because the typical time between collisions at altitudes corresponding to the peak ionic density of planetary ionospheres is no longer than 1 s [28]. For such a reason, a stable doubly charged ion can be lost in such environments through one of the following reactions: (i) chemical reaction with a neutral partner; (ii) recombination with a thermal electron; or (iii) Coulomb explosion [6,13,28]. On the other hand, molecular dications can be formed in a metastable state having a lifetime shorter than 1 s, as discussed in the next section. In such a case, they can undergo many dissociation reactions towards fragment ions with a high kinetic energy content because they are unstable with respect to Coulomb explosion. Among these ions, some may experience collisions and may release their energy, but some of them may reach the top of the atmosphere and may escape into space, in particular, those metastable dications formed at around the exobase level (above this altitude, which is about 200 km on Mars with a ground pressure of 636 Pa, the pressure force does not play any role because of rare collisions, and the particles have an escape probability equal to 1/e) [28,35]. A very interesting and recent study by Lilensten et al. [35] investigated the role of CO_2^{2+} molecular dications in the atmospheric escape of Mars, showing that, in general, dications may constitute a source of the escape of planetary atmospheres, which had not been taken into account until now. Actually, the escape probability of molecules depends on gravitational forces. However, sometimes a simple thermal distribution cannot explain the escape rate of some species. This indicates that some chemical processes must be responsible for high kinetic energy products, like charge exchange or photodissociation phenomena. Lilensten et al., retrieving the measured electron escape flux, were able to perform a first attempt in the evaluation of the escape flux of fast ions due to dissociation of the unstable CO_2^{2+} dication. In the case of Mars atmosphere, with a mean solar flux $F_{10.7} = 120$ (for details, see the Appendix A in [35]), they obtained an escape flux of fast ions of about 8.4×10^{23} s^{-1}, indicating that, in general, this escape process should account for up to about 6% of the total, with the escape of thermal ions accounting for the remainder. However, their calculations extended to Venus and HD 209458 b (one of the three observed exoplanets, i.e., "hot Jupiters" gas giants) do not agree with experimental observations, and, in the latter case, cannot explain the measured escape flux of C^+ [35]. We consider that this should be an indication that the role of metastable molecular dications in the production of fast fragment ionic species able to escape into space is still underestimated in the kinetic modeling of planetary ionospheres.

The aim of this paper is to stimulate the research in the field of chemical characterization of planetary atmospheres in order to focus the attention of the scientific community on the role of molecular dications inducing Coulomb explosion and fragment ion formation with a high kinetic energy content. For such a reason, molecular dications are considered as exotic species, and when they are formed in planetary ionospheres, the possibility of generating dissociative products with a kinetic energy of several eV allows these ionic fragments to reach sufficient velocity to escape into space. Therefore, double ionization processes can, in principle, contribute to the continuous erosion of the atmosphere of some planets of the Solar System, like Mars and Titan (the largest satellite of Saturn), as discussed in the following sections.

2. EXPERIMENTS

The data presented and discussed here have been recorded by double photoionization experiments performed at the ELETTRA Synchrotron Light Laboratory (Trieste, Italy) using the ARPES (Angle-Resolved Photoemission Spectroscopy) end station of the Gas Phase Beamline. Details on the beamline and the end station have already been reported elsewhere [36], and the apparatus used for the experiment discussed here has also been described previously [37,38]. Therefore, only some features relevant to the present investigation are outlined here.

As observed in Figure 1, the monochromatic energy from the selected synchrotron light beam crosses an effusive molecular beam of CO_2, N_2O and C_2H_2 neutral precursors, and the product ions are then detected in coincidence with photoelectrons.

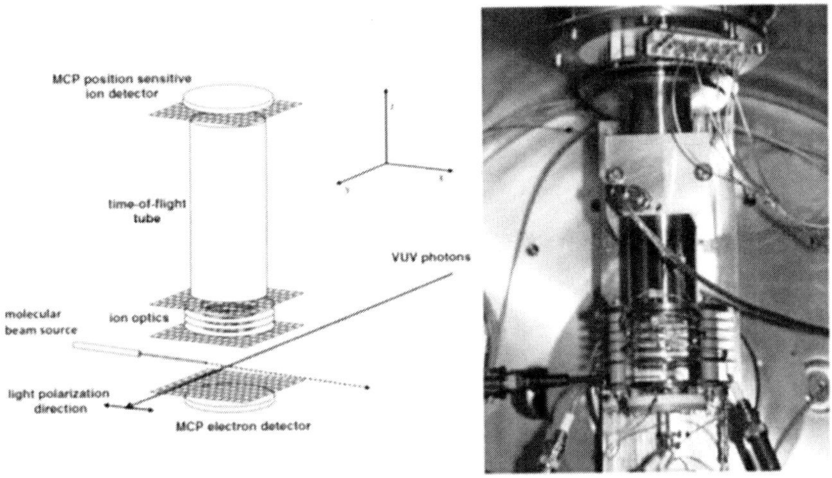

(a) (b)

Figure 1. The electron-ion extraction and detection system used for the photoelectron-photoion coincidence measurements. (**a**) a scheme of the set-up and (**b**) a picture of such a device.

The molecular beam of CO_2, N_2O and C_2H_2 crosses at right angles the VUV light beam, having the light polarization vector parallel to the synchrotron ring plane and perpendicular to the time-of-flight direction of detected ions. Incident photon fluxes and the gas pressure are monitored, and the ion yield has been corrected for flux changes of pressure and photons, when the photon energy was scanned. The gas inlet effusive source is supplied with a mixture of the molecule under study (CO_2, N_2O and C_2H_2) with helium, allowing the normalization of all ion signals (measured at each photon energy) to the total ion yield of helium at that energy. In order to record photoions in coincidence with photoelectrons, we used the electron-ion-ion coincidence technique. The coincidence electron-ion-ion extraction and detection system was built following the model described in detail in [18] and its scheme can be seen in Figure 1. This device consists of a time-of-flight (TOF) spectrometer equipped with an ion position sensitive detector (stack of three micro-channel-plates with a multi-anode array arranged in 32 rows and 32 columns). It has been especially designed in order to properly measure the spatial momentum components of the dissociation ionic products [37]. Finally, by the analysis of distribution maps, the kinetic energy of the two products, released into the two ionic fragments, can be extracted. Such kinetic energy is obtained by a simple analysis of the ion intensity maps based on the method suggested by Lundqvist et al. [39,40]. In particular, this target can be reached by looking at the dimensions and shapes of the peak for each ion pair in the coincidence spectra measured at all investigated photon energies. Through analysis of coincidences distribution as a function of the arrival time differences ($t_2 - t_1$) of fragment ions (to the ion position sensitive detector, being a multi-channel-plate (MCP)) generated by Coulomb explosion of the molecular dication under study, we are able to calculate the lifetime of such metastable species by using the procedure proposed by Field and Eland [41,42,43]. All experimental components were controlled by a computer used to record experimental data. The incident photon flux and the gas pressure were monitored and stored in separate acquisition channels [39,44].

Carbon dioxide, nitrous oxide and acetylene, from a commercial cylinder at room temperature, were supplied to a needle effusive beam source. The used CO_2, N_2O and C_2H_2 gases had a 99.99% and ~99.0% (in the case of acetylene) nominal purity. Carbon dioxide and nitrous oxide were used without any further treatment, whereas acetylene was used after a ~193 K cold trap purging to remove acetone impurities. The performance of the cold trap was verified by recording mass spectra. An adjustable leak valve along the input gas pipe line was used in order to control the gas flow, which was monitored by checking the pressure in the main vacuum chamber.

3. RESULTS AND DISCUSSION

Carbon dioxide, nitrous oxide and acetylene are simple molecules of interest for interstellar medium ISM and planetary atmospheres (not only for the Earth but also for other planets of the Solar System like Mars, Venus and also Titan). The presence of CO_2, N_2O in the ISM has been demonstrated by microwave spectroscopy, whereas C_2H_2 was detected by IR spectroscopic measurements [33]. Acetylene is also found as a minor component in the atmosphere of gas giants like the planet Jupiter, in the atmosphere of Saturn's satellite Titan, and in comets, where photochemical experiments have demonstrated that this simple hydrocarbon is a likely precursor of C_2, a widely observed component in such environments [45,46,47,48].

On the other hand, the importance of carbon dioxide and nitrous oxide are well known in the Earth's atmosphere where both species are important greenhouse gases and N_2O can participate in ozone depletion [49,50]. Moreover, CO_2 is the main component of the Mars (95.3%) and Venus (96.5%) atmospheres and, together with acetylene molecules, has been detected as a minor component in Titan's atmosphere (2–4 ppm and about 10 ppb, respectively) [51]. It has to be noted that the presence of VUV light photons in these environments makes it highly probable that double photoionization of these molecular species occurs with their subsequent dissociation into ionic fragments having a high kinetic energy content of several eV. This translational energy is sufficient for some of them to escape from the upper atmosphere of Mars and Titan into space, as we discuss in the following subsections.

3.1. The Double Photoionization of CO_2 and the Lack of O^+ Concentration in Mars Atmosphere

The double photoionization of CO_2 by VUV synchrotron light (34–50 eV energy range) is in good agreement with previous data [21,51,52,53] and indicates that four two-body dissociation processes are possible with their measured threshold energies:

$$CO_2 + h\nu \rightarrow (CO_2^+)^* + e^- \rightarrow CO^+ + O^* \rightarrow CO^+ + O^+ + e^- \qquad h\nu \geq 35.6\,\text{eV} \tag{1}$$

$$CO_2 + h\nu \rightarrow CO_2^{2+} + 2e^- \qquad h\nu \geq 37.3\,\text{eV} \tag{2}$$

$$CO_2 + h\nu \rightarrow (CO_2^{2+})^*_{\text{long lived}} + 2e^- \rightarrow CO^+ + O^+ \qquad h\nu \geq 38.7\,\text{eV} \tag{3}$$

$$CO_2 + h\nu \rightarrow (CO_2^{2+})^*_{\text{short lived}} + 2e^- \rightarrow CO^+ + O^+ \qquad h\nu \geq 39.0\,\text{eV} \tag{4}$$

In our mass spectrometric determinations, no clear evidence was found for possible three-body dissociation reactions despite that Eland and coworkers [22] were able to detect the final ion products generated from the three body dissociative double photoionization processes of some triatomic molecules (SO_2, CS_2, OCS, ICN, CO_2, and N_2O) at a photon energy higher than 40 eV. The reason for such a discrepancy is probably due to the low cross-section values characterizing the three body fragmentation channels in the energy range investigated in our experiment. In

Reactions (3) and (4), several electronic states of the intermediate molecular (CO_2^{2+}) dication are involved with different lifetimes τ. Reaction (3) occurs through a long-lived dication with $\tau \geq 3.1$ μs corresponding to the formation of CO_2^{2+} ions in the ground $X^3\Sigma^+_g$ electronic state with an internal energy below the threshold towards the Coulomb explosion or in excited singlet states undergoing slow spin forbidden intersystem crossing to the triplet ground state, followed by fast dissociation over the ground state potential energy surface [21,53]. Reaction (4) involves a short-lived dication having $\tau \leq 50$ ns corresponding to the formation of the ground $X^3\Sigma^+_g$ electronic state with an internal energy above the threshold for dissociation. Reactions (1), (3) and (4) form the same CO^+ and O^+ product ions. In particular, reaction (1) is an indirect process occurring below the double photoionization threshold of 37.34 eV, via the formation of an excited state of the $(CO_2^+)^*$ monocation, followed by the production of an intermediate autoionizing oxygen atom O*.

By analysing coincidence spectra obtained for each investigated photon energy, we can evaluate the kinetic energy release (KER) for the CO^+ and O^+ production in the "fast" fragmentation process by looking at the dimensions and shapes of the dot intensity for each ion pair peak. With such a procedure, applied to the CO^+/O^+ coincidences signal related to the Reactions (1) and (4) fragmentation processes, we have measured kinetic energy distributions of each product ions for different values of the investigated photon energy (36.0, 39.0, 41.0, 44.0 and 49.0 eV) as shown in Figure 2.

From the kinetic energy distributions reported in Figure 2, it is evident that the maximum KER value for the O^+ ions ranges between 2.2 and 3.8 eV at all investigated photon energies. This is a consequence of the general behavior of doubly charged molecular ions coming from their instability towards the Coulomb explosion, with the subsequent formation of dissociative ionic fragments with a kinetic energy of several eV, as already mentioned above. In the case of O^+ formation by VUV double photoionization of carbon dioxide molecules, this energy is large enough for the atmosphere of Mars and Titan to allow these fragments to reach enough velocity to escape into space, contributing to the continuous erosion of these atmospheres.

In Table 1 are reported the measured KER values for fragments ions produced by VUV dissociative double photoionization of CO_2, C_2H_2 and N_2O in comparison with the typical escape energy for various ions at the exobase in the atmosphere of Earth, Venus, Mars and Titan [28]. In particular, considering the ionosphere of Mars, the possibility to form carbon dioxide CO_2^{2+} dications (due to UV light double photoionization of CO_2 neutral precursors) in this environment followed by its Coulomb explosion and subsequent production of O^+ fragments having a high translational energy (ranging between 2.2 and 3.8 eV—see Figure 2), could explain the lack in the O^+ expected concentration in Mars' atmosphere.

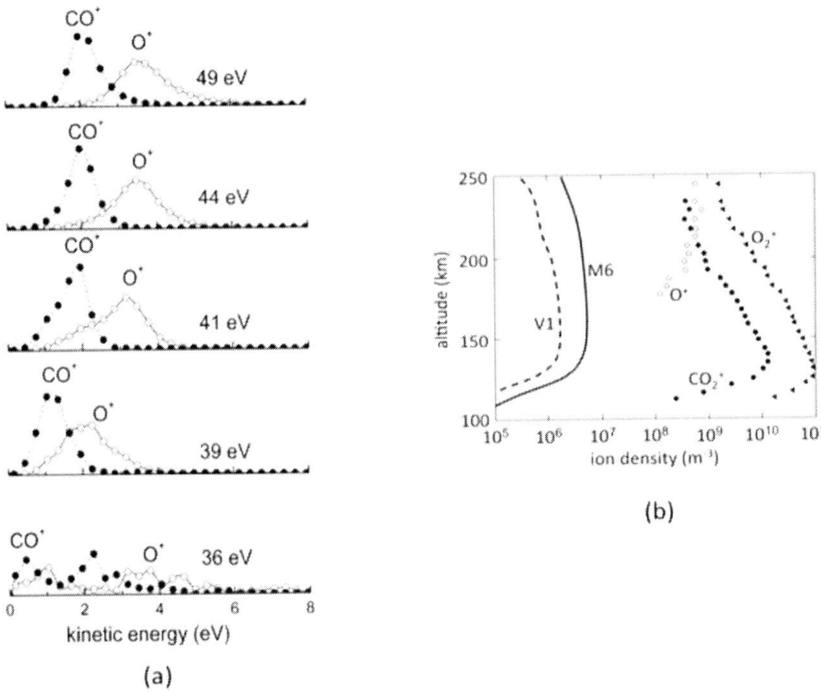

Figure 2. (**a**) Kinetic energy distributions for CO^+ and O^+ fragment ions originating by Coulomb explosion of CO_2^{2+} dication produced in the double VUV photoionization experiment of CO_2, as a function of the photon energy. (**b**) Calculation of CO_2^{2+} ion density profiles for Viking 1 lander (V1) and Mariner 6 spacecraft (M6) conditions by Witasse et al. [33]. The O^+, CO_2^+, and O_2^+ densities profiles (on the right part of the panel) measured by Viking 1 lander are also plotted.

In fact, by looking at O^+, we can argue that the observed behavior of its density profile measured by Viking 1 lander in the upper atmosphere of Mars, compared with the CO_2^{2+} density profile as calculated by Witasse et al. for Viking 1 lander and Mariner 6 spacecraft geophysical conditions (see Figure 2) [33], could be explained by invoking the dissociative double photoionization of CO_2 induced by VUV photons. These authors found a maximum in the CO_2^{2+}density of about 5×10^6 ions·m^{-3} at an altitude of 160 km where the O^+ concentration (measured by Viking 1 with an ion density decreasing from 6 ×10^8 ions·m^{-3} at 250 km to 1 × 10^8 ions·m^{-3} at 175 km of altitude) disappears, becoming undetectable [33]. This surprising behaviour could be explained by our experimental observations discussed here: by double VUV photoionization of CO_2 molecules, when the photon energy is higher than 48 eV, the stable molecular dication formation (Reaction (2)) and the "fast" fragmentation process (Reactions (1) and (4)) achieve the same importance and their cross sections gain the maximum value, so the O^+ formation with a KER sufficient to escape from the atmosphere of Mars becomes very high and its experimental detectability very difficult. An analogous situation could characterize Titan's ionosphere, where the low typical escape energy values for ions (see Table 1) allow for the high probability of the escape process into space, not only for O^+, but also for CO^+ ions, having typical escape energy smaller than 1.0 eV.

Table 1. The range of measured values of KER distributions (eV) for fragments ions produced by VUV dissociative double photoionization of CO_2, N_2O, C_2H_2 in the range of 28–65 eV compared with the typical escape energy (eV) for various ions in the atmosphere of Earth, Venus, Mars and Titan [28].

Ionic Species	Measured Range of KER Distributions (eV)	Typical Escape Energy (eV) for Various Ions in the Atmosphere (at the Exobase) of Some Planets of the Solar System [28]			
		Earth	Venus	Mars	Titan
H^+	2.8 ÷ 6.0 [(a)]	0.62	0.53	0.13	0.02
C^+	1.7 ÷ 3.3 [(b)]	7.4	6.4	1.5	0.28
CH^+	1.3 ÷ 3.2 [(c)]	8.0	6.9	1.6	0.30
CH_2^+	1.6 ÷ 2.9 [(b)]	8.6	7.5	1.8	0.32
N^+	2.3 ÷ 5.2 [(d)]	8.6	7.5	1.8	0.32
O^+	1.0 ÷ 5.2 [(e)] 2.2 ÷ 3.8 [(f)]	9.8	8.6	2.0	0.37
C_2H^+	0.1 ÷ 0.4 [(a)]	15.4	13.3	3.1	0.58
CO^+	0.4 ÷ 2.6 [(f)]	17.3	14.9	3.5	0.65
N_2^+	0.4 ÷ 2.9 [(e)]	17.3	14.9	3.5	0.65
NO^+	1.0 ÷ 2.9 [(d)]	18.5	16.0	3.75	0.70

[(a)] formed by Reactions (8) + (9)—see text; [(b)] formed by Reaction (11)—see text; [(c)] formed by Reaction (10)—see text; [(d)] formed by Reaction (6)—see text; [(e)] formed by Reaction (5)—see text; [(f)] formed by Reactions (1) + (4)—see text.

3.2. The Coulomb Explosion of Nitrous Oxide Dication

When nitrous oxide molecules are photoionized by using VUV synchrotron radiation in the 28–40 eV energy range, the production of N_2O^{2+} molecular dications in a metastable state is possible. After its formation, N_2O^{2+} can evolve towards the formation of N_2^+/O^+ and N^+/NO^+ ionic fragments via Coulomb explosion (see Reactions (5) and (6) reported below, respectively). According to a previous study [49], to which we refer for an overview of the experimental and theoretical works already performed by several research groups, no stable N_2O^{2+} molecular dications have been observed in our experiment, indicating that the lifetime of the produced molecular dications is never longer than 10^{-6} s, which was the typical average time-of-flight of the ions in our detection device. In the whole investigated energy range, we observed the two dissociation channels reported below with a threshold energy of about 32.2 eV:

$$N_2O + h\nu \rightarrow (N_2O^{2+})^* + 2e^- \rightarrow N_2^+ + O^+ + 2e^- \tag{5}$$

$$N_2O + h\nu \rightarrow (N_2O^{2+})^* + 2e^- \rightarrow N^+ + NO^+ + 2e^- \tag{6}$$

No evidence for three-body dissociation reactions was found in our investigated photon energy range, in agreement with previous results by Eland and coworkers, who found product ions originating from the three-body fragmentation processes of the N_2O^{2+} dication at a photon energy higher than 40 eV [22]. In Reactions (5) and (6), $(N_2O^{2+})^*$ indicates an intermediate "short-lived" molecular dication. In such an experiment, product ions were detected as a function of the photon energy, in coincidence with one of the two photoelectrons and the recorded delay times. Recorded correlation time diagrams have shown that, for energies below the vertical threshold for the N_2O^{2+} dication formation (35.8 eV), autoionizing states are formed, followed by dissociative autoionization finally leading to ion products of both Reactions (5) and (6). By increasing the photon energy, N_2O^{2+} can be formed above the threshold, and also in this case, it dissociates through the same channels by a direct Coulomb explosion [49]. Now, we are able to analyze our data using the same procedure already discussed for CO_2, providing the total kinetic energy of ionic fragments as a function of the investigated photon energy. In Figure 3 are reported the maximum values of the peak for the total kinetic energy of the two couples of product ions $N^+ + NO^+$ and $O^+ + N_2^+$ originating from Reactions (5) and (6), respectively, as a function of the photon energy.

From Figure 3, it is possible to see that ions produced by Reaction (5), $O^+ + N_2^+$, are formed with a total KER of about 5.0 eV, while $N^+ + NO^+$ originating from Reaction (6) posses a total KER of about 6.0 eV. Looking at the typical escape energy for various ions reported in Table 1, we can argue that the measured kinetic energy content (ranging between 1.0

and 2.3–5.2 eV, respectively) for both O^+ and N^+, produced by dissociative double VUV photoionization of nitrous oxide is, in principle, compatible with their possible escape from the upper atmosphere of Mars and Titan

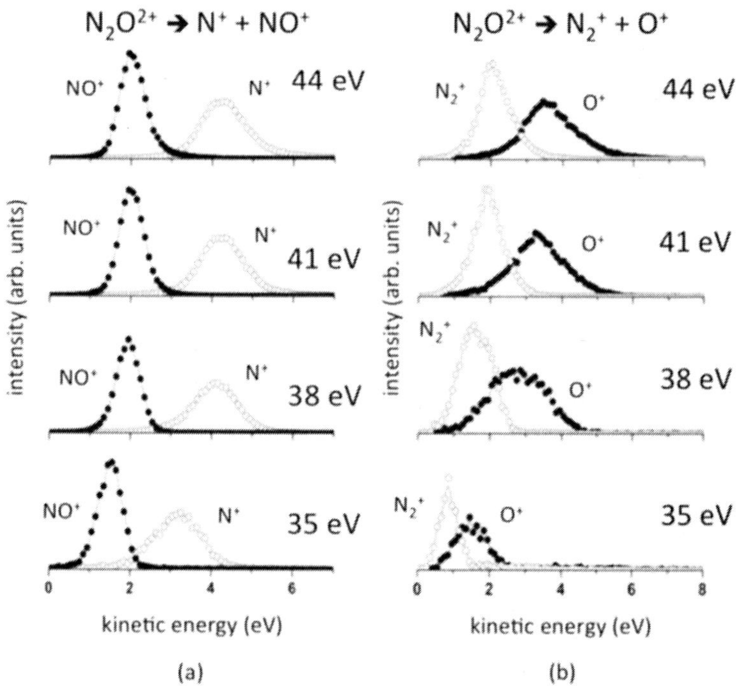

Figure 3. (**a**) Kinetic energy distributions for each fragment ion N^+ and NO^+ (see Reaction (6) on the text) originating from the Coulomb explosion of N_2O^{2+} dication, produced in the double VUV photoionization experiment of nitrous oxide (see text), as a function of the investigated photon energy. (**b**) measured kinetic energy distributions for N_2^+ and O^+ ions (see Reaction (5) on the text) originating from the Coulomb explosion of N_2O^{2+}, and produced in the same experiment discussed in the text.

3.3. The Coulomb Explosion of Acetylene Dication

The double photoionization of acethylene by using VUV synchrotron radiation in the 32–65 eV photon energy range produces acetylene dication with subsequent dissociation reactions. Our previous measurements [44], in agreement with experimental determinations by

Eland and coworkers [22], show the following accessible channels with their respective threshold energies:

$$C_2H_2 + h\nu \rightarrow C_2H_2^{2+} + 2e^- \quad h\nu \geq 31.7\,\text{eV} \tag{7}$$

$$C_2H_2 + h\nu \rightarrow (C_2H_2^{2+})^*_{\text{long lived}} + 2e^- \rightarrow H^+ + C_2H^+ \quad h\nu \geq 33.8\,\text{eV} \tag{8}$$

$$C_2H_2 + h\nu \rightarrow (C_2H_2^{2+})^* + 2e^- \rightarrow H^+ + C_2H^+ \quad h\nu \geq 34.0\,\text{eV} \tag{9}$$

$$C_2H_2 + h\nu \rightarrow (C_2H_2^{2+})^* + 2e^- \rightarrow CH^+ + CH^+ \quad h\nu \geq 34.0\,\text{eV} \tag{10}$$

$$C_2H_2 + h\nu \rightarrow (C_2H_2^{2+})^* + 2e^- \rightarrow (H_2CC^{2+})^* \rightarrow C^+ + CH_2^+ \quad h\nu \geq 34.0\,\text{eV} \tag{11}$$

where $C_2H_2^{2+}$ indicates a stable molecular dication formation, $(C_2H_2^{2+})^*_{\text{long-lived}}$ refers to a molecular dication with a lifetime of 108 ± 22 ns, as we determined in our recent work [44], and $(C_2H_2^{2+})^*$ indicates a doubly charged intermediate molecular species with a lifetime shorter than the characteristic window time of our apparatus for the probed experimental conditions, that is ~50 ns. Reactions (8) and (9) indicate two different mechanisms for the same deprotonation process, involving intermediate "long-lived" and "short-lived" dication species, respectively. Reaction (11) is a proton transfer rearrangement reaction which occurs via formation of the intermediate vinylidene dication, $(H_2CC^{2+})^*$, with subsequent dissociation towards the C^+ and C_2H^+ ion products [22,44].

In the present experiment, we have focused our attention on the two-body dissociation Reactions (8)–(11) originating from Coulomb explosion of $C_2H_2^{2+}$ dications, induced by 39.0 eV photons, and using photoelectron-photoion-photoion coincidence and ion imaging techniques. The recorded coincidence plot at this energy is reported in Figure 4a. By analyzing the density distribution of the recorded coincidences, we are able to extract a lifetime of 108 ± 22 ns for the intermediate "long-lived" $C_2H_2^{2+}$ molecular dication of Reaction (8), and the kinetic energy distributions of each product ion (see Figure 4b) by using the same procedure already discussed in the case of carbon dioxide.

In the Figure 4a are also shown the total kinetic energy distributions related to product ion pairs of Reactions (8)–(11). Such total KER distributions show: (i) for the deprotonation process (Reactions (8) and (9)), a maximum at about 4.3 eV with a full width at half maximum (FWHM) of about 60%; (ii) for the symmetric dissociation (Reaction

(10)) a maximum at 5.2 eV with a FWHM of 44%; and (iii) for the proton transfer rearrangement process via formation of the intermediate vinylidene dication (Reaction (11)) a maximum in the KER distribution at 4.5 eV with a FWHM of about 28%.

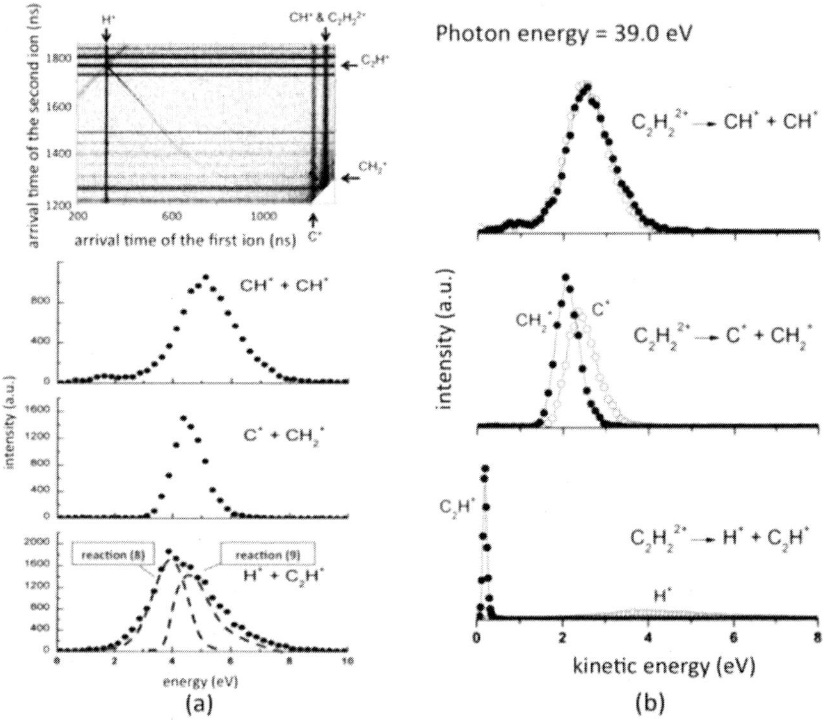

Figure 4. (**a**) on the top is the photoion-photoion coincidence plot as measured at a photon energy of 39 eV; on the bottom are the total kinetic energy distributions of product ions for the three measured two-body dissociation reactions indicating in the case of H^+ + C_2H^+ ions, formation of a bimodal distribution due to the presence of different reaction mechanisms (see text). (**b**) Kinetic energy distributions for each two-body fragment ion originating from the Coulomb explosion of the $C_2H_2^{2+}$ molecular dication (see Reactions (7)–(11) and text), as measured at a photon energy of 39 eV.

One interesting characteristic of such distributions is that the one referred to the deprotonation process leading to the formation of H^+ and C_2H^+ ions shows bimodal behavior (see Figure 4a) originated by the two possible mechanisms contributing to this reaction: (i) a "slow" dissociation component, Reaction (8), more important at low photon energy ($h\nu \leq 35$ eV) where $C_2H_2^{2+}$ is formed mainly in its ground $^3\Sigma^-_g$ electronic state with a measured lifetime of about 108 ns; and (ii) a "fast" dissociation component, Reaction (9), via formation of a "short-lived" $C_2H_2^{2+}$ intermediate dication having a lifetime shorter than 50 ns and coming from the $^3\Pi_u$ excited electronic state whose accessible Franck-Condon zone occurs at about 37 eV. This state, accordingly to the potential energy curves calculated by Thyssen et al. [22], correlates to ground state C_2H^+ and H^+ products along an entirely repulsive curve with an energy release of about 7–8 eV. This "fast" component for the deprotonation reaction is expected to play an important role as the photon energy becomes higher than 36.0 eV and should be characterized by a higher KER for the product C_2H^+ and H^+ ions with respect to the "slow" component which involves the formation of the $^3\Sigma^-_g$ dicationic metastable state. The presence of two different mechanisms for this process is confirmed by the quite broad FWHM value (~60%) of the related total KER distribution (see Figure 4b). In order to clarify the relative importance of the two mechanisms discussed above (see Reactions (8) and (9)), we are planning to extend our photoelectron-photoion-photoion coincidence and ion imaging study to a wider photon energy range. Finally, it has to be noted that the translational energy content of each H^+, C^+, CH^+ and CH_2^+ product ion is quite large, with an average value of 4.0, 2.5, 2.0 and 2.5 eV, respectively. This kinetic energy is sufficient to allow this species to participate in atmospheric escape from Mars and Titan (see Figure 4 and Table 1).

4. CONCLUSIONS

The dissociative double photoionization processes, induced by VUV light photons and producing fragment ions with a high kinetic energy content, could be a generally important way for ionic species to escape from the atmosphere of some planets of the Solar System, like Venus, Mars and Titan. In fact, these processes occur via formation of intermediate molecular dications that could dissociate by Coulomb explosion towards the formation of two ionic fragment species having a kinetic energy release of several eV, and therefore, much larger than the limiting thermal escape velocity characterizing some planetary atmospheres. In the case of the double VUV photoionization of CO_2, C_2H_2 and N_2O molecules (in the photon energy range of 28–65 eV), the fragment product ions O^+, CO^+, N^+, CH_2^+, CH^+, C^+, H^+ are characterized by a translational energy ranging between 1.0 and 6.0 eV that is large enough to allow their escape process from the upper atmospheres of Mars and Titan. Moreover, these studies could be helpful in understanding important details about the chemistry of planetary ionospheres, like in the case of Mars, where we were able to explain the observed behavior of the O^+ and CO_2^{2+} ion density profiles.

ACKNOWLEDGMENTS

Financial contributions from MIUR (Ministero dell'Istruzione, dell'Università e della Ricerca) through the PRIN 2009 (Grant 2009W2W4YF_002) project is gratefully acknowledged. The authors also gratefully thank the "Fondazione Cassa di Risparmio di Perugia" for partial support (Project code: 2014.0255.021).

CONFLICTS OF INTEREST

The authors declare no conflict of interest.

REFERENCES

1. Balucani, N.; Bartocci, A.; Brunetti, B.G.; Candori, P.; Falcinelli, S.; Pirani, F.; Palazzetti, F.; Vecchiocattivi, F. Collisional autoionization dynamics of Ne*(3P2,0)-H2O. Chem. Phys. Lett. 2012, 546, 34–39.
2. Brunetti, B.G.; Candori, P.; Cappelletti, D.; Falcinelli, S.; Pirani, F.; Stranges, D.; Vecchiocattivi, F. Penning Ionization Electron Spectroscopy of water molecules by metastable neon atoms. Chem. Phys. Lett. 2012.
3. Falcinelli, S.; Bartocci, A.; Cavalli, S.; Pirani, F.; Vecchiocattivi, F. The stereo-dynamics of collisional autoionization of ammonia by helium and neon metastable excited atoms through molecular beam experiments. J. Chem. Phys. 2015.
4. Falcinelli, S.; Bartocci, A.; Cavalli, S.; Pirani, F.; Vecchiocattivi, F. Stereo-dynamics in collisional autoionization of water, ammonia, and hydrogen sulfide with metastable rare gas atoms: Competition between intermolecular halogen and hydrogen bonds. Chem. Eur. J. 2016, 22, 764–771.
5. Falcinelli, S.; Pirani, F.; Vecchiocattivi, F. The possible role of penning ionization processes in planetary atmospheres. Atmosphere 2015.
6. Alagia, M.; Balucani, N.; Candori, P.; Falcinelli, S.; Pirani, F.; Richter, R.; Rosi, M.; Stranges, S.; Vecchiocattivi, F. Production of ions at high energy and its role in extraterrestrial environments. Rend. Lincei Sci. Fis. Nat. 2013.
7. Vuitton, V.; Dutuit, O.; Smith, M.A.; Balucani, N. Chemistry of Titan's atmosphere. In Titan: Surface, Atmosphere and Magnetosphere; Wodarg, M., Müller-Wodarg, I., Griffith, C.A., Lellouch, E., Cravens, T.E., Eds.; Cambridge University Press: Cambridge, UK, 2003; pp. 224–284.
8. Falcinelli, S.; Rosi, M.; Candori, P.; Vecchiocattivi, F.; Farrar, J.M.; Pirani, F.; Balucani, N.; Alagia, M.; Richter, R.; Stranges, S. The escape probability of some ions from Mars and Titan ionospheres. In ICCSA 2014, Part I, Lecture Notes in Computer Science LNCS; Springer: Berlin/Heidelberg, Germany, 2014; Volume 8579, pp. 554–570.

9. Larsson, M.; Geppert, W.D.; Nyman, G. Ion chemistry in space. Rep. Prog. Phys. 2012, 75, 066901.
10. Lombardi, A.; Faginas Lago, N.; Laganà, A.; Pirani, F.; Falcinelli, S. A bond-bond portable approach to intermolecular interactions: Simulations for N-methylacetamide and carbon dioxide dimers. In ICCSA 2012, Part I, Lecture Notes in Computer Science LNCS; Springer: Berlin/Heidelberg, Germany, 2012; pp. 387–400.
11. Pei, L.; Carrascosa, E.; Yang, N.; Falcinelli, S.; Farrar, J.M. Velocity map imaging study of charge-transfer and proton-transfer reactions of CH3 radicals with H3+. J. Phys. Chem. Lett. 2015, 6, 1684–1689.
12. Candori, P.; Falcinelli, S.; Pirani, F.; Tarantelli, F.; Vecchiocattivi, F. Interaction components in the hydrogen halide dications. Chem. Phys. Lett. 2007.
13. Sabzyan, H.; Keshavarz, E.; Noorisafa, Z. Diatomic dications and anions. J. Iran. Chem. Soc. 2014, 11, 871.
14. Falcinelli, S.; Fernandez-Alonso, F.; Kalogerakis, K.; Zare, R.N. Mass spectrometric detection of alkaline earth monohalide dications. Mol. Phys. 1996, 88, 663–672.
15. Tosi, P.; Correale, R.; Lu, W.; Falcinelli, S.; Bassi, D. Production of the molecular dication ArN2+ in the reaction Ar2+ + N2. Phys. Rev. Lett. 1999, 82, 450–452.
16. Teixidor, M.M.; Pirani, F.; Candori, P.; Falcinelli, S.; Vecchiocattivi, F. Predicted structure and energetics of HCl2+. Chem. Phys. Lett. 2003, 379, 139–146.
17. Alagia, M.; Brunetti, B.G.; Candori, P.; Falcinelli, S.; Teixidor, M.M.; Pirani, F.; Richter, R.; Stranges, S.; Vecchiocattivi, F. Threshold-photoelectron-spectroscopy-coincidence study of the double photoionization of HBr. J. Chem. Phys.2004, 120, 6985–6991.
18. Alagia, M.; Biondini, F.; Brunetti, B.G.; Candori, P.; Falcinelli, S.; Moix, T.M.; Pirani, F.; Richter, R.; Stranges, S.; Vecchiocattivi, F. The double photoionization of HCl: An ion-electron coincidence study. J. Chem. Phys. 2004, 121, 10508–10512.
19. Alagia, M.; Candori, P.; Falcinelli, S.; Pirani, F.; Mundim, M.S.P.; Richter, R.; Rosi, M.; Stranges, S.; Vecchiocattivi, F. Dissociative double photoionization of benzene molecules in the 26–33 eV energy range. Phys. Chem. Chem. Phys.2011, 13, 8245–8250.
20. Price, S.D.; Eland, J.H.D.; Fournier, P.G.; Fournier, J.; Millié, P. Electronic states and decay mechanisms of the N2O2+dication. J. Chem. Phys. 1988, 88, 1511–1515.
21. Slattery, A.E.; Field, T.A.; Ahmad, M.; Hall, R.I.; Lambourne, J.; Penent, F.; Lablanquie, P.; Eland, J.H. Spectroscopy and metastability of CO22+ molecular ions. J. Chem. Phys. 2005.
22. Thissen, R.; Delwiche, J.; Robbe, J.M.; Duflot, D.; Flament, J.P.; Eland, J.H.D. Dissociations of the ethyne dication C2H22+. J. Chem. Phys. 1993, 99, 6590–6599.
23. Hochlaf, M.; Bennett, F.R.; Chambaud, G.; Rosmus, P. Theoretical study of the electronic states of CO22+. J. Phys. B At. Mol. Opt. Phys. 1998, 31, 2163–2175.
24. Sharma, V.; Bapat, B.; Mondal, J.; Hochlaf, M.; Giri, K.; Sathyamurthy, N. Dissociative double ionization of CO2: Dynamics, energy levels, and lifetime. J. Phys. Chem. A 2007, 111, 10205–10211.
25. Pauling, L. The normal state of the helium molecule-ions He2+ and He2++. J. Chem. Phys. 1933, 1, 56–59.
26. Nicolaides, C.A. Energy generation from volcanic ground-states—Application to cold He22+. Chem. Phys. Lett. 1989, 161, 547–553.
27. Stauber, P.; Doty, S.D.; van Dishoeck, E.F.; Benz, A.O. X-ray chemistry in the envelopes around young stellar objects. Astron. Astrophys. 2005, 440, 949–966.
28. Thissen, R.; Witasse, O.; Dutuit, O.; Wedlund, C.S.; Gronoff, G.; Lilensten, J. Doubly-charged ions in the planetary ionospheres: A review. Phys. Chem. Chem. Phys. 2011, 13, 18264–18287.
29. Schio, L.; Li, C.; Monti, S.; Salén, P.; Yatsyna, V.; Feifel, R.; Alagia, M.; Richter, R.; Falcinelli, S.; Stranges, S.; Zhaunerchyk, V. NEXAFS and XPS studies of nitrosyl chloride. Phys. Chem. Chem. Phys. 2015, 17, 9040–9048.
30. Alagia, M.; Bodo, E.; Decleva, P.; Falcinelli, S.; Ponzi, A.; Richter, R.; Stranges, S. The soft X-ray absorption spectrum of the allyl free radical. Phys. Chem. Chem. Phys. 2013, 15, 1310–1318.
31. Falcinelli, S.; Rosi, M.; Candori, P.; Vecchiocattivi, F.; Farrar, J.M.; Pirani, F.; Balucani, N.; Alagia, M.; Richter, R.; Stranges, S. Kinetic energy release in molecular dications

fragmentation after VUV and EUV ionization and escape from planetary atmospheres. Plan. Space Sci. 2014, 99, 149–157.
32. Lilensten, J.; Witasse, O.; Simon, C.; Soldi-Lose, H.; Dutuit, O.; Thissen, R.; Alcaraz, C. Prediction of a N2++ layer in the atmosphere of Titan. Geophys. Res. Lett. 2005, 32, L03203.
33. Witasse, O.; Dutuit, O.; Lilensten, J.; Thissen, R.; Zabka, J.; Alcaraz, C.; Blelly, P.L.; Bougher, S.W.; Engel, S.; Andersen, L.H.; et al. Prediction of a CO2++ layer in the atmosphere of Mars. Geophys. Res. Lett. 2002.
34. Gronoff, G.; Lilensten, J.; Simon, C.; Witasse, O.; Thissen, R.; Dutuit, O.; Alcaraz, C. Modeling dications in the diurnal ionosphere of Venus. Astron. Astrophys. 2007, 465, 641–645.
35. Lilensten, J.; Simon Wedlund, C.; Barthélémy, M.; Thissen, R.; Ehrenreich, D.; Gronoff, G.; Witasse, O. Dications and thermal ions in planetary atmospheric escape. Icarus 2013, 222, 169–187.
36. Blyth, R.R.; Delaunay, R.; Zitnik, M.; Krempasky, J.; Krempaska, R.; Slezak, J.; Prince, K.C.; Richter, R.; Vondracek, M.; Camilloni, R.; et al. The high resolution gas phase photoemission beamline, Elettra. J. Electron Spectrosc. Relat. Phenom. 1999, 101–103, 959–964.
37. Alagia, M.; Candori, P.; Falcinelli, S.; Lavollèe, M.; Pirani, F.; Richter, R.; Stranges, S.; Vecchiocattivi, F. Dissociative double photoionization of CO2 molecules in the 36–49 eV energy range: Angular and energy distribution of ion products. Phys. Chem. Chem. Phys. 2010.
38. Alagia, M.; Candori, P.; Falcinelli, S.; Lavollée, M.; Pirani, F.; Richter, R.; Stranges, S.; Vecchiocattivi, F. Anisotropy of the angular distribution of fragment ions in dissociative double photoionization of N2O molecules in the 30–50 eV energy range. J. Chem. Phys. 2007.
39. Lavollée, M. A new detector for measuring three-dimensional momenta of charged particles in coincidence. Rev. Sci. Instrum. 1990.
40. Lundqvist, M.; Baltzer, P.; Edvardsson, D.; Karlsson, L.; Wannberg, B. Novel time of flight instrument for doppler free kinetic energy release spectroscopy. Phys. Rev. Lett. 1995, 75, 1058–1061.
41. Field, A.; Eland, J.H.D. Lifetimes of metastable molecular doubly charged ions. Chem. Phys. Lett. 1993, 211, 436–442.
42. Alagia, M.; Candori, P.; Falcinelli, S.; Mundim, K.C.; Mundim, M.S.; Pirani, F.; Richter, R.; Stranges, S.; Vecchiocattivi, F. Lifetime and kinetic energy release of metastable dications dissociation. Chem. Phys. 2012, 398, 134–141.
43. Alagia, M.; Candori, P.; Falcinelli, S.; Mundim, M.S.P.; Pirani, F.; Richter, R.; Rosi, M.; Stranges, S.; Vecchiocattivi, F. Dissociative double photoionization of singly deuterated benzene molecules in the 26–33 eV energy range. J. Chem. Phys. 2011.
44. Alagia, M.; Callegari, C.; Candori, P.; Falcinelli, S.; Pirani, F.; Richter, R.; Stranges, S.; Vecchiocattivi, F. Angular and energy distribution of fragment ions in dissociative double photoionization of acetylene molecules at 39 eV. J. Chem. Phys. 2012.
45. Brooke, T.Y.; Tokunaga, A.T.; Weaver, H.A.; Crovisier, J.; Bockelee-Morvan, D.; Crisp, D. Detection of acetylene in the infrared spectrum of comet Hyakutake. Nature 1996, 383, 606–608.
46. Cernicharo, J.; Heras, A.M.; Pardo, J.R.; Tielens, A.G.G.M.; Pardo, J.R.; Herpin, F.; Guélin, M.; Waters, L.B.F.M. Infrared space observatory's discovery of C4H2, C6H2, and benzene in CRL 618. Astrophys. J. 2001, 546, L123–L126.
47. Woods, P.M.; Millar, T.J.; Zijlstra, A.A.; Herbst, E. The synthesis of benzene in the proto-planetary Nebula CRL 618. Astrophys. J. 2002, 574, L167–L170.
48. Skouteris, D.; Balucani, N.; Faginas-Lago, N.; Falcinelli, S.; Rosi, M. Dimerization of methanimine and its charged species in the atmosphere of Titan and interstellar/cometary ice analogs. Astron. Astrophys. 2015.
49. Alagia, M.; Candori, P.; Falcinelli, S.; Lavollée, M.; Pirani, F.; Richter, R.; Stranges, S.; Vecchiocattivi, F. Double photoionization of N2O molecules in the 28–40 eV energy range. Chem. Phys. Lett. 2006, 432, 398–402.

50. Brunetti, B.; Candori, P.; Falcinelli, S.; Lescop, B.; Liuti, G.; Pirani, F.; Vecchiocattivi, F. Energy dependence of the Penning ionization electron spectrum of Ne*(3P2,0) + Kr. Eur. Phys. J. D 2006, 38, 21–27.
51. Cravens, T.E.; Robertson, I.P.; Waite, J.H., Jr.; Yelle, R.V.; Kasprzak, W.T.; Keller, C.N.; Ledvina, S.A.; Niemann, H.B.; Luhmann, J.G.; McNutt, R.L.; et al. Composition of Titan's ionosphere. Geophys. Res. Lett. 2006, 33, L07105.
52. Franceschi, P.; Thissen, R.; Zabka, J.; Roithová, J.; Herman, Z.; Dutuit, O. Internal energy effects in the reactivity of CO22+ doubly charged molecular ions with CO2 and CO. Int. J. Mass Spectrom. 2003, 228, 507–516.
53. Alagia, M.; Candori, P.; Falcinelli, S.; Lavollée, M.; Pirani, F.; Richter, R.; Stranges, S.; Vecchiocattivi, F. Double photoionization of CO2 molecules in the 34–50 eV Energy range. J. Phys. Chem. A 2009, 113, 14755–14759.

CHAPTER 8

XUV excitation followed by ultrafast non-adiabatic relaxation in PAH molecules as a femto-astrochemistry experiment

A. Marciniak1 , V. Despre´1 , T. Barillot1 , A. Rouze´e2, M.C.E. Galbraith2, J. Klei2, C.-H. Yang2, C.T.L. Smeenk2, V. Loriot1 , S. Nagaprasad Reddy4, A.G.G.M. Tielens3, S. Mahapatra4, A.I. Kuleff5, M.J.J. Vrakking2 & F. Le´pine1

1 Institut Lumie`re Matie`re, Universite´ Lyon 1, CNRS, UMR 5306, 10 rue Ada Byron, 69622 Villeurbanne Cedex, France.

2 Max-Born-Institut, Max Born Strasse 2A, D-12489 Berlin, Germany.

3 Leiden Observatory, Leiden University, PO Box 9513, NL-2300RA Leiden, The Netherlands.

4 School of Chemistry, University of Hyderabad, Hyderabad 500 046, India.

5 Theoretische Chemie, PCI, Universita¨t Heidelberg, Im Neuenheimer Feld 229, D-69120 Heidelberg, Germany

ABSTRACT

Highly excited molecular species are at play in the chemistry of interstellar media and are involved in the creation of radiation damage in a biological tissue. Recently developed ultrashort extreme ultraviolet light sources offer the high excitation energies and ultrafast time-resolution required for probing the dynamics of highly excited molecular states on femtosecond (fs) (1 fs=10^{-15}s) and even attosecond (as) (1 as=10^{-18} s) timescales. Here we show that polycyclic aromatic hydrocarbons (PAHs) undergo ultrafast relaxation on a few tens of femtoseconds timescales, involving an interplay between the electronic and vibrational degrees of freedom. Our work reveals a general property of excited radical PAHs that can help to elucidate the assignment of diffuse interstellar absorption bands in astrochemistry, and provides a benchmark for the manner in which coupled electronic and nuclear dynamics determines reaction pathways in large molecules following extreme ultraviolet excitation.

1. INTRODUCTION

When a molecule interacts with a single UV or vacuum ultraviolet (VUV) photon, it can respond by promoting one of its electrons to an excited bound state or to the ionization continuum, leaving a hole in the electron density. At its simplest, this excitation process can be theoretically described using a single-particle picture. While such independent electron approximations are often sufficient in low-frequency light fields, they rapidly fail when excitation at higher photon energies (that is, in the extreme ultraviolet (XUV) domain) is considered. In this case, electron correlation cannot be neglected, and multi-electronic effects in the ionization are important, leading to processes such as Auger decay and shake-up/shake-off, where several electrons are ionized or excited as a result of the absorption of a single photon1.

Nuclear dynamics can occur following an electronic excitation. The typical case considered in femtochemistry experiments involves the excitation of single, well-defined electronic states that display time-dependent structural dynamics. This situation applies when the timescale for electronic motion (given by the inverse electronic level spacing) is much faster than that of the nuclear motion, and allows the use of the well-known Born–Oppenheimer approximation. In the Born–Oppenheimer approximation, the electrons instantaneously adapt to the nuclear geometry and the atoms move under the influence of forces that derive from potential energy surfaces describing the electronic energy as a function of nuclear geometry. The Born–Oppenheimer approximation breaks down when the timescales of the electronic and nuclear motion become comparable. In this case, the electronic and nuclear motion are strongly coupled, leading to ultrafast non-adiabatic electronic relaxation that typically takes place on a (sub)-picosecond time scale2.

The breakdown of both independent electron picture and Born–Oppenheimer picture upon excitation with high-energy photons was revealed in recent real-time investigations in simple molecules such as N_2

(ref. 3), O_2 (ref. 4) and N_2O (ref. 5). The possible occurrence of multi-electronic mechanisms is even more pertinent in complex molecular species such as proteins, where they determine the production of slow electrons that are responsible for bond breaking, that is, radiation damage6,7. Likewise, non-Born–Oppenheimer effects at conical intersections (CIs) are prevalent in larger molecules, and are thought to be responsible for the photostability of DNA8. Understanding multi-electronic and non-adiabatic processes in large species at a microscopic level by means of time-resolved investigations is a major challenge within current ultrafast photochemistry research, made difficult by the rapidly increasing number of reaction pathways that occur when the size of the molecule increases.

In this article, we have investigated polycyclic aromatic hydrocarbon (PAH) molecules that are well known in astrochemistry, as they account for 10% of the elemental carbon in galaxies. Their photochemical fate is key to understand the origin of biologically relevant molecules, eventually leading to the emergence of life in the universe9. Interstellar large cationic PAHs (composed of several 10 C-atoms) are often considered to be prime candidates for being the carriers of the diffuse interstellar absorption bands (DIBs)—a set of some 400 absorption bands generally ascribed to (hitherto unidentified) interstellar molecules. Experimentally measured line widths of visible absorption bands of small PAH cations have been interpreted in terms of non-radiative relaxation pathways governed by an energy gap law9. This energy gap law has been taken to imply that for high-lying states and for large PAHs, where the density of electronic states is expected to be very high, the width of (visible) absorption lines becomes very large and incompatible with those observed for the DIBs. The linewidths of the DIBs imply that the lifetimes of the involved electronic states typically lie in the range of a few 10 s of fs to a few ps. Consequently, only the lowest electronic transitions are deemed to be relevant candidates for DIBs10, but the intensity of these transitions is too weak to support this hypothesis. As a result, the question of the origin of the DIBs remains puzzling. A definitive assignment of the DIBs to PAH molecules would be crucial as they illustrate the chemical complexity of the 'molecular universe' (ref. 11 and references therein). In that context, quantitative determination of lifetimes of both small and larger, highly excited PAH cationic species is compulsory.

PAH molecules are also archetypal systems for which XUV-induced multi-electronic effects have been described12. Compared with more fragile molecules of the same size, they have the advantage of being rather chemically stable in the XUV spectral region. Therefore, from a fundamental point of view, they provide a situation where ionization involving many electrons, vibronic couplings and energy relaxation can be studied using high-level theoretical descriptions that need not take fragmentation into account13.

In this paper, we present the first experimental evidence for a general ultrafast mechanism that follows the absorption of an XUV photon by PAH molecules. This mechanism is sketched in Fig. 1a. An XUV photon ionizes the molecule creating a hole in the valence shell. As a result of electron correlation, the exit of the photoelectron may lead to excitation of another electron, that is, the formation of an excited 'shake-up' state. The excited molecular cationic state relaxes through a non-adiabatic mechanism. The goal of our work is to unravel this mechanism and to determine its timescale, and thereby to establish the typical relaxation times of the excited radical PAHs.

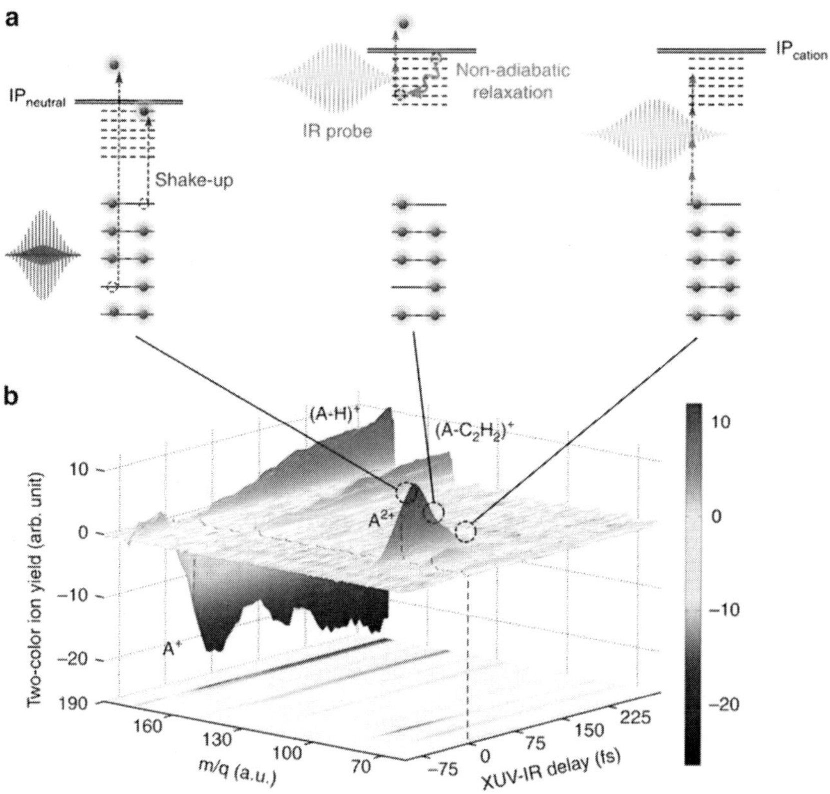

Figure 1: Schematic of the experiment.(**a**) Schematic of the XUV-induced dynamics in PAH molecules studied in this paper. Excited states are created in the valence shell of the cation through one of two possibilities, namely the formation of a single-hole configuration or the formation of a 2hole-1particle configuration (involving a shake-up process) (left) (IP stands for Ionization potential). The cation can be further ionized by the IR probe laser, provided that non-adiabatic relaxation has not taken place yet

(middle). After relaxation, the IR probe cannot ionize the cation anymore (right). (**b**) Two-colour XUV+IR ion signals measured in the case of anthracene, as a function of the detected mass-to-charge ratio and the XUV-IR delay. XUV-only and IR-only signals have been subtracted. The XUV pump and IR probe pulses overlap at zero delay (black dashed line). A red colour corresponds to a signal increase, while a blue colour signifies depletion. For positive XUV-IR delays, a very fast dynamics is observed for the doubly charged anthracene ion (A^{2+}, m/q=89). As explained in the text, the measurement reflects non-adiabatic relaxation in the anthracene cation (A^{+}). The dynamics observed in the first fragment ($A\text{-}C_2H_2^{+}$) is not discussed in this article.

2. RESULTS

2.1 Experiments

The mechanism was experimentally demonstrated using a fs XUV pump—fs IR (infrared) probe arrangement (Supplementary Figs 1 and 2, and Supplementary Note 1). A 25-fs long XUV pulse, containing photon energies between 17 and 35 eV, ionized neutral PAH molecules, and was followed at a variable time delay by a 35-fs long, weak (typically 10^{11} W cm^{-2}) IR pulse that probed the relaxation dynamics. The probing used ionization of the excited cationic states, and exploited the fact that rapid energy relaxation increases the energy that needs to be absorbed to doubly ionize the molecule, thus removing the molecules from the probe laser observation window. Experimentally, the formation of doubly charged cations was measured as a function of the XUV-IR pump-probe delay, as shown in Fig. 1b, where, for the case of anthracene, a 2-colour signal (ΔS) is shown, which is defined as

$$\Delta S = S_{XUV+IR} - (S_{XUV} + S_{IR}) \quad (1)$$

where S_{XUV+IR} corresponds to the ionization signal measured when both the XUV and IR lasers are present, and S_{XUV} and S_{IR} correspond to the signals measured when only the XUV laser or the IR laser present, respectively. As the main finding of our work, a rapid decay of the doubly charged parent ion signals was observed as a function of XUV-IR delay, for all four PAH molecules that were investigated: naphthalene (N, $C_{10}H_8$), anthracene (A, $C_{14}H_{10}$), pyrene (P, $C_{16}H_{10}$) and tetracene (T, $C_{18}H_{12}$).

To extract lifetimes from our measurements, the time-dependent doubly charged ion signals presented in Fig. 1b were fitted to an exponential decay convoluted with a Gaussian (see Fig. 2a):

$$\Delta S_{\mathrm{Fit}} = \mathrm{Heaviside}(t - t_0) * A_{\mathrm{Fit}} \exp\left(-(t - t_0)/\tau_{\mathrm{decay}}\right) \otimes \exp\left(-4\ln(2)(t/\tau_{\mathrm{crossco}})^2\right) \quad (2)$$

where A_{Fit} is the amplitude of the signal, t_0 defines the time overlap of the XUV and IR pulses, τ_{decay} is the lifetime of the excited state, and τ_{crossco} is the full-width at half-maximum of a Gaussian function that accounts for the cross-correlation time between the XUV and IR pulses. In Fig. 2a results are shown for A^{2+}, yielding an excited state lifetime of 32±4 fs. At delays >100 fs, the two-colour signal completely vanishes, since the IR intensity is no longer sufficient to ionize the cation A^+. (For further details about the data analysis, see the Supplementary Note 2.)

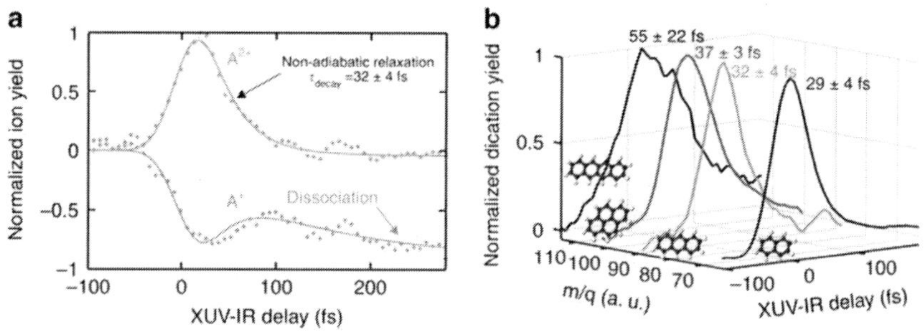

Figure 2: Time-dependent measurements.(**a**) Experimental time-dependent di-cationic (A^{2+}, orange) and cationic (A^+, grey) signals measured in XUV pump-IR probe experiments on anthracene (see Supplementary Note 2 for details about curve fitting). (**b**) Time-dependent di-cation signals measured for four PAH molecules: naphthalene (blue), anthracene (orange), pyrene (red) and tetracene (black).

In Fig. 2a the yield of the parent ion A^+ is also plotted as a function of the XUV-IR pump–probe delay. For short XUV-IR delays (<100 fs), the time dependence of the A^+ parent ion yield mirrors that of the A^{2+}doubly charged ion. This is consistent with our interpretation of the A^{2+}formation in terms of an IR-induced depopulation of excited cationic states, producing an A^{2+} ion and inhibiting the detection of an A^+ ion. In other words, unless ionized by the IR laser, the XUV-induced excited states relax through internal conversion within a few tens of fs. At later delays a slow (>100 fs) reduction of the A^+ yield appears, which reflects the occurrence of IR-induced dissociation of the vibrationally hot A^+ions.

The experiment was carried out for the four PAH molecules mentioned. As shown in Fig. 2b, decay times of 29±4 fs for naphthalene, 32±4 fs for anthracene, 37±3 fs for pyrene and 55±22 fs for tetracene were measured. Moreover, these measured decay times were independent of the IR intensity and the gas used in the XUV high-harmonic generation process (Supplementary Fig. 3). The latter changed the high-energy cutoff of the XUV spectrum from H17 (26.4 eV) in Xe to H23 (35.6 eV) in Ar. The robustness of the measured decay times to changes in the XUV spectral cutoff can be explained by the XUV absorption cross-section of the PAH molecules, which peaks at 17 eV (reaching 400 Mbarn) and decreases substantially above this photon energy[14,15]. Hence, only low-order harmonics (H11 (17.05 eV), H13 (20.15 eV) and H15 (23.25 eV)) will be efficiently absorbed and contribute to the observed mechanism. In addition, the cationic states formed are expected to be very similar for all harmonics, with the excess energy being carried off by the ejected photoelectron. Measurements also showed that the decay times barely vary with the IR intensity (Supplementary Figs 3 and 4), meaning that the IR probe intensity does not affect the measured dynamics, and that the relaxation rates are similar for the highly excited A^+ ions that can be ionized by one or two IR photons and less excited A^+ ions that require more IR photons. We note that the production of A^{2+} favours ionization processes via the absorption of a limited number of IR photons, since absorption at higher intensity of a larger number of IR photons leads to more extensive fragmentation, and does not contribute to the A^{2+}signal.

The ionization thresholds of the neutral PAH molecules used are 8.1 eV (naphthalene), 7.4 eV (anthracene), 7.6 eV (pyrene) and 7 eV (tetracene). The double ionization threshold of neutral anthracene is ~20 eV and can therefore be reached directly by H13, H15 and up, but also by, for example, absorption of H11 plus 2 IR photons. However, the latter requires that following the XUV ionization the ion has to remain in an electronically excited state that is close enough to the ionization threshold to be further ionized. Ultrafast non-adiabatic relaxation leads to energy flow from electronic to vibrational degrees of freedom, and therefore increases the number of IR photons that are needed for producing A^{2+}, thus causing the observed time dependence.

2.2 Theoretical model

In order to shed light onto the multi-electronic and coupled electronic and nuclear dynamics, high-level theoretical calculations were carried out (Supplementary Notes 3 and 4). First, a multi-electronic approach was used to compute both the single-ionization spectrum and the multidimensional potential energy surfaces of the PAH molecules, using the third-order *ab initio* algebraic diagrammatic construction (ADC(3)) Green's function method16,17. Nineteeen ADC calculations were performed for each selected normal mode (corresponding to 19 geometries). This calculation was followed by wave packet propagation employing the multiconfigurational time-dependent Hartree (MCTDH)18approach, to compute the non-adiabatic relaxation of the states determined by the ADC method. These theoretical studies were confined to naphthalene because of the huge computational effort required. In Fig. 3a, we show the high-energy part of the calculated ionization spectrum of naphthalene at 23 eV (Supplementary Fig. 5). Each line represents a cationic eigenstate and is located at the corresponding ionization energy. The height of the line reflects the contribution of a particular one-hole (1h) configuration to the ionic states (accounting for the process of removal of an electron from a particular molecular orbital). Hence, a peak height of 1 reflects a fully mono-electronic process. Reduced peak heights reflect the effect of electron correlation and, more specifically, the involvement of two-hole-one-particle (2h-1p) configurations (accounting for the removal of an electron from a particular orbital accompanied by the excitation of another electron) and higher order configurations. While the ionization below 10 eV is described essentially by 1h configurations, electron correlation effects start to play a very important role above 10 eV, and shake-up excitation and higher order processes dominate above 15 eV (Fig. 3a and Supplementary Fig. 5). Note that a photon with a given energy may populate all states with a lower binding energy (for instance, in our experiment, harmonic H13 populates all states below 20.15 eV). In the experiment, and in agreement with earlier photoelectron spectroscopy measurements19, the excitation accesses a spectral domain where correlation cannot be neglected and where excited cations are created due to correlation. The calculation shows that nine states are efficiently populated in the energy range of 16–20 eV (shake-up region) close to the ionization threshold of the cation. This defines the first step of the observed process.

Next, the excited cations relax through non-adiabatic-coupled electron-nuclear dynamics. A model diabatic Hamiltonian20 restricted to the nine dominant electronic states (covering the spectral range of interest), calculated using the ADC method described above, was constructed in terms of the dimensionless normal displacement coordinates (Q) of the reference electronic ground state of neutral naphthalene (N) (for technical details, see Supplementary Fig. 6, Supplementary Table 2, Supplementary Notes 3 and 4). In Fig. 3b one-dimensional cuts of the multidimensional diabatic potential energy surface of N^+ are shown along the normal coordinate of the symmetric inter-ring C=C stretching mode Q7 (ν=1,051 cm^{-1}). From this diagram it can be seen that there are numerous curve crossings among this group of states, which develop into multiple multidimensional CIs. The location of the energy minimum of the seam of these CIs with respect to the equilibrium position of a state (Supplementary Table 1) primarily governs the dynamics.

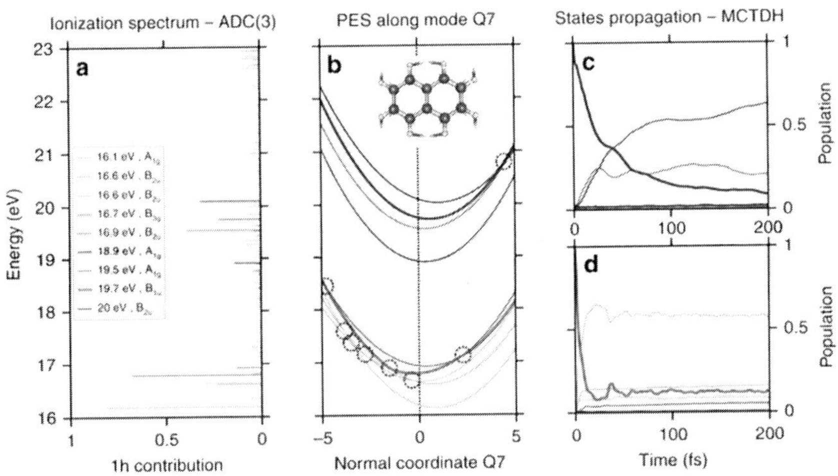

Figure 3. Multi-electronic ADC(3) and non-adiabatic MCTDH propagation calculations performed on naphthalene. (**a**) Zoom of the ADC spectrum in the relevant high energy range, indicating the nine ionic states selected as initial states for the non-adiabatic dynamics simulations (the same colour code indicated in the inset is used in the other parts of the figure). (**b**) Cut through the multidimensional potential energy surface along the normal coordinate of the symmetric inter-ring C=C stretching mode Q7 (see inset), for the nine selected electronic states. The seams (dashed circle) are all located close to the equilibrium geometry (dashed line) and the high complexity of their multidimensional topology governs the depopulation dynamics. (**c**,**d**) Time-dependent evolution of the diabatic population of the electronic states when beginning with population=1 in the B_{1u} (19.7 eV) and B_{3g} (16.7 eV) states, calculated using the MCTDH method.

First principles nuclear dynamics calculations were carried out employing the vibronic Hamiltonian constructed as sketched above (Supplementary Note 4). For a selected state, the initial wave packet is defined as a Franck–Condon projection starting from the ground vibronic state of the neutral molecule. The wave packet is propagated over 200 fs in the coupled manifold of 9 N^+ states with 29 relevant vibrational modes. The 29 chosen modes are the ones that induce noticeable coupling strength in the electronic states according to the symmetry selection rules. Note that the remaining discarded modes have extremely low coupling strength and high frequency and are thus less important in the dynamics. As a typical example, computed time-dependent diabatic electronic populations starting from initial population of the states B_{1u} (19.7 eV) and B_{3g} (16.7 eV) are shown in Fig. 3c,d (the time-dependent populations starting from other states is shown in the Supplementary Fig. 7), and clearly evidence rapid non-adiabatic relaxation dynamics from higher-lying to lower-lying states of the cation. The B_{3g} state (16.7 eV) shows very fast dynamics characterized by a ~10 fs decay time. The quasi-degeneracy of the seam minima with the equilibrium minimum of this state and moderate interstate coupling causes its ultrafast non-radiative decay. The decay of the B_{1u} state (19.7 eV) occurs within 40 fs, and can again be understood in relation with the location of the seam minima. The discussion above can be generalized to the other electronic states including very low-lying states as presented in the Supplementary Fig. 7and emphasizes the link between the timescale of the relaxation dynamics and the location of the CIs with respect to the equilibrium position. We note that our model includes only part of all the cationic states present in this energy range. However, as discussed above the decay is determined by topology of the crossings rather than the density of states. The timescale observed is a signature of the rigid bonds of the carbon backbone of the molecule. As a consequence, lower states will only weakly affect the depopulation of states such as B_{1u}, as they will lead to similar crossings.

Although modelling of the non-adiabatic relaxation for larger PAH species is computationally out of reach, we expect that the same process will take place in all PAH molecules due to the similarity of their vibrational modes21. As shown in Supplementary Fig. 5, the larger PAH molecules have very similar ionization spectra. The relaxation time for larger PAHs is a competition between two effects: on the one hand, the increase in the density of states facilitates rapid energy transfer (the energy gap law). On the other hand, the wave packet has to travel through more intermediate CIs before reaching states from which the system can no longer be ionized by the IR pulse, slowing down the dynamics. This is exactly what we observe experimentally. However, we also note that the evolution of the relaxation lifetime with the size is modest, and even for larger species, it remains in the sub-picosecond regime.

3. DISCUSSION

The study reported here involves states at higher energies (typically up to 12 eV above the ground state, that is, binding energy of 20 eV) than those involved in the DIBs (for example, typically below 8 eV above the ground state). Nevertheless, our results also demonstrate that, despite the existence of small energy gaps between the highly excited states probed here, lifetimes in the range of 30–60 fs are observed, and even increasing for large species. Moreover, our calculations show that this decay rate is general and is found for states in the 0–12 eV energy range. This is because of the topology of the CIs. It follows that the energy gap laws cannot be directly extended to large PAHs or higher energy states, suggesting that low-lying states which potentially could be involved in the DIB problem are likely to have much longer lifetimes than previously assumed. As electronic transitions remain the best probe to identify the specific PAH molecules present in space, further studies on the link between PAH cation species and the DIBs are warranted. The use of short XUV pulses to prepare these cation species could have significant implications for unraveling the extent of molecular complexity in the universe.

In conclusion, we have identified for the first time a general process in PAH molecules under conditions where both the single active electron and Born–Oppenheimer approximations breakdown. The XUV ionization is followed by a very fast relaxation of the excited cationic states due to non-Born–Oppenheimer couplings. The present study offers the first direct observation of ultrafast relaxation pathways in a family of large molecular systems under XUV irradiation. The rapid relaxation is due to the favourable topology of the seams of the CIs corresponding to slight distortions of the rigid molecular backbone which relax within a fraction of the vibrational period. The slight evolution of the decay with molecular size indicates that the relaxation remains time confined in the same range even for larger species.

Ultrafast XUV science offers a perfect platform for studying the importance of very short time-scale dynamics in natural processes involving potentially very large molecules. Here we show that, beyond statistical molecular relaxation, other fundamental aspects of quantum mechanics, such as multi-electronic excitation and non-Born–Oppenheimer vibronic couplings play an important role in the context of astrochemistry and have to be accurately included in models. Our work also has possible implications for photoinduced coherent charge transport such as attosecond hole migration following electron removal, as discussed in the recent literature 22,23,24,25,26.

REFERENCES

1. Becker, U. & Shirley, D. A. VUV and soft X-ray photionization 47–80Springer (1996).Domcke, W. & Yarkony, D. R. Role of conical intersections in molecular spectroscopy and photoinduced chemical dynamics. Annu. Rev. Phys. Chem 63, 325–352 (2012).Gagnon, E. et al. Soft X-ray-driven femtosecond molecular dynamic. Science 317,

1374–1378 (2007).Timmers, H., Shivaram, N. & Sandhu, A. Ultrafast dynamics of neutral superexcited oxygen: a direct measurement of the competition between autoionization and predissociation. Phys. Rev. Lett. 109, 173001 (2012).Zhou, X. et al. Probing and controlling non-Born-Oppenheimer dynamics in highly excited molecular ions. Nat. Phys. 8, 232–237 (2012).

2. Persson, P., Lunell, S., Szoke, A., Ziaja, B. & Hajdu, J. Shake-up and shake-off excitations with associated electron losses in X-ray studies of proteins. Protein Sci. 10, 2480–2484 (2001).Boudaïffa, B., Cloutier, P., Hunting, D., Huels, M. A. & Sanche, L. Resonant formation of DNA strand breaks by low-energy (3 to 20 eV) electrons. Science 267, 1658–1660 (2000).Satzger, H. et al. Primary processes underlying the photostability of isolated DNA bases. Proc. Natl Acad. Sci. USA 103, 10196–10201 (2006).Tielens, A. G. G. M. Interstellar polycyclic aromatic hydrocarbon molecules. Annu. Rev. Astron. Astrophys. 46, 289–337 (2008).
3. Reddy, V. S., Ghanta, S. & Mahapatra, S. First principles quantum dynamical investigation provides evidence for the role of polycyclic aromatic hydrocarbon radical cations in interstellar physics. Phys. Rev. Lett. 104, 111102 (2010).Tielens, A. G. G. M. The molecular universe. Rev. Mod. Phys. 85, 1021 (2013).
4. Deleuze, M. S., Trofimov, A. B. & Cederbaum, L. S. Valence one-electron and shake-up ionization bands of polycyclic aromatic hydrocarbons. I. Benzene, naphthalene, anthracene, naphthacene, and pentacene. J. Chem. Phys. 115, 5859 (2001).
5. Jankala, J. et al. Double photoionization in ring molecules: search of the cooper pair formation. Phys. Rev. Lett. 112, 143005 (2014).Malloci, G., Mulas, G., Cappellini, G. & Joblin, C. Time-dependent density functional study of the electronic spectra of oligoacenes in the charge states -1, 0, +1, and +2. Chem. Phys. 340, 43–58 (2007).
6. Malloci, G., Cappellini, G., Mulas, G. & Mattoni, A. Electronic and optical properties of families of polycyclic aromatic hydrocarbons: a systematic (time-dependent) density functional theory study. Chem. Phys. 384, 19–27 (2011).
7. Schirmer, J., Cederbaum, L. S. & Walter, O. New approach to the one-particle greens-function for finite Fermi systems. Phys. Rev. A 28, 1237 (1983).
8. Schirmer, J., Trofimov, A. B. & Steller, G. A non-Dyson third-order approximation scheme for the electron propagator. J. Chem. Phys. 109, 4734 (1998).
9. Meyer, H.-D., Manthe, U. & Cederbaum, L. S. The multi-configurational time-dependent Hartree approach. Chem. Phys. Lett. 165, 73–78 (1990).
10. Bally, T., Carra, C., Fulscher, M. P. & Zhu, Z. Electronic structure of the naphthalene radical cation and some simple alkylated derivatives. J. Chem. Soc., Perkin Trans. 2 8, 1759–1766 (1998).Köppel, H., Domcke, W. & Cederbaum, L. S. in. Advances in Chemical Physics, Vol. 57 (eds Prigogine, I. & Rice, S. A.) (Wiley, 1984).Bauschlicher, C. W. et al. The NASA ames polycyclic aromatic hydrocarbon infrared spectroscopic database: the computed spectra. Astrophys. J. Suppl. Ser. 189, 341 (2010).Calegari, F. et al. Ultrafast electron dynamics in phenylalanine initiated by attosecond pulses. Science 346, 336–339 (2014).Cederbaum, L. S. & Zobeley, J. Ultrafast charge migration by electron correlation. Chem. Phys. Lett. 307, 205 (1999).
11. Remacle, F., Levine, R. D., Schlag, E. W. & Weinkauf, R. Electronic control of site selective reactivity: a model combining charge migration and dissociation. J. Phys. Chem. A 103, 10149 (1999).
12. Kuleff, A. & Cederbaum, L. S. Ultrafast correlation-driven electron dynamics. J. Phys. B 47, 124002 (2014).
13. Lépine, F., Ivanov, M. Y. & Vrakking, M. J. J. Attosecond molecular dynamics: fact or fiction? Nat. Photon. 8, 195–204 (2014).

CHAPTER 9

Integrated Science Investigation of the Sun (ISIS): Design of the Energetic Particle Investigation

D. J. McComas12, N. Alexander1, N. Angold1, S. Bale3, et al.

1.Southwest Research InstituteSan AntonioUSA

2.University of Texas at San AntonioSan AntonioUSA

3.University of California at BerkeleyBerkeleyUSA

ABSTRACT

The Integrated Science Investigation of the Sun (ISIS) is a complete science investigation on the Solar Probe Plus (SPP) mission, which flies to within nine solar radii of the Sun's surface. ISIS comprises a two-instrument suite to measure energetic particles over a very broad energy range, as well as coordinated management, science operations, data processing, and scientific analysis. Together, ISIS observations allow us to explore the mechanisms of energetic particles dynamics, including their: (1) Origins—defining the seed populations and physical conditions necessary for energetic particle acceleration; (2) Acceleration—determining the roles of shocks, reconnection, waves, and turbulence in accelerating energetic particles; and (3) Transport—revealing how energetic particles propagate from the corona out into the heliosphere. The two ISIS Energetic Particle

Instruments measure lower (EPI-Lo) and higher (EPI-Hi) energy particles. EPI-Lo measures ions and ion composition from ~20 keV/nucleon–15 MeV total energy and electrons from ~25–1000 keV. EPI-Hi measures ions from ~1–200 MeV/nucleon and electrons from ~0.5–6 MeV. EPI-Lo comprises 80 tiny apertures with fields-of-view (FOVs) that sample over nearly a complete hemisphere, while EPI-Hi combines three telescopes that together provide five large-FOV apertures. ISIS observes continuously inside of 0.25 AU with a high data collection rate and burst data (EPI-Lo) coordinated with the rest of the SPP payload; outside of 0.25 AU, ISIS runs in low-rate science mode whenever feasible to capture as complete a record as possible of the solar energetic particle environment and provide calibration and continuity for measurements closer in to the Sun. The ISIS Science Operations Center plans and executes commanding, receives and analyzes all ISIS data, and coordinates science observations and analyses with the rest of the SPP science investigations. Together, ISIS' unique observations on SPP will enable the discovery, untangling, and understanding of the important physical processes that govern energetic particles in the innermost regions of our heliosphere, for the first time. This paper summarizes the ISIS investigation at the time of the SPP mission Preliminary Design Review in January 2014.

Keywords: Solar Probe Plus, ISIS Solar, energetic particles, SEPs, CMEs, Corona Particle acceleration

1. INTRODUCTION

1.1 Science Background and Objectives

NASA's Solar Probe Plus (SPP) mission is designed to plunge repeatedly into the innermost regions of the solar system where the Sun's million-degree atmosphere, or corona, begins its outward expansion to produce the supersonic solar wind. The outward-flowing solar wind interacts with the Earth's magnetosphere and other objects in the solar system and ultimately inflates a bubble in the interstellar medium known as the heliosphere, which engulfs and protects the Earth and the planets from galactic cosmic radiation. SPP is designed to survey the birthplace of the solar wind and its embedded magnetic and electric fields, explore the origins of large-scale disturbances created during powerful explosions known as solar flares and coronal mass ejections (CMEs), and reveal how these conspire to accelerate solar energetic particles (SEPs)—suprathermal and energetic particles from ~few keV up to GeV. By making direct in-situ measurements of the inner heliospheric environment from <10 to >60 solar radii (R_S), SPP is poised to redefine Solar and Heliospheric physics as we currently understand them (see McComas et al. 2007).

Solar Probe Plus (Fox et al. 2014) carries a complement of four state-of-the-art scientific instruments/instrument suites: a solar wind plasma suite—the Solar Wind Electrons Alphas and Protons (SWEAP) investigation (Kasper et al. 2014); an electric and magnetic field suite—"FIELDS" (Bale et al. 2014); a wide field imager—"WISPR" (Howard et al. 2014); and the energetic particle suite—the Integrated Science Investigation of the Sun (ISIS). ISIS is designed to provide comprehensive measurements of the energetic particle environment, including energy spectra, anisotropy, and composition of suprathermal and solar energetic ions from ~0.02–200 MeV/nucleon (nuc), as well as the energy spectra and arrival direction of ~0.025–6 MeV electrons. Within ISIS, observations are made with two complementary instruments: Energetic Particle Instrument-Low energy (EPI-Lo) and Energetic Particle Instrument-High energy (EPI-Hi), which measure the lower and higher energy energetic particles, respectively. This paper describes the science goals and instrumentation of the ISIS investigation as planned and designed as of SPP's Preliminary Design Review (PDR) in January 2014.

The overall science objective of ISIS is to explore the physical mechanisms that produce, accelerate, and transport energetic particles in the inner heliosphere. ISIS achieves this objective by addressing the following three questions:

1) What is the origin or seed population of solar energetic particles (SEPs)?

2) How are these SEPs and other particle populations accelerated?

3) What mechanisms are responsible for transporting the different particle populations into the heliosphere?

ISIS additionally contributes to addressing SPP's other two overarching scientific objectives, namely, (1) Trace the flow of energy that heats and accelerates the solar corona and solar wind and (2) Determine the structure and dynamics of the plasma and magnetic fields at the sources of the solar wind through its measurements of suprathermal particle (~2–400 keV/nuc H–Fe ions) tails. The ISIS design is also set to reveal previously unknown features of the inner heliosphere, which stretches from the Sun's corona to the orbit of Mercury. Thus, we also anticipate great discovery science from ISIS and more generally from the SPP mission.

The ISIS investigation is carried out by a world-class team of scientists, engineers, technicians, and business and support personnel; Fig. 1 provides ISIS' organizational chart and identifies the key team members leading in each area and Table 1 lists the Science Team members. The team was assembled by combining top people and institutions that have been highly successful across a broad range of prior missions. ISIS leadership works to foster an environment where everyone works together to produce even greater synergies and uses the combined experience and capabilities of the team to make ISIS the best possible science investigation for SPP.

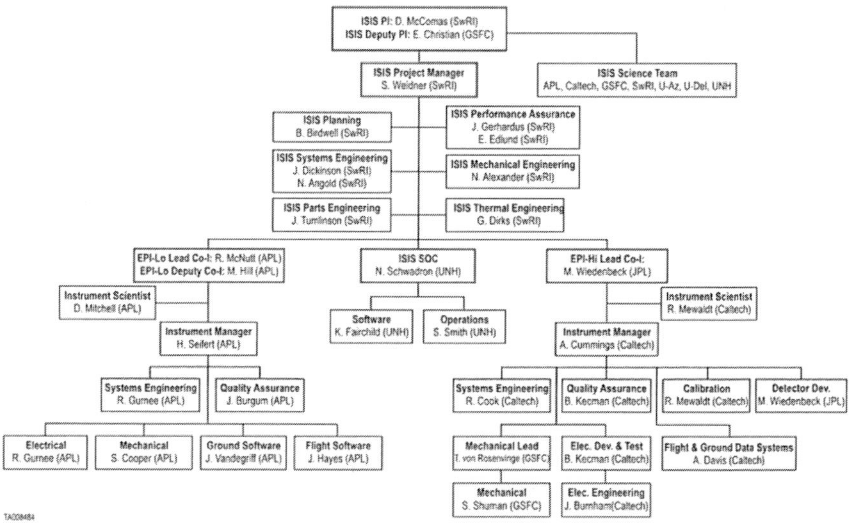

Figure 1. Organizational chart for the Integrated Science Investigation of the Sun

The remainder of Sect. 1 provides a scientific justification for suprathermal and energetic particle measurements on SPP and discusses each of the above science questions in terms of the several particle populations that SPP will encounter. The focus is on particular topics that only the ISIS investigation can address. Section 2 provides an overview of ISIS, including suite philosophy, the ISIS suite viewing, mechanical and electrical aspects of the overall design, and plans for science data and operations. Sections 3 and 4 provide details of the EPI-Lo and EPI-Hi instruments, respectively, including overall design, planned and expected performance vs. requirements, electrical and mechanical design, and calibration plans; Sect. 5 describes planned science operations, data processing, and data products in more detail, while Sect. 6provides a brief summary.

Table 1. ISIS Science Team members

Title	Name	Organization
PI	Dave McComas	SwRI
DPI	Eric Christian	GSFC
Co-I	Alan Cummings	Caltech
Co-I	Mihir Desai	SwRI
Co-I	Joe Giacalone	U Arizona
Co-I	Matthew Hill	JHU/APL
Co-I	Stefano Livi	SwRI
Co-I	Bill Matthaeus	U Delaware
Co-I	Ralph McNutt	JHU/APL
Co-I	Dick Mewaldt	Caltech
Co-I	Don Mitchell	JHU/APL
Co-I	Nathan Schwadron	UNH
Co-I	Tycho von Rosenvinge	GSFC
Co-I	Mark Wiedenbeck	JPL
SSM[a]	Robert Gold	JHU/APL
SSM[a]	Stamatios Krimigis	JHU/APL
SSM[a]	Edmond Roelof	JHU/APL
SSM[a]	Ed Stone	Caltech

[a]SSM is "Senior Science Mentor": Non-Co-I science team members who provide senior-level guidance and advice to the rest of the team

1.2 Solar Energetic Particles, Solar Flares, and Coronal Mass Ejections

1.2.1 Two Sources of Solar Energetic Particles

Prior to the mid-1980s, solar flares were believed to produce much of the solar energetic particle population observed near Earth orbit. However, this viewpoint was challenged when coronal mass ejections (CMEs) were first observed and then recognized as powerful alternative means by which the Sun expels considerable matter and energy into the heliosphere and causes geomagnetic disturbances (e.g., Gosling 1993, and references therein). As illustrated in Fig. 2, it is now widely accepted that the Sun accelerates charged particles (ions and electrons) from ~few keV up to near-relativistic energies in at least two ways (e.g., Reames 2013). Mechanisms associated with magnetic reconnection occur during solar flares and are largely responsible for producing the smaller "impulsive" SEP events that are enriched in electrons, ^{3}He, and heavier ions such as Ne–Fe (e.g., Mason 2007). In contrast, large-scale shock waves and related processes, driven by fast CMEs plowing through the ambient corona and the solar wind are associated with the larger "gradual" SEP events (Reames 1999).

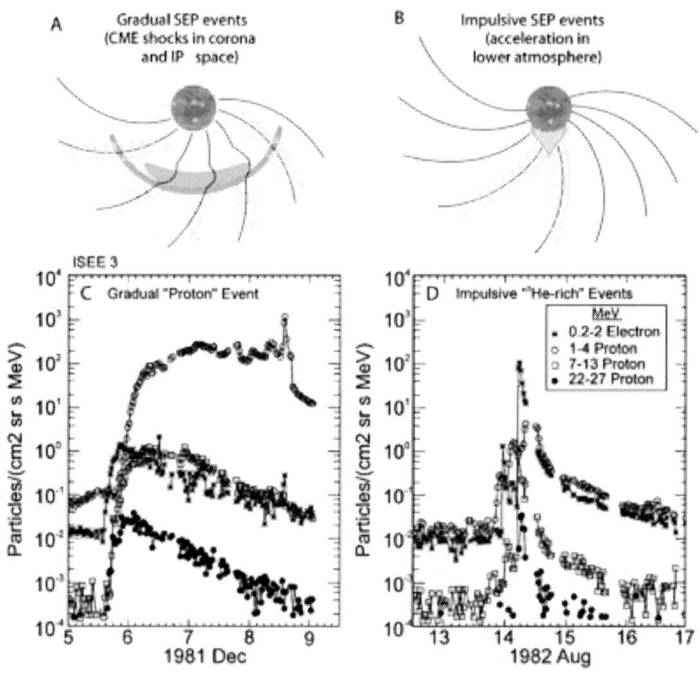

Figure 2. Two sources of SEP events. (**A**) A large gradual event is produced by an expanding CME-driven shock wave that populates interplanetary magnetic field (IMF) lines with SEPs over a broad longitudinal extent. (**B**) A solar flare produces an impulsive event that populates well-connected IMF lines, leaving nearby IMF lines relatively empty. Intensity-time profiles of electrons and protons in (**C**) a large gradual and (**D**) a small impulsive SEP event (adapted from Reames 1999)

Though the acceleration mechanisms in the two cases are distinct, and both are highly efficient, their effects are often intertwined. The net result is that charged particles are frequently energized up to GeV energies within a few seconds to minutes (Rice et al. 2003).

1.2.2 Why Do We Need to Make Measurements Closer to the Sun?

An impediment to understanding the origin and acceleration mechanisms and developing and testing physics-based models using near-Earth observations is that by the time the SEPs reach 1 AU, particles from multiple sources (flares, CME shocks etc.), having been accelerated by different physical processes (e.g., reconnection, shock acceleration), are completely mixed. Figure 3 shows a clear example of the need for SEP observations in the inner heliosphere. For these observations, the Helios-1 spacecraft located at ~0.4 AU detected at least 5 separate impulsive-like injections of electrons and He ions, while IMP-8 at 1 AU observed a single particle event. The dramatically different time-intensity profiles could result from a combination of one or more of the following processes: (1) scattering and diffusion during transport between 0.4 and 1 AU; (2) distinct magnetic connections of the spacecraft to the particle sources; and (3) the spatio-temporal evolution of the particle sources.

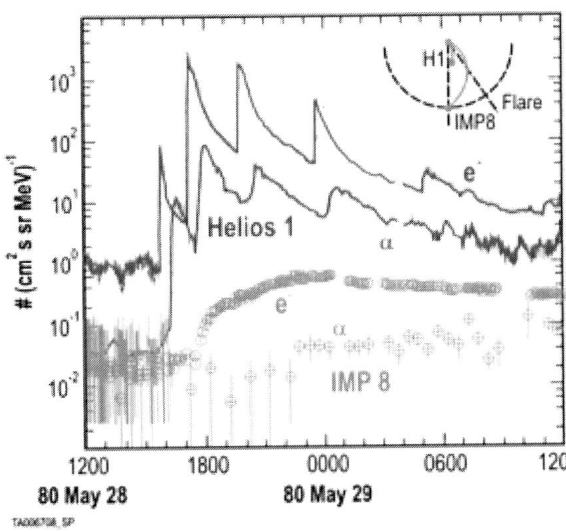

Figure 3. Electron (e: 0.3–0.8 MeV) and He (α: 2–4 MeV/nuc) time profiles from Helios-1 (0.3 AU) and IMP-8 (1 AU) during five impulsive SEP events in 1980 (from Wibberenz and Cane 2006). Magnetic connections to the flare site are indicated at *upper right*. Helios-1 observed five injections that had merged into a single event by the time they reached IMP-8

These mechanisms smear the time profiles at 1 AU, making it difficult to identify the source locations and therefore understand the nature of the acceleration mechanisms. By making SEP measurements close to and, in some cases, within the acceleration region and at various radial distances along the way, ISIS allows us to understand and untangle the effects of transport, including velocity dispersion, for the first time.

1.3 Identify the Origins of Energetic Particles in the Inner Heliosphere

1.3.1 Impulsive SEP Events

Combined observations of the energy spectra and onset times of ^{3}He-rich impulsive SEP events and >5 keV electrons with imaging of Type III radio bursts and Extreme ultraviolet (EUV) observations of the flaring regions indicate that many impulsive SEP events are associated with jet-like ejections that occur around the solar particle release times (Pick et al. 2006). From extrapolations using Potential Field Source Surface models, as in Wang et al. (2006), the source regions of these particles are seen to be adjacent to coronal holes containing Earth-directed open field lines. In this scenario, the jets are signatures of magnetic reconnection or exchange of connectivity between closed field lines in the source regions and Earth-directed open field lines, and the discrete ^{3}He-rich particle events seen in Fig. 4 are observed when such foot point exchanges occur. While this model could account for two discrete ^{3}He-rich increases seen in the time-intensity profiles in Fig. 4 on October 12, 2005, it is unclear how it could also explain the presence of ^{3}He for a period of more than ~3 days thereafter. Mason (2007) suggested that a nearly continuous emission process associated with magnetic reconnection between closed and open field lines may be associated with the acceleration and subsequent release of ^{3}He into the interplanetary medium. In addition, the three physical mechanisms discussed in Sect. 1.2.2could also play important roles in producing long duration ^{3}He-rich intervals.

Because the expected intensities and therefore the effective sensitivity will be significantly higher in the inner heliosphere, SPP will measure discrete injections of ^{3}He during numerous SEP events that are smaller than currently detected from Earth-orbit. In addition, by combining observations of energetic electrons, Type-III radio bursts and other imaging observations, SPP will also be able to identify and associate periods of continuous ^{3}He emission with nearly steady coronal processes involving foot point exchange and magnetic reconnection.

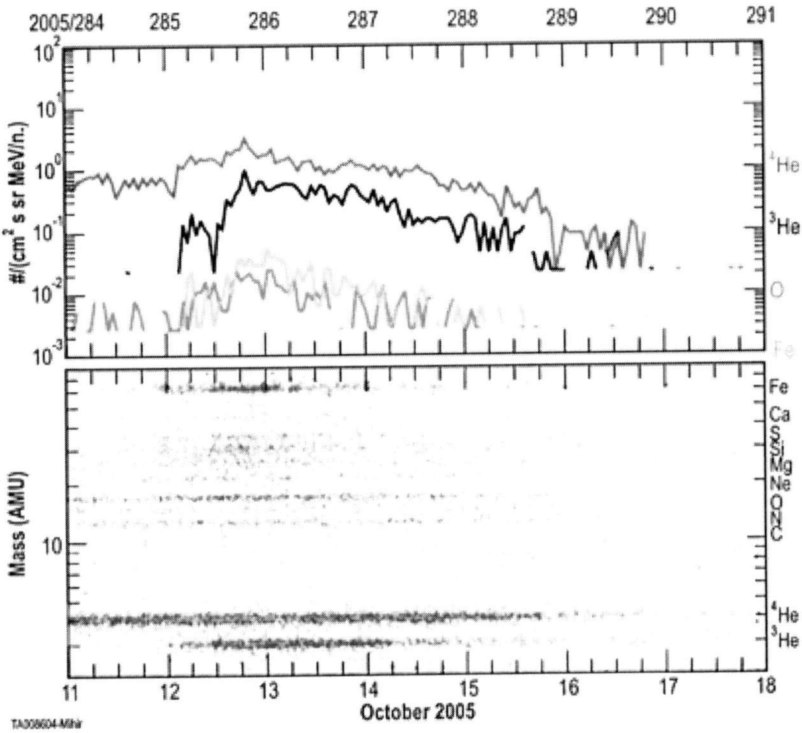

Figure 4. *Top*: Time-intensity profiles of ^{3}He, ^{4}He, Fe, and O and mass spectrogram of ions with mass between 3 and 60 AMU during a 7-day period (taken from Mason 2007)

These ISIS measurements, in conjunction with FIELDS and SWEAP, will also contribute to another SPP scientific objective, to trace the flow of energy that heats and accelerates the solar corona and solar wind. In particular ISIS, FIELDS and SWEAP enable studies to distinguish among heating mechanisms, such as reconnection, ion-cyclotron waves, and turbulent dissipation and associated stochastic heating (Cranmer and van Ballegooijen 2010; Isenberg and Vasquez 2013; Cranmer et al. 2007; Chandran et al. 2013). Direct parallel acceleration of suprathermals by reconnection up to several times the Alfvén speed in the low beta corona may be diagnosed by ISIS measurement of pitch angle distributions and energy spectra of electrons and energetic ions. Stochastic acceleration associated with Fermi processes and propagating waves, a process familiar in flares and astrophysical applications (Liu et al. 2004), will produce bidirectional beaming distributions of energized ions and widely varying energy spectra features. In contrast, stochastic acceleration of the betatron type may occur near (but outside of) reconnecting current sheets

(Dmitruk et al. 2004; Dalena et al. 2014; Teaca et al. 2014) and may produce higher energy suprathermals with perpendicular pitch angle distributions. Finally, ion cyclotron resonance can produce distinctive pitch angle signatures associated with a population of resonant waves of the required polarization (Isenberg and Vasquez 2013). ISIS measurement of pitch angle anisotropies and energy spectra together with FIELDS magnetic field and electric field spectral energy density, polarization and helicities, and SWEAP ion velocities and temperatures, will allow direct and correlative measurements that will distinguish among these mechanisms. These studies will be done at various heliocentric distances approaching and within the Alfvén radius, enabling an evaluation of the radial variation of the relative effects of these distributed heating mechanisms.

1.3.2 Large Gradual SEP Events

As the outward moving CMEs expand and slow down, the associated shocks weaken, implying that most CME shocks are very efficient at accelerating particles when they are close to the Sun and produce most of the higher energy ions within ~20 R_S. Since CMEs propagate into the ambient corona and the slow solar wind, the associated shocks were generally believed to inject and accelerate coronal or solar wind material (e.g., Reames 1999). However, the ion composition in many large gradual CME-associated SEP events was enriched in the rare isotope ^{3}He and in heavy ions such as Ne–Fe, leading some researchers (see e.g., Mason et al. 1999) to suggest that CME shocks encounter and re-accelerate suprathermal flare material that is frequently present in the interplanetary medium (Wiedenbeck et al. 2008). In contrast, Cane et al. (2006) suggested that flares make a direct contribution to many large gradual SEPs, especially at energies above ~10 MeV/nuc. By making the first-ever in-situ measurements of the properties of CME shocks, suprathermal ions, and the accelerated SEPs simultaneously within 20 R_S combined SPP measurements will help reveal how flares contribute to large gradual SEPs, either by providing the suprathermal seed population or by providing the higher energy SEPs themselves. A key element in unraveling the balance of these effects will be ISIS analyses of transport effects, discussed further below.

1.3.3 Suprathermal Ions

The presence of rare (~10^{-4} of ^{4}He) solar wind ion species such as ^{3}He and He^+ in substantial amounts in the gradual SEP population indicates that the associated CME-driven shocks draw much of their seed population from a pool of suprathermal ions with speeds above that of the ambient solar wind ions (e.g., Desai et al. 2001), at least for these species, and possibly others as well. Ions from multiple sources can contribute to a commonly observed suprathermal tail (~1.5–20 times the solar wind speed). Suprathermal ions are defined here as ~2–400 keV/nuc H–Fe ions. These sources can be highly variable in space and time, and include gradual and impulsive SEP events, CME-driven interplanetary shocks, and corotating interaction regions (CIRs). Seed particles may also be extracted from the pool of heated and accelerated solar wind ions, and interstellar and inner source pickup ions.

Figure 5 demonstrates this connection to more energetic particles that provides an important motivation to understand the physical mechanisms responsible for the production of suprathermal (ST) ions. The left panel shows that the >12 MeV/nuc Fe fluence near Earth in large SEP events is likely to be significantly larger if the ST ion intensities ~1 day prior to the occurrence of the corresponding SEP event were also high (Mewaldt et al. 2012b). The right panel shows that the >30 keV/nuc ST Fe intensities vary by three orders in magnitude as compared with the factor of ~10 variations in the solar wind densities (Mason et al. 2005). Such variations in the ST ion populations are expected to play a critical role in determining what types of seed populations are available for acceleration by CME shocks.

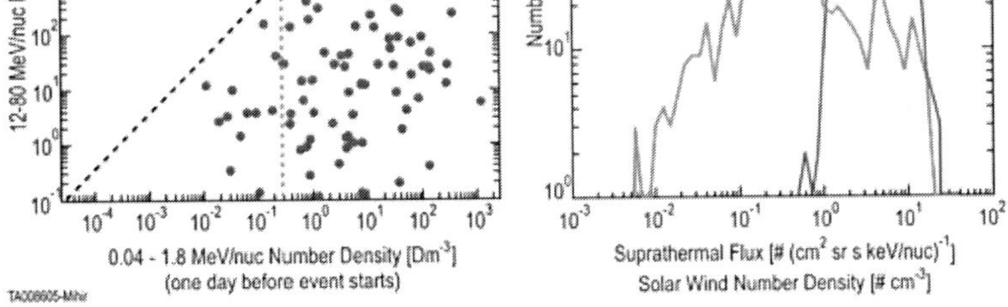

Figure 5. *eft*: Fluence of >12 MeV/nuc Fe in large SEP events vs. suprathermal Fe density one day prior to the onset of the corresponding SEP event (after Mewaldt et al. 2012b). *Right*: Hourly averaged intensity of suprathermal ~30 keV/nuc Fe (*red*) and number density of solar wind Fe (*blue*) for a 100 day period in 2004 (taken from Mason et al. 2005)

Existing observations of ST ion populations, as shown in Figs. 6 and 7, tell somewhat conflicting stories (Mason and Gloeckler 2012). For instance,

Fig. 6 shows that the ST proton tails between ~1.5–6 times the solar wind speed are characterized by an energy power-law with a nearly constant spectral index of −1.5 (i.e., $E^{-1.5}$, corresponding to a distribution function $f(v) \sim v^{-5}$, where v is particle velocity) under a variety of solar wind conditions, with most enhancements in the ST tail density being associated with extended compression regions (Fisk and Gloeckler 2012). In contrast, Fig. 7 shows that the ST tail between ~6–20 times the solar wind speed has highly variable spectral indices and that the ion composition varies with solar activity cycle, implying that the contributions from various sources (e.g., CIRs, SEPs etc.) also vary in time and space (Dayeh et al. 2009). Giacalone (2012) showed that strong interplanetary shocks are significant sources of suprathermal ions and that the peak enhancement of the particles occurred at the same time as the passage of nearly all the shocks studied. This suggests that shocks may also be an important a source for the ever-present, but variable, suprathermal particle tail given that energetic particles, being quite mobile, can remain within the inner heliosphere for some time.

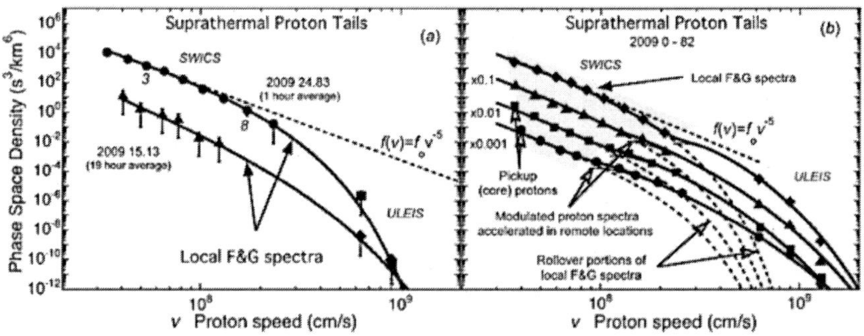

Figure 6. Proton phase space densities exhibit a common spectra shape under a variety of solar wind conditions (taken from Fisk and Gloeckler 2012)

Present theoretical models regarding the origin and acceleration of suprathermal ions can be grouped into two categories: (1) ST tails result from a distributed, continuous acceleration process. In interplanetary space this could be any second order Fermi process with (the problematic) constraint that acceleration and escape times are essentially equal. Specific suggestions have been made to achieve this, including distributed reconnection embedded in solar wind turbulence (Le Roux et al. 2002) and acceleration in extended compression regions (Fisk and Gloeckler 2012). SPP also could discover that ST tails originate in the solar corona due to micro- or nano-flares, a possibility that once again would have implications for observed signatures of transport; or (2) ST tails are lower energy portions of material accelerated in energetic particle events such as CIRs, transient interplanetary shocks (Giacalone 2012), SEPs etc. (Livadiotis and McComas 2009; Jokipii and Lee 2010; Schwadron et al. 2010).

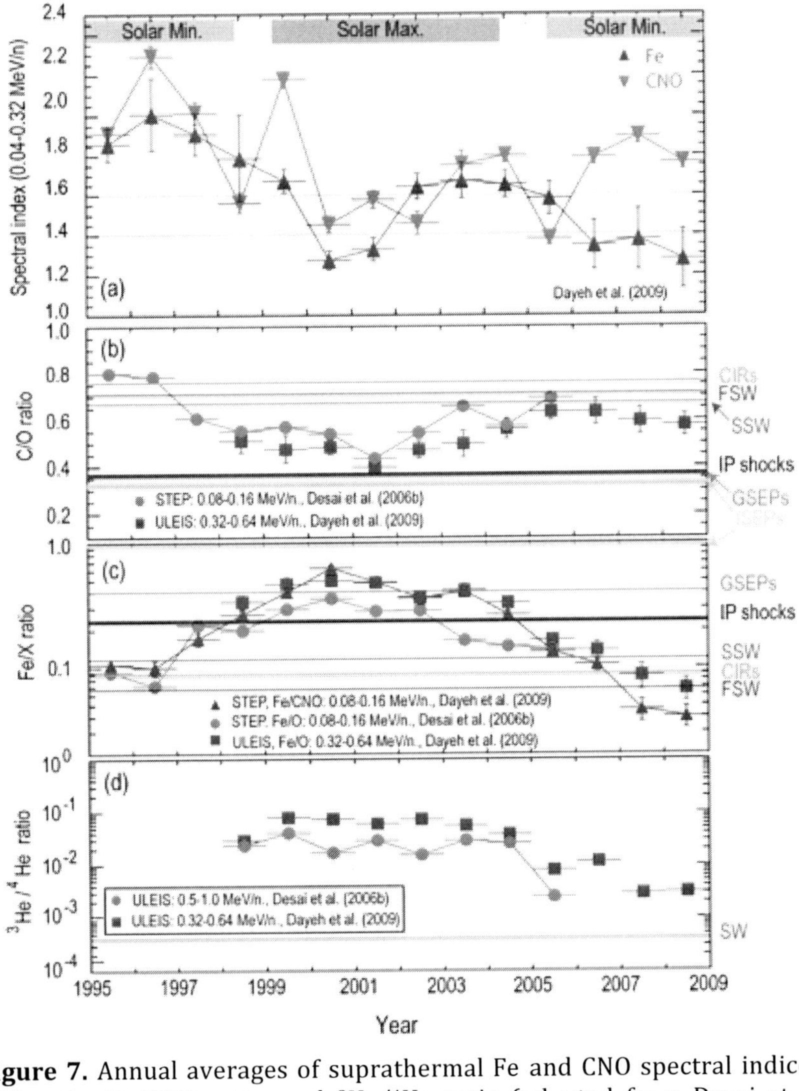

Figure 7. Annual averages of suprathermal Fe and CNO spectral indices, C/O ratio, Fe/CNO ratio, and $^{3}He/^{4}He$ ratio (adapted from Desai et al. 2006; Dayeh et al. 2009)

Adding to these controversies is the fact that, under typical solar wind conditions, current observations either do not have the capability, or lack the sensitivity, to measure the composition of suprathermal ions below ~50 keV/nuc to discern contributions from various sources on shorter timescales (less than a day). If the suprathermal ion tails are produced by CME shocks and solar flare-reconnection driven processes (e.g., nano flares), then the tail fluxes are likely to be significantly higher in the inner heliosphere; therefore ISIS observations that can simultaneously measure both suprathermal protons and heavy ions up to Fe will determine how the contributions from different sources vary on shorter timescales and with distance.

1.4 Understand SEP Acceleration Mechanisms

1.4.1 Impulsive SEP Events

Magnetic reconnection-associated mechanisms during powerful solar flares accelerate electrons and ions to near-relativistic energies; these may involve first order Fermi processes as well as second order (stochastic) acceleration, or a mixture of first and second order processes (Ambrosiano et al. 1988). Acceleration via reconnection can be complex, involving direct acceleration, trapping particles in collapsing magnetic islands (Drake et al. 2006), stochastic acceleration (Le Roux et al. 2002; Chandran et al. 2010), and perpendicular betatron acceleration (Dmitruk et al. 2004; Dalena et al. 2014). Stochastic and reconnection-related acceleration mechanisms may be driven by low frequency fluctuations associated with MHD turbulence, or by plasma waves that may preferentially heat the ^{3}He nuclei and heavy ions, while also preferentially occurring near disturbances such as shocks. Various theoretical models (e.g., Aschwanden 2002; Petrosian et al. 2009; Kartavykh et al. 2008) are partially successful in accounting for some observed features in impulsive SEP events, but none has so far explained all of the key observations simultaneously. These observed features include:

1) ^{3}He/^{4}He abundance ratios significantly enhanced relative to the ambient corona (Mason 2007).

2) Event-to-event variability in spectral forms; spectra for all species are either power-laws or broken power-laws with nearly constant abundance ratios (Fe/O and ^{3}He/^{4}He) with energy or they exhibit curvature for only ^{3}He and Fe, leading to dramatic energy-dependent abundances (left panel in Fig. 8, taken from Mason et al. 2002).

3) Power-law behavior of the Q/M-dependent enhancements in the abundances of heavy and ultra-heavy ions with mass >60 AMU (see Fig. 8 right panel, taken from Mason et al. (2004).

4) Ionization states of heavy ions that increase with increasing energy, and significantly higher average ionization states compared with the solar wind (Klecker et al. 2006).

5) Timing of suprathermal electrons that indicates source regions located in the upper corona, while the associated ions with higher ionization states need to be stripped if they passed through the denser lower corona (Kartavykh et al. 2008).

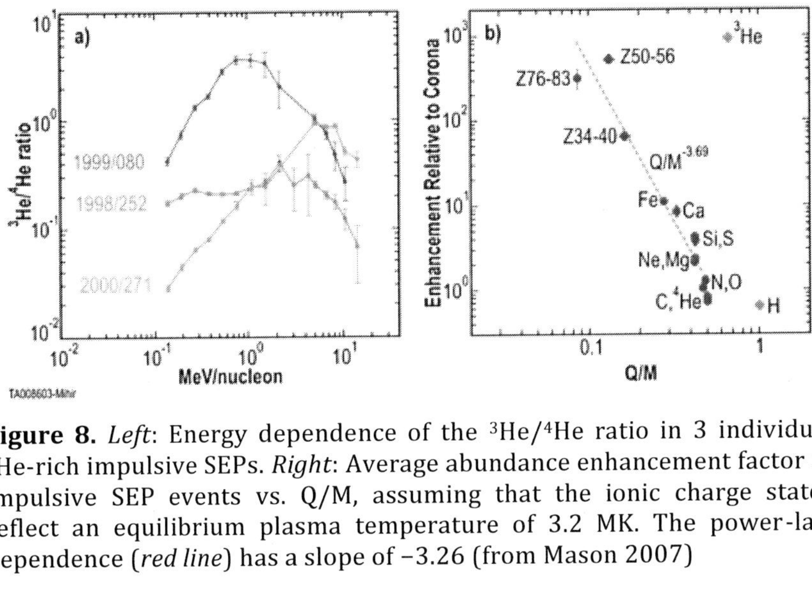

Figure 8. *Left*: Energy dependence of the $^3He/^4He$ ratio in 3 individual ^3He-rich impulsive SEPs. *Right*: Average abundance enhancement factor in impulsive SEP events vs. Q/M, assuming that the ionic charge states reflect an equilibrium plasma temperature of 3.2 MK. The power-law dependence (*red line*) has a slope of −3.26 (from Mason 2007)

By making simultaneous measurements of >25 keV electrons and a number of ion species including ^3He and Fe over a broad energy range with far greater sensitivity, compared to 1 AU observations, in several individual impulsive SEP events, ISIS will provide powerful constraints on existing acceleration models.

1.4.2 Diffusive Shock Acceleration at CME Shocks

Diffusive shock acceleration (DSA) is the process by which energetic charged particles undergo acceleration through a combination of spatial transport and drift across the large plasma compression associated with a collisionless shock, such as that driven by a CME. DSA comprises first-order Fermi acceleration when applied to shocks that move along the magnetic field (parallel shocks), and shock-drift acceleration when applied to shocks that move nearly normal to the magnetic field (quasi-perpendicular shocks) (e.g., Lee 2005). Although the basic physics involved in these processes is well understood, several factors can contribute and cause the large variability that is often seen in key properties such as peak intensities, time-intensity profiles, energy spectra and composition of SEP events. Some important factors are variable seed populations (Mason et al. 1999), geometry of the shock (Tylka and Lee 2006), the presence or absence of a preceding CME from the same active region (Gopalswamy et al. 2004), scattering by ambient turbulence or by self-generated Alfvén waves during acceleration and transport (Giacalone 2005; Ng et al. 2003; Mason et al. 2012), and the presence of flare accelerated material (Cane et al. 2006).

A complete understanding of the contributions from each of these processes is needed to answer the following questions:

1) Why are CMEs with the same speed associated with peak proton intensities that vary by four orders in magnitude (Kahler et al. 2000)?
2) Why do CMEs with nearly the same kinetic energies produce SEP events having vastly different amounts of energy in accelerated particles (Mewaldt et al. 2008a)?
3) Why do some SEP events exhibit spectral breaks that vary strongly as a function of the ion's charge-to-mass (Q/M) ratio (see Fig. 9 and Mewaldt et al. 2005)?
4) 4.What roles do self-excited Alfvén waves play during intense SEP events and how do they affect the streaming limits (Reames and Ng 1998; Ng et al. 2003)?
5) Do preceding CMEs generate turbulence in their wakes and/or produce suprathermal seed populations that affect particle acceleration by following CMEs (Gopalswamy et al. 2012)?
6) What factors determine the maximum energies in SEP events and why are only a small fraction of CMEs associated with the so-called Ground Level Events (GLEs) (Mewaldt et al. 2012a)?

CME shocks weaken and expand as they propagate through the inner heliosphere, therefore their properties, the conditions, and the seed populations they encounter close to the Sun are likely to be vastly different than those further out in interplanetary space. It is perhaps because of this that comparisons between 1 AU data and predictions of shock acceleration models have been unsatisfactory (e.g., van Nes et al. 1984; Desai et al. 2004). As SPP moves into the region where the CME-shock acceleration processes are most efficient and the self-excited Alfvén waves should become more prevalent, it will also identify the seed populations that are available and sample the physical environment into which different CMEs are launched. Simultaneously, ISIS will measure SEP heavy ion spectra and composition over a broad energy range between ~10 keV/nuc to ~100 MeV/nuc. These observations will facilitate direct comparisons with Q/M-dependent fractionation effects and predictions of various shock acceleration models, thereby providing answers to fundamental questions concerning SEP acceleration at CME shocks.

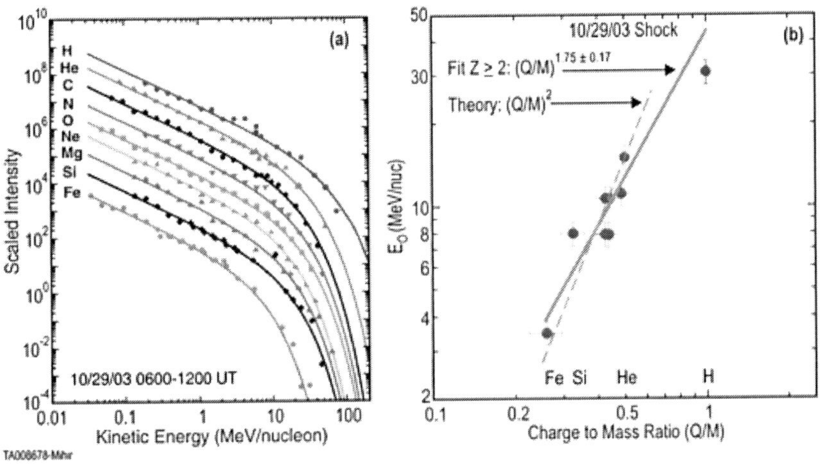

Figure 9. Heavy ion energy spectra in a CME-associated SEP event are fitted with Ellison and Ramaty (1985) spectral form for differential intensity $j = j_0 E^{-\gamma} \exp(-E/E_0)$. All elements have the same spectral index of $\gamma = 1.3$ (Mewaldt et al. 2005). Spectra of different elements are offset for clarity. (*Right*): Values of the e-folding energies E_0 versus the ion's Q/M ratio. Fits to the values of E_0 yield a $(Q/M)^{1.75}$ dependence, which according to Li et al. (2009) is appropriate for a strong quasi-parallel shock

1.5 Disentangle the Role of Transport

1.5.1 Impulsive SEP Events

Flare particles accelerated on closed field lines require access to open field lines to escape into the heliosphere (in the absence of cross-field transport). These escaping impulsive SEPs sometimes propagate out to Earth-orbit without being scattered (see e.g., Mason et al. 1989; Haggerty and Roelof 2002). The scatter-free propagation provides accurate information about the release times of the SEPs, and also yields powerful diagnostics of the magnetic field configuration near the source regions and of the conditions under which the acceleration and escape occur. Among the mechanisms that could allow the flare-accelerated particles to escape are interchange reconnection (Wang et al. 2006) between open and closed field lines, which can occur at neutral points (Fisk 1996), and even more frequently at component reconnection sites at coronal hole boundaries (Rappazzo et al. 2012). Perpendicular diffusion (Matthaeus et al. 2003) and field-line random walk (Giacalone et al. 2000) also contribute to escape. Particles in individual SEP events that escape via reconnection are also affected by field-line meandering, as empty flux tubes (flux dropouts) are occasionally observed in between filled ones (see Fig. 10and Mazur et al. 2000). Thus, time-intensity variations during impulsive SEP events may reflect the magnetic field topology near the source regions, as well as magnetic topological trapping boundaries that form in evolving interplanetary flux tubes (Ruffolo et al. 2003; Seripienlert et al. 2010).

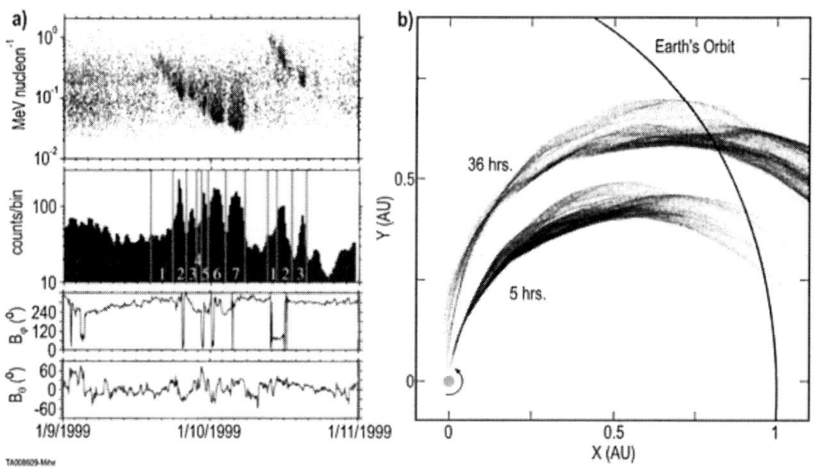

Figure 10. *Left*: Energy of H–Fe ions vs. arrival time at 1 AU for two impulsive SEP events showing flux dropouts (from Mazur et al. 2000). *Right*: Results of a simulation illustrating particle positions projected onto the ecliptic plane at two different times during an impulsive SEP event. Particles are allowed to diffuse along their field line. Groups of field lines were allowed to meander at their base (from Giacalone et al. 2000)

As SPP ventures closer to the Sun and observes many individual SEP events, ISIS will accurately determine the travel path-lengths and particle release times by measuring the onset times of ions and electrons with vastly different speeds, and by using pitch-angle distributions to pinpoint the arrival directions and the source locations. These along with other supporting measurements allow SPP to determine the coronal magnetic field configuration and conditions under which impulsive SEPs are accelerated, transported, trapped, and released into the interplanetary medium.

1.5.2 Large Gradual SEP Events

Particles accelerated in large gradual SEP events undergo diffusive transport from the Sun to Earth, which affects properties such as spectral breaks, time-intensity profiles, and abundance variations. However, these scattering and transport processes are poorly understood because these effects are mixed with those that result from variations in the suprathermal seed populations, CME shock properties and acceleration efficiencies, and other magnetically influenced transport effects that may be correlated with shocks (Tessein et al. 2013). Nonetheless, temporal variations in the abundance ratios (e.g., Fe/O) have provided powerful clues about the Q/M-dependent scattering during interplanetary transport (Tylka et al. 1999; Mason et al. 2006, 2012). Figure 11 shows an example of how the Fe/O ratio in a CME-shock associated SEP event is affected by scattering. The left panel shows that the Fe/O ratio at the same kinetic energy-per-nuc (~30 MeV/nuc) exhibits dramatic variations; it is close to ~1 at the start of the event and then drops to <0.1 around the shock arrival time. This effect led some researchers (e.g., Cane et al. 2006) to suggest that this large gradual SEP event was dominated by flare-accelerated material at the onset and later by shock-accelerated material. However, as shown in the right panel, these variations completely vanish and the Fe/O ratio becomes nearly constant when the ~30 MeV/nuc Fe intensity is compared with the normalized O intensity at roughly twice the kinetic energy-per-nucleon. The same effect is observed in many large SEP events and is now attributed to Q/M-dependent scattering of particles as they propagate through the corona and the interplanetary medium (Mason et al. 2012). Nonetheless, the details of where and how this scattering process occurs, and why similar effects are not observed in all large SEP events, are not well understood. By making such measurements in many SEP events that occur along its trajectory at various radial distances, ISIS will determine how and where scattering effects start dominating SEP properties.

1.6 Additional Science Opportunities

In addition to obtaining required measurements of energetic electrons, ions from H through Fe, and He isotopes, the EPI-Hi telescope designs lend themselves to additional important measurements with no modification of the hardware. These measurements provide additional science opportunities by separating ^{20}Ne and ^{22}Ne isotopes and by enabling serendipitous observations of energetic neutral atoms (ENAs) from large eruptions on the Sun.

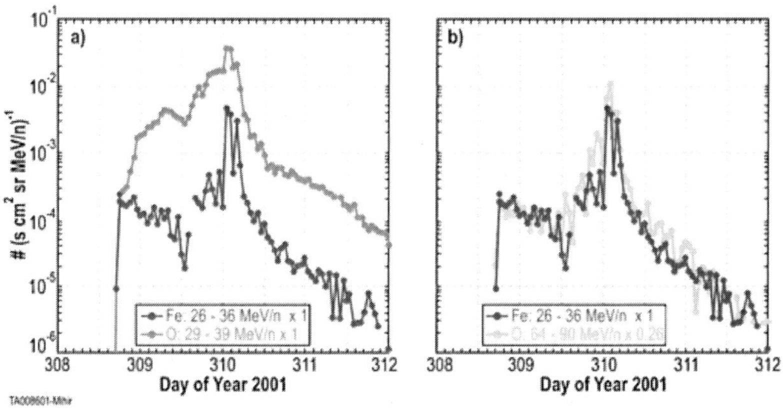

Figure 11. *Left*: Fe and O intensity-time profiles during the November 21, 2003 large SEP event at ~35 MeV/nuc. *Right*: Time-intensity profiles of Fe and O are compared with O at ~twice the kinetic energy of Fe (adapted from Mason et al. 2006, 2012)

Interactions of accelerated ions and electrons with the solar atmosphere produce emissions that include gamma rays and neutrons, which can be uniquely identified in EPI-Hi coincidence measurements (see Sect. 4.9). Solar neutron intensities will be orders of magnitude greater than near Earth because most solar neutrons decay long before they reach 1 AU. The response of HET to secondary neutrons produced by SEP interactions in spacecraft material will be calibrated against 1–200 MeV proton intensities measured in LET and HET. SPP's proximity to the Sun greatly benefits these observations. The direct detection of neutrons will be complementary to other measurements that EPI-Hi makes of the charged products of neutron decay.

In the 6 December 2006 solar event, which occurred at E79°, the STEREO/LET instruments observed H ENAs arriving from within ±10° of the Sun hours before energetic ions arrived (Mewaldt et al. 2009). EPI-Hi may similarly detect ENAs from the charge-exchange of protons accelerated by shocks in the solar corona in cases where ENAs arrive earlier than the accelerated ions, which can occur, for example, when the spacecraft is poorly connected to the acceleration region.

The $^{22}Ne/^{20}Ne$ ratio in SEP events varies by a factor ~4 from one solar event to another (Leske et al. 2007) because of charge-to-mass (Q/M) dependent fractionation during solar energetic particle acceleration and/or transport. Thus, by assuming that both isotopes have the same mean ionic charge, we can use $^{22}Ne/^{20}Ne$ measurements to determine the extent of Q/M fractionation and, in some cases, obtain insight into the origins of event-to-event variations of elemental abundance ratios such as Fe/O.

1.7 ISIS Science Requirements and Performance

ISIS provides comprehensive measurements of the energy spectra, anisotropy, and composition of suprathermal and solar energetic ions from ~0.02–200 MeV/nuc, as well as the energy spectra and arrival direction of ~0.025–6 MeV electrons (energy ranges indicate expected performance, see Fig. 12). The ISIS driving requirements were identified in the original ISIS proposal to NASA and have been vetted and agreed to through the development of various SPP Requirements documents. By combining measurements with other SPP instruments and from instruments on other spacecraft, ISIS addresses key questions concerning the origin, acceleration, and transport of different types of inner heliospheric particle populations, including those associated with CIRs, ACRs, and GCRs. During the course of the SPP mission, ISIS will measure a sufficient number of impulsive and gradual SEP events (see Fig. 13) to meet all of its scientific objectives. Driving requirements for key instrument functional parameters are provided in Table 2; we note that while meeting these requirements is adequate to achieve all of our scientific objectives, the expected performance is even better.

Figure 13 shows that the numbers of gradual and impulsive SEP events expected to be observable with EPI-Hi during the SPP prime mission, as a function of particle fluence and heliocentric radius, are more than sufficient to address all the science questions discussed above and to enable great discovery science on the SPP mission.

2. ISIS SUITE OVERVIEW

2.1 Introduction

The ISIS Energetic Particle Suite measures energetic electrons, protons, and heavy ions across a broad range of energies. To provide this wide coverage, ISIS includes two instruments that contain multiple sensors with detectors optimized for various parts of the energy measurement range. The suite combines EPI-Lo and EPI-Hi mounted together on the ISIS Bracket (Fig. 14). ISIS is mounted at the aft end of the ram side of the SPP spacecraft providing an open FOV toward the direction of the nominal Parker spiral over much of the solar encounter phase, while staying within the umbra of the SPP Thermal Protection System (TPS).

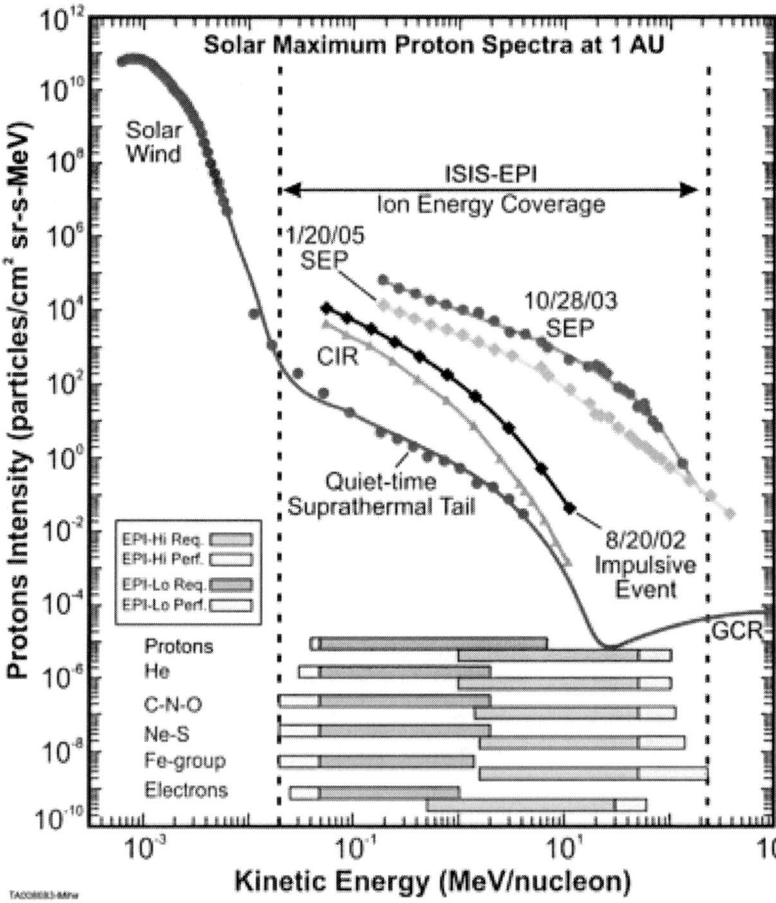

Figure 12. Ion energy spectra of different inner heliospheric particle populations that SPP will encounter and the required energy range coverage for EPI-Lo (*green*) and EPI-Hi (*blue*) as well as the broader expected performance (*lighter shades*) for ISIS' overall energy coverage. Also shown is the energy range coverage for electrons

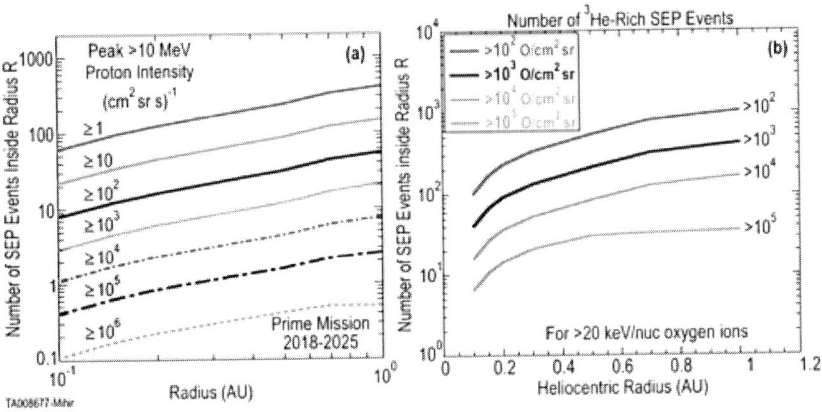

Figure 13. Number of gradual (*left*) and impulsive (*right*) SEP events during the SPP prime mission inside a given heliocentric radius as a function of event size. These plots include the relative amount of time spent at each radial distance. The estimates for gradual events are based on NOAA/GOES data from 1976–2008, an assumed 11-year solar cycle, and a radial gradient of $R^{-2.4}$ based on Lario et al. (2006). Estimates for impulsive ^{3}He-rich events are based on 1998–2006 data from ACE/ULEIS, measurements of impulsive electron events by Wind/3DP (Wang 2010), and on an assumed fluence radial gradient of R^{-2}

Table 2. Measurement requirements for ISIS-EP suite

Functional parameter	Measurement requirements
Energy range	$e-$: <0.05 to >3 MeV
	$p+$/ions: <0.05 to >50 MeV
Energy binning (resolution)	>6 bins/decade
Cadence	$e-$: <1 s for selected electron rates
	$p+$/ions: 5 s for selected ion rates
Field of view	$>\pi/2$ ster coverage in both sunward and anti-sunward hemispheres, including coverage within 10 degrees of the nominal Parker spiral field direction at perihelion
Angular sectoring	$e-$: <45° sectors
	$p+$/ions: <30° sectors
Composition	At least H, He, ^{3}He, C, O, Ne, Mg, Si, Fe
Max intensity <1 MeV	$>10^6$ particles cm^{-2} sr^{-1} s^{-1}
Max intensity >1 MeV	$>5 \times 10^5$ particles cm^{-2} sr^{-1} s^{-1}

EPI-Lo (Fig. 15) measures energetic ions from 0.02 MeV/nuc to ~15 MeV total energy and energetic electrons from 25–1000 keV. To provide a large FOV with hemispherical coverage, EPI-Lo has eight wedges. Each wedge has 10 apertures that collimate energetic particles into pathways through the electro-optics, which allows their speed and energy to be measured while also registering which aperture they passed through. The eight sensor wedges are serviced by an electronics box that contains the Event Board and the Power Board. The Event Board contains the analog and digital processing circuits to record the events and communicates to the spacecraft through command and telemetry channels. The Power Board contains both the low voltage power converters as well as the high voltage power required for the sensors.

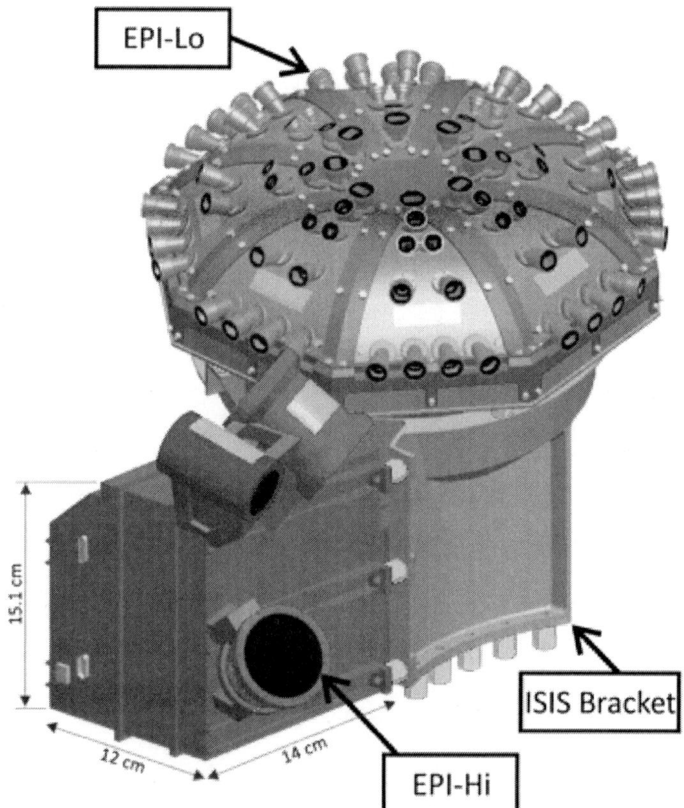

Figure 14. ISIS suite

EPI-Hi (Fig. 16) measures energetic protons and He nuclei from ~1 to ~100 MeV/nuc (and higher energies for heavier elements) and energetic electrons from ~0.5 to ~6 MeV. To cover this energy range, and to provide wide FOV coverage, EPI-Hi has three telescopes, a double-ended high energy telescope (HET), a double-ended low energy telescope (LET1), and a single-ended low energy telescope (LET2). These telescopes are mounted on the EPI-Hi Electronics Box, which contains an analog and digital processing electronics board for each telescope; a detector bias power supply; a digital processing unit for the instrument to coordinate its operations and communicate by command and telemetry channels with the spacecraft; and a low voltage power supply.

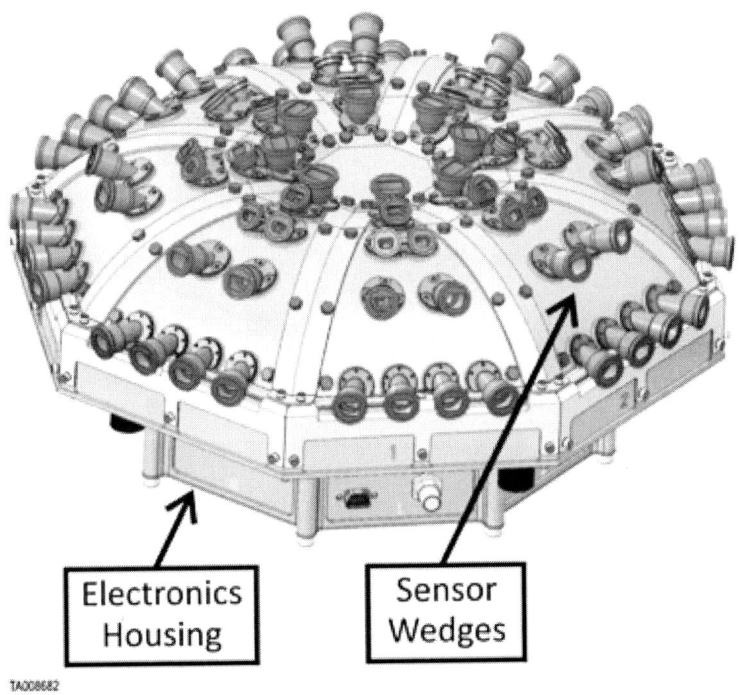

Figure 15. EPI-Lo mechanical design

2.2 ISIS Fields of View

2.2.1 Location on Spacecraft

ISIS is located as far aft on the spacecraft body as possible, on the ram side, just inside the umbra line (Fig. 17). This provides protection from direct solar heating, but still allows ISIS to view within 10° of the Sun-Probe line, thereby providing access to the nominal direction of the Parker Spiral magnetic field over much of the solar encounter phase.

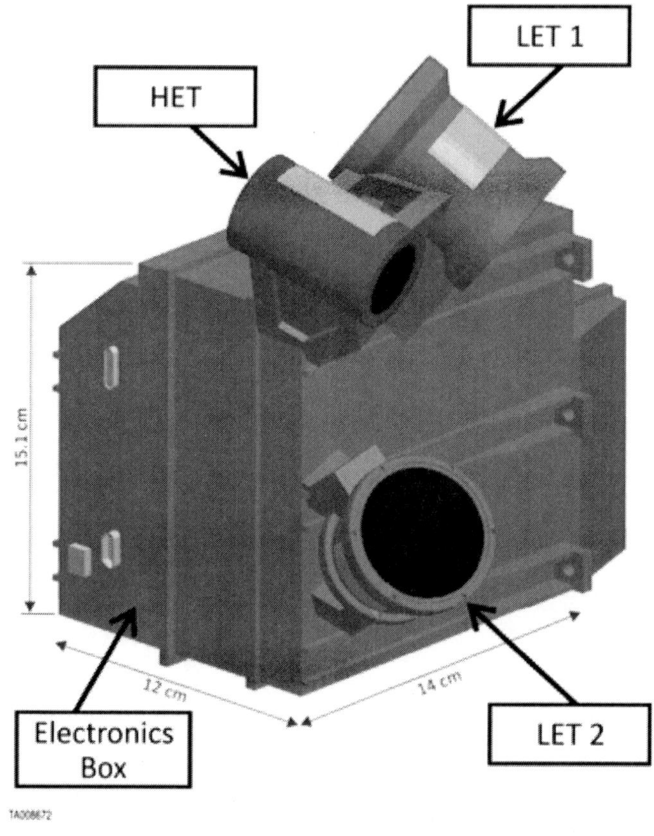

Figure 16. EPI-Hi mechanical design

2.2.2 FOV Maps

Because ISIS must remain in the umbra, its FOV has several blockages from the spacecraft itself. The TPS is the predominant FOV blockage (see Figs. 18 and 19). At the extremes of ISIS' FOV, there are small blockages due to the deployed solar arrays and the solar limb sensors. The EPI-Hi FOVs are a series of five overlapping 45° half-angle cones, three of which are provided by the low energy telescopes (the double-ended LET1, and the single-ended LET2). The other two 45° half-angle cones are provided by the double-ended, higher energy telescope (HET). The EPI-Hi telescope FOVs are overlapped to provide full energy coverage in the sunward and anti-sunward direction in two, nearly-complete 45° half-angle cones. Further coverage is provided in the directions that are not blocked by the spacecraft by the low energy telescopes alone. The EPI-Lo FOV is an array of 80 apertures, ten on each of eight wedges. These apertures are arranged to sample a hemispherical FOV that includes viewing in the sunward and anti-sunward hemispheres as well as coverage near the direction of the nominal Parker spiral.

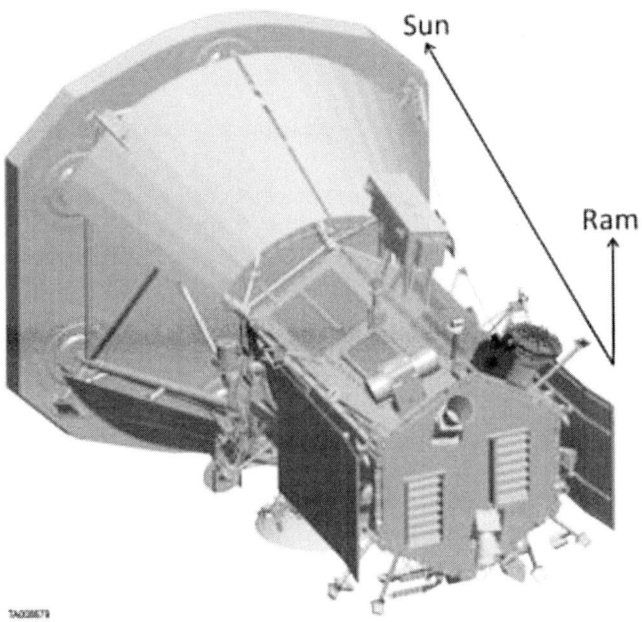

Figure 17. Location of ISIS on SPP spacecraft

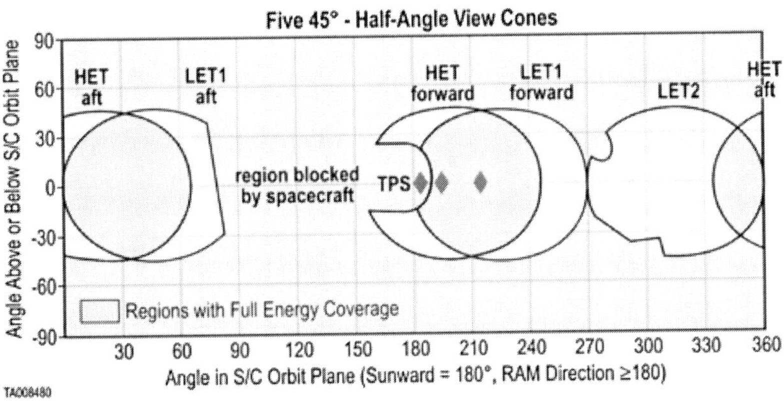

Figure 18. EPI-Hi FOV Map. *Three blue diamonds* indicate the locations of the average Parker spiral magnetic field for a solar wind velocity of 400 km/s at heliocentric distances of 0.05, 0.25, and 0.7 AU (*left to right*)

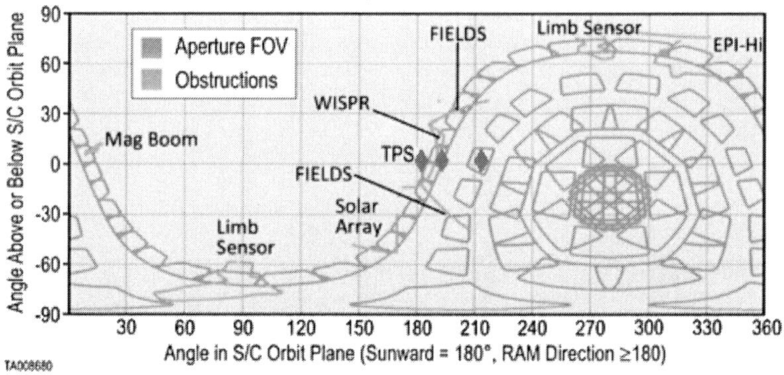

Figure 19. EPI-Lo FOV Map. *Three blue diamonds* indicate the locations of the average Parker spiral magnetic field for a solar wind velocity of 400 km/s at heliocentric distances of 0.05, 0.25, and 0.7 AU (*left to right*)

2.3 ISIS Bracket

The ISIS suite is integrated onto a single, combined ISIS bracket. The bracket is designed to allow flexibility and position the ISIS instruments as close to the allowable umbra-line as possible since the final position of the SPP TPS may be shifted slightly during spacecraft Integration and Testing (I&T). The nominal bracket design holds ISIS just within the allowed limit with ~2° safety margin to the actual umbra line, which keeps the ISIS FOVs as close as possible to the Sun-Probe line. However, the SPP spacecraft center-of-mass must be very close to the center-of-pressure on the TPS in order to minimize pointing perturbations. To do this, the TPS, which is initially oversized, will be trimmed to its final configuration during spacecraft I&T. By design, the TPS will not be trimmed below the limit we have designed to, but it could be trimmed less, which would move the actual umbra line farther away from ISIS and reduce viewing toward the Sun-Probe line. To prevent this reduction in our FOV in this key region of interest, the ISIS bracket is designed to accommodate a late TPS modification by extending the suite farther away from the spacecraft panel, thus restoring ISIS to the position as close as possible to the umbra.

2.4 Spacecraft Accommodation

2.4.1 Mass, Power, and Telemetry

ISIS has been optimized for the SPP mission. Mass is the most constrained resource for the SPP spacecraft, and power is second because of the mass penalty for cooling the solar arrays. The ISIS team has worked diligently to remain within limits set by the mission design. The resource estimates come from a detailed Master Equipment List in which the estimates for each item are carefully maintained and revised over the course of development. Given the heritage of the electronics, the power and telemetry estimates have remained steady during the course of the project development to date. To minimize overall SPP risk, the ISIS team was given a small mass increase to reduce development risks during the risk-reduction SPP activities in Phase B. The Current-Best-Estimates (CBEs) as of the time of the ISIS and mission PDRs, as well as uncertainty estimates are listed in Table 3.

Table 3. ISIS resource table

	EPI-Hi			EPI-Lo			ISIS bracket			ISIS
	CBE	Uncertainty	Total	CBE	Uncertainty	Total	CBE	Uncertainty	Total	Total
Mass [kg]	3.628	0.692	4.320	3.435	0.656	4.091	0.817	0.156	0.973	9.384
Instrument power [W]	5.810	0.960	6.770	4.170	0.830	5.000	NA	NA	NA	11.770
Operational heaters [W]	0.480	0.070	0.550	NA	NA	NA	NA	NA	NA	0.550
Survival heaters [W]	3.810	0.570	4.380	2.450	0.370	2.820	NA	NA	NA	7.200

	EPI-Hi			EPI-Lo			ISIS
	Raw	Burst	Total	Raw	Burst	Total	Total
Uncompressed data volume [Gbit/orbit]	3.660	0.000	3.660	11.320	0.200	11.520	15.180
Compressed data volume (75 %) [Gbit/orbit]	2.745	0.000	2.745	8.490	0.200	8.690	11.435
Packetized data volume (105 %) [Gbit/orbit]	2.882	0.000	2.882	8.915	0.210	9.125	12.007

Data volume formula: 105 % [75 % * (EPI-Hi Raw + EPI-Lo Raw) + EPI-Hi Burst + EPI-Lo Burst]

2.4.2 Thermal Design

On a spacecraft that is designed to fly much closer to the Sun than any other mission in history, one might reasonably expect that the thermal environment would be extreme and that the limiting hot-case for the thermal design would be at perihelion. However, the limiting hot-case for the thermal design of ISIS is actually near 1 AU, just after launch when the SPP spacecraft has to perform various maneuvers that allow direct solar illumination of ISIS. After those early operations, ISIS remains in the umbra of the TPS at all other times, which provides a stable environment that is "in-family" with past, heritage missions for this type of instrumentation.

The thermal design of the SPP mission requires that the instruments be thermally isolated from the SPP spacecraft. The ISIS bracket configuration provides thermal isolation from the spacecraft deck, from the instruments, between the instruments, and from the bracket, thermally isolating the instruments themselves. This is accomplished by including 0.5 in (1.27 cm) ULTEM® spacers at all of the mounting bolt locations on both sides of the bracket. To provide electrical conductivity, while still maintaining thermal isolation, thin straps of copper bridge the gap. These provide a very small cross-section to minimize thermal conductivity, but have a large surface area for good, high-frequency electrical grounding. Multi-layer insulation (MLI) covers the majority of ISIS, except for the apertures and thermal radiators, to minimize radiant heat transfer. EPI-Hi and EPI-Lo have independent survival heaters controlled by the spacecraft, and EPI-Hi also has an operational heater controlled by the instrument itself.

2.4.3 Electrical Interfaces

EPI-Hi and EPI-Lo have completely independent electrical interfaces to the SPP spacecraft. Each has separate command and telemetry interfaces to the SPP Command and Data Handling (C&DH) unit as well as separate power interfaces for instrument power to the SPP Power Distribution Unit (PDU). The SPP spacecraft also provides independent temperature sensors and survival heaters that are used to control the EPI-Hi and EPI-Lo temperatures.

Each instrument has separate A-side/B-side command and telemetry interfaces. Low Voltage Differential Signaling (LVDS) is used to establish a high-speed serial communication protocol with Interface Transfer Frames (ITFs) as the packet format. Command ITFs are used to send spacecraft time and status as well as instrument-command packets to each instrument. ISIS uses the spacecraft time and status information to configure itself autonomously. The instruments return instrument-status and telemetry packets in Telemetry ITFs. The instrument-status can be interpreted on-board the spacecraft to make real-time autonomy decisions. The telemetry packets are stored on the SPP data recorder for down-link to ground stations at the appropriate time. To maintain time synchronization, a 1 Pulse-Per-Second (PPS) signal is provided by the spacecraft. Instead of using a separate line, a "virtual PPS" is provided by carefully controlling the timing of the start-bit of the first byte of the Command ITFs.

EPI-Hi and EPI-Lo have independent power connections to the SPP PDU, each with their own switches. At the SPP mission level (including measurements from the other instruments), mission success can be achieved with either EPI-Hi or EPI-Lo, so additional power redundancy is not required. The spacecraft provides a standard +28 V power bus. ISIS uses a common Low Voltage Power Supply (LVPS) design, which meets the Electromagnetic Interference/Compatibility (EMI/EMC) requirements of the SPP Mission. The LVPS design is customized for each instrument and provides reverse voltage protection, in-rush current limiting, EMI filtering, and isolation between the spacecraft power return and the ground returns of the secondary power supply rails that are developed within each instrument.

2.4.4 Alignment and Fields of View Blockages

The alignment requirements of ISIS are relatively straightforward to achieve. We require an overall accuracy of 1° alignment between ISIS' and the spacecraft's coordinate systems. This is achieved through close tolerances on mounting holes. The budget for the tolerances is distributed through each of the components and tracked by systems engineering at the ISIS level. Besides our alignment with the spacecraft, we also require knowledge of blockages in our FOV. These are primarily caused by the TPS, the solar arrays, and the solar limb sensors. As described in Sect. 2.3, ISIS' FOV is sensitive to the final location of the TPS and we have made provisions to adjust the ISIS bracket in response to shifts in the TPS during spacecraft I&T.

2.4.5 Nitrogen Purge

In order to maximize the performance and lifetime of the detectors within ISIS, high-purity gaseous nitrogen (GN2) purge must be maintained during instrument I&T, during spacecraft I&T, and up until launch. Before integration onto the SPP spacecraft, ISIS Ground Support Equipment (GSE) is used to provide purge. Once integrated to the spacecraft, purge is provided through the flight purge lines built into the spacecraft. The instrument designs include provisions for venting to prevent over-pressure due to the purge and during launch.

2.4.6 Covers

In order to protect the sensitive apertures of the instruments, the ISIS team provides "red-tag" remove-before-flight covers. During I&T, these covers remain installed except for specific times for tests that require that they be removed. In those cases, special procedures must be used to protect the apertures. No launch covers are required. The ability to withstand the launch environment without covers will be qualified and verified through component-level Engineering Model (EM) and full Flight Model (FM) testing.

2.4.7 Safe Arm Plug

The EPI-Lo instrument has a red-tag remove-before-flight Safe Plug that limits high voltage within the instrument to air-safe levels. This red-tag Safe Plug is removed before full high voltage in vacuum testing and before launch. The EPI-Hi instrument does not require a Safe Plug because the detector voltages and conductor spacing is air-safe even at full voltage levels.

2.5 Science Data and Operations

2.5.1 ISIS Science Operations Center

The ISIS Science Operations Center (SOC) processes, distributes, and archives all ISIS data. This integrated approach provides an efficient and effective method for verifying all data products and coordinating and combining data not just between EPI-Lo and EPI-Hi, but with data from the other SPP instrument teams as well as ancillary data.

The ISIS SOC coordinates the development of instrument command sequences based on science activity plans developed across the ISIS team. The ISIS SOC is tasked with verifying and validating the command sequences and coordinating with the SPP Mission Operations Center (MOC) to send the commands to the SPP spacecraft. It also maintains command and telemetry databases and develops processing pipelines using algorithms generated by the ISIS team. The health and safety of the instruments are monitored, and alerts to the team are issued if alarms are raised. The housekeeping and science data are processed and distributed within the team and to the broader community on a timely basis to support the SPP mission. Thoroughly vetted and validated data products will be transmitted to the NASA-designated repository for permanent archiving.

2.5.2 Integrated Science Team

The ISIS science team includes scientists from the ISIS Leadership, EPI-Hi, EPI-Lo, and SOC groups as well as Senior Science Mentors and theory and modeling team members who are leading experts in energetic particle science (see Table 1). These scientists are integrated into the on-going instrument development process as well as the operations planning and data analysis efforts. The ISIS science team works in close coordination with the other SPP instrument investigation teams. It also coordinates data collection and analysis with other missions (e.g. Solar Orbiter) and ground-based observations. This close coordination inside the ISIS science team and cooperation with the broader science community provides a truly outstanding integrated science investigation of the Sun.

3. ENERGETIC PARTICLE INSTRUMENT—LOW ENERGY (EPI-LO)

EPI-Lo measures energetic particles in the lower portion of the ISIS energy range. As discussed in Sect. 2, the two ISIS instruments complement one another in their energy ranges and their sky coverage in order to obtain the comprehensive set of observations needed to understand solar energetic particle sources, acceleration and transport close to the Sun.

Table 4. EPI-Lo instrument required and projected performance

Parameter	Required	Goal (expected)	Comment/Heritage
Electron energies	50–500 keV	25–1000 keV	Electron capability from JEDI, RBSPICE
Ion energies	50 keV/nuc– 15 MeV Total E	~20 keV/nuc– 15 MeV Total E ~85 MeV Total E for Fe	Capability based on that of RBSPICE. Maximum energy ~1.5 MeV/nuc for Fe
Energy resolution	45 % for required energy range	11 % for required energy range	Telemetry limited
Time sampling	5 sec	1 sec	Telemetry and/or statistics limited
Angle resolution	$<30° \times <30°$	Ions, $\sim15° \times 12°$ to $<30° \times <30°$ $e-$, 45°	Varies with elevation
Pitch Angle (PA) coverage	0°–90° or 90°–180°, some samples in both hemispheres	0°–90° or 90°–180°, some samples in both hemispheres	
Time for full PA	1–5 sec	1–5 sec	Telemetry limited
Ion composition	H, ^{3}He, ^{4}He, C, O, Ne, Mg, Si, Fe	H, ^{3}He, ^{4}He, C, O, Ne, Mg, Si, Fe	^{3}He/^{4}He ~50 to 1000 keV/nuc
Electron sensitivity, geometric factor, counting rate, and background drivers	$<10^6$ cm^{-2} s^{-1} sr^{-1} ($G \sim 0.05$ cm^2 sr; measure event rates to >50 kHz)	10^2–10^7 cm^{-2} s^{-1} sr^{-1} ($G > 0.05$ cm^2 sr; measure event rates to >700 kHz; background rate <1 Hz)	G = Geometric factor (cm^2 sr) 8 pixels/sensor; background rate is spectral-slope dependent
Ion sensitivity, geometric factor, counting rate, and background drivers	10^1–10^6/cm^{-2} s^{-1} sr^{-1} ($G \sim 0.05$ cm^2 sr; measure event rates to >50 kHz; at minimum intensity require accidental rate <~50 kHz)	10^0–10^7/cm^{-2} s^{-1} sr^{-1} ($G > 0.07$ cm^2 sr; measure event rates to >700 kHz; at minimum intensity require accidental rate <~5 kHz)	80 pixels/sensor

3.1 EPI-Lo Overview

EPI-Lo is a novel, light-weight, high-heritage, time-of-flight (TOF) based, mass spectrometer that measures energetic electron (25–1000 keV) and ion (~0.04–7 MeV for protons and ~0.02–2 MeV/nuc for heavier ions) spectra and resolves all major heavy ion species and ^{3}He and ^{4}He over much of this energy range in multiple directions. ISIS thus covers the critical energy range from suprathermal energies (~20 keV/nuc) up to the lower portion of the EPI-Hi energy range with a single instrument. The EPI-Lo characteristics and projected performance are summarized in Table 4.

EPI-Lo (Figs. 15 and 20) consists of eight sensor wedges mounted above an electronics box. It has 80 separate entrances (10 on each of eight wedges) densely sampling over half of the sky. This configuration permits full angular distributions without articulation or duty cycle, and allows for measuring the first-arriving, field-aligned ions at the spacecraft for a broad range of vector magnetic (**B**) field directions.

Understanding coronal acceleration processes requires ion-mass resolution sufficient to measure separately the intensities of the isotopes ^{3}He and ^{4}He (Fig. 21), and the elements C, O, Ne, Mg, Si, and Fe, while simultaneously providing angular coverage with at least 45 % energy resolution.

EPI-Lo achieves this resolution, including separation of ^{3}He from ^{4}He, over most of its angular coverage from 200 keV to 4 MeV total particle energy. EPI-Lo also returns high-resolution energy spectra (64 log-spaced energy bins) at reduced time resolution (necessitated by downlink limitations) and measures the required ion composition and pitch angle distributions from 30 keV/nuc to 0.3–1.0 MeV/nuc every 5 or 30 s. In this way ion acceleration histories are completely characterized with no uncertainty owing to an insufficient energy range or composition misidentification. EPI-Lo rejects background by requiring coincidence between the start and stop pulses for the TOF measurement, along with an energy measurement between appropriate TOF time gates. This rejection is non-linear, and is very effective (resulting in inconsequential rates of false valid events) for background counting rates in individual detectors (singles rates) below ~10^{6} s^{-1}; projected singles rates from combined background and foreground sources are ≤3×10^{5} s^{-1}. Incoming ion velocities are determined by measuring the TOF between two thin (100 nm Start, 65 nm Stop) carbon-polyimide-aluminum (and palladium, Start only) foils. An ion passing through each of the foils produces secondary electrons, which are deflected toward a microchannel plate (MCP) producing "Start" and "Stop" pulses. The ion entrance angles are determined from the position where the Start electrons strike the MCP and are unique for each entrance foil location. The ion energy deposited in the SSD, together with velocity from the TOF, determines species through an onboard table lookup.

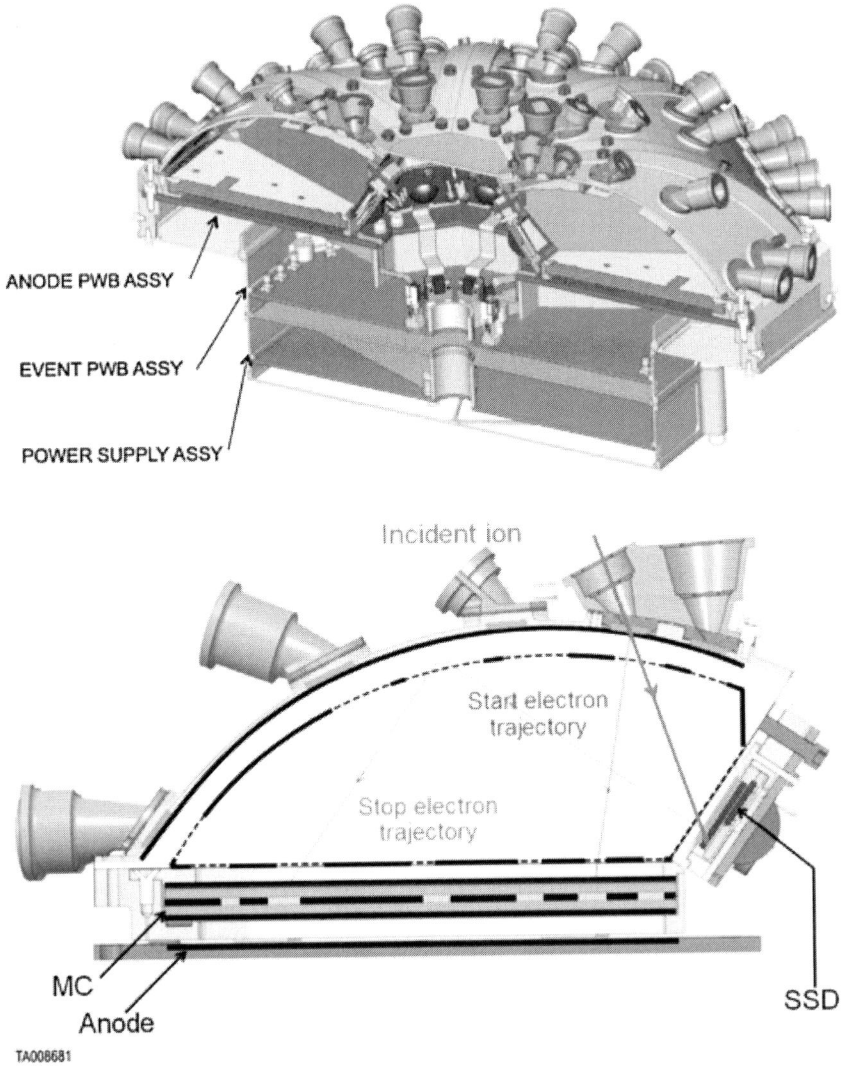

Figure 20. EPI-Lo comprises eight wedges mounted on a common electronics box. Ions generate start/position electrons as they transit thin foils in the entrance apertures, and then strike a stop foil and solid-state detector (SSD), yielding angle and $E \times$ TOF. Energetic electrons enter the same apertures as the ions, and are detected in a second set of SSDs located behind light- and ion-rejecting cover foils

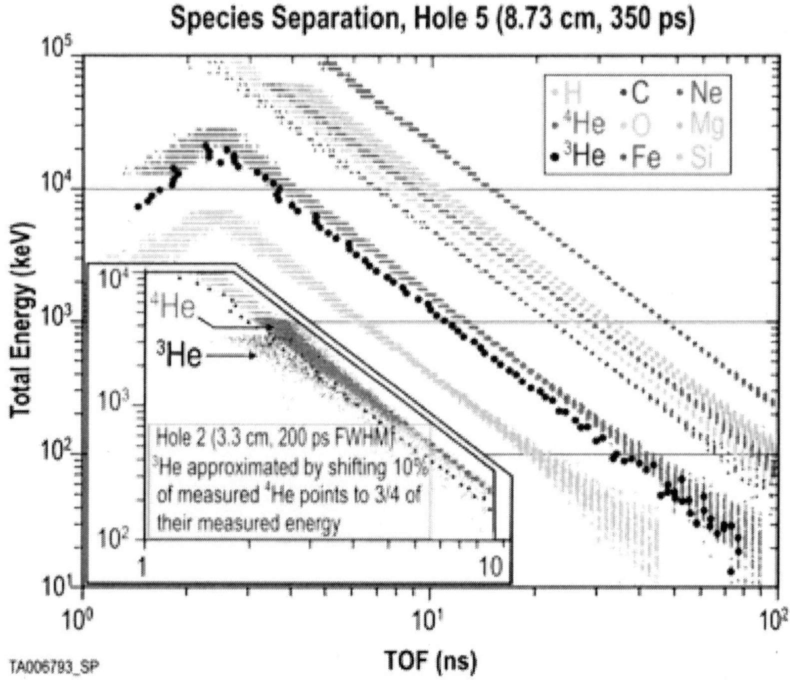

Figure 21. Representative species separation for EPI-Lo holes 2 (22.5° from the instrument normal) and 5 (the holes about the instrument perimeter, 90° from instrument normal) based on measured TOF resolution performance on test model. Data taken using an alpha source in hole 2 is overlaid on the hole 2 simulated data. ^{3}He is well distinguished from ^{4}He at 1:100 abundance ratio to ≥1.5 MeV/nuc. The best performance for species identification is in the 64 holes at the 3, 4 and 5 positions, at 45°, 67.5°, and 90° from the instrument normal

EPI-Lo also detects electrons from ~25 keV to 1000 keV. EPI-Lo uses solid-state detectors (SSDs) shielded by aluminum flashing of ~2 μm thickness as also used in multiple current and upcoming missions, e.g., Cassini-MIMI (Krimigis et al. 2004), MESSENGER EPS (Andrews et al. 2007), New Horizons PEPSSI (McNutt et al. 2008), Juno JEDI (Mauk et al. 2013), Van Allen Probes RBSPICE (Mitchell et al. 2013), and MMS EIS (Mauk et al. 2014). The relatively thick aluminum flashing naturally suppresses light, which is a very important feature for this intrinsically single parameter measurement. While the primary electron measurement does not identify which entrance aperture an electron enters through, each EPI-Lo wedge contains independent electron SSDs, so the sector of the sky is identified over an angular coverage similar to that for the ions.

For the small subset of the electrons that produce secondary electrons as they transit the entrance foils, an additional mode that requires a Start pulse along with the SSD pulse identifies the specific entrance aperture that the electron entered through. This mode also provides better background rejection.

MCP detectors and SSD's have been optimized for use on this mission. We also produced a prototype wedge (Fig. 22), which has been used for extensive testing and development of the EPI-Lo concept and design. Finally, lead development engineers and instrument scientists on JEDI, RBSPICE, and EIS are all participating in the EPI-Lo effort, as well the corresponding leads that developed and continue to operate PEPSSI.

Figure 22. EPI-Lo wedge prototype

3.2 EPI-Lo Expected Performance

The detectors in EPI-Lo are sensitive to charged particles and to light. The MCPs are sensitive to ultraviolet (UV) light, including both EUV and shorter far ultraviolet (FUV) wavelengths whereas the SSDs are sensitive to everything from X-rays to visible wavelengths. However, the mechanisms are different. MCPs respond to charged particles impacting their surface, and UV produces secondary electrons (photoelectrons) on surfaces (either the surface of the MCP itself, or the Start and Stop foils designed to produce secondary electrons as ions penetrate them, or other surfaces inside the sensor volume), which can then enter channels and be amplified similarly to signals from any other particle.

SSDs respond to particles as they deposit energy in the bulk material of the detector. They free bound electrons to create electron-hole pairs and so increase conductivity. In SSDs, light can also energize bound electrons sufficiently that they become free and increase conductivity. Energetic X-rays may deposit enough energy to create free electrons that then raise the energy of sufficient numbers of bound electrons so that the resulting current spike is above threshold. Lower energy photons will not do so, but enough of them arriving within an SSD event time constant (~1 microsecond) can free sufficient numbers of electrons to register as events, or to raise the electronic noise of the detector unacceptably.

In both cases, light (both UV and visible) must be reduced to a level where these effects cannot degrade the particle measurements. Controlling the amount of light entering the sensor volume involves a variety of techniques: (1) eliminating stray-light leak paths into the instrument; (2) collimating the entrance apertures to eliminate exposure to trajectories that are not useful for particle analysis; (3) coating surfaces from which light may be reflected into the detectors with low-reflectance coatings; (4) using high-work-function surfaces in locations where UV photons may strike surfaces from which photoemission is undesirable; (5) designing the Start and Stop foils using materials and thicknesses that reflect and/or filter the UV and visible light, reducing the light that can directly enter the sensor volume to acceptable levels; (6) taking precautions against failure of this filtering approach as a consequence of small numbers of pinholes in the foils (either from dust impact or launch vibration damage); and, (7) designing the sensor timing constraints such that controlled rates of photoelectron production from UV photons do not impact particle measurements.

Briefly, the instrument design carefully considers: (1) stray light paths and avoids them; (3) coating surfaces as a second-order consideration relative to direct paths for light and UV; and (7) timing constraints fundamental to the instrument design: an approach successfully used on numerous past flight programs (the "hockey pucks"[1] on MESSENGER, New Horizons, Van Allen Probes, Juno, and MMS as well as the ENA imagers on Cassini and IMAGE).

All of these programs also rely on collimation and filtering by foils, and all were subject to pinhole degradation at various levels, so these considerations are not new; however, the SPP environment is sufficiently unique that careful consideration to these approaches is required for EPI-Lo.

The environment in which EPI-Lo will operate is one never before experienced. It has been extensively modeled, and there is considerable knowledge of the light environment. The SPP team has provided their best estimate of the worst-case photon environment as a function of solar elongation angle at perihelion. This, and the other expected SPP environments are specified in an internal Project-provided environmental design and test requirements document. Light at the ISIS location is dominated by Thomson-scattered photons, to first order a process that preserves the shape of the solar spectrum. The intensity of the scattered light is a fairly strong function of elongation angle, dominated primarily by the line-of-sight integrated electron density, which varies with elongation (Fig. 23). While dust-scattered light dominates at large elongations it is much weaker than the Thomson-scattered light at smaller elongations, so the foil filtering design required by the Thomson-scattered light at smaller elongations is more than adequate for the dust-scattered light.

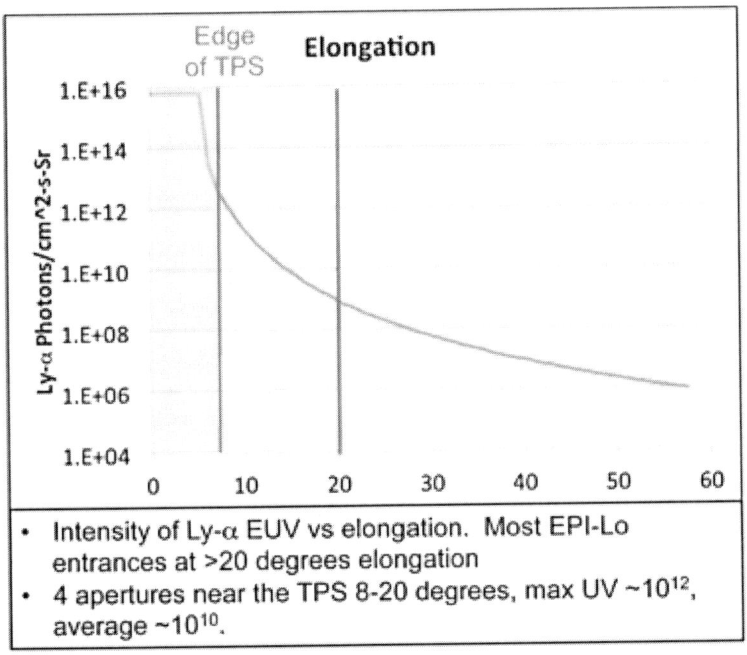

Figure 23. Intensity of Ly-α EUV vs. solar elongation. Most EPI-Lo entrances include angles no closer than 20 degrees elongation (which implies a maximum Ly-α intensity of $\sim 10^9 cm^{-2} s^{-1} sr^{-1}$). For four apertures close to the TPS, the maximum is $\sim 10^{12}$ $cm^{-2} s^{-1} sr^{-1}$, but average only $\sim 10^{10}$ $cm^{-2} s^{-1} sr^{-1}$

We know the elongation angles of the various apertures on EPI-Lo. In considering these, we treat the EPI-Lo apertures in two distinct groups: those that contain sky elements within ~12° elongation angle (the edge of the TPS is at about 8°), and those that do not. For those with elongation angles less than 12°, the rough order-of-magnitude (ROM) for the UV and light environment is approximately equivalent to having the disk of the Sun directly in the FOV of that aperture at 1 AU. This then means that the filtering requirements for those apertures are approximately the same as what would be required for EPI-Lo at 1 AU with the Sun directly in the aperture FOV.

Several approaches contribute to determining the required filtering properties of the foils. For visible light (which primarily concerns the SSDs), we have used flight experience from various instruments in Earth orbit. For example, the ISEE-1 Energetic Particle Detector (EPD) had 30 $\mu g/cm^2$ of Al covering its SSD, and suffered from visible light contamination when the detector viewed the Sun directly. The IMP-8 EPD had 40 $\mu g\,cm^{-2}$ of Al covering its SSD and did not respond to direct Sun in the visible, but it did respond to solar X-ray events. From this we conclude that 40 $\mu g\,cm^{-2}$ of Al or the equivalent would be sufficient for EPI-Lo for elongations <12°. For these small elongation foils we are baselining a Start foil of 24 $\mu g\,cm^{-2}$ of Al plus 18 $\mu g\,cm^{-2}$ of palladium (Pd), which combined with the Stop foil containing 7.3 $\mu g\,cm^{-2}$ of Al, yields a predicted noise level on the SSD of ~0.04 keV μs^{-1}, well below the electronic noise level of ~7 keV μs^{-1}.

For the larger elongation apertures we have baselined a total of 7.3 $\mu g\,cm^{-2}$ of Al plus 18 $\mu g\,cm^{-2}$ of Pd on the Start foil, and 7.3 $\mu g\,cm^{-2}$ of Al on the Stop foil. The predicted noise level for this combination is 0.7 keV μs^{-1} in the SSD. We plan to test at this level, and increase the thickness of the aluminum layers if necessary. It should be noted that these are ROM estimates and not really directly comparable. The Thomson scattered light is diffuse, whereas sunlight at 1 AU is collimated. Appropriate descriptions in terms of photons at a particular wavelength are number of photons $cm^{-2}\,s^{-1}$ for sunlight at 1 AU but number $cm^{-2}\,s^{-1}\,sr^{-1}$ for the scattered light at 10 R_S (~perihelion for SPP). The way we employ our model is by using the EPI-Lo geometric factor for the apertures, which lie within a given elongation angle range we are modeling, and calculating the number of photons s^{-1} that hit the aperture foils, so the calculation is correspondingly conservative. We then run the filter model with a fraction of the full Sun at 1 AU determined such that we match that number of photons/s on the same aperture foils.

The foils employed as filters result in energy loss of ions entering the instrument before they reach the TOF section and subsequently the Stop foil and SSD. These energy losses and straggling affect the minimum energy and energy resolution at the lower end of the EPI-Lo energy range. However, based on TRIM and GEANT4 simulations, as well as experience with similar foils on previous instruments, the required foils will still permit EPI-Lo to provide its full, required measurements.

We calculated the transmission of visible light through Al and Pd. Aluminum is quite efficient at filtering visible light, but palladium is not, so we rely primarily on Al for filtering in the visible wavelength range with palladium included to filter short-wavelength EUV.

For UV transmission we have employed a filter model that we have maintained throughout the development of the Cassini INCA and IMAGE HENA, instruments and all of the "hockey pucks," which incorporates the solar EUV spectrum from about 10 to 140 nm and uses Henke atomic scattering tables (http://henke.lbl.gov/optical_constants/asf.html). This model has been quite successful in modeling the expected responses of foil-based time-of-flight by energy (TOF × E) instruments to the UV environments of interplanetary space, Earth, Jupiter, and Saturn. Applying this model to the UV environment provided by the SPP team, the ISIS team determined that the predicted Start foil rates for an EPI-Lo *quadrant* from UV photo-electrons is $\sim 1\times10^{3}$ s^{-1}. The predicted Stop rate is also $\sim 1\times10^{3}$ s^{-1}. The predicted accidental TOF × E rate is $\ll 1.0$ s^{-1} for the quadrant and, hence, not significant.

The foil materials are sufficiently thick that neither photoelectrons nor solar wind electrons (nor solar wind ions) will penetrate them. However, particles can enter through holes in the foils (either pre-existing, from launch, or produced by dust impacts in flight), so we also calculate and account for the susceptibility to backgrounds from photoelectrons and solar wind ions and electrons. The highest photoelectron intensities will be in the vicinity of the Sun-facing surface of the spacecraft TPS where the plasma sheath is predicted to have a thickness[2] on the order of centimeters, therefore those electrons will not reach EPI-Lo (which is in the umbra, ~3 meters from the TPS edge). EPI-Lo must be able to tolerate the photoelectron flux produced by the UV flux striking surfaces relatively close to the EPI-Lo apertures, and therefore umbral UV intensities. Those intensities are small (both in number flux and in energy) relative to solar wind electron intensities, therefore we consider only the latter.

Solar wind electron flux at perihelion is expected to rise to $\sim 2\times10^{12}$ $cm^{-2}\,s^{-1}$ (with a typical thermal energy of ~100 eV; numbers provided by the SPP Project). Dividing by 2π sr, this amounts to about 3.2×10^{11} $cm^{-2}\,s^{-1}\,sr^{-1}$. Solar wind electrons entering through pinholes in the EPI-Lo aperture foils at relatively low energy (≤100 eV) will behave like Start electrons inside the instrument; they will be focused onto the MCP areas that register Start events. These need to be limited to less than $\sim 1\times10^{6}$ s^{-1}, preferably much less, but at this level the instrument will still function as designed; for example, the RBSPICE instruments on the Van Allen Probes mission routinely run with Start rates of 2 to $4\times10^{6}\ s^{-1}$ and still return well-behaved, calibrated TOF × E data.

The largest solid angles viewed by the Start foils are ~0.2 sr, so those foils would see $\sim6\times10^{10}$solar wind electrons $cm^{-2}\ s^{-1}$. If we wish to limit the intensity to $\sim10^{6}\ s^{-1}$ for a quadrant, this requires that the total pinhole area through the 20 foils (of the quadrant, there being 10 apertures to the octant) be $\sim2\times10^{-5}\ cm^{2}$. The foil support grids are 180 lines/inch, or ~70 lines/cm. This corresponds to a grid element width of 1/70 cm, or 143 µm. The grid wires are 12 µm in width, so a grid element has an area of $\sim130\times130\ \mu m^{2}$, or about $1.7\times10^{-4}\ cm^{2}$. This implies that with only one grid-element sized pinhole in any one of the 20 foils in a quadrant, we would already exceed our limit by about one order of magnitude ($\sim6\times10^{10}\times1.7\times10^{-4}\ cm^{2}\sim10^{7}\ s^{-1}$).

In tests of high velocity dust impacting flight-like foils at the dust facility at the University of Colorado, only micron-sized holes were produced by the dust. Therefore, even though we do not expect a dust grain to be able to take out a whole grid element, during Phase B, we decided to take an even more conservative approach and use two foils in the collimators, separated by 0.5 cm minimum (effectively turning a single hole into an aperture-constrained "telescope"). Then the geometric factor for pinholes in both foils is $\Omega\times1.7\times10^{-4}\ cm^{2}$ where Ω is the solid angle defined by the two pinholes. This geometric factor is $\sim6\times10^{-8}\ cm^{-2}\ sr^{-1}$. Multiplying by $3.2\times10^{11}\ cm^{-2}\ s^{-1}\ sr^{-1}$, we obtain a rate of $\sim2\times10^{4}\ s^{-1}$. Even for one such pair of holes in every entrance of a quadrant, the total rate would amount to only $\sim4.0\times10^{5}\ s^{-1}$, which more than meets our criterion.[3]

It should be pointed out that the need for double foils for mitigation of solar wind electrons does not require both foils be of the same design. We only require the collimator foils to reduce the susceptibility to pinholes of the electron-blocking capacity. Therefore, our collimator foil is much simpler than the start foil, using a single layer (1000 Å of polyimide with 50 Å C flashing on the inward-facing side) sufficiently thick that pinholes on such a foil supported by a 180 line-per-inch grid become very unlikely.

3.3 EPI-Lo Electrical

The EPI-Lo electronics box contains all the electronics to run the instrument other than the energy and timing preamps, which are located in the sensor wedges. The box contains two octagonal boards mounted into metal frames. The boards stack one on top of the other, with an internal connector providing electrical interconnects between the boards. The functional breakup between the two boards minimizes the number of interconnects needed. See Fig. 24 for the block diagram.

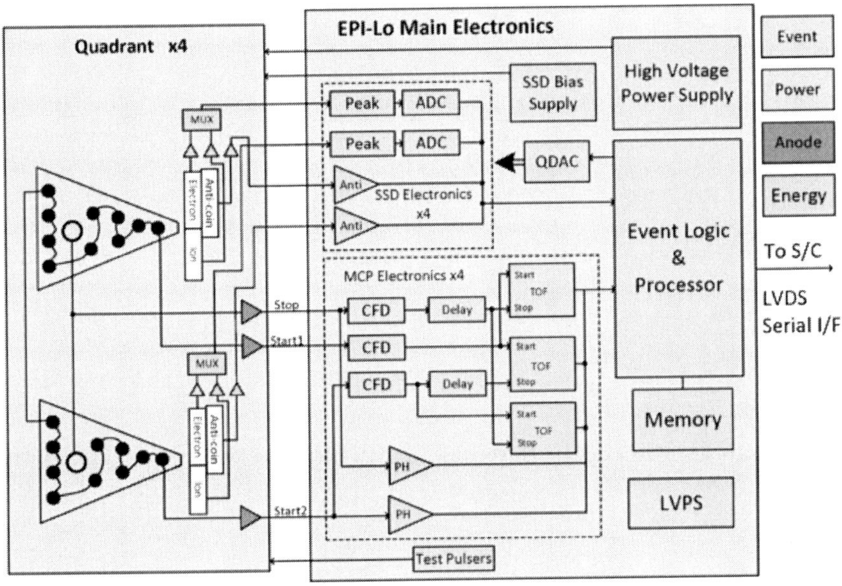

Figure 24. EPI-Lo block diagram

The electronics are designed to handle solar energetic particle event intensities up to at least 10^6 particles $cm^{-2}\,sr^{-1}\,s^{-1}$. This includes handling electron counts rates ≥70,000 counts per second and total ion count rates (SSD and MCP valid coincident event) of ≥70,000 counts per second. In both cases, these events can be either evenly distributed over the entire instrument or concentrated in one wedge. The term "handle" is used to mean that the incoming particles are processed in the instrument such that the particle types, directions, and rates can be determined. Note that ground software rate correction will be necessary when rates are sufficiently high (e.g. ≥10^6 starts per second or ≥40,000 total ion count rates per second). With the current mission and environment assumptions, we expect that rates will be sufficiently high only ~5 % of the time to require such corrections.

3.3.1 Event Board

An ion entering the sensor through one of the collimator apertures will deposit energy in the SSD and produce secondary electrons in the Start and Stop foils, which are amplified by the MCP and collected in distinct positions on the anode. This collection of time-correlated signals is called an event for the purpose of the description of the electronics that follows. The Event board directly processes the sensor SSD and anode preamp output signals, and contains all necessary analog and digital circuitry to process and store event information on an event-by-event basis. The energy signals from the eight SSD preamplifier sets are processed in parallel peak-detect/discriminator circuit/ADC chains. The MCP anode signals are processed via constant-fraction discriminators (CFDs) and time-to-digital conversion (TDC) circuitry; these measured time differences are converted into event look direction and particle velocity in the field-programmable gate array (FPGA). FPGA-based event logic also determines which signals comprise valid ion and electron events and coordinates all event hardware processing timing. A soft-core processor (i.e. a processor implemented in VHDL) is also embedded in the FPGA to provide all command, control, telemetry, and data processing functions of the instrument. SRAM, MRAM, and PROM memory storage is provided on the board to support the processor.

EPI-Lo uses a pulse-width technique to determine the energy deposited in each detector for energies above ~1 MeV (Paschalidis 2008). This method, used on the Juno JEDI, the RBSPICE, and MMS EIS, allows the energy dynamic range to be extended from ~1.5 MeV to a total energy ~15–20 MeV. In order to cover fully iron composition with no gaps between EPI-Lo and EPI-Hi, the maximum energy measured will be extended to ~85 MeV for iron (1.5 MeV/nuc for Fe nuclei). Three separate approaches to this extension of the iron energy range have been identified, and the preferred approach will be finalized and tested in the engineering model prior to instrument CDR. Each sensor uses an existing, flight-qualified, application-specific integrated circuit (ASIC) containing preamplifier/shaper circuits to amplify the SSD and APD (analog peak detect) signals, shape the pulse, and generate timing triggers on the rise and fall of each pulse. These signals feed into the EPI-Lo FPGA where the coincidence logic and other digital processing begin. The EPI-Lo FPGA-based processor accumulates events into rates, and packetizes the telemetry products.

3.3.2 Power Board

The Power board contains both the low and high voltage power supplies. The low voltage portion takes spacecraft primary power on a single 9 pin connector and generates 1.5 V (for the FPGA core), 3.3 V (primarily for digital interface logic, memories, and TDCs), and 5 V (primarily for analog functions). A 15 V output powers the high voltage power supply (HVPS). The HVPS generates the necessary high voltage outputs for the sensor MCP and electron optics, with a maximum voltage output of 3300 V. It can independently control high voltage to each of the four quadrants. A bias supply provides up to 250 V for the SSDs.

3.3.3 Anode Board

The four anode boards are located directly under the MCPs and each board covers two octants (sensor wedges). Each anode board (a "quadrant") has 20 start pick-up anodes configured as a delay line, and two stop pick-up anodes tied together. One pre-amplifier reads out each of the two sides of the delay line, and one pre-amplifier reads out the combined stop anodes. Locating the pre-amplifiers directly on the anode board reduces system noise.

3.3.4 Energy Boards

Each of the eight SSD assemblies has an energy board that contains the pre-amplifiers and shapers for each of the electron, ion, and anti-coincidence detectors. The energy board is mounted directly to the back of the SSD to minimize noise. The electron and ion SSDs have pulse height analysis electronics on the event board while the anti-coincidence detector is monitored by a comparator with a programmable threshold to give a simple yes/no particle event result. The energy board also supports test inputs, both external and from the event board, and filtering for the bias voltage.

3.3.5 EMI/EMC Design Considerations

Of principal concern for EMC design are the power supplies. These are controlled to a frequency window centered at $n\times50$ kHz with $n\geq3$ and 500 ppm wide overall operating conditions and time. The LVPS is synchronized to 200 kHz by a 400 kHz clock provided by the digital boards. EPI-Lo has a 40 MHz oscillator and EPI-Hi has a 58.8 MHz oscillator; both evenly divide to 400 kHz. To minimize interference, transformers and large inductors are placed as far from the box walls as possible, and stable currents are employed to minimize changes in magnetic emissions. EPI-Lo does have nickel grids, and concerns associated with those are mitigated with careful handling, use of non-magnetic tools, and testing.

All electronic parts are selected for proven radiation tolerance: total ionizing dose (TID) >100 krad, no latch-up, and acceptable single-event upset (SEU) performance. Triple module redundancy (TMR) and error correction code (ECC) are used on vulnerable registers inside the FPGA. PROM-based boot code is used to ensure reliable memory loading and checking capabilities. Analog parts are selected with low dose effects in mind. Parts comply with Goddard Space Flight Center's (GSFC's) "EEE-INST-002" Level II requirements.

3.4 EPI-Lo Mechanical

EPI-Lo uses a symmetric design to enable the wide field of view in a compact, low-mass package. Parts consist of the eight wedge assemblies and their closeouts, a top close out, the "spider" frame, which holds the wedges, and the event- and power-board slices, which comprise the components of the electronics box (Fig. 25). The common wedge design is shown in Fig. 26. Each contains an MCP assembly, an SSD assembly, and a collimator set. Alignment of all of the pieces enables the coverage and functionality of the detector overall. Preliminary design analysis has been done with a finite element model (FEM) using 89,077 nodes and 70,645 elements.

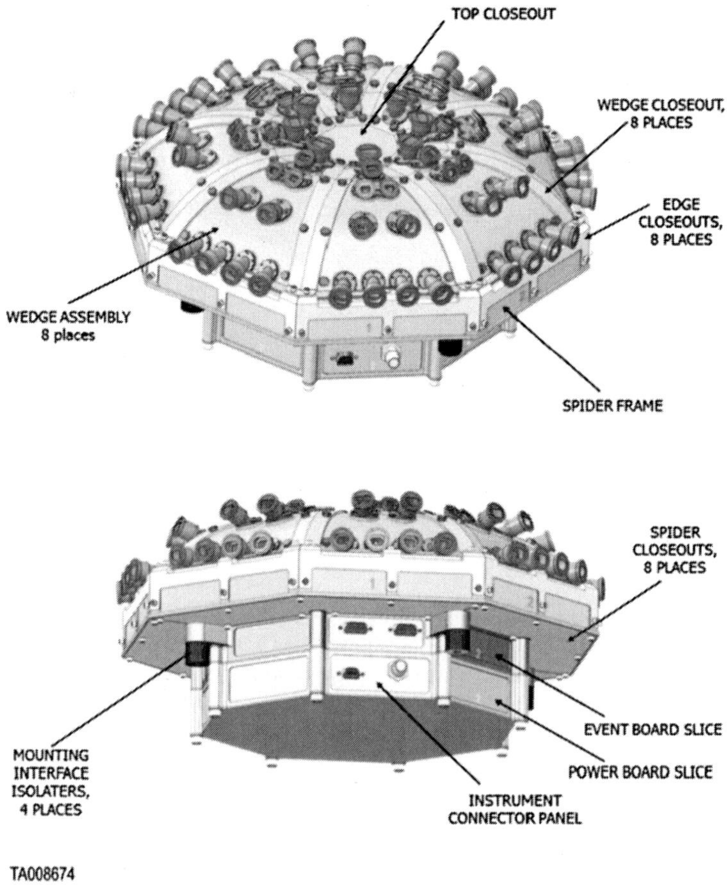

Figure 25. EPI-Lo mechanical assembly

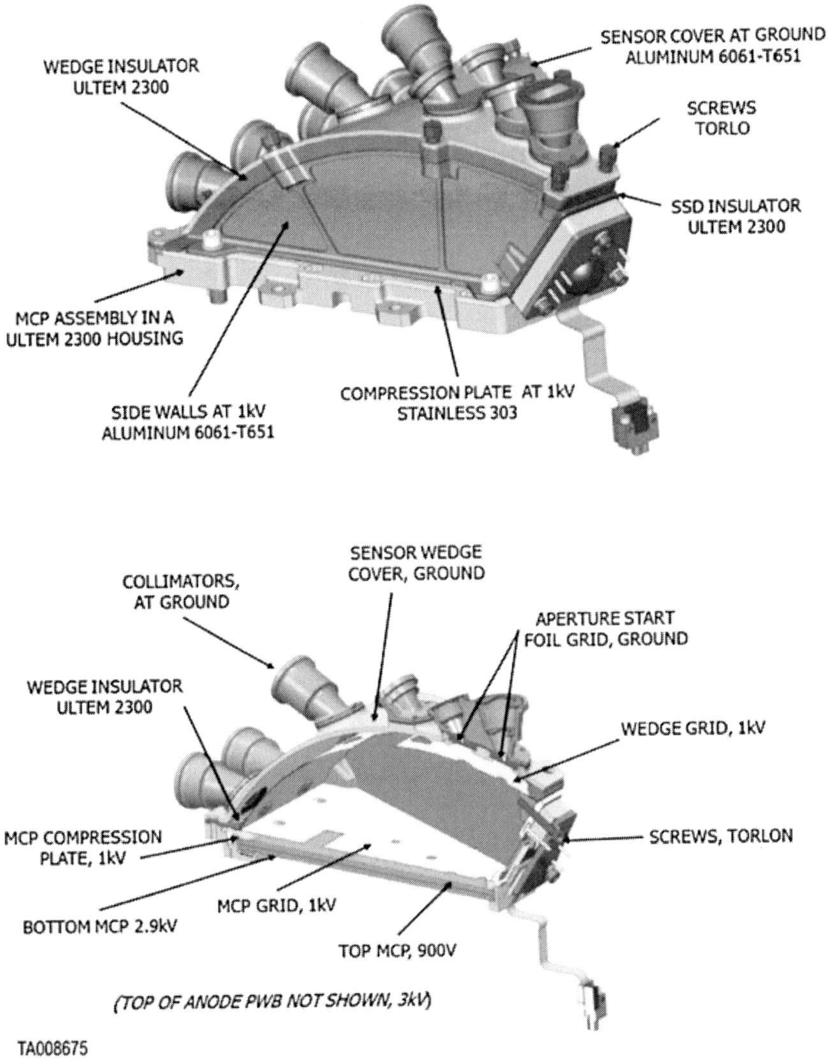

Figure 26. Wedge design

The preliminary structural analysis of the baseline design was performed using MSC Nastran, MAYA SATK®, and Femap for analysis. The model was simplified wherever possible to reduce solution time. Printed wiring assemblies (PWAs) were modeled as plate elements with uniform stiffness, thickness, and density, and the instrument model was oriented to the ISIS bracket configuration in relation to the spacecraft panels.

We performed a mechanical modal analysis to determine mechanical resonances. The analysis showed the first mode to be 304 Hz (for the Event PWA) and 553 Hz for the overall instrument. The analysis environmental input levels per spacecraft requirements were performed for all three orthogonal axes relative to the spacecraft panel. The 3-σ acceleration random response enveloped the static load requirement as desired for EPI-Lo displacements, stresses, and forces. The random-vibration PWA displacement response may be relatively high for electrical, electronic, and electromechanical (EEE) part solder/lead-wire fatigue resistance, and further analysis is planned after EEE parts placement is finalized. All margins of safety are positive for the current model configuration, and a detailed analysis will be carried out for the flight configuration to confirm that the flight design has positive margins and meets the minimum frequency requirement.

3.5 EPI-Lo Software

EPI-Lo software consists of instrument Common Software (reused from other projects), SPP Host Software, EPI-Lo Application Software, and EPI-Lo Boot Software. The Common software includes packet telemetry, command handling, macros (stored command sequences), memory management, monitoring and alarm generation, and status reporting. About 50 % of the application software and >90 % of the instrument boot code is reused.

EPI-Lo software identifies ion species and energies on-board by referencing look up tables stored in non-volatile memory that define carefully tailored regions ("boxes") in the TOF vs. Energy parameter space. In the instrument's ion-composition mode, each incoming ion event is associated with a 32-bit species and energy accumulation bin that will record the total number of events per integration interval within the defined box. The accumulation bins are arranged in packets and sent to the SSR. A selection of raw event data is also sent to the SSR, allowing more detailed compositional studies (but with decreased statistics due to downlink data volume constraints). We have used these techniques successfully many times on past missions (most recently, and with very similar software, on PEPSSI, RBSPICE, and JEDI).

In the spacecraft interface software, command and telemetry use the 115,200-baud Universal asynchronous receiver/transmitter (UART) protocol with 8 data bits plus odd parity. There is a redundant (side A and B) interface, for which the command arrival determines the active side; telemetry is sent only to that side. Dynamic side switching is supported. The system is interrupt driven: per-byte interrupt for command and telemetry, and side A and B 1PPS interrupts.

3.6 Calibration Plan

EPI-Lo measurements are intended to generate the information needed to derive differential intensity j in $(cm^2\,sr\,s\,keV/nuc)^{-1}$. The goal of EPI-Lo characterization and calibration efforts is to develop the quantitative procedures for converting the count rates R (s^{-1}) reported by EPI-Lo into estimates of j for the various defined ranges of energies, particle species, and arrival angles.

Calibration for a particle instrument like EPI-Lo means determining the following:

1) Transfer function from counts into flux (engineering units to physical units)
2) Correct rates for effects such as dead times and hardware or software saturation
3) Response to visible and ultraviolet light
4) Response to high-energy (out of band) particle backgrounds

Calibration will be done primarily using the APL particle accelerator (e.g., McNutt et al. 2008), a versatile system capable of producing a broad range of ion species at energies from 20 to 170 keV. The system includes an electron-impact ionization source, extraction gap, Einzel Lens and Wien filter mounted in the insulated terminal structure along with all associated power supplies. The system produces beams of H, He, O, and noble gas ions with intensities over the range of 10^2–10^6 particles $cm^{-2}\,s^{-1}$ at the target position (~mm^2 to cm^2). These tests will be supplemented with a variety of radioactive sources (e.g., ^{241}Am) as stimuli.

In addition, calibration runs will be made using the accelerator at Goddard Space Flight Center. A proton beam will be used to scan both angles to complete the characterization of the transfer function. The sensor response to electrons will be characterized from ~100 keV to 1 MeV, and heavy ions (He, O, and Ar) will also be used to characterize the instrument response. Calibrations of the response to both foreground and background ions and electrons at very high energies (up to ~20 MeV) require additional specialized facilities. These include the accelerator at Lawrence Berkeley National Laboratory (Berkeley Lab) and an accelerator at Idaho National Laboratory (INL).

Calibration will begin on the prototype unit to validate the instrument design and performance. This activity will also be used to establish testing procedures for the flight unit. Most sensor-level calibration for EPI-Lo occurs sequentially, with subsystems sequenced through fixed calibration setups. Sensor wedges are integrated with the electronics, followed by instrument-level testing and pre-qualification calibration, which establishes a calibration baseline. The integrated EPI-Lo undergoes EMI/EMC, vibration, and thermal vacuum testing. Final calibration is performed to ensure no changes have occurred during environmental testing. In addition to ground calibration and pre- and post-environmental qualification, in-flight calibration is used to finalize knowledge of the full instrument response. For instance, data will be used to monitor MCP gain, and efficiencies can be tracked separately for each wedge by using relative Start, Stop, and SSD rates. There are also on-board pulsers to calibrate the SSD and anode electronics. Such in-flight calibration times will also naturally allow cross-calibration between EPI-Lo and EPI-Hi.

3.7 Radiation/Contamination Effects

All EPI-Lo electronics parts have been used in recent flight instruments, are currently available, and have a minimum tolerance of >100 krad. The electronics box thickness meets that dose level with a factor of two margin. EPI-Lo contains detectors sensitive to certain contaminants, e.g. hydrocarbons; however, all detectors reside within cavities behind cover foils protected pre-launch by GN2 purge. EPI-Lo is located out of direct impact by thruster plumes; high voltage can be turned down for thruster firings if analysis indicates temporary pressure increases from such events.

3.8 Data Collection Strategy and Products

EPI-Lo delivers telemetry in Consultative Committee for Space Data Systems (CCSDS) packets using the CCSDS file Delivery Protocol (CFDP). Most data are losslessly compressed (using the FAST algorithm) prior to packetization; compression factors of >1.6 have been achieved with similar instruments on recent deep-space missions (e.g., MESSENGER and New Horizons). Sensor-event collection, command-and-telemetry processing, and instrument control functions are implemented by the instrument processor, an approach successfully used for the previously mentioned instruments now in operation or development.

The telemetry formats are flexible and adjustable by ground command. EPI-Lo has low average data rates fitting the nominal allocation suggested in the original SPP Announcement of Opportunity (NASA Science Mission Directorate 2014) and high data rates to allow for higher time resolution and additional event data, sometimes called pulse height analysis (PHA) data. EPI-Lo has ion counting rates for up to 64 energy bins, at least 9 species, and 80 angular "pixels" (one for each aperture). Event data include high-resolution TOF and energy PHA values. Energy-spectra data are collected and averaged from the selected detectors. High-resolution electron rates are also collected.

Data products conform to time, angle, and energy-resolution requirements that address the science objectives. Rate data for ions are generated nominally at 30 and 5 s time-resolution (time resolution can also be adjusted by ground command between 1 s and several hours).

High-resolution energy spectra are sent every few minutes so that fast developing features can be followed. The high level of flexibility enables maximal use of telemetry allocation and permits adjusting the operating parameters in response to evolving flight experience if needed.

4. ENERGETIC PARTICLE INSTRUMENT—HIGH ENERGY (EPI-HI)

EPI-Hi measures energetic particles in the upper portion of the ISIS energy range. As discussed in Sect. 2, the two ISIS instruments are complementary in their energy range and sky coverage in order to obtain the comprehensive set of observations needed to fully understand solar energetic particle sources, acceleration, and transport close to the Sun.

4.1 EPI-Hi Overview

EPI-Hi measures energetic electrons, protons, and heavy ions in the MeV energy range using the dE/dx versus total energy technique in a sensor system based entirely on ion-implanted silicon SSDs. It builds on a heritage of more than 40 years of SSD-based energetic particle instruments, and most directly on the Low Energy Telescope (LET, Mewaldt et al. 2008b) and High Energy Telescope (HET, von Rosenvinge et al. 2008) that are part of the IMPACT instrument suite on the twin STEREO spacecraft; these instruments have been providing multi-point measurements of energetic particles in the heliosphere since late 2006.

The sophistication of SSD instruments has steadily advanced as capabilities have improved with high performance, low power, and miniaturized, front-end electronics in the form of ASICs. The EPI-Hi requirements and expected performance are summarized in Table 5.

Table 5. EPI-Hi instrument required and expected performance

Parameter	Required	Goal (expected)	Comment/Heritage
Electron energies	0.5–3 MeV	0.5–6 MeV	STEREO/HET
Ion energies	~1 to ≥50 MeV/nuc	p, He: 1 to 100 MeV/nuc $Z \geq 6$: 1.5 to >100 MeV/nuc	STEREO/LET & HET
Energy binning	≥6 bins per decade	12 bins/decade	Logarithmic bins
Cadence	Fastest: e 1 s, p 5 s	Fastest: e 1 s, p 1 s most data products: 1 min angular distributions: 5 min	Large energy bins best energy resolution
Fields of view	≥$\pi/2$ steradians in sunward and anti-sunward hemisphere	Five 45° half-angle view cones covering sunward and anti-sunward hemispheres	View cones overlap to provide full energy coverage near Parker spiral to within 10° of spacecraft-Sun line
Angular sectoring	e: ≤45°; ions: ≤30°	45° half-angle cones subdivided into 25 overlapping sectors	Overlapping sectors provide improved angular resolution for deriving pitch angle distributions
Elemental composition	H, He, C, O, Ne, Mg, Si, Fe	H, He, C, N, O, Ne, Na, Mg, Al, Si, S, Ar, Ca, Cr, Fe, Ni	STEREO/LET
Isotopic composition	^{3}He/^{4}He	^{3}He/^{4}He	In selected viewing directions STEREO/LET
Intensity range	Up to 3×10^6 protons cm^{-2} sr^{-1} s^{-1}	Normal mode: up to ~1×10^6 protons cm^{-2} sr^{-1} s^{-1} Pixel mode: up to ~4×10^7 protons cm^{-2} sr^{-1} s^{-1}	>10 MeV protons
Geometrical factor	N/A	5 view cones, each with $A\Omega \approx 0.5$ cm^2 sr	Value at energy with maximum $A\Omega$

Figure 27 shows a block diagram of EPI-Hi. The sensor system consists of three detector stacks, commonly called "particle telescopes", each controlled by a dedicated electronics board that provides front-end analog signal processing, PHA, coincidence determination, data accumulation and analysis, and formatting for readout. We have tailored the telescope designs for measurements in two different but overlapping energy ranges. A pair of Low Energy Telescopes (LET1 and LET2) measure protons and heavy ions from ~1 to >20 MeV/nuc and electrons from ~0.5 to ~2 MeV while a single High Energy Telescope (HET) extends the energy coverage for electrons up to ~6 MeV and for protons and heavy ions to at least 100 MeV/nuc. LET1 and HET analyze particles incident from both ends of their detector stacks while LET2 analyzes particles incident from only one end because the other end is blocked by the spacecraft. Each of the five active telescope ends has an angular acceptance cone with a half-angle of 45° centered on the symmetry axis of the telescope.

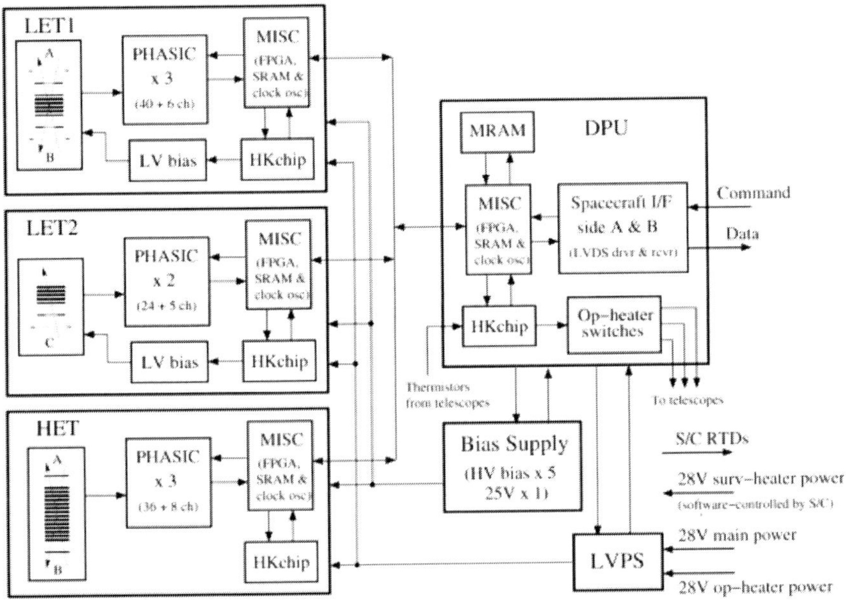

Figure 27. EPI-Hi block diagram

In addition to the three telescope boards, EPI-Hi includes a Data Processing Unit (DPU) board that is responsible for the overall control of the instrument, including managing the boot-up of processors on the telescope boards, collecting data from those boards and merging them into a single data stream that is passed to the spacecraft, and receiving commands from the spacecraft and passing them to the processor for which they are intended.

The remaining two electronics boards contain the LVPS that produces the required regulated, filtered voltages needed by the instrument and the bias supply that produces programmable voltages for biasing SSDs.

A metal enclosure, designated the "E-Box", holds the electronics boards and provides a degree of shielding against radiation and high-speed dust impacts. We also use the E-Box, which is attached to the ISIS bracket (see Sect. 2.3), as a platform on which we mount the detector telescopes in the orientations needed to achieve the required viewing directions, as illustrated in Fig. 28.

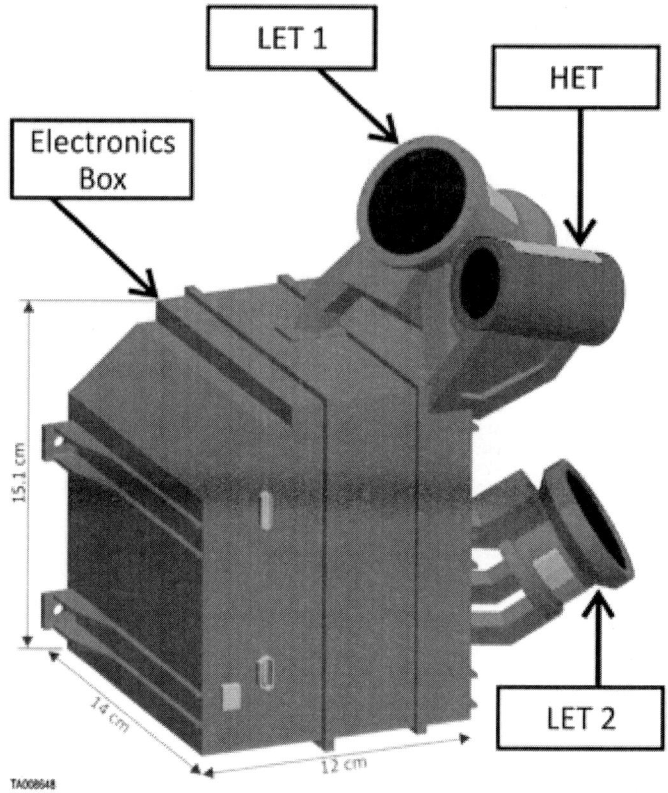

Figure 28. EPI-Hi mechanical configuration including E-Box and telescopes

4.2 High-Energy Telescope

Figure 29 shows cut-away views of the telescopes. HET has a thick central stack of detectors together with a single detector spaced farther away at each end. A coincidence between this front detector and the central stack defines the FOV of the telescope. The front detectors (H1) and the outermost detectors at each end of the central stack (H2) have five main active elements, consisting of a central bull's-eye and four quadrants of a surrounding annular region, with all elements having equal areas, as illustrated in Fig. 30a. Identifying which detector segment was hit by an incident particle in each of the two detectors allows us to subdivide the FOV into 25 overlapping sectors, thus enabling measurements of the distribution of particle incidence directions.

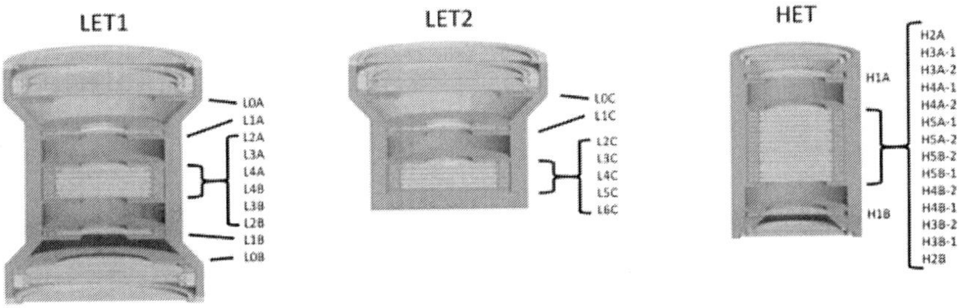

Figure 29. Cut-away illustration of the EPI-Hi telescope designs and detector naming

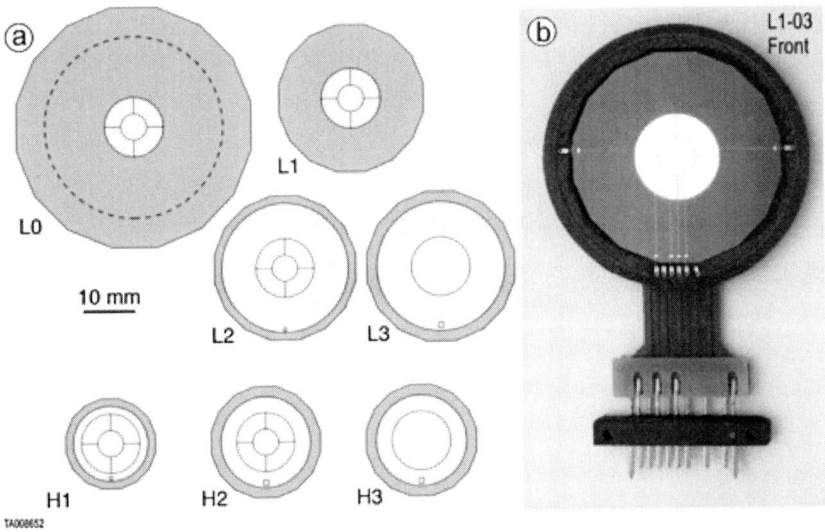

Figure 30. (**A**) Illustration of the detector segmentation. Detectors L0 through L3 are used in LET; detectors H1 through H3 are used in HET. The *dashed line* in the L0 detector drawing indicates the diameter inside which the silicon thickness is 12 μm. The L4, L5, and L6 detectors are identical to L3 while the H4 and H5 detectors are identical to H3. (**B**) Photograph of a prototype L1 detector

The detectors other than H2 in the central HET stack have a central segment with an active area equivalent to the combined areas of the five segments in H1 and H2. The combination of the thickness of the central stack and the active diameter of the detectors is such that some particles incident at large angles from the telescope axis will exit through the sides of the stack. To identify and reject these side-exiting particles, the detectors include an annular "guard" segment that is used as anti-coincidence in the analysis. The guard also allows identification of particle trajectories that clip the detector edge resulting in reduced energy-loss signals. This feature is important because the detector thickness is a non-negligible fraction of the detector diameter and thus edge-clipping trajectories are not improbable. The HET detectors (as well as the LET detectors other than L0 and L1) also include one additional segment in the form of a small pixel that is used to extend the dynamic range in particle intensities to values above which the larger detector segments will saturate, as discussed in Sect. 4.5.

The individual silicon detectors used in the HET telescopes have thicknesses of either 500 μm (H1) or 1000 μm. However, for H3, H4, and H5, we use two successive 1000 μm detectors connected to the same front-end electronics to produce what is effectively a 2000 μm detector without requiring the high bias voltage that an individual detector of this thickness would need.

4.3 Low-Energy Telescope

The conceptual design of the LETs is similar to that of HET, using a central detector stack in conjunction with position-sensitive detectors that define the FOV and subdivide it into a number of sectors that we use for measuring particle angular distributions. Similar to HET, the LET detectors, other than L0 and L1, have thicknesses of either 500 μm (L2) or 1000 μm. In order to achieve a low threshold energy for EPI-Hi and to minimize the energy gap between EPI-Hi and EPI-Lo, we have developed a process for fabricating very thin silicon detectors. At the front of LET we have L0 and L1 detectors with thicknesses of 12 μm and 25 μm, respectively (Fig. 29). A 1 MeV proton has energy just sufficient to penetrate the L0 detector and the thin windows in front of the telescope (~3 μm silicon equivalent) and provide the 2-parameter measurement required for particle identification. The thin detectors are sufficiently uniform in thickness to allow the required species separation, including distinguishing between the isotopes of He as shown in Sect. 4.4. Like the H1 and H2 detectors, L0, L1, and L2 are all position sensitive with a central bull's-eye surrounded by an annulus subdivided into four quadrants. Figure 30b shows a photograph of a prototype L1 detector.

Protons with energy greater than ~8 MeV will have energy losses in the L0 detector that fall below the detection threshold of the front-end electronics (see Sect. 4.7). For this reason, in LET we identify events either by a coincidence between L0 and L1 or a coincidence between L1 and L2. It is possible for an event to satisfy the L1•L2 coincidence without triggering L0 either because the signal fell below threshold in L0 or because the particle trajectory did not pass through the active area of this device. We have designed the L0 detector to have its 1 cm^2 active area located at the center of a large (~9 cm^2) thin silicon membrane that covers the entire field of view defined by the L1•L2 coincidence. This ensures that the analyzed particles will have passed through a consistent amount of material, thereby enabling accurate calculations of their incident energies. The L2 detector has a large annular guard region so that we can determine whether an event defined by a coincidence between L0 and the center element of L1 but lacking a signal in the central five segments of L2 is due to a particle stopping in L1 or to a trajectory passing outside the central part of L2.

Unlike in HET, the central stack detectors in LET (L3 and L4) have an active diameter large enough to intercept even the widest-angle trajectories defined by the L1•L2 coincidence. Thus, the LET detectors do not require guard segments to identify side-exiting particles. We have also segmented these detectors into a central bull's-eye surrounded by a wide annular region so that we can dynamically adjust the LET geometrical factor and thereby increase the dynamic range in particle intensities that EPI-Hi can measure (see Sect. 4.6).

4.4 Species and Energy Coverage

Figure 31 shows the energy range covered by LET and HET for elements with atomic numbers in the range $1 \leq Z \leq 30$. The primary measurements involve the analysis of particles that trigger two position-sensitive detectors and then stop in the central detector stack, as indicated by the absence of a signal in the detector at the opposite end of the stack. In the case of HET we extend the coverage to higher energies by also analyzing events due to particles that penetrate the entire thickness of the stack.

Figure 32 shows results from a Monte Carlo simulation of the HET response to a population of particles with heavy-ion composition typical of a large SEP event. The element response tracks have distinct branches that correspond to stopping and penetrating particles. For events on the stopping branch we derive the particle energy by simply adding the energy losses in all of the detectors and making a small correction for energy lost in the front windows. On the penetrating branch the detected energy is less than the incident energy, but we are able to use the location along the track to obtain a reasonably accurate value for the incident energy up to about twice the energy required to penetrate the stack. This allows us to measure energy spectra up to at least 100 MeV/nuc for the more abundant elements.

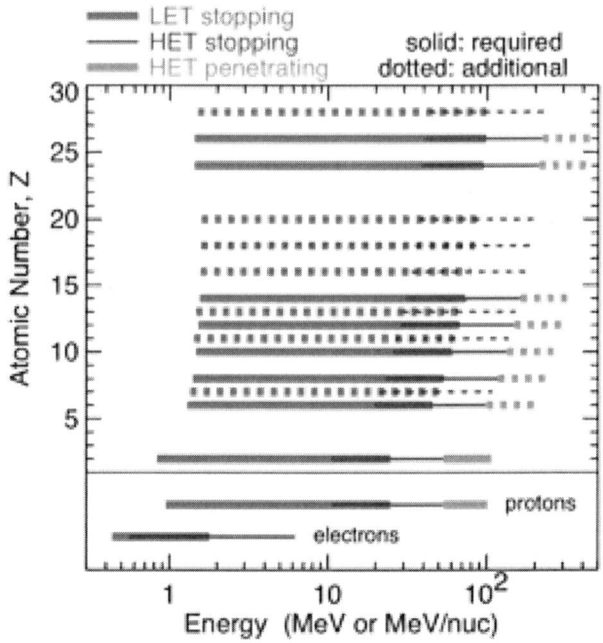

Figure 31. Species and energy coverage by LET and HET telescopes

Figure 33 shows results from a Monte Carlo simulation of the LET response to particles that satisfy the L1•L2 coincidence and stop in L2. The figure shows L1 and L2 energy losses with a simple multiplicative correction that depends on the mean secant of the incidence angle in the assigned sector (Mewaldt et al. 2008b). This correction, which removes a significant fraction of the track broadening caused by variations in the angle at which the particles penetrate the L1 detector, allows us to display all of the events in well resolved tracks on a single plot.

In addition to separating the response tracks for major elements, we have designed the EPI-Hi telescopes to distinguish helium isotopes. A large enhancement of ^{3}He relative to ^{4}He, sometimes by factors of 10^3 or greater relative to the solar coronal value of $\sim 4\times10^{-4}$, is a distinguishing characteristic of impulsive SEP events in which particle acceleration is thought to be associated with magnetic reconnection rather than due to coronal and interplanetary shocks (see Sect. 1). For He isotope measurements we use portions of the FOV that have small angles, θ, relative to the telescope axis, thus minimizing the uncertainty in the dE/dxmeasurement due to the range of particle path-lengths through the dE/dx detector (secθvariation). The separation of He isotopes is most challenging at the lowest energies that we measure with LET since the effects of detector thickness non-uniformities and of Bohr/Landau fluctuations (e.g., Ahlen 1980) in the particle energy losses are most significant there. Because ^{4}He in this energy range can be obtained from radioactive alpha-particle sources, we were able to test the performance in the laboratory using an L0 detector to measure dE/dx and an L1 detector to measure residual energy. Figure 34 shows the ^{4}He response track obtained from these measurements and compares it with track locations calculated from a simple parameterization of the range-energy relation in silicon. As shown in the inset, the ^{4}He resolution is sufficient to distinguish ^{3}He from ^{4}He down to ratios as small as a few percent.

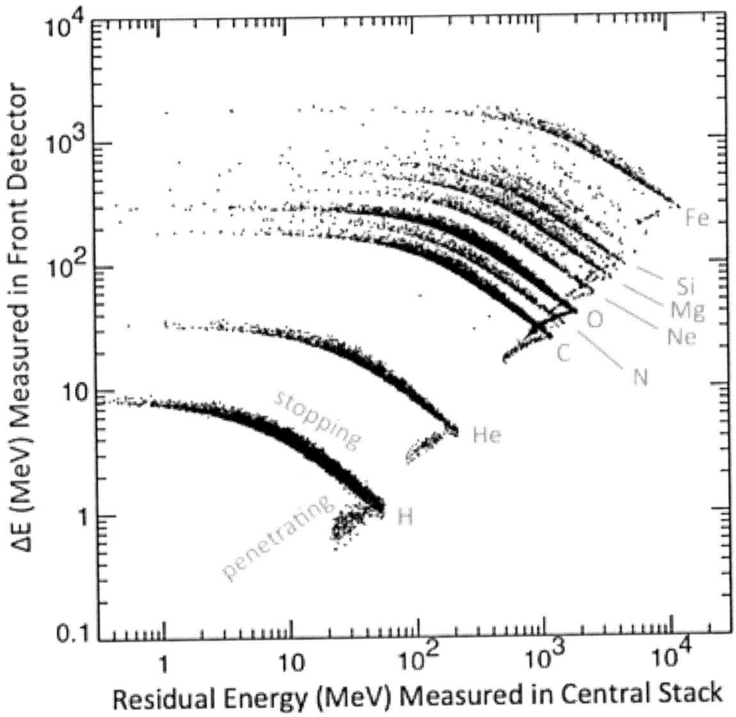

Figure 32. Monte Carlo simulation of HET response measuring ΔE using the front detector (H1A) and the residual energy, E', using all of the detectors in the central stack. Relative abundances are typical of a large SEP event except that He has been suppressed by a factor of 200 and H by 2000

The energy range for electron measurements in HET extends from the energy required to penetrate the H1 detector and front windows up to a maximum set by the active thickness of the entire HET detector stack. In LET, electron energy losses in the L0 and L1 detectors always fall below threshold so we use coincidences among L2 and the following stack detectors for electron detection. In this case we do not have trajectory information (other than the telescope end that the electron entered), so we use a model of the instrument response to derive electron energy spectra from the data. Data from the LETs are important for obtaining electron energy spectra measurements over a broad field of view as well as for determining electron event onset times in cases where the first-arriving electrons do not fall within the HET field of view. The EPI-Hi fields-of-view provide excellent coverage close the Parker spiral direction over the full orbit and typically will allow measurements of the first-arriving particles.

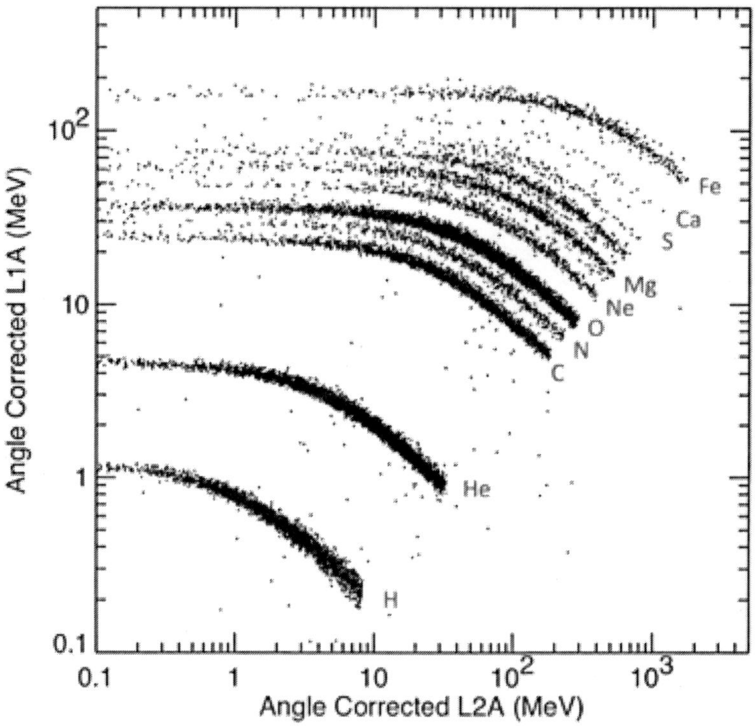

Figure 33. Monte Carlo simulation of the LET response for particles stopping in the L2A detector. A simple multiplicative correction factor depending on the angular sector in which each particle was detected has been applied to the ΔE (L1A) and E' (L2A) values in order map all of the events onto a common response matrix

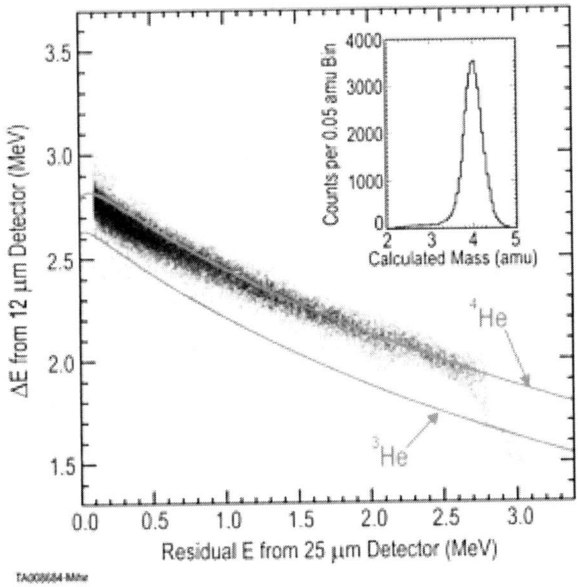

Figure 34. Measured response of 2-detector telescope consisting of one L0 detector and one L1 detector to ^{4}He particles from a ^{244} Cm source. A piece of fine fabric was interspersed between the source and the detectors to produce a spread of energies. The *inset* shows a mass histogram accumulated over the residual energy interval from 1.0–2.5 MeV. *Red curves* are approximate tracks for ^{3}He and ^{4}He calculated from a simple range-energy relation

4.5 Viewing Directions and Angular Sectioning

Figure 18 shows the locations of the EPI-Hi 45° half-angle conical FOVs in the sky. Although these fields are largely unobstructed, there are several places where portions of the spacecraft or of other instruments block them. The most significant blockage is by the spacecraft's TPS. The forward view cones of HET and LET1 are intentionally oriented so that the TPS cuts through them, thus allowing detection of particles arriving from as close as possible to the radial direction over the full EPI-Hi energy range. The combination of the EPI-Hi telescopes will allow measurement of nearly complete pitch angle distributions for nominal Parker spiral magnetic fields over the entire orbit except for blockage by the TPS, which affects less than 10° even near perihelion. A number of the EPI-Lo apertures (Fig. 19) are also oriented for viewing within the overlap region between the HET and LET1 forward fields of view, thus allowing excellent spectral coverage in the region where the average Parker spiral magnetic field will often be located when the spacecraft is near the Sun.

We assign detected particles to one of 25 sectors based on which of the 5 main detector segments they hit in two position-sensitive detectors (Fig. 30). By orienting one of the two detectors with its quadrant sectors rotated 45° relative to those on the other detector we are able to identify 25 unique mean viewing directions, as illustrated in Fig. 35a. The angular width of the individual sectors is broad, extending as much as 25° from the mean direction in some cases. However, ≥80 % of the geometrical acceptance of the sectors fall within 15° of the mean. We rely on the significant overlaps among the sectors to provide information about particle angular distributions, as illustrated in Fig. 35b, where we show the accuracy with which the arrival direction of a narrow beam of particles can be determined as a function of that direction and of the number of particles counted.

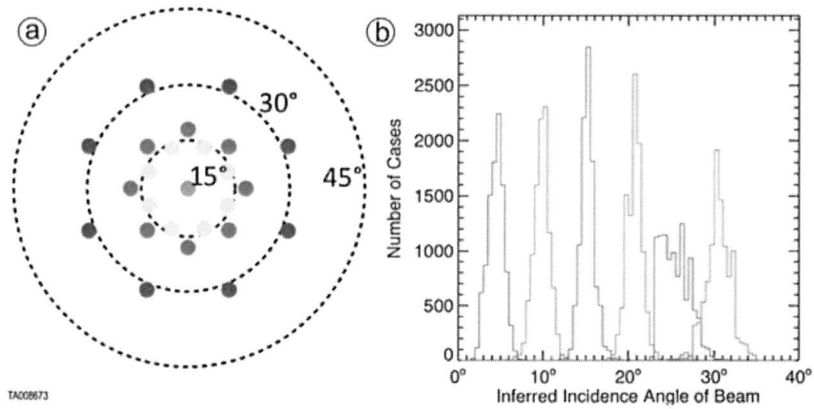

Figure 35. (**A**) Mean viewing directions of each of the 25 sectors. *Dashed circles* indicate angles of 15°, 30°, and 45° from the telescope axis. (**B**) Simulted distributions of derived incidence directions for parallel beams incident at 5°, 10°, 15°, 20°, 25°, and 30° from the telescope axis when determining the direction using 200 detected particles

4.6 Geometrical Factors and Dynamic Range in Particle Intensities

We have designed EPI-Hi to measure MeV particles over a large dynamic range of intensities, since this range must encompass the combined effects of varying numbers of accelerated particles, varying radial distances from the Sun, and varying magnetic connection to the acceleration region, as well as intrinsic differences in the source abundances among different species. To accommodate this large range, the telescopes have relatively large geometrical factors to cover the lowest intensities of interest but at the same time use detectors that are small and closely spaced to minimize the geometrical factor for out-of-geometry particles that can cause backgrounds and dead time. We further reduce the effect of these unwanted particles by using a relatively thick housing for the telescopes that serves as a passive shield to absorb low energy particles. In addition, by using segmented detectors we are able to reduce the geometrical acceptance when particle intensities are high and thereby help avoid saturating the front-end electronics.

Figures 36a and 36b show the dependence of the geometrical factors of the HET and LET telescopes on incident particle energy and on particle arrival direction relative to the telescope axis, respectively. We calculated these geometrical factors for a single LET or HET entrance aperture and have not included corrections for blockages of the fields of view or for the geometrical factor reductions that we impose when intensities are high.

The EPI-Hi front-end electronics can handle count rates up to $\sim 10^5$ particles s^{-1} from a single telescope before the analysis dead time approaches 100 % (see Sect. 4.7). In order to keep the dead-time percentage significantly below this value, we employ a system of "dynamic thresholds" based on an approach that we successfully used in the STEREO/LET (Mewaldt et al. 2008b) and HET (von Rosenvinge et al. 2008) instruments. This approach takes advantage of the fact that the count rates are typically dominated by protons, electrons, and He nuclei, which have relatively small energy losses in the detectors. When the intensity gets high enough to cause excessive dead time, the thresholds are raised automatically on some segments in a detector so that they will no longer be sensitive to these low pulse heights but will still remain sensitive to heavy ions. By raising thresholds in this way for successively deeper detectors in the stack we are able to reduce the geometrical factor for electrons, protons, and He in several stages. We base the determination of when thresholds need to be raised (and when they subsequently need to be lowered as intensities decrease) on the count rate in a detector segment for which we do not raise the threshold. We take the threshold changes into account when we derive particle intensities from the measured event rates. Based on historical data and the assumption that SEP intensities will scale as $1/r^3$ (an extreme assumption), we expect only one event during the prime mission that will have a peak intensity too high to measure with the HET telescope using the dynamic thresholds (Lario and Decker 2011).

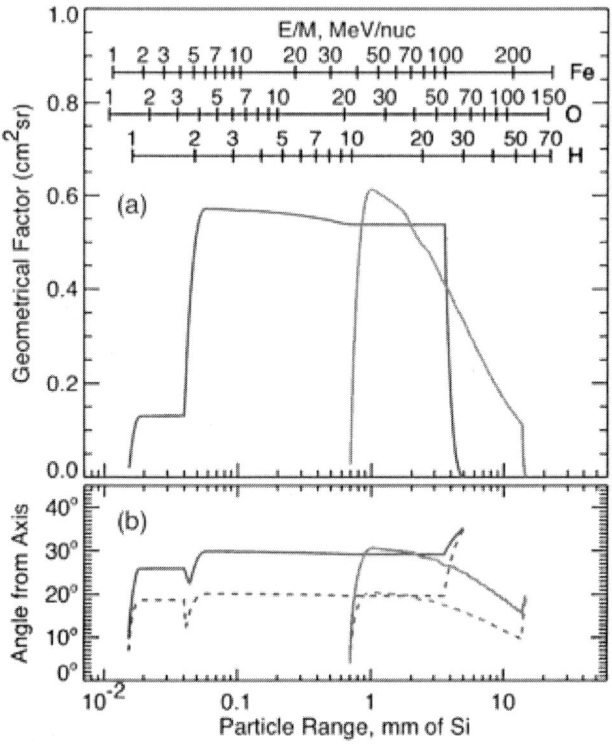

Figure 36. (**A**) LET and HET geometrical factors as a function of the particle range in silicon. Energy scales for H, O, and Fe are also shown. *Blue curves* are used for LET, *red curves* for HET. Protons above ~25 MeV have energy losses in the L1 detector that fall below threshold and thus do not produce a coincidence in LET. Geometrical factors are shown for one end of each telescope. Plot is based on stopping particles; penetrating particle analysis will allow modest extensions to higher energies. (**B**) *Solid* (*dashed*) *curves* show the angles inside which 90 % (50 %) of the trajectories from an isotropic particle distribution fall as a function of the particle range in silicon

Estimates of the proton intensities that could be encountered if a large solar energetic particle event occurs when the spacecraft is close to perihelion and well connected to the acceleration region are extreme (Lario and Decker 2011), with a 5 % chance that we could experience an event with a proton integral intensity above 1 MeV exceeding 1.5×10^{8} $cm^{-2}\,sr^{-1}\,s^{-1}$. If Solar Probe Plus should experience a rare event approaching this intensity, the count rate in even one of the detector segments will be high enough to saturate the front-end circuitry. In order to obtain a rough measure of intensity under such circumstances we have designed all of the thick detectors to include an extra small "pixel" segment with an area of 1 mm^2 (1000 μm detectors) or 0.36 mm^2 (500 μm detectors), which is ≤5 % of the area of the smallest normal segment (see Fig. 30a). We use pixels on detectors at a few depths in the telescopes to measure singles count rates of energy losses exceeding a selected threshold. By setting the threshold slightly below the energy that will be deposited by a proton stopping in the full thickness of the detector (~12 MeV for 1000 μm detectors and ~8 MeV for 500 μm detectors we can use the measured count rate to infer an approximate proton integral intensity above an energy that is set by the material above the pixel. Using a simulation of the instrument response we can combine measurements from pixels located at a few different depths to derive a rough energy spectrum for extreme events. From the correlation between intensity measurements made using the pixels and those obtained in the normal operating mode of the instrument we are able to calibrate the pixel measurements using data that we obtain before and after the highest intensity periods when normal operation is not possible. Referring to Fig. 13, we estimate that in the LET telescopes the pixels will be used for measurements when the integral intensity of protons above 10 MeV exceeds $\sim 2\times10^4$ cm^{-2} sr^{-1} s^{-1} and in HET when the intensity exceeds $\sim 10^6$ cm^{-2} sr^{-1} s^{-1}.

4.7 Front-End Electronics

The telescope boards process the signals from the silicon detectors using an ASIC called the "Pulse Height Analysis System Integrated Circuit" (PHASIC). A first generation version of this part is flying in the STEREO/LET and HET instruments. Solar Probe Plus requires a higher radiation tolerance than STEREO, so we have made several modifications to the PHASIC design to facilitate the production of a version that can tolerate a dose of at least 100 krad and also to improve several key aspects of its performance.

The PHASIC contains 16 complete dual-gain PHA chains. Figure 37 shows a schematic diagram of one of these chains. It consists of a charge-sensitive preamplifier that drives two independent post-amplifiers that shape the pulse and provide additional gain. By having two parallel analysis chains we are able to make precise pulse height measurements over the preamplifier's entire dynamic range. The shaped pulse, which is bipolar with a peaking time of 1.9 μs, drives a peak detector that contains a linear gate and discriminator. The discriminator controls the linear gate and prevents subsequent pulses from contaminating the detected peak level until it has been digitized. The PHASIC uses Wilkinson rundown ADCs, which determine the pulse height by measuring the time interval required for discharging a hold capacitor using a constant current. The PHASIC's on-chip digital circuitry also counts the number of triggers of the high- and low-gain analysis chains and provides several programmable-input OR-gates that are used to perform the lowest level of event coincidence logic and thereby enable rapid rejection of some uninteresting triggers without requiring readout of the digitized pulse heights. The PHASIC developed for EPI-Hi also incorporates shift registers to record the time-history of the discriminator outputs and aid in identification of crosstalk. Each dual-gain PHA includes a precision test pulser for use in the functional testing and calibration of the circuit.

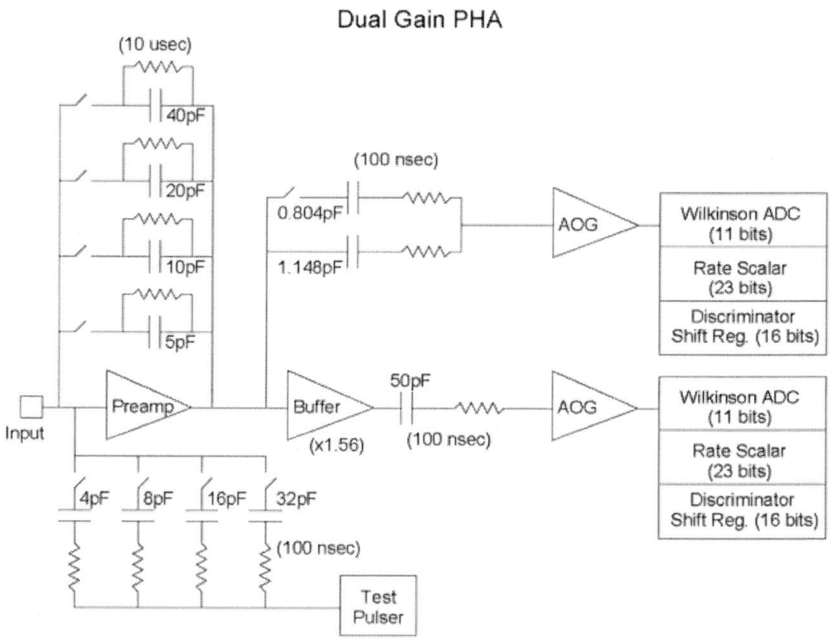

Figure 37. Schematic diagram of one PHASIC channel

The PHASIC design includes several parameters that are programmable by setting bits in a digital control register. This programmability provides sufficient flexibility so that the PHASIC can meet the analysis requirements of all of the different detectors used in EPI-Hi. By programming the feedback capacitance in the preamplifier we select the gain appropriate to the range of signals for a given thickness of detector. We can also select between two options for the gain ratio between the high- and low-gain analysis chains. Other programmable parameters include the discriminator thresholds and the value of the capacitor used by the test pulser for injecting a calibrated charge into the preamplifier.

Table 6 summarizes the PHASIC specifications. As illustrated in Fig. 38, we install the ASIC in a hybrid circuit containing a number of passive components that support the operation of the chip.

These include a precision resistor for each PHA chain that sets the ADC's rundown current and provides excellent stability over temperature. The design of the ASIC itself relies on ratios of component values for setting key parameters, since such ratios are much more predictable and stable than absolute component values.

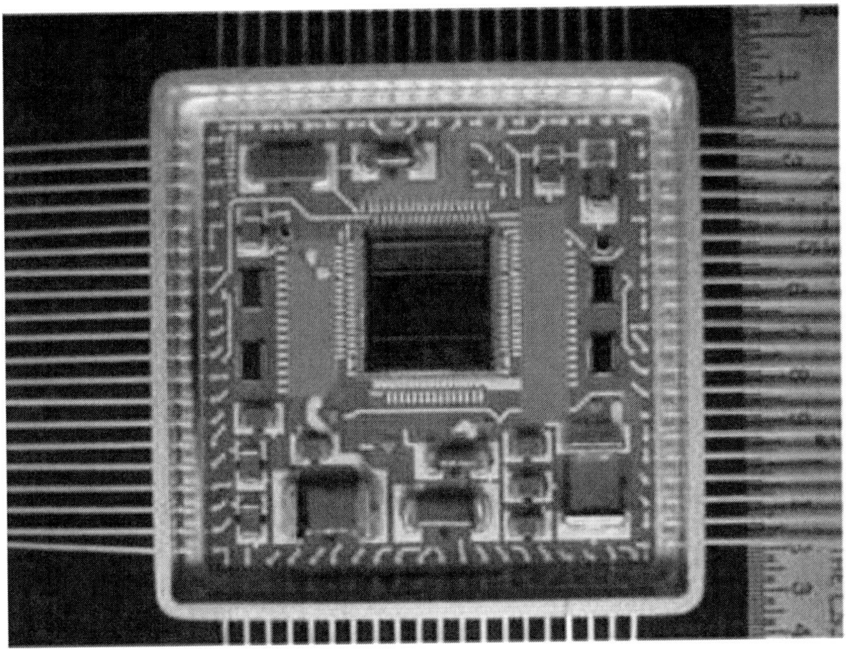

Figure 38. Photograph of PHASIC chip installed in hybrid. *Centimeter scale at right* indicates size

We have fabricated and tested a prototype version of the EPI-Hi PHASIC using the commercial ON-Semi C5N CMOS process that will allow the addition of Aeroflex processing steps to improve the total tolerance to >100 krad.

4.8 Digital Control and Data Processing

We use individual "Minimal Instruction Set Computers" (MISCs) to control each of the three telescope boards and also the DPU board. The STEREO and NuSTAR missions previously used the same MISC design, which is implemented in a portion of an FPGA and runs a Forth operating system (Mewaldt et al. 2008b). The logic circuitry required by each of the boards also fits in the RTAX250 FPGA, with significant margin. Each MISC has an associated SRAM chip to provide read-write memory and, in addition, the DPU MISC has non-volatile magnetoresistive random-access memory (MRAM) that contains the code and tables needed by all of the MISCs. Working from this stored copy of the software, the DPU MISC supports the booting of the MISCs on the telescope boards ("telescope MISCs") when needed.

Table 6. PHASIC chip specifications

Table 6 PHASIC chip specifications

Number of dual-gain PHAs	16
Chip size	7.4 by 7.4 mm
Power	11 mW per active PHA
Dynamic range	Up to 23000 (full scale/trigger threshold)
Integral non-linearity	<0.05 % of full scale
Differential non-linearity	<1 %
High/Low gain ratio	68 or 40, configurable
ADC type	Wilkinson
ADC resolution (each gain)	11 bits, 12th bit overflow
Shaping	Bipolar, 1.9 μsec to peak
Preamp feedback capacitance	5–75 pF, programmable in 5 pF steps
Preamp full scale output swing	4.0 Volts
Cross-talk between PHAs	<0.2 %
Radiation tolerance	>100 krad, no latchup below 80 MeV/(mg/cm^2)
Gain temperature coefficient	<50 ppm/°C
Offset temperature coefficient	<0.1 channel/°C
Operating temperature range	−30 to +50 °C
Leakage current balancing	Up to 32 μA with 10-bit resolution
Threshold programmability	Up to 6 % of full scale (each gain) with 10 bit resolution
Deadtime per trigger	4–67 μs (pulse height dependent)

The telescope MISCs handle the sorting of events that satisfy the coincidence requirements implemented in the PHASIC and FPGA logic into a number of different categories depending on the type of analysis they require. These MISCs determine the species, energy, and incidence direction of accepted particles, and they count the numbers of events that occur with various combinations of these parameters using selected accumulation times ranging from 1 s to 1 hour as appropriate for the expected count rates and the data storage allocated to EPI-Hi on the spacecraft recorder.

The flight software derives the atomic number of each detected ion by using the measured energy deposited in the detector in which the particle stopped (E') and in the preceding detector (ΔE) together with the mean thickness penetrated in the ΔE detector, which depends on both the direction of incidence and the measured thickness characteristics of the detector. We determine the regions occupied by the tracks for each of the elements of interest, and also for electrons, based on the calculated instrument response validated using preflight accelerator calibrations and refined based on data collected in flight (see Sect. 4.10). We determine the incident energies of stopping particles by summing up the energies deposited in each detector and adding a small correction for the unmeasured energy loss in the telescope window. For ions we use the mass of the most abundant isotope having the derived atomic number in order to calculate the incident energy per nucleon, which is the quantity used for assigning the particle to a particular energy bin. For He nuclei having incidence directions for which the mass resolution is sufficient for resolving ^{3}He from ^{4}He, we also calculate the mass of the detected isotope.

In addition to counts of different categories of events accumulated at various time cadences, we also accumulate the corresponding efficiency factors and live-times needed to correct the measured count rates. These rates and associated correction factors are the primary data returned by EPI-Hi.

We employ a prioritization scheme for selecting a sample of events to be returned with their complete set of measured pulse heights. We use these event data for validating the on-board analysis and for addressing specialized science topics. The number of events to be recorded is flexible and we adjust it depending on the volume of data that can be downlinked for a particular orbit.

We also measure and return a variety of housekeeping parameters including temperatures, voltages, and detector leakage currents. We have developed a second ASIC called the "HK Chip" for measuring these parameters. This chip also handles assorted other support functions and allows a significant reduction in the parts count on the telescope and DPU boards. For example, the HK Chip includes a high-precision DAC that provides the reference voltage required by the test pulsers in the PHASICs. As with the PHASIC, we have already fabricated and tested a non-rad-hard version of the HK Chip to validate the design that will be used for production of rad-hard flight parts by Aeroflex Inc.

4.9 Gamma Rays and Neutrons

The HET telescope can detect gamma rays and neutrons (see Sect. 1.6) since the central regions of H3A through H3B (see Fig. 39) provide a 1.2 cm^3 sensitive volume of silicon almost completely surrounded by active material consisting of their guard regions together with H2A and H2B. As a result, triggers of one or more central segments (with the surrounding detector elements in anti-coincidence) can be caused by ~0.5–6 MeV gamma-rays that Compton-scatter or pair-produce as well as by ~2–20 MeV neutrons that undergo nuclear reactions such as ^{28}Si(n,p)^{28}Al or ^{28}Si(n,α)^{25}Mg (Mewaldt et al. 1977).

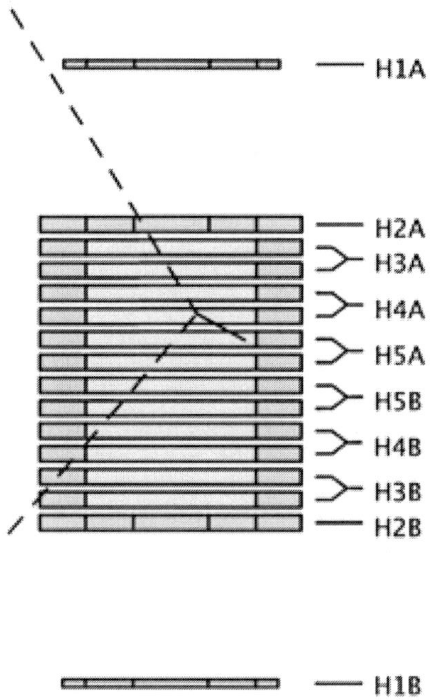

Figure 39. Cross section of the HET silicon detector active areas illustrating how detectors are used for detection of gamma rays and neutrons. For these measurements, all of the orange-shaded detector segments are used in anticoincidence to reject incident charge particles. The *green-shaded segments* are used to record energy losses from charged particles produced when neutrals interact in the interior of the telescope. *Lines* indicate a neutral particle (*upper dashed line*) that scatters producing a charged particle (*solid line*) plus a second neutral (*lower dashed line*) that escapes from the telescope. Only the charged particle is detected

4.10 Calibration Plan

Onboard determinations of the composition and energy spectra of ions ranging from H to Ni and of electrons require precise calibrations of the response of the LET and HET detectors and electronics over a broad range of operating temperatures and incident particle energies, directions, and intensities. To carry out the particle identification procedure described in Sect. 4.8 (see Fig. 33) we must know precisely the mean thickness of all detector segments, the transfer function from pulse-height to energy in each of the PHASIC circuits, and range-energy relations for all of the ions of interest.

We performed an initial test of several LET L0 and L1 detectors in October 2013 at Berkeley Lab's 88-inch cyclotron. The measured narrowness of the tracks in Fig. 40 demonstrates that both L0 and L1 have the required energy resolution and thickness uniformity to identify elements up to at least Kr (Z=36). We also measure the thickness characteristics of the 12 and 25 μm detectors in the laboratory with collimated alpha-particle sources. Our experience with SSDs on STEREO indicates that the 500 and 1000 μm detectors for EPI-Hi will be sufficiently uniform to preclude the need to carry out thickness uniformity studies of these devices. From Fig. 40 we conclude that existing heavy-ion range-energy relations require only minor adjustments to calculate the locations of the element tracks.

The on-board particle identification system also requires accurate pre-launch and in-flight measurements of the gain and offset of each PHASIC's dual-gain PHA. Since EPI-Hi may experience temperature variations over the orbit, we will measure the temperature coefficients of each flight PHASIC. Fortunately, the PHASICs are very stable over temperature, with gain temperature coefficients of <50 ppm/°C, and offset temperature coefficients of <0.1 channel/°C. In addition, we will perform periodic in-flight calibrations of each ADC using the test pulsers built into the PHASICs.

For the electron calibrations for LET and HET, we model the telescope responses using the Geant4 simulation package (http://geant4.web.cern.ch/geant4/) and spot-check the simulation results using laboratory beta-decay sources such as ^{106}Ru, which has a maximum electron energy of 3.54 MeV.

We are planning a final end-to-end test of EPI-Hi at the Michigan State University National Superconducting Cyclotron Laboratory (NSCL), which can accelerate ^{58}Ni ions up to 160 MeV/nuc. By fragmenting ^{58}Ni and a lighter beam like Ne in a thick target, we can measure the ΔE versus E' response tracks for elements from H to Ni (Mewaldt et al. 2008b). We can also test and calibrate the instrument's performance under high-rate conditions at NSCL.

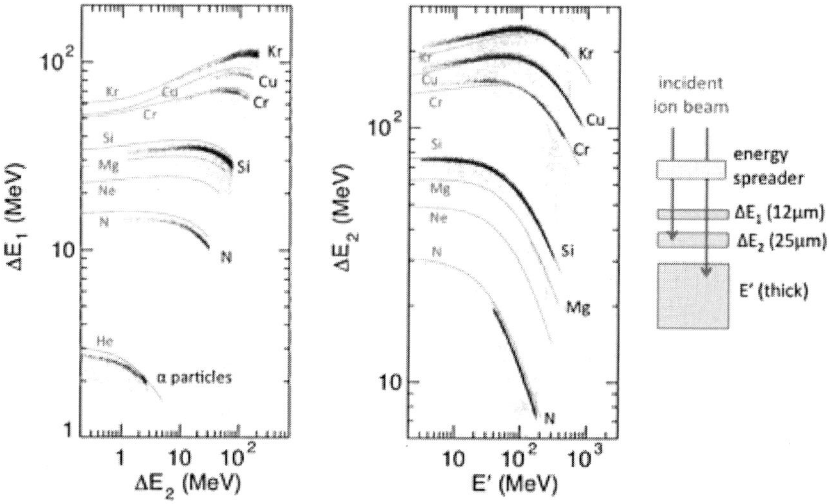

Figure 40. Data from a heavy-ion test of a detector telescope consisting of one L0 detector and one L1 detector in front of a thick residual energy detector, as illustrated in the *sketch at the right*. A "cocktail beam" containing a variety of heavy ions was obtained from the 88-inch cyclotron at the Lawrence Berkeley National Laboratory. *Red curves* show the response calculated from a published range-energy relation. Accelerator data of the type shown can be used to better calibrate the locations of the response tracks

Following launch, we will collect and telemeter energetic particle data to evaluate the performance of the LET and HET on-board particle identification and update the calibration parameters as necessary. Data collected in flight will also be used for intercalibrating EPI-Hi with EPI-Lo.

4.11 Environmental Tests

We have subjected prototypes of the thin silicon detectors, L0 and L1, to selected environmental tests since these detectors are not only a new technology development but will also experience some particularly harsh conditions. We constructed a mechanical model of the L0 detector having a thin silicon membrane of the same diameter but only 80 % of the L0 thickness, and subjected it to a severe acoustic test that went up to an overall sound pressure level of 143 dB. The model came through this test undamaged.

Although the radiation specification used for most components on SPP is relatively modest, ~20 to 40 krad for total ionizing dose and $\sim 1\times10^{11}$ to 2×10^{11} protons/cm^2 fluence for displacement damage, the thin silicon detectors are, of necessity, protected by much less shielding than a typical electronic subassembly.

We tested prototype thin silicon detectors up to 10 Mrad total dose using the ^{60}Co source in the Jet Propulsion Laboratory (JPL) high-dose-rate test facility and up to a proton fluence of 10^{13} cm^{-2} at the University of California Davis cyclotron. The devices survived these extreme tests and showed negligible degradation of their response when tested with alpha particle sources.

The increase in the detector leakage currents due to the displacement damage was significant, with the L0 current going from ~5 nA to ~0.7 μA after the proton exposure, but was in good agreement with the expected increase (Lutz 1999). This change is well within the design limits of the EPI-Hi bias system and front-end electronics.

The characteristics of the population of high-speed dust particles orbiting close to the Sun are poorly known, but these particles could present a significant risk of damaging sensitive exposed surfaces such as the thin windows and thin front detectors in LET. Because of this, we have already carried out initial tests of prototypes of the polyimide windows at dust accelerators in Heidelberg, Germany and at the University of Colorado.

Dust particles with kinetic energies approximating those expected close to the Sun were able to penetrate the few-μm thicknesses of these windows, but did not cause any windows to rupture. We also performed a preliminary test of the effect of dust impacts on a thin silicon detector and determined that the detector was still operable after the test. More detailed dust testing is continuing.

We have also designed the coincidence logic for the LET telescopes to improve their robustness against damage due to dust impacts. As discussed above (Sect. 4.3), we can select events for analysis in LET on the basis of a coincidence between L1 and L2, without requiring a signal from L0.

Thus if the L0 detector were to fail due either to a direct dust impact or to a large noise increase as might occur if the windows were to suffer enough damage to allow a significant flux of light to impinge on L0, operation could continue with a somewhat higher LET threshold energy.

5. SCIENCE OPERATIONS, DATA PROCESSING, AND DATA PRODUCTS

5.1 Software/Instrument Modes

The SPP prime mission is composed of 24 highly elliptical, heliocentric orbits with decreasing orbital periods from 168 days for orbit 1, settling into an ~88 day orbit period midway through the mission. Each orbit is broken into two distinct periods, the Solar Encounter period (inside 0.25 AU) and the Cruise/Downlink period (outside 0.25 AU). For each orbit, the solar encounter duration is ~10–11 days. During the solar encounters all instruments across the SPP payload are powered on and continuously collecting science data (for more on instrument operations, see below). Although limited commanding is supported during the encounter period, spacecraft commanding is not nominally planned during this period. All normal spacecraft activities during the encounter period will execute via time-tagged or event-tagged onboard command sequences.

The ISIS instruments are planned to operate similarly in two operational modes:

- Normal Science Mode
- Spacecraft-Sun Distance: $R \leq 0.25$ AU (R is heliocentric radius)
- Full nominal power
- High data collection rate and burst data (EPI-Lo)
- Low-rate Science Mode
- Spacecraft-Sun Distance: $R > 0.25$ AU
- Full power when not downlinking and whenever possible
- Reduced data collection rate (fits within ISIS telemetry allocation)
- Commanding window scheduled late in the series of telemetry passes, although it may not be used every orbit

The instruments are capable of switching between these routine modes autonomously, based on information received once-per-second from the spacecraft. EPI-Hi has no "Burst" mode. During "Burst" periods EPI-Lo sends telemetry to the spacecraft at higher rates.

At least three special operational modes are currently envisioned:

- Software upload mode

- Calibration Science mode
- Safe mode

During Calibration Science mode, samples of PHA event data are accumulated and returned to validate onboard assignments of species, energy, and incidence angle, and for assessing instrument backgrounds. This mode is activated outside 0.25 AU by command several times early in the mission when sufficiently high intensities of solar energetic particles are present. Following ground analysis of data gathered in Calibration Science mode, onboard tables and parameters will be updated as needed. These tables and parameters are pre set, based on the pre-flight calibrations, but it is likely that some fine-tuning will be needed. We expect to upload new tables about 3 weeks into the statistics gathering period, and then use the remaining time to verify the new tables or make further updates as needed. In safe mode, the ISIS instruments return to a "quiet" state, awaiting commands from the ground. This mode is reserved for instrument I&T, commissioning, and instances where the flight software encounters a fault or out-of-limit condition.

Particle events are sporadic so the goal is to approach 100 % coverage. Whenever possible, ISIS instruments should be on and taking data.

The ISIS team streamlines instrument operation by minimizing commanding and mode changes. The spacecraft telemetry file system is used to prioritize data downlink:

- First priority: Housekeeping for health and safety
- Second priority: Snapshot data to identify important time periods (immediately relayed to other SPP instruments)
- Third priority: Full Science Data (rates at different cadences and event data)

Because solar encounters occur frequently, the team has a three-step hierarchy of planned commands for near-term and longer-term changes. Specifically, command loads will be planned for solar passes N, $N+1$, and $N+2$ where pass N corresponds to the next solar pass. This pro-active planning process will allow for any needed adjustments based on recent data, while ensuring that plans are in place well in advance of imminent passes.

5.2 ISIS Science Operations Center

The ISIS SOC is housed at the University of New Hampshire (UNH) with heritage based on the Interstellar Boundary Explorer (IBEX) Mission. The primary functions of the ISIS SOC functions include planning, commanding, monitoring instrument health and safety, and producing and archiving data products. The key requirements of the ISIS SOC are:

- Provide public "quick look" data 60 days after downlink (a less stringent requirement of 6 months for quick look data applied for the first 3 orbits)
- Provide public science data 6 months after downlink
- Share science/engineering/ancillary data among the SPP team
- Archive all telemetry, data, software and documents for mission + 1 year
- Place all software and data in a final, deep archive by 12 months after the end of the mission
- Implement SPP project approved security on SOC computers
- Receive "remote SOC notification" of instrument fault conditions as detected by the spacecraft

The ISIS SOC has a common framework that flows from Ground Support Equipment Operations Systems (GSEOS) through development of the engineering models, flight models, I&T, and flight operations. The common framework allows the ISIS SOC to develop as the team builds and learns to operate the instruments. It is important to have cross-coordination and cross-calibration, to the greatest degree possible. The ISIS SOC participates throughout instrument characterization, calibration, I&T, and flight. The ISIS SOC comprises a Command, Telemetry and Ground (CTG) group, a Science Data group, the pipeline and other analysis software, a Command and Telemetry Archive, and a Distributed Science Archive (Fig. 41). The connections between the SOC and MOC are illustrated in Fig. 42.

We apply the concept of "test as you fly, and fly as you test". The intention of "test as you fly" is to avoid improvisation during the critical flight phase. Operations plans, including instrument modes, will be modified based on experience from I&T, commissioning, and early ops. However, we emphasize commonalities among all those phases rather than planning from the start to operate in very different modalities during testing and flight.

Instrument Ground Support Equipment (GSEs) fulfils two critical roles by providing a direct connection to instruments for ground testing and flight-like modes so that the spacecraft simulator can be used to test operations. The GSE flight-mode operation will be incorporated into the SOC GSEOS system to provide a nearly identical "look and feel" to the instrument/SOC teams during testing, I&T, commissioning and normal flight operations.

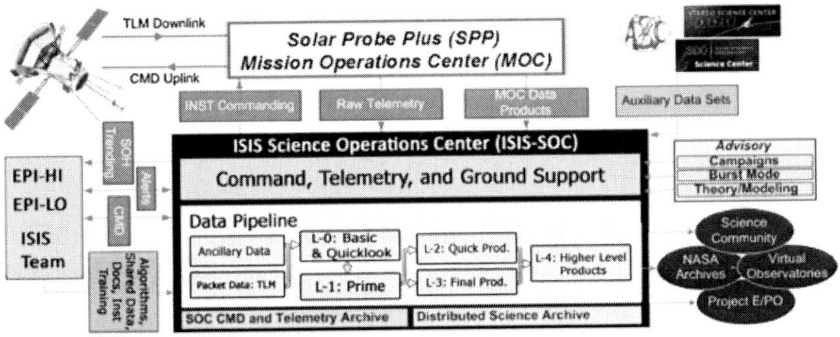

Figure 41. Data flow through the ISIS-SOC housed at UNH. The ISIS SOC runs the data pipeline to produce data products (Levels 1–4) and houses the commanding center for ISIS. The ISIS and SPP teams aid the ISIS SOC for algorithm development, input concerning commanding, alert checking, and state-of-health trending. The data at the output of the ISIS SOC pipeline is delivered to the science community, NASA archives, virtual observatories and SPP project Education and Public Outreach centers

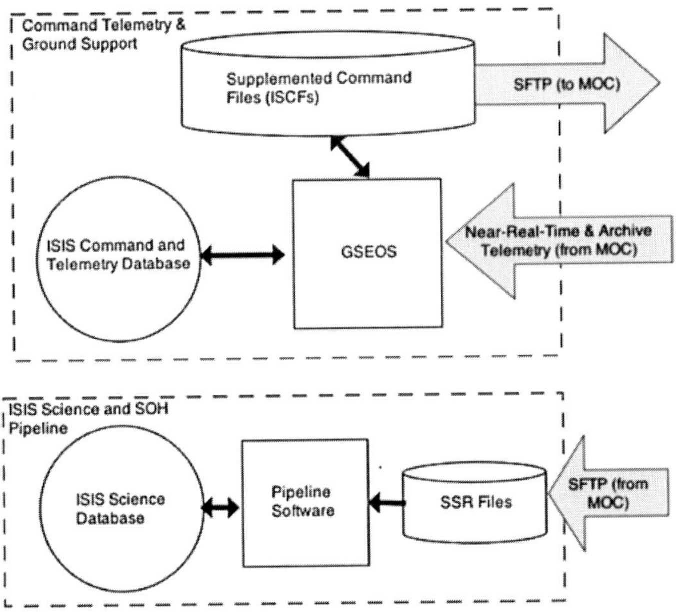

Figure 42. Secure ftp (SFTP) connections to the SPP Mission Operations Center (MOC) at APL provides for transfer to the MOC of command files and transfer from the MOC of instrument telemetry. A near-real-time connection between the SOC and MOC to GSEOS provides for near-real-time instrument commanding and instrument monitoring

The ISIS SOC CTG group is responsible for uploading instrument commands, validating command requests from the instrument teams, generating coordinated command sequences, receiving science and ancillary data from the MOC, monitoring instrument state-of-health, responding to alerts in coordination with the SPP instrument teams, processing raw science and ancillary data (including housekeeping), producing time-information for instrument data and housekeeping, and supporting ground testing and routine operations from pre-flight to post-flight. The ISIS SOC is staffed by a small team responsible for day-to-day operations, with operators brought on-board early in the project and operating the instruments during I&T. The command and telemetry database developed during I&T is used for flight operations.

The SOC CTG group checks instrument health and safety limits throughout testing, I&T, commissioning, and flight. GSEOS is used at ISIS SOC for health and safety monitoring during contacts. The ISIS SOC also plans to develop specific ISIS software for health and safety checking, and SPP-accessible web pages with state-of-health checks and long-term trending. Housekeeping and status data are processed as soon as possible after receipt (within 3 days) and posted for viewing for the ISIS team through an ISIS SOC website. The housekeeping and status data are downloaded first after passes, and provide critical information from which to build command loads in subsequent passes.

The ISIS SOC Science Data (SDG) group manages the repository for processing software and supports the SPP team throughout the project by processing Level 0, 1, 2, 3 and 4 (L-0 to L-4) data products; developing data analysis tools in coordination with the instrument teams and SPP communities; generating summary and quick-look plots; consolidating SPP data products with auxiliary data sets; serving data and analysis software to SPP communities; providing a center for scientific interchange meetings throughout the project; and developing a database of models and derived quantities.

The library of analysis software is maintained in a version control system downloadable to other sites maintained within the team. The analysis software is written in platform-independent languages such as C, IDL and Java for the pipeline software. Tools such as Mission Independent Data Layer (MIDL; http://sd-www.jhuapl.edu/MIDL/) and IDL will be available for visualization and analysis. Most processes within the ISIS SOC will be partially automated using Unix/Shell scripts that launch executables with clearly specified I/O.

Three types of software support the creation and analysis of science data for ISIS: software for integrating and testing of the instruments, ISIS-SOC components for processing telemetry into calibrated data products, and visualization tools for displaying and manipulating science data products.

Data processing algorithms incorporate straightforward routines for decommutation, particle identification and application of calibration data. All formal pipeline processing is performed at the ISIS SOC.

The SOC pipeline executes telemetry processing modules to be developed in coordination with the EPI-Lo and EPI-Hi teams. The processing modules convert telemetry into Level 1 data products, and the modules interact with the pipeline through a simple interface. UNH hosts the SOC web site, which serves calibrated data to science data centers, virtual observatories, and the public. The SOC will ensure that the team has rapid access to data.

The ISIS team is fully committed to an open and timely data release to maximize the usage and usability of the ISIS measurements from the SPP mission. The availability of all ISIS data will be fully compliant with the science data policy defined in the NASA Heliophysics Science Data Management Policy. After completion of in-orbit commissioning and cross-calibration, the SOC delivers data products.

The ISIS team retains the data internally for only the time period necessary to ensure that the data are properly processed, calibrated, and verified. The ISIS Team and SOC strive to make this period as short as possible after receipt of the original science telemetry and auxiliary orbit, attitude and spacecraft status information.

The ISIS SOC also centrally creates metadata and handles data archiving deliveries. Crucial to making the archiving cost effective is ensuring that all ISIS data products are created archive-ready. This will involve some data design work by the teams early on to coordinate data formats, layouts, and metadata standards.

This will enable the data to be released to the long-term archive with no changes. Solar Probe Plus falls under the purview of the Heliophysics Data Policy, which recommends the use of the CDF data format and the Space Physics Search Archive and Extract (SPASE) metadata standard. The public archive for the mission is the Space Physics Data Facility (SPDF) at the Goddard Space Flight Center.

5.3 ISIS Data Products

EPI-Hi and EPI-Lo data products are detailed in Table 7. Level 0 data products are essentially raw, whereas Level 1 data products are validated and released to the science community after the validation process is complete. Level 2 data products are being defined.

Table 7. Data products level, description, latency and any dependencies

Data level	Description	Latency	Users
L0 status data	Snapshots of EPI-Lo and EPI-Hi data through the orbit	Downloaded first following each orbit	ISIS instrument teams
L0 housekeeping	EPI-Hi and EPI-Lo voltages, currents, temperatures, rate monitors	Downloaded first following each orbit	ISIS instrument teams
L0 command response data	Data products that detail results of specific commands	Data delivered asap after tests	ISOC and ISIS instrument teams
L0 memory dumps	EPI-Hi and EPI-Lo will perform slow (bit/s) memory dumps periodically	Data delivered after each pass	ISOC and ISIS instrument teams
L0 EPI-Hi events (Z & E)	Sample of nuclear charge (Z) and kinetic energy E of individual particles at processing rates of >1000 particles/sec	Data delivered after each pass	ISOC and ISIS instrument teams
L0 EPI-Hi Z vs. E matrices	Each of the processed "events" is sorted into several Z vs. E matrices	Data delivered after each pass	ISOC and ISIS instrument teams
L0 EPI-Hi Z vs. E vs. direction matrices	Z vs. E matrices accumulated over multiple look directions	Data delivered after each pass	ISOC and ISIS instrument teams
L0 EPI-Lo electron rates	Electron rates in angle and energy bins	Data delivered after each pass	ISOC and ISIS instrument teams
L0 EPI-Lo ion rates	Proton, He and heavy ion rates in angle and energy bins	Data delivered after each pass	ISOC and ISIS instrument teams
L0 EPI-Lo PHA events	Complete information on select events. Cadence of 0.1 s < 0.25 AU and 5 s > 0.25 AU	Data delivered after each pass	ISOC and ISIS instrument teams
L0 EPI-Lo TOF-only events	TOF information allowing detailed species separation	Data delivered after each pass	ISOC and ISIS instrument teams
L1 EPI-Lo events, rates	Products similar to EPI-Lo L0 data products, however validated and time-sorted with redundancies removed	Data accumulated after each pass (N) and made available to public in pass ($N + 1$)	SPP science community
L1 EPI-Lo particle intensities	Absolute intensities (in units of particles ($cm^{-2}\ sr^{-1}\ s^{-1}\ [MeV/nuc]^{-1}$) for ion species and in electrons ($cm^{-2}\ sr^{-1}\ s^{-1}\ MeV^{-1}$) for electrons). These are produced from the L0 data	Data accumulated after each pass (N) and made available to public in pass ($N + 1$)	SPP science community
L1 EPI-Hi events, rates, matrices	Products similar to EPI-Hi L0 data products, however validated and time-sorted with redundancies removed	Data accumulated after each pass (N) and made available to public in pass ($N + 1$)	SPP science community
L1 EPI-Hi LET and HET particle intensities	Absolute intensities (in units of particles ($cm^{-2}\ sr^{-1}\ s^{-1}\ [MeV/nuc]^{-1}$) for ion species and in electrons ($cm^{-2}\ sr^{-1}\ s^{-1}\ MeV^{-1}$) for electrons)	Data accumulated after each pass (N) and made available to public in pass ($N + 1$)	SPP science community

L1 EPI-Hi expanded event data	These data are produced for a sample of the individual ions/electrons that trigger the instrument. A priority system will ensure that all species, energies, and directions are sampled	Data accumulated after each pass (*N*) and made available to public in pass (*N* + 1)	SPP science community
L2 data sets	Higher level products that combine various data products. Examples include data products time-ordered with solar wind and magnetic field data, element abundance ratios, fits to particle anisotropy data, and particle intensities associated with particular solar events	Data accumulated after each pass (*N*) and made available to public in pass (*N* + 1)	SPP science community

The ISIS SOC also provides to the public a quick-look processed version of science data within 6 months of downlink for the first three orbits after launch and within 60 days of downlink thereafter, including:

- EPI-Lo spectrograms of <1 MeV electron counts/data-interval as a function of time, at less than 1 hour cadence, with 4 or more energy bins.
- EPI-Lo spectrograms of <1 MeV/nuc total ion or single species counts/data-interval as a function of time, at a <1 hour cadence, with 4 or more energy bins.
- EPI-Hi plots of 1-hour averages of four rates: 1–5 MeV electrons; 2–10 MeV protons; 10–50 MeV protons; and 4–40 MeV/nuc HiZ ($6 \leq Z \leq 28$).

Data verification and validation are partly automated, but also require evaluation by ISIS scientists. Verification and validation processes are coordinated by the ISIS Science Data group. After the 3-orbit period, all ISIS data products will be made available to the public no later than 6 months after downlink. The ISIS team also defines and produces additional high-level products as needed. For example, Solar Proton Event lists and Co-rotating Interaction Region (CIR) event lists will likely be produced. The list of such products will evolve throughout the mission as new regimes are discovered. The ISIS team will fully engage with the SPP team as a whole to define and provide new high-level products throughout the mission. These products along with the most accurate data and a complete set of ISIS data products will be regularly released to the SPDF.

The coordination of data with other SPP teams and with information from outside missions and ground-based data will be carried out through the SOC. A series of Level 2 data products will be generated including coordinated SPP products (solar wind, magnetic field, energetic particles), SPP products coordinated with 1 AU data sets (e.g., STEREO, ACE, L1 observers, GOES), data sets from spacecraft beyond 1 AU, and data sets from other missions inside 1 AU such as Solar Orbiter. Also critical to understanding the coronal environment are data sets from remote and ground based observations. Potential field models, for example, provide important information about the connection between features observed by SPP and possible source regions on the Sun. ISIS Level 2 data sets therefore incorporate not only data but also results of data-based models such as potential field models. Level 3 and higher level data products are developed based on scientific needs, incorporation of models, and physical context of the environment.

6. SUMMARY AND CONCLUSIONS

The ISIS investigation on the Solar Probe Plus mission has been optimized to explore the sources of energetic particles near the Sun and mechanisms that accelerate and transport them as their distributions evolve outward into interplanetary space. ISIS will finally answer the fundamental questions of the origins, acceleration, and transport of solar energetic particles and provide new discovery science.

ISIS accomplishes this with its two electrically independent instruments that together measure energetic particles over the broad energy range from ~0.02 to ~200 MeV/nuc for ions and ~0.025–6 MeV for electrons. The instruments are mounted together on a common ISIS bracket that allows viewing as close to the spacecraft heat shield umbra as possible, which ensures broad pitch angle observations across these energy ranges. Science planning and operations are carried out through the ISIS Science Operation Center, which also produces the combined ISIS science products and coordinates data with the other SPP investigation science centers, including science teams from many other space missions, such as Solar Orbiter, ground-based observers, and others. The data products will optimize the scientific productivity and output not just from the SPP investigation teams, but from the entire Heliophysics community.

EPI-Lo is a high-heritage, TOF-based mass spectrometer that measures energetic electron (25–1000 keV) and ion spectra (~0.02 MeV/nuc to a maximum energy of 15 MeV for most elements). By resolving all major heavy ion species and ^{3}He and ^{4}He over much of this energy range in multiple directions, EPI-Lo provides the required data across the critical energy range from suprathermal energies (~20 keV/nuc) up to the lower portion of EPI-Hi energy range with a single instrument.

EPI-Hi measures energetic particle spectra, composition, and angular distributions using the dE/dx vs. E technique in a sensor system consisting of a double-ended High Energy Telescope (HET), and two Low

Energy Telescopes (LETs), one double-ended (LET1) and one single-ended (LET2). Together these EPI-Hi telescopes cover ~1 to at least 100 MeV for protons, and higher for heavy elements, and ~0.5 to 6 MeV for electrons.

Together, coordinated observations from EPI-Hi and EPI-Lo, along with the centralized ISIS SOC and broader ISIS science team, are configured to revolutionize our understanding of the critical energetic particle environment that SPP will encounter as it explores the inner heliosphere progressively closer to the Sun.

In conclusion, Solar Probe Plus is the key Heliophysics mission for understanding the solar corona and inner heliosphere. Only by repeatedly traveling very close to the Sun is it possible to observe the regions where solar energetic particles are born, energized, and injected into the interplanetary environment. Only by making these critical observations is it possible to understand the origins, energization and transport of solar energetic particles, which drive key aspects of space weather and pose a significant risk for human exploration of space and for our increasingly technological space-based infrastructure. The Integrated Science Investigation of the Sun will make these unique observations and develop the critical scientific advances and knowledge needed to understand them.

FOOTNOTES

1 The Energetic Particle Spectrometer (EPS) on the MErcury Surface, Space ENvironment, GEochemistry, and Ranging (MESSENGER) mission to Mercury, the Pluto Energetic Particle Spectrometer Science Investigation (PEPSSI) on the New Horizons mission to Pluto, the three JEDI instruments on the Juno mission to Jupiter, the Radiation Belt Storm Probes Ion Composition Experiment (RBSPICE) on the twin Van Allen Probes in the Earth's radiation belts and the Energetic Ion Spectrometer (EIS) on the Magnetospheric Multiscale mission (MMS) spacecraft to be launched into Earth orbit.

2 The Debye length sets the scale length for the plasma sheath. With respect to the solar wind, the electron temperature is expected to have a characteristic temperature of ~100 eV and at 10 R_S, the density should be typically no more than $\sim(215/10)^2 \times 10$ $cm^{-3} \sim 4600$ cm^{-3}; the corresponding Debye length is ~1 meter. Photoelectron temperatures are much less ~1 eV, and the densities will be much larger, so the sheath thickness will be much less near the sunlit parts of the spacecraft.

3 Work by the SPP project indicates about 0.1 hits/foil over the mission (~0.25–0.4 hits/mission in the direction viewing the Sun). This is the geometry for a hit through both foils, but it is under the assumption that the particle will survive the first foil hit. So EPI-Lo will likely accumulate holes but at a tolerable rate.

ACKNOWLEDGEMENTS

We are deeply indebted to all of the outstanding men and women who have made the Solar Probe Plus Mission and ISIS investigation a reality. These include members of the Solar Probe/Solar Probe Plus Science and Technology Definition Team (STDT), NASA headquarters personnel who have supported and continue to support the mission, project team members at APL and GSFC, instrument team members at a variety of universities and other institutions across the country and around the world, and other members of the community who support this critical mission through their advisory work for the NRC and informally through many other means. We also thank Wendy Mills for the great job she did in assembling and editing this paper for the ISIS Team.

REFERENCES

1. S.P. Ahlen, Theoretical and experimental aspects of the energy loss of relativistic heavily ionizing particles. Rev. Mod. Phys. **52**(1), 121 (1980)
2. J. Ambrosiano, W.H. Matthaeus, M.L. Goldstein, D. Plante, Test particle acceleration in turbulent reconnecting magnetic fields. J. Geophys. Res. **93**, 14383–14400 (1988).
3. G.B. Andrews, T.H. Zurbuchen, B.H. Mauk, H. Malcom, L.A. Fisk, G. Gloeckler, G.C. Ho, J.S. Kelley, P.L. Koehn, T.W. Lefevere, S.S. Livi, R.A. Lundgren, J.M. Raines, The energetic particle and plasma spectrometer instrument on the MESSENGER spacecraft. Space Sci. Rev. **131**(1–4), 523–556 (2007).
4. M.J. Aschwanden, Particle acceleration and kinematics in solar flares—a synthesis of recent observations and theoretical concepts (Invited review). Space Sci. Rev. **101**(1), 1–227 (2002).
5. Bale et al., Space Sci. Rev. (2014, this issue)
6. H.V. Cane, R.A. Mewaldt, C.M.S. Cohen, T.T. von Rosenvinge, Role of flares and shocks in determining solar energetic particle abundances. J. Geophys. Res. **111**(A6), A06S90 (2006).
7. B.D.G. Chandran, B. Li, B.N. Rogers, E. Quataert, K. Germaschewski, Perpendicular ion heating by low-frequency Alfvén-wave turbulence in the solar wind. Astrophys. J. **720**(1), 548–554 (2010).
8. B.D.G. Chandran, D. Verscharen, E. Quataert, J.C. Kasper, P.A. Isenberg, S. Bourouaine, Stochastic heating, differential flow, and the alpha-to-proton temperature ratio in the solar wind. Astrophys. J. **776**(1), 45 (2013).
9. S.R. Cranmer, A.A. van Ballegooijen, Can the solar wind be driven by magnetic reconnection in the Sun's magnetic carpet? Astrophys. J. **720**(1), 824–847 (2010).

10. S.R. Cranmer, A.A. van Ballegooijen, R.J. Edgar, Self-consistent coronal heating and solar wind acceleration from anisotropic magnetohydrodynamic turbulence. Astrophys. J. Suppl. **171**(2), 520–551 (2007).
11. S. Dalena, A.F. Rappazzo, P. Dmitruk, A. Greco, W.H. Matthaeus, Test-particle acceleration in a hierarchical three-dimensional turbulence model. Astrophys. J. **783**, 143 (2014).
12. M.A. Dayeh, M.I. Desai, J.R. Dwyer, H.K. Rassoul, G.M. Mason, J.E. Mazur, Composition and spectral properties of the 1 AU quiet-time suprathermal ion population during solar cycle 23. Astrophys. J. **693**(2), 1588–1600 (2009).
13. M.I. Desai, G.M. Mason, J.R. Dwyer, J.E. Mazur, C.W. Smith, R.M. Skoug, Acceleration of 3He nuclei at interplanetary shocks. Astrophys. J. **553**(1), L89–L92 (2001).
14. M.I. Desai, G.M. Mason, M.E. Wiedenbeck, C.M.S. Cohen, J.E. Mazur, J.R. Dwyer, R.E. Gold, S.M. Krimigis, Q. Hu, C.W. Smith, R.M. Skoug, Spectral properties of heavy ions associated with the passage of interplanetary shocks at 1 AU. Astrophys. J. **611**(2), 1156–1174 (2004).
15. M.I. Desai, G.M. Mason, R.E. Gold, S.M. Krimigis, C.M.S. Cohen, R.A. Mewaldt, J.E. Mazur, J.R. Dwyer, Heavy-ion elemental abundances in large solar energetic particle events and their implications for the seed population. Astrophys. J. **649**(1), 470–489 (2006).
16. P. Dmitruk, W.H. Matthaeus, N. Seenu, Test particle energization by current sheets and nonuniform fields in magnetohydrodynamic turbulence. Astrophys. J. **617**(1), 667–679 (2004).
17. J.F. Drake, M. Swisdak, H. Che, M.S.A. Shay, Electron acceleration from contracting magnetic islands during reconnection. Nature **443**(7111), 553–556 (2006).
18. D.C. Ellison, R. Ramaty, Shock acceleration of electrons and ions in solar flares. Astrophys. J. **298**, 400–408 (1985).
19. L.A. Fisk, Motion of the footpoints of heliospheric magnetic field lines at the Sun: implications for recurrent energetic particle events at high heliographic latitudes. J. Geophys. Res. **101**(A7), 15547–15554 (1996).
20. L.A. Fisk, G. Gloeckler, Particle acceleration in the heliosphere: implications for astrophysics. Space Sci. Rev. **173**(1–4), 433–458 (2012).
21. Fox et al., Space Sci. Rev. (2014, this issue)
22. J. Giacalone, Particle acceleration at shocks moving through an irregular magnetic field. Astrophys. J. **624**(2), 765–772 (2005).
23. J. Giacalone, J.R. Jokipii, J.E. Mazur, Small-scale gradients and large-scale diffusion of charged particles in the heliospheric magnetic field. Astrophys. J. **532**(1), L75–L78 (2000).
24. J. Giacalone, Energetic charged particles associated with strong interplanetary shocks. Astrophys. J. **761**(1), 28 (2012).
25. N. Gopalswamy, H. Xie, S. Yashiro, S. Akiyama, P. Mäkelä, I.G. Usoskin, Properties of ground level enhancement events and the associated solar eruptions during solar cycle 23. Space Sci. Rev. **171**(1–4), 23–60 (2012).
26. N. Gopalswamy, S. Yashiro, S. Krucker, G. Stenborg, R.A. Howard, Intensity variation of large solar energetic particle events associated with coronal mass ejections. J. Geophys. Res. Space Phys. **109**(A12), A12105 (2004).
27. J.T. Gosling, The solar flare myth. J. Geophys. Res. **98**(A11), 18,937–18,949 (1993).
28. D.K. Haggerty, E.C. Roelof, Impulsive near-relativistic solar electron events: delayed injection with respect to solar electromagnetic emission. Astrophys. J. **579**(2), 841–853 (2002).
29. Howard et al., Space Sci. Rev. (2014, this issue)
30. P.A. Isenberg, B.J. Vasquez, Kinetic models of fast solar wind driven by imbalanced ion cyclotron dissipation, in *SOLAR WIND 13: Proceedings of the Thirteenth International Solar Wind Conference*. AIP Conference Proceedings, vol. 1539 (2013), pp. 22–25.
31. J.R. Jokipii, M.A. Lee, Compression acceleration in astrophysical plasmas and the production of $f(v) \propto v^{-5}$ spectra in the heliosphere. Astrophys. J. **713**(1), 475–483 (2010).
32. S.W. Kahler, A.H. McAllister, H.V. Cane, A search for interplanetary energetic particle events from solar posteruptive arcades. Astrophys. J. **533**(2), 1063–1070 (2000).

33. Y.Y. Kartavykh, W. Dröge, B. Klecker, L. Kocharov, G.A. Kovaltsov, E. Möbius, Charge state formation of energetic ultraheavy ions in a hot plasma. Astrophys. J. **681**(2), 1653–1659 (2008).
34. Kasper et al., Space Sci. Rev. (2014, this issue)
35. B. Klecker, E. Möbius, M.A. Popecki, L.M. Kistler, H. Kucharek, M. Hilchenbach, Observation of energy-dependent ionic charge states in impulsive solar energetic particle events. Adv. Space Res. **38**(3), 493–497 (2006).
36. S.M. Krimigis, D.G. Mitchell, D.C. Hamilton, S.A. Livi, J. Dandouras, S. Jaskulek, T.P. Armstrong, J.D. Boldt, A.F. Cheng, G. Gloeckler, J.R. Hayes, K.C. Hsieh, W.-H. Ip, E.P. Keath, E. Kirsch, N. Krupp, L.J. Lanzerotti, R.A. Lundgren, B.H. Mauk, R.W. McEntire, E.C. Roelof, C.E. Schlemm, B.E. Tossman, B. Wilken, D.J. Williams, Magnetospheric imaging instrument (MIMI) on the Cassini mission to Saturn/Titan. Space Sci. Rev. **114**(1–4), 233–329 (2004).
37. D. Lario, R.B. Decker, Estimation of solar energetic proton mission-integrated fluences and peak intensities for missions traveling close to the Sun. Space Weather **9**, S11003 (2011).
38. D. Lario, M.-B. Kallenrode, E.C. Roelof, S.M. Krimigis, A. Aran, B. Sanahuja, Radial and longitudinal dependence of solar 4–13 MeV and 27–37 MeV proton peak intensities and fluences: Helios and IMP-8 intensities. Astrophys. J. **653**(2), 1531–1544 (2006).
39. M.A. Lee, Coupled hydromagnetic wave excitation and ion acceleration at an evolving Coronal/Interplanetary shock. Astrophys. J. Suppl. Ser. **158**(1), 38–67 (2005).
40. J.A. le Roux, G.P. Zank, W.H. Matthaeus, Pickup ion acceleration by turbulent field-aligned electric fields in the slow low-latitude solar wind. J. Geophys. Res.., Space Phys. **107**(A7), SSH 9-1 (2002).
41. R.A. Leske, R.A. Mewaldt, C.M.S. Cohen, A.C. Cummings, E.C. Stone, M.E. Wiedenbeck, T.T. von Rosenvinge, Solar isotopic composition as determined using solar energetic particles. Space Sci. Rev. **130**(104), 195–205 (2007).
42. G. Li, G.P. Zank, O. Verkhoglyadova, R.A. Mewaldt, C.M.S. Cohen, G.M. Mason, M.I. Desai, Shock geometry and spectral breaks in large SEP events. Astrophys. J. **702**(2), 998–1004 (2009).
43. S. Liu, V. Petrosian, G.M. Mason, Stochastic acceleration of stochastic acceleration of ^{3}He and ^{4}He by parallel propagation plasma waves. Astrophys. J. **613**(1), L81–L84 (2004).
44. G. Livadiotis, D.J. McComas, Beyond kappa distributions: exploiting Tsallis statistical mechanics in space plasmas. J. Geophys. Res. **114**(A11), A11105 (2009).
45. G. Lutz, *Semiconductor Radiation Detectors* (Springer, Munich, 1999)
46. G.M. Mason, ^{3}He-rich solar energetic particle events. Space Sci. Rev. **130**(1–4), 231–242 (2007)
47. G.M. Mason, M.I. Desai, J.E. Mazur, J.R. Dwyer, Energetic particles accelerated by shocks in the heliosphere: what is the source material? in *The Physics of Collisionless Shocks: 4th Annual IGPP International Astrophysics Conference*. AIP Conference Proceedings, vol. 781 (2005), pp. 219–226.
48. G.M. Mason, G. Gloeckler, Power law distributions of suprathermal ions in the quiet solar wind. Space Sci. Rev. **172**(1–4), 241–251 (2012)ADS
49. G.M. Mason, D.K. Haggerty, E.C. Roelof, M.I. Desai, C.M. Cohen, R.A. Mewaldt, J.R. Dwyer, Injection and interplanetary propagation of solar energetic particles observed on ACE. Am. Geophys. Union **43**, SH43B-01 (2006)
50. G.M. Mason, G. Li, C.M.S. Cohen, M.I. Desai, D.K. Haggerty, R.A. Leske, R.A. Mewaldt, G.P. Zank, Interplanetary propagation of solar energetic particle heavy ions observed at 1 AU and the role of energy scaling. Astrophys. J. **761**(2), 104 (2012).
51. G.M. Mason, J.E. Mazur, J.R. Dwyer, ^{3}He enhancements in large solar energetic particle events. Astrophys. J. **525**(2), L133–L136 (1999).
52. G.M. Mason, J.E. Mazur, J.R. Dwyer, J.R. Jokipii, R.E. Gold, S.M. Krimigis, Abundances of heavy and ultraheavy ions in ^{3}He-rich solar flares. Astrophys. J. **606**(1), 555–564 (2004).
53. G.M. Mason, C.K. Ng, B. Klecker, G. Green, Impulsive acceleration and scatter-free transport of about 1 MeV per nucleon ions in (He-3)-rich solar particle events. Astrophys. J. **339**, 529–544 (1989).

54. G.M. Mason, M.E. Wiedenbeck, J.A. Miller, J.E. Mazur, E.R. Christian, C.M.S. Cohen, A.C. Cummings, J.R. Dwyer, R.E. Gold, S.M. Krimigis, R.A. Leske, R.A. Mewaldt, P.L. Slocum, E.C. Stone, T.T. von Rosenvinge, Spectral properties of He and heavy ions in 3He-rich solar flares. Astrophys. J. **574**(2), 1039–1058 (2002).
55. W.H. Matthaeus, G. Qin, J.W. Bieber, G.P. Zank, Nonlinear collisionless perpendicular diffusion of charged particles. Astrophys. J. **590**(1), L53–L56 (2003).
56. B.H. Mauk, J.B. Blake, D.N. Baker, J.H. Clemmons, G.D. Reeves, H.E. Spence, S.E. Jaskulek, C.E. Schlemm, L.E. Brown, S.A. Cooper, J. Craft, J.F. Fennell, R.S. Gurnee, C.M. Hammock, J.R. Hayes, P.A. Hill, G.C. Ho, J.C. Hutcheson, A.D. Jacques, S. Kerem, D.G. Mitchell, K.S. Nelson, N. Paschalidis, E. Rossano, M.R. Stokes, J.H. Westlake, The energetic particle detector (EPD) investigation and the energetic ion spectrometer (EIS) for the magnetospheric multiscale (MMS) mission. Space Sci. Rev. (2014).
57. B.H. Mauk, D.K. Haggerty, S.E. Jaskulek, C.E. Schlemm, L.E. Brown, S.A. Cooper, R.S. Gurnee, C.M. Hammock, J.R. Hayes, G.C. Ho, J.C. Hutcheson, A.D. Jacques, S. Kerem, C.K. Kim, D.G. Mitchell, K.S. Nelson, C.P. Paranicas, N. Paschalidis, E. Rossano, M.R. Stokes, The Jupiter energetic particle detector instrument (JEDI) investigation for the Juno mission. Space Sci. Rev. (2013).
58. J.E. Mazur, G.M. Mason, J.R. Dwyer, J. Giacalone, J.R. Jokipii, E.C. Stone, Interplanetary magnetic field line mixing deduced from impulsive solar flare particles. Astrophys. J. **532**(1), L79–L82 (2000)
59. D.J. McComas, M. Velli, W.S. Lewis, L.W. Acton, M. Balat-Pichelin, V. Bothmer, R.B. Dirling Jr., W.C. Feldman, G. Gloeckler, S.R. Habbal, D.M. Hassler, I. Mann, W.H. Matthaeus, R.L. McNutt Jr., R.A. Mewaldt, N. Murphy, L. Ofman, E.C. Sittler Jr., C.W. Smith, T.H. Zurbuchen, Understanding coronal heating and solar wind acceleration: case for in situ near-sun measurements. Rev. Geophys. **45**(1), RG1004 (2007).
60. R.L. McNutt, S.A. Livi, R.S. Gurnee, M.E. Hill, K.A. Cooper, G.B. Andrews, E.P. Keath, S.M. Krimigis, D.G. Mitchell, B. Tossman, F. Bagenal, J.D. Boldt, W. Bradley, W.S. Devereux, G.C. Ho, S.E. Jaskulek, T.W. LeFevere, H. Malcom, G.A. Marcus, J.R. Hayes, G.T. Moore, M.E. Perry, B.D. Williams, P. Wilson, L.E. Brown, M.B. Kusterer, J.D. Vandegriff, The Pluto Energetic Particle Spectrometer Science Investigation (PEPSSI) on the new horizons mission. Space Sci. Rev. **140**(1–4), 315–385 (2008)
61. R.A. Mewaldt, E.C. Stone, R.E. Vogt, Neutral particle background in cosmic Ray telescopes composed of silicon solid state detectors. Space Sci. Instrum. **3**, 231–242 (1977)
62. R.A. Mewaldt, C.M.S. Cohen, G.M. Mason, A.W. Labrador, M.L. Looper, D.E. Haggerty, C.G. Maclennan, A.C. Cummings, M.I. Desai, R.A. Leske, G. Li, J.E. Mazur, E.C. Stone, M. Wiedenbeck, Solar energetic particle spectral breaks, in *The Physics of Collisionless Shocks: 4th Annual IGPP International Astrophysics Conference*. AIP Conference Proceedings, vol. 781 (2005), pp. 227–232.
63. R.A. Mewaldt, Solar energetic particle composition, energy spectra, and space weather. Space Sci. Rev. **124**(1–4), 303–316 (2006).
64. R.A. Mewaldt, C.M.S. Cohen, J. Giacalone, G.M. Mason, E.E. Chollet, M.I. Desai, D.K. Haggerty, M.D. Looper, R.S. Selesnick, A. Vourlidas, How efficient are coronal mass ejections at accelerating solar energetic particles? in *Particle Acceleration and Transport in the Heliosphere and Beyond: 7th Annual International Astrophysics Conference*. AIP Conference Proceedings, vol. 1039 (2008a), pp. 111–117.
65. R.A. Mewaldt, C.M.S. Cohen, W.R. Cook, A.C. Cummings, A.J. Davis, S. Geier, B. Kecman, J. Klemic, A.W. Labrador, R.A. Leske, H. Miyasaka, V. Nguyen, R.C. Ogliore, E.C. Stone, R.G. Radocinski, M.E. Wiedenbeck, J. Hawk, S. Shuman, T.T. von Rosenvinge, K. Wortman, The low-energy telescope (LET) and SEP central electronics for the STEREO mission. Space Sci. Rev. **136**(1–4), 285–362 (2008b)
66. R.A. Mewaldt, R.A. Leske, E.C. Stone, A.F. Barghouty, A.W. Labrador, C.M.S. Cohen, A.C. Cummings, A.J. Davis, T.T. von Rosenvinge, M.E. Wiedenbeck, STEREO observations of energetic neutral hydrogen atoms during the 2006 December 5 solar flare. Astrophys. J. **693**(1), L11–L15 (2009).
67. R.A. Mewaldt, M.D. Looper, C.M.S. Cohen, D.K. Haggerty, A.W. Labrador, R.A. Leske, G.M. Mason, J.E. Mazur, T.T. von Rosenvinge, Energy spectra, composition, and other

properties of ground-level events during solar cycle 23. Space Sci. Rev. **171**(1–4), 97–120 (2012a).

68. R.A. Mewaldt, G.M. Mason, C.M.S. Cohen, The dependence of solar energetic particle fluences on suprathermal seed-particle densities, in *Space Weather: The Space Radiation Environment.* AIP Conference Proceedings, vol. 1500 (American Institute of Physics, Melville, 2012b), pp. 128–133
69. D.G. Mitchell, L.J. Lanzerotti, C.K. Kim, M. Stokes, G. Ho, S. Cooper, A. Ukhorskiy, J.W. Manweiler, S. Jaskulek, D.K. Haggerty, P. Brandt, M. Sitnov, K. Keika, J.R. Hayes, L.E. Brown, R.S. Gurnee, J.C. Hutcheson, K.S. Nelson, N. Paschalidis, E. Rossano, S. Kerem, Radiation belt storm probes ion composition experiment (RBSPICE). Space Sci. Rev. **179**(1–4), 263–308 (2013).
70. NASA Science Mission Directorate, Solar probe plus announcement of opportunity, Solicitation: NNH10ZDA002O. http://nspires.nasaprs.com/external/solicitations/summary.do?method=init&solId={BC6CAAB3-F512-B60F-5F11-2942EAACFA8B} &path=closedPast. Accessed 4 June 2014
71. C.K. Ng, D.V. Reames, A.J. Tylka, Modeling shock-accelerated solar energetic particles coupled to interplanetary Alfvén waves. Astrophys. J. **591**(1), 461–485 (2003).
72. N.P. Paschalidis, The power energy ASIC, with voltage and time-over-threshold read outs: application to space science particle instruments, in *IEEE Nuclear Science Symposium and Medical Imaging Conference, N02-5, Dresden, Germany* (2008)
73. V. Petrosian, Y.W. Jiang, S. Liu, G.C. Ho, G.M. Mason, Relative distributions of fluences of 3He and 4He in solar energetic particles. Astrophys. J. **701**(1), 1–7 (2009)
74. M. Pick, G.M. Mason, Y.-M. Wang, C. Tan, L. Wang, Solar source regions for 3He-rich solar energetic particle events identified using imaging radio, optical, and energetic particle observations. Astrophys. J. **648**(2), 1247–1255 (2006).
75. A.F. Rappazzo, W.H. Matthaeus, D. Ruffolo, S. Servidio, M. Velli, Interchange reconnection in a turbulent corona. Astrophys. J. Lett. **758**(1), L14 (2012).
76. D.V. Reames, Particle acceleration at the Sun and in the heliosphere. Space Sci. Rev. **90**(3/4), 413–491 (1999).
77. D.V. Reames, The two sources of solar energetic particles. Space Sci. Rev. **175**(1–4), 53–92 (2013).
78. D.V. Reames, C.K. Ng, Streaming-limited intensities of solar energetic particles. Astrophys. J. **504**, 1002 (1998).
79. W.K.M. Rice, G.P. Zank, G. Li, Particle acceleration and coronal mass ejection driven shocks: shocks of arbitrary strength. J. Geophys. Res. **108**(A10), SSH 5-1 (2003).
80. D. Ruffolo, W.H. Matthaeus, P. Chuychai, Trapping of solar energetic particles by the small scale topology of solar wind turbulence. Astrophys. J. **597**(2), L169–L172 (2003).
81. N.A. Schwadron, M.A. Dayeh, M.I. Desai, H. Fahr, J.R. Jokipii, M.A. Lee, Superposition of stochastic processes and the resulting particle distributions. Astrophys. J. **713**(2), 1386–1392 (2010).
82. Seripienlert, D. Ruffolo, W.H. Matthaeus, P. Chuychai, Dropouts in solar energetic particles: associated with local trapping boundaries or current sheets? Astrophys. J. **711**(2), 980–989 (2010).
83. Teaca, M.S. Wield, F. Junco, R. Schlickeiser, Acceleration of particles in imbalanced magnetohydrodynamic turbulence (March 2014).
84. J. Tessein, W.H. Matthaeus, M. Wan, K.T. Osman, D. Ruffolo, J. Giacalone, Association of suprathermal particles with coherent structures and shocks. Astrophys. J. Lett. **776**(1), L8 (2013).
85. A.J. Tylka, M.A. Lee, A model for spectral and compositional variability at high energies in large, gradual solar particle events. Astrophys. J. **646**(2), 1319–1334 (2006).
86. A.J. Tylka, D.V. Reames, C.K. Ng, Observations of systematic temporal evolution in elemental composition during gradual solar energetic particle events. Geophys. Res. Lett. **26**(14), 2141–2144 (1999).
87. P. van Nes, R. Reinhard, T.R. Sanderson, K.-P. Wenzel, R.D. Zwickl, The energy spectrum of 35- to 1600-keV protons associated with interplanetary shocks. J. Geophys. Res. **89**, 2122–2132 (1984).

88. T.T. von Rosenvinge, D.V. Reames, R. Baker et al., The high energy telescope for STEREO. Space Sci. Rev. **136**(1–4), 391–435 (2008)
89. L. Wang, Personal communication (2010)
90. Y.-M. Wang, M. Pick, G.M. Mason, Coronal holes, jets, and the origin of 3He-rich particle events. Astrophys. J. **639**(1), 495–509 (2006).
91. G. Wibberenz, H.V. Cane, Multi-spacecraft observations of solar flare particles in the inner heliospherc. Astrophys. J. **650**(2), 1199–1207 (2006).
92. M.E. Wiedenbeck, C.M.S. Cohen, A.C. Cummings et al., Persistent energetic 3He in the inner heliosphere, in *Proceedings of the 30th International Cosmic Ray Conference, July 3–11*, vol. 1 (2008), pp. 91–94

INDEX